International Archives of the History of Ideas Archives internationales d'histoire des idées

Volume 253

International Archives of the History of Ideas/Archives internationales d'histoire des idées is a series which publishes scholarly works on the history of ideas in the widest sense of the word. It covers history of philosophy, science, political and religious thought and other areas in the domain of intellectual history. The chronological scope of the series extends from the Renaissance to the Post-Enlightenment. Founded in 1963 by R.H. Popkin and Paul Dibon, the International Archives of the History of Ideas/Archives internationales d'histoire des idées, edited by Guido Giglioni and John Christian Laursen, with assistance of Former Director Sarah Hutton, publishes, edits and translates sources that have been either unknown hitherto, or unavailable, and publishes new research in intellectual history, and new approaches within the field. The range of recent volumes in the series includes studies on skepticism, astrobiology in the early modern period, as well as translations and editions of original texts, such as the *Treatise of the Hypochondriack and Hysterick Diseases* (1730) by Bernard Mandeville. All books to be published in this Series will be fully peer-reviewed before final acceptance.

Peter Cheyne
Editor

Matter and Life in Coleridge, Schelling, and Other Dynamical Idealists

Springer

Editor
Peter Cheyne
British Literature and Culture
Shimane University
Matsue, Japan

ISSN 0066-6610 ISSN 2215-0307 (electronic)
International Archives of the History of Ideas Archives internationales d'histoire des idées
ISBN 978-3-031-78314-2 ISBN 978-3-031-78315-9 (eBook)
https://doi.org/10.1007/978-3-031-78315-9

This Springer imprint is published by the registered company Springer Nature Switzerland AG
The registered company address is: Gewerbestrasse 11, 6330 Cham, Switzerland

Contents

About the Editor and Contributors

About the Editor

Peter Cheyne is professor in the School of Letters at Hiroshima University; life member of Clare Hall, Cambridge; and visiting fellow in philosophy at Durham University. He is the author of *Coleridge's Contemplative Philosophy* (Oxford UP, 2020), editor of *Imperfectionist Aesthetics in Art and Everyday Life* (Routledge, 2023) and *Coleridge and Contemplation* (Oxford UP, 2017), and co-editor of *The Philosophy of Rhythm: Aesthetics, Music, Poetics* (Oxford UP, 2020).

Contributors

Lydia Azadpour Department of Philosophy, University of Nottingham, Nottingham, Nottinghamshire, UK

Peter Cheyne School of Letters, Hiroshima University, Hiroshima, Japan

Andrew Cooper Philosophy Department, Warwick University, Warwick, Warwickshire, UK

Allison Dushane Department of English and Modern Languages, Angelo State University, San Angelo, TX, USA

Alexander J. B. Hampton Department for the Study of Religion, University of Toronto, Toronto, ON, Canada

Philipp Höfele Freiburg Institute for Advanced Studies, University of Freiburg, Freiburg, Baden-Württemberg, Germany

Charity Ketz English Department, Pennsylvania State University, University Park, State College, PA, USA

J. C. C. Mays School of English Drama Film, University College Dublin (emeritus), Dublin, Ireland

Tim Milnes English Department, University of Edinburgh, Edinburgh, Midlothian, UK

Cecilia Muratori Dipartimento di Studi Umanistici, University of Pavia, Pavia, Italy

Lara Ostaric Philosophy Department, Temple University, Philadelphia, PA, USA

Kimberley Page-Jones Faculty of Law and Economics, University of Western Brittany, Brest, Brittany, France

Tilottama Rajan Department of English and Writing Studies, University of Western Ontario, London, ON, Canada

Dale E. Snow Philosophy Department, Loyola University, Baltimore, MD, USA

Dillon Struwig Department of English Studies, Sophia University, Tokyo, Tokyo Prefecture, Japan

Andrea Timár Department of English Studies, Eötvös Loránd University, Budapest, Hungary

James Vigus English Department, Queen Mary University of London, London, Greater London, UK

Daniel Whistler Department of PIRP, Royal Holloway, University of London, Egham, Surrey, UK

Thomas Wormald Centre for the Study of Theory and Criticism, University of Western Ontario, London, ON, Canada

Chapter 1
Introduction to Dynamical Idealists on Matter and Life

Peter Cheyne

Abstract The first section one gives the rationale for the book by (i) criticizing still surviving tendencies to support outmoded versions of materialism and (ii) outlining the surprising focus of a diverse group of idealist thinkers from Böhme to Schelling, who argued for energy-based theories of matter and life. This book identifies these thinkers as 'dynamical idealists'. This diverse group arguably provided truer, more unified, and cogent accounts of matter in the context of force and energy than corpuscularian materialism, which is still sometimes uncritically adopted in twenty-first-century literary studies (e.g. new materialism) and by related humanities scholars who denigrate 'idealism' as unembodied and detached from material concerns. Emphasizing the more fundamental place of energy, from which they argued that matter emerges, dynamical idealists identified the dimension of reality that underlies phenomena which support materialist views of matter. Discussion of recent physics and ecological thinking are given, arguing that the fundamentals of dynamical idealism are surprisingly consistent with twentieth- and twenty-first-century science on matter and life. The second section outlines this book, grouping the chronologically arranged chapters into three parts. A short third, final section appeals for an end to crude materialist–idealist polarization, to engage a broader and philosophically and scientifically informed energism that can meet the increasing need to pursue ecological thinking and the understanding of matter and life within systems of energy flow.

P. Cheyne (✉)
School of Letters, Hiroshima University, Hiroshima, Japan

P. Cheyne (ed.), *Matter and Life in Coleridge, Schelling, and Other Dynamical Idealists*, International Archives of the History of Ideas Archives internationales d'histoire des idées 253, https://doi.org/10.1007/978-3-031-78315-9_1

1.1 The Dynamical Idealists

This book examines the interrelationship between dynamical conceptions of matter and philosophical and scientific theories of life in the period 1612–1843.[1] The volume contributors' approaches to theories of matter and life are intended to illuminate the study of early-modern to romantic philosophical idealism, literature, and intellectual culture. Uniting research in philosophy and literature with the history of physics and biology, this book examines important themes that occupied an historically and philosophically diverse group of thinkers whom this volume identifies as 'dynamical idealists', thinkers of the idealist and dynamical theories of matter and life, in seventeenth- to nineteenth-century Europe, especially in Germany and Britain. Not just a contribution to philosophy and literature, the current volume intends also to be a useful work of history and philosophy of science that interrelates important strands in idealist and dynamical theories of matter and the life sciences of the period.

While the chapters in Part One discuss theories from the first two-hundred years of that period, the majority (Parts Two and Three) of the book focusses on thinkers around the years 1800–50, including G. W. F. Hegel, and especially Friedrich Schelling and Samuel Taylor Coleridge. Besides introducing the new genealogical category of the dynamical idealists, including Coleridge among such philosophical thinkers is a novel and perhaps a controversial aspect of this book. Schelling's centrality in the dynamical philosophy of nature is highlighted but also Coleridge's role in adapting, wrangling with, and importing German philosophical and scientific insights into the domain of British philosophy and science runs through the current volume. Coleridge is an influential interdisciplinary figure who offers, with Schelling, a unifying role. Placing Coleridge alongside philosophers such as Hegel and Schelling, the current volume is distinctive in considering Coleridge's role in transposing Germanic ideas into the British philosophical landscape, demonstrating the remarkable alternative to British empiricism that the programme of dynamical idealism offered.

Throughout, the volume consistently refers to the development of 'dynamical idealist' theories in earlier thinkers as diverse as the alchemically tinged theosopher Jacob Böhme, Cambridge Platonists Henry More and Ralph Cudworth, rationalist and monadologist G. W. Leibniz, and transcendental idealist Immanuel Kant.[2] Kant (building on Leibniz) is the salient and central figure here regarding dynamical idealism, and his *Metaphysical Foundations of Natural Science* (1786) was rightly seen

---

[1] This book was supported by JSPS Kakenhi grant number 19 K00107 and by a Shimane University SDG grant for my Living Ideas Cross-faculty Research project. The work also benefitted from a sabbatical visiting fellowship at Clare Hall, Cambridge, and seminars at the history and philosophy of science group there and at philosophy of religion meetings at the divinity faculty.

[2] Friedman (2013, esp. ch. 2, 'Dynamics', 97–274), is recommended for its thorough and insightful reading of Kant's dynamical theory of matter.

as a pillar of the dynamical account of matter by Schelling, the other philosophers of nature (the *Naturphilosophen*), Coleridge, and related romantic-era thinkers.

The term 'dynamical' is primarily used in this book with the sense of physical force or energy, such as, for instance, is meant in the science of dynamics. Figurative senses, applied psychologically, for example, or within the 'dynamics' of theories or systems, cultural or psychological, are also conveyed by the term 'dynamic', depending on the context. Rather than require from the authors in this volume a uniform view on what the book identifies as dynamical idealism in theories of matter and life, patterns and tendencies of interpretation remain open for development. Nonetheless, a working notion of the term 'dynamical idealism' was shared for consideration throughout the writing of this book. That sense of the term connotes a theory of matter as a state of relative stability of energy, but not as a basic (fundamental), corpuscularian given. As Schelling put it, concerning his dynamical theory, 'corporeal matter is not an original constitution' (2000 [*c.*1815], 61).

The dynamical idealists, as this book claims, are a group of divergent thinkers of various kinds of idealism (e.g. hermetic, rationalist, transcendental, absolute, objective) who accept the existence of matter yet see it as an emergence and not as a foundation. In the broad view, open to interpretation, of dynamical theories of matter, matter comes into and passes out of being and is preceded by, succeeded by, and always remains, energy. Life may then be seen as an energic, emergent development of matter that is itself energic and emergent. Coleridge, for instance, held that nature has its own power of self-transcendence that progresses through evolution, or as he put it: 'Nature leaves nothing *behind* but still takes up the lower into the higher, still refining and ennobling what it elevates' (1957–2002, 4: §4517 [April 1819]). Around a year later, when he was about 48 years old, he expressed his continuing joy, in his 'Sonnet: To Nature', in facing 'all created things | Deep, heartfelt, inward joy that closely clings' and of finding in nature 'Lessons of love and earnest piety' (2001 [1820?], 1: 92–3).

The titular 'dynamical idealists' of the current volume represent the thinkers covered in this book, from Böhme to Schelling, as proponents of varieties the dynamical theory of matter—i.e. the theory that matter arises from oppositional constellations of energy—and as adherents of various forms of philosophical idealism. This volume thereby proposes a new genealogy in modern philosophy that accommodates literary and cultural figures, such as Novalis and Coleridge, according to a worldview that holds matter and life to be evolving emergences from an underlying, pre-material and sustaining background of forces, powers, and universal or cosmic ideas.

The focus on theories of matter and life provides a concentrated approach to the study of early-modern to romantic science, literature, and culture as well as of philosophical idealism. In essence, the book examines the development of dynamical theories of matter, especially Leibniz's and Kant's accounts of matter as fundamentally energic rather than corpuscular,[3] and explores how these influenced theories of matter and life in Europe, with special attention to Schelling and Coleridge. While Coleridge's appreciation of Kant's dynamical theory of matter was fostered through others such as Thomas Beddoes, Henrik Steffens, and Schelling, he developed his understanding of it through careful study of Kant's *Metaphysical Foundations of Natural Science*, a book to which, as Dillon Struwig's chapter in the current volume explores, Coleridge added extensive and insightful commentary.

During the period from Böhme to Schelling, natural science and philosophy grew increasingly empiricist. The more rationalist Cambridge Platonists, who posited a universal 'plastic principle', or 'shaping spirit', adapted to yet generally resisted this modern current. By the late stage of the period examined, Friedrich Schelling and the *Naturphilosophen* were combining rationalist and transcendental philosophy and 'romantic science' with empirical methodology.[4] Most intriguingly, Schelling combined a detailed knowledge of experimental science with an aesthetic and religious sense of nature as divinely revelatory, and he went so far as to argue that matter, perceived aright in its scintillating beauty, could be viewed as being in reality the appearance of a spiritual essence, the 'clothing'—as Thomas Carlyle (2008 [1831]) had his Professor of 'Things in General' at 'Weissnichtwo University' proclaim—of a transformative divinity. According to Schelling:

> If we consider the wondrous transformations by which matter is subjugated in the organic world with the human eye, out of which shine the spirit, the intellect, and the will in an uncomprehending yet sensuously sensitive fashion, then one is allowed to consider all of matter as mere appearance, that is, simply as a deferred image of the essence that really lies at the basis. It then sees all bodies only as the clothing or masks that hide that inner point of transfiguration from us. (Schelling 2000 [1815], 61)

In this period, too, Hegel was constructing a more fully rationalist system of objective idealism that was nevertheless replete with empirical particularities. Coleridge, relatedly, was synthesizing such ideas with physician and Kant scholar Beddoes's similar German importations and Richard Saumarez's new Hunterian theories on physiology and the 'life principle'. In this, Coleridge closely followed the chemical experiments of Humphry Davy and recent developments such as animal magnetism, while trying to fuse these developments within a rational-physical evolutionary schema. Connecting back to the beginning of the period addressed by this book, Hegel's, Schelling's, and Coleridge's different logics of transformation shared a

[3] Holden (2004) remains the most detailed and comprehensive study of the main philosophical problems related to the theory of matter from Galileo to Kant, esp. in the context of developing dynamical theories.

[4] Berger (2023) compares the philosophies of nature of Schelling and Hegel and relates them to twenty-first-century issues around understanding matter and life. Grant (2006) covers Schellingian and post-Schellingian physics, along with some zoology and geology, in relation to different aspects of *Naturphilosophie*. See also Dunham, Grant, and Watson (2011) for a useful book with aims and historical coverage near to the current volume. Beiser (2008), esp. pt. 4 on 'Schelling and Absolute Idealism', including chap. 5 on 'Theory of Life and Matter' gives a useful intellectual history relevant to the themes of the current book.

derivation from Böhme's theocosmology of the universe arising from opposed, interdependent energies, or spiritual qualities (dark–light; rage–love; bitter–sweet; etc.) that pre-exist, give birth to, and evolve the material world.

Written by philosophers, intellectual historians, and literary theorists, the chronological series of chapters in this book provides an interdisciplinary assessment of issues surrounding energy-based theories of matter and developmental theories of life. These issues relating empirical and theoretical science to philosophical idealism and materialism continue to influence current thinking in philosophy, literature, and cultural studies. The book aims to provide a more sophisticated exposition of idealism and materialism than is often given in works of literary and cultural theory and also aims to stimulate further thinking in the intellectual history of dynamical idealism. It demonstrates that idealist thinkers including Böhme, Leibniz, Kant, Hegel, Schelling, and Coleridge were committed to the existence of matter through their different but related theories of material existence as emergent or sublated from a prior substrate of powers or beings such as energies, forces, and monads. The simplistic, polarized view that these idealists denied the reality of matter is rejected, as the sophisticated histories of the dynamical idealists' theories of matter and organic life are explicated.

The main question motivating this volume asks how a varied group of philosophical and literary thinkers from 1612 to 1843, who were idealists of one form or another, conceived of physical matter and the development of organic life. The date commencing that period is when Böhme wrote *Aurora* (1612), and a key work at the end of that period is Schelling's final statement on nature philosophy, his 'Presentation of the Process of Nature' (1843). The volume contributors' findings include, to reiterate with emphasis, what is to some the surprising realization that these idealist thinkers were prepossessed with theories of physical matter and the development of organic life and that far from denying the existence of matter, as some outmoded views still maintain, these thinkers, in their various ways, propounded or supported sophisticated theories of matter as essentially equivalent to energy. Differing in many ways, their theories, to clarify, can nonetheless be considered together in their acceptance of the existence of matter while holding that matter itself is not a basic level of existence but is something which arises from manifestations or oppositions of energy. In this respect, the titular 'dynamical idealists' represented in this book are found to have more in common with Einsteinian and post-Einsteinian theories of physics than do the rival theorists of classical, corpuscularian materialism. This connection makes the current book timely, since it brings together philosophy, history of science, anthropology, and literature to address historical issues in the physics of matter and the development of biological organisms that are still living issues (in progressed forms) today, as the possible existence of dark energy and dark matter is being interrogated and as the understanding of quantum-physical underpinnings of matter and life is making astonishing advances.[5]

[5] Al-Khalili and McFadden (2014), esp. chs 1 and 2, discusses relevant issues in current history and philosophy of science of physics and biology.

A major objective in creating this volume is not only to identify dynamical idealism but also to show emphatically that the existence of the dynamical idealists proves that there are forms of idealism that neither deny nor denigrate matter and biology, but which instead commence with the interplay of fundamental powers and forces that are prior to matter, these forces being physical yet non-material. Matter, according to dynamical idealists including Böhme, More, Cudworth, Leibniz, Kant, Hegel, Schelling, and Coleridge, arises from this dynamical interplay, whether as the 'pause' between opposed forces, or as fields of force surrounding points in space, which Newton, tutored by More, conceived as the 'divine sensorium',[6] in his metaphysics of gravity and the creation and persistence of matter. That is, dynamical idealists are so far from denying matter that they were in fact especially careful to develop physical and metaphysical theories, often, as with Leibniz, involving mathematical theory too, to account for matter as arising from more fundamental forces. Those forces, these thinkers generally agreed, despite many differences in the details of their theories, derive from powers, potencies, which are in some sense positively related to what are sometimes known as objective ideas, whether as universal laws of nature that have an objective reality or as 'living ideas' in the universal mind of God. The idealism of these thinkers does not refer merely to ideas formed in the human mind but, variously for each of these thinkers, to rational ideals of perfection; principles such as the transformation of energies from one quality or spirit into another; a biological formational drive; the pre-material ideality of universal laws; and a cosmic, developmental nisus.

The dynamical idealists, unlike materialist corpuscularians, who required invisible, corpuscular effluences, could naturally admit non-material, physical forces (e.g. gravitation and electromagnetism) and universal, physical but non-material laws of nature. What 'physical' means in such contexts is often very metaphysical, with relations (of energy or 'spirits') preceding material expression or resolution. Their shared commitment to varying arguments that ideas and powers precede physical existence is crucial to the dynamical idealist view of matter and physical forces. Ideas and powers, in this view, precede matter, but they do not swallow it up or usurp its place: the physical remains real. Relating this to later physics, the dynamical idealists stand near the Einsteinians who maintain the substantial reality of laws of nature, as opposed to those who argue that the laws of nature are changeable and have no reality beyond the material phenomena that they only appear to order.[7]

[6] Coleridge seems to have drawn from this view when he formulated a view of mental directedness such that when it ascends (becomes more general and intellectually intuitive) it becomes progressively humane and spiritual and when it descends (becomes local, particular, and more physically sensory) it is rendered 'Receptivity', prompting him to 'call the Reason the sensorium of the Spirit' (1957–2002, 4: §4938, May 1823). A staunch corpuscularian and 'mechanical philosopher', however, Newton was no dynamical idealist.

[7] Ellis (2001) is relevant to both sides of the debate, esp. ch. 4, 'Realism and Essentialism in Science', section 4.7 focussing on physics and 4.8 on biology; pt. 4, 'Laws of Nature', esp. ch. 6, 'Theories of Laws of Nature', provides an excellent overview of the philosophy of laws of nature.

Inverting, though probably without knowing it, the Galilean-Cartesian-Lockean primary–secondary quality distinction, Böhme theorized matter and all extended qualities as emerging from an original cosmic genesis of subjective qualities and spirits, or energies (*Quellgeisten*). Hegel, Schelling, and Coleridge were all deeply influenced by Böhme's radically different and voluminously thought-through worldview. Hegel, for instance, was implicitly following this Behmenist principle when he wrote of 'The transition from Quality to Quantity' as one 'not found in our ordinary way of thinking', which is indeed quite opposite to it, yet 'quantity just means quality superseded and absorbed: and it is by the dialectic of quality … that this supersession is effected' (1975 [1830], §98, 145).

In Böhme's view, the temporal world, as well as heaven and hell, and everything between, are formed in the dynamic and interpenetrative circulation of powers, qualities (spirits), and beings. In this process, God is eternally becoming tripartite in self-reflective self-knowing. The dark, wrathful, divine abyss reflects itself in the light, warm, loving Christ, and the perichoresis, or connecting rotation, of the Holy Spirit transforms and materializes the divine energies, such that 'the Spirit manifesteth the Deity in Nature: he *spreadeth forth* the Glance of the Majesty, so that it is seen in the Wonders of Nature' (Böhme 1764a–81 [1620, *The Threefold Life of Man*], 2.1: ch.4, 41). This is precisely Schelling's view when he writes, in the third draft of his *The Ages of the World*, that 'a matter known to be meshed with the wheel of the cosmos, nay, with its innermost and highest relationships, also demands in itself holy awe' (2000 [*c.*1815], 67). Böhme's God is the original and universal 'Genetrix' or mother, the cosmic matrix. Böhme took up St Paul's notion (1 Corinthians 15) of the 'spiritual body' versus the 'natural' or 'physical body', as when the Lusatian dialectical theosopher wrote that, 'the true real Body which is hidden In the Grossness is a spiritual Body' (1764b–81 [1623], 3: ch.16, §3). Böhme's metaphysical, theophanic cosmogenesis represents the often subterranean, hidden road that a branch of thinking took, where qualities and 'spirits' are primary and where matter and extended things arise out of them. That submerged path is in opposition to the dominant inverse view, which has had outstanding and still accelerating technological success, where qualities are secondary, sometimes even epiphenomenal, to material quantity.

In a genealogy of thought that made dynamical idealism clearer, more modern, and more philosophically respectable in general, this and related ideas progressed through More, shed their alchemical, hermetic vestiges with Cudworth, and found an influential expression in the monadology of Leibniz. Kant took this up in the 'Dynamics' chapter of his *Metaphysical Foundations of Natural Science* (1786), where he argued for the dynamical theory whereby matter arises from the opposition of fundamental forces of attraction and repulsion. The theory, unlike corpuscularian mechanism, denies fundamental status to matter. Instead, it sees matter as arising from the opposed forces of expansion and attraction, anticipating Roger

---

For the opposite view to Ellis, Ladyman, Ross, Spurrett, and Collier (2007) gives robust and quite aggressive analytic criticisms of recent analytic metaphysics and philosophy of science.

Boscovich's theory, following shortly after, of matter having an impenetrability due to fields of force and not from any putative atomic solidity. Later, Hegel honed more emphatically on the dialectic process of production through opposition that powered the dynamical construction of matter. Hegel commends, in his *Science of Logic* (1975 [1830], §98, 144), that 'To Kant we owe the completed theory of matter as the unity of repulsion and attraction. The theory is correct… Still, this dynamical construction of matter… tak[es] for granted, instead of deducing, attraction and repulsion'. Hegel regrets that while 'German physicists for some time accepted this pure dynamic', scientists 'in modern times have found it more convenient to return to the Atomic point of view … which, for its utter unintelligence, there would be sufficient reason to guard against' (144).

In his 1843 restatement of nature-philosophy 'Presentation of the Process of Nature', Schelling affirms the anti-corpuscularian view he held over decades, reiterating his view that 'matter … is maybe the hardest of all metaphysical concepts' (1856–61, 10: 310). That position follows necessarily from his dynamical idealism, and he earlier expressed it in his 1813 second-draft fragment of the *Ages of the World*, writing that 'the idea of matter [as] in … itself spiritual and incorporeal will appear completely unfathomable to many people' (1997 [1813], 150). His commitment to the dynamical theory of matter goes back further still, to at least 1800 and his *System of Transcendental Idealism*, where he affirms in the 'Deduction of Matter' that 'the two activities [expansive/unlimited and attractive/liminant], which maintain equilibrium in the [phenomenal, material] product, can appear only as fixed, static activities, that is as *forces*' (1978 [1800], 83). He continues there to give his version of the dynamical idealist 'construction of matter' over several pages, affirming that 'Matter is indeed nothing else but mind viewed in an equilibrium of its activities' (92). Schelling explicitly argues, then, that matter is 'spiritual' and is 'mind' in a kind of balance, or oppositional equilibrium.

While this idealism soon becomes complicated in his *Identitätsphilosophie*, affirming a mind–matter equivalence, his 1809 *Freiheitsschrift* (*Essence of Human Freedom*) and his unfinished and much redrafted *Weltalter* (*Ages of the World*) clearly reaffirm the dynamical idealist construction of matter. In the second-draft *Weltalter*, for instance, he maintains the already 'well-known theory of the construction of matter out of forces, which states that the inner essence of all matter is spiritual in a broad sense, since forces are undeniably something incorporeal, and to that extent spiritual' (Schelling 1997 [1813], 150). A twentieth- or twenty-first-century philosopher would qualify that remark by noting that while the energic theory of matter cannot strictly be called materialist, in the sense of materialism asserting matter to be fundamental and irreducible, it is tenable within the broader tenets of physicalism.

Schelling's third draft of *Ages of the World* alludes to Kant's endeavour, building on Leibniz, 'to present matter as the product of two forces'. These forces combined are also the power behind life, being for him 'the primordial forces of all life: the attractive and the expansive'. Yet how, he asks, does 'their clash' result in 'something graspable' (2000 [*c*.1815], 33)? Instead of the opposed forces 'sublimat[ing] each other reciprocally', with 'nothing perceptible remaining anywhere', nor with one vanquishing the other, corporeality must arise, he argues, as a product. This production of matter distinguishes 'a prior and posterior … between the forces', so that 'each force [expansive and attractive] stays in its Being and its essence', requiring that 'a product must come forth', and that product, 'like matter, stands between total restriction and complete expansion, stopped, so to speak, in the middle' (33).

Since the twentieth century, physicists again hold that matter arises from energy, abandoning the corpuscular materialism that dominated the Enlightenment and much of the nineteenth century. If matter is ultimately to be understood as energy, then mind–matter dualism gives way to a nearer relation—perhaps even identity—of mind and energy, in what twenty-first-century physicist Carlo Rovelli (2014, 12) calls, 'A reality that seems to be made of the same stuff that our dreams are made of'. Rather than caricaturize thinkers such as Cudworth, Schelling, or Coleridge as Quixotic idealists jousting romantically against matter and embodiedness, it should be recognized that they and other dynamical idealists were prepossessed with the origin and nature of matter and physical forces. Those thinkers include philosophers Cudworth, Leibniz, Kant, Hegel, and physicists Albert Einstein, who explained the equivalence of matter and energy, Max Planck, and David Bohm, as well as recent theoretical physicists such as Rovelli, all of whom proposed or supported profound theories of matter arising in the interplay of opposed forces.

Note that Einstein's original article stated mass–energy equivalence with the formula $m = E/c^2$ (equivalent to the later and popularly known $E = mc^2$), showing that mass is a property exhibited by energy. The following explanation from Brian Greene (2005), a Columbia University professor of mathematics and physics, expresses this clearly:

> Einstein showed [that] mass is a frozen form of energy that can be converted into the more familiar energy of motion….
>
> A little bit of mass can thus yield enormous energy. The destruction of Hiroshima and Nagasaki was fueled by converting less than an ounce of matter into energy; the energy consumed by New York City in a month is less than that contained in the newspaper you're holding. Far from having no energy, the baseball that has come to rest on the china shop's floor contains enough energy to keep an average car running continuously at 65 m.p.h. for about 5,000 years.

In the context of this book, it is most intriguing to note that Greene's explanation that 'mass is a frozen form of energy' parallels Coleridge's description (following More, as I shall detail in Chap. 15) of matter as the 'coagulum …, the pause … of opposite energies' (1956–71, 4: 775, to Charles Augustus Tulk, September 1817). Matter as such is energy-in-waiting. As Einstein's formula requires, mass–energy equivalence applies not only to nuclear explosions and power stations. $E = mc^2$ is constantly at work: when cars burn petrol to drive, when music players deplete their batteries to create sound waves, and, as the physicist Greene adds, when the eye moves and brain synapses fire to create perception and thought in 'chemical reactions that interchange mass and energy, once again in accord with Einstein's formula'.

Importantly, for thinking forward, when theological content is bracketed as neither rejected nor affirmed, the dynamical idealists are much nearer to twentieth- and twenty-first-century physics than either is to classical, corpuscularian materialism. Dynamical idealism, unsurprisingly, also stands near to the philosophical stance of physicalism, which is committed only to the existence of entities and forces required by the current and future science of physics. Some physicalists even oppose the purported reality of matter with quantum theory.

Matter, generally agreed to be a mode of energy, is a tightly bonded system of fundamental forces and particles. Matter also has kinetic energy in the form of motion relative to any other location. In calling the notion that matter is a relative fixing of energy the 'pause' or 'coagulation' of energy, Coleridge drew from the theories of Böhme, More, Cudworth, Leibniz, Schelling, and, especially, Kant. Matter as coagulated or extinguished and inverted energy or spirit was similarly expressed by Schelling in his *First Outline of a System of the Philosophy of Nature* (2004 [1799], 132):

> What is matter other than extinguished spirit? All duplicity is cancelled in it; its state is a state of absolute identity and of rest. In the transition from homogeneity to duplicity a world already dawns, and with the restoration of duplicity the world itself opens up, and what is this world other than visible spirit?

Schelling's description here is more in the terms of neo-Platonic emanation, where universal mind (spirit) or reason-principles emanate, according to Plotinus, from the One (equivalent to the Absolute of the post-Kantians), through nous, the divine intellect that contains all ideas or forms and energies, then into soul, which manifests life and the world-soul, before bottoming out into the most basic matter, where everything spiritual or intellectual is at its most attenuated. The spirit or intellect that for Plotinus was 'attenuated' in basic matter, Schelling, in this period, called 'dormant'. Hence, matter is 'extinguished spirit' for Schelling in 1799. The French philosopher, German idealist-influenced Victor Cousin (2025 [1826], 75–6) should also be counted among the dynamical idealists, especially for his view, in his 1826 preface to his *Philosophical Fragments*, that the almost total predominance of 'forces and laws' in modern science accounts for why 'physics ... became spiritualist':

> Philosophical considerations which reduce the concept of the external world to that of force have already made some progress and govern modern physics without [physics] knowing it. What physicist, since Euler, seeks anything in nature but forces and laws? Who speaks today of atoms? And who now takes molecules, derivative of atoms, as anything but a hypothesis? ... I can conclude rigorously that physics, whether it knows it or not, is no longer materialist and that it became spiritualist ... The physical sciences have themselves started down the great path of spiritualism properly understood.

Readers today can rightly criticize Cousin's brushing off atoms and molecules as mere hypotheses. However, the atomic hypotheses of his day included John Dalton's now odd-seeming view of atoms as entirely solid, unbreakable spheres, with molecules schematized by his ball-and-stick model where the sticks can break and reform in other patterns while the atomic balls remain eternally solid.[8] The molecular part of his theory remains, at that level, sound and pedagogically useful, but the atomic component is entirely misleading and obsolete to the degree that it now appears as a caricature of materialism.

The dynamical idealist view of matter, by contrast, as coagulated or attenuated ('extinguished') spirit or energy, when modernized and given in its most general terms, has held up rather well. Current cosmology supports the matter-as-relative-'pause'-of-energy theory that commenced with the dynamical idealists. Matter in the form of stable atoms arose in the universe about 380,000 years after the Big Bang, cosmologists now say, when the universe had cooled enough, by expansion, to allow electrons to form clouds around nuclei of the protons and neutrons that formed within the first second of the Big Bang, almost 14 billion years ago.[9] Before those atoms formed, the universe contained other forms of energy manifestations, including massless particles such as photons, light waves and other modes of electromagnetic radiation and fields of force, but it contained no matter. In these respects, then, matter is an emergence and a relative stabilizing of energic phenomena, as the dynamical idealists proposed.

1.2 Outline of the Current Book

The interconnecting chapters of the current book construct a genealogy of how theories of matter and of living matter developed from alchemical and theosophical intuitions in early modern philosophy and mysticism to the emergent holism and concerns about system in the romantic era. Throughout, questions of the nature of matter and the evolution of life are shown to be central concepts of dynamical idealist thinkers—who developed these ideas with extraordinary sophistication—and not merely the preserve of the rival materialist schools. The chapters are arranged chronologically and into three parts. The four chapters in Part 1 (numbered Chaps. 2–5, since this introduction is Chap. 1) move from Böhme through Cudworth, Leibniz, and Shaftesbury to Novalis and the age of romanticism.[10] The six chapters of Part 2 then focus on Schelling, with some reference to Hegel and Coleridge. Part 3 then gives eight chapters on Coleridge, with foci also on Blumenbach and

---

[8] His famous Dalton's balls were produced as three-dimensional models in 1810 and were used to illustrate principles in Dalton, Wollaston, and Thomson (1893a [1802–8]) and Dalton, Gay-Lussac, and Avogadro (1893b [1808–11]).

[9] Barrow (2009) is an authoritative and highly readable source of information and discussion on current cosmology.

[10] Beiser (2002), pt. 3 covers early German absolute idealism and pt. 4 continues with a focus on Schelling. Part 3, chs. 2–4 explore Hölderlin, Novalis, and Schlegel as literary figures who arguably made philosophical contributions to absolute idealism. That approach overlaps with some aspects of this volume, in terms of recovering and rethinking major contributions by philosophically neglected figures who are often supposed not quite to count as philosophers.

Schopenhauer. It should be clear that despite a traditional, disciplinarian neglect in philosophy, the current book adds to recent efforts, some by contributors to this volume, to read the work of literary-philosophical figures such as Goethe, Novalis, and Coleridge from a broader intellectual-historical and philosophical perspective.

Chapter 2, by Cecilia Muratori and James Vigus, explores Coleridge's engagement with Böhme's works, focusing on his critical adoption of the German mystic's ideas through the language of contemporary *Naturphilosophie*, especially Henrik Steffens. Coleridge interprets Böhme's account of light using the concept of polarity and, following Schelling, brings in the modern term 'gravitation'. The chapter also examines the influence of the 'Teutonic theosopher' on Coleridge's later thought, particularly his 'spiritual chemistry', and discusses Coleridge's ambivalent stance on what he saw as Böhme's struggles with pantheism. In Chap. 3, Charity Ketz examines Leibniz's concept of force in the composition of matter, comparing it to Coleridge's views. Both saw matter as shaped by the conditions of possibility imposed upon actuality. While Leibniz explored force to identify individuality in action, Coleridge focused on consciousness to reveal dynamic relations. Coleridge's treatment of divine influence on nature and the interplay of opposition and incorporation reflects a Leibnizian notion of force in his analysis. In Chap. 4, Tom Wormald traces the concept of plasticity from its kernel in Cudworth's plastic nature through Shaftesbury to eighteenth- and nineteenth-century thinkers including Herder and Hegel in Germany and Coleridge in Britain to reveal the paradoxical presence of the idealist concept plastic 'life' at the heart of contemporary materialist thought. Reference is also made to plasticity in the work of French philosopher Catherine Malabou, as one of the most interesting and rich concepts in contemporary materialist thought. Chap. 5, by Alex Hampton, examines Novalis's philosophical, linguistic, and religious writings to illuminate how he understood the logos of nature and expressed it through poetry. Through an examination of key works including *Fichte Studien*, *Monolog*, and *Die Christenheit oder Europa*, poetry emerges not as a subjective expression of one's impression of nature, but as an external metaphysical force, the constitutive logos of nature, and the ideal structuring principle of society.

The Schellingian Part 2 opens with Tilottama Rajan examining Schelling and Hegel alongside Coleridge and J. H. Green. Green's concept of 'physiogony' extends Schelling's *Stufenfolge* into an aspirational discipline consistent with Hegel's reading of nature through a gradation of disciplines in which, however, nature talks back to mind. Where Hegel is caught in the agon between thought and matter, for Schelling they enfold each other, rendering nature a force that generatively shapes thought or (for Hegel) deforms and re-forms it. Philipp Höfele's Chap. 7 argues that Schelling, following Goethe, sought a concept of nature and life addressing dynamics, self-creation, and purposefulness. They shared a methodological focus on holistic phenomena, opposing a Newtonian transcendental philosophy of nature. Both aimed to redefine the ontological and methodological understanding of nature, anticipating process-philosophical environmental ethics and emphasizing the interconnectedness of life and the environment. In Chap. 8, Lara Ostaric traces Schelling's principle of life to Kant's antinomy of teleological judgment. She contends that while Schelling closely follows Kant, he ontologically extends his own solution to the antinomial conflict by positing a unity of mechanical and teleological causes in the principle of life, the 'world-soul', which has a constitutive and not a regulative status and which permeates both inorganic and organic nature. This unity is possible only insofar as both inorganic and organic nature exhibit the prior unity of mind and world in the Absolute. Dan Whistler, in Chap. 9, describes a set of post-Kantian accounts of meaning (particularly attending to Schelling), according to which: (i) meaningful properties are formal ones; (ii) these properties should be described in terms of natural processes; (iii) meanings should therefore be considered natural forms. This set of naturalist positions typically construes meaning as 'living form', but equally—at its most radical—exceeds this organicist framework to think inorganic meanings.

Chapter 10, by Dale Snow, studies Schelling's and Coleridge's evolving theories of matter. She examines how Coleridge and Schelling share the idea that matter, as reflected in natural entities, has an educative, revelatory function. Discussing the neo-Platonic roots of this idea, she shows how Schelling and Coleridge both argued for an understanding of nature which could serve to aid in our recovery from our alienation from it. In Chap. 11, Tim Milnes highlights the matter of effluence within the fluidity of energy. Following Slavoj Žižek, he focusses on Schelling's 'indivisible remainder', an 'incomprehensible base of reality in things'. Milnes takes this volatile nub in Coleridge's philosophical chiasmus to idealize organic life *and* to enliven and 'organicize' thought. Milnes argues that Coleridge's '*caput mortuum*' corresponds with Schelling's void, an ontological excess that threatened to erupt volcanically within his 'living' ideas. Milnes concludes that the central problem of Coleridge's 'Life-Philosophy' remains one of what to do with God's waste. The problematic recalls lines in the poem that Jim Mays reads at the end of this book, 'Human Life Contemplated on the Denial of Immortality' (1811–15), which portrays and assails the atheistic view of human being as:

> Surplus of Nature's dread activity,
>
> Which, as she gazed on some nigh-finished vase,
>
> Retreating slow, with meditative pause,
>
> She formed with restless hands unconsciously!
>
> Blank accident! nothing's anomaly!
>
> (2001 [composed 1811–15], 1: 887)[11]

Lydia Azadpour commences the Coleridgean Part 3. Her Chap. 12 examines Coleridge's ambivalence about analogy in his philosophy of nature. She untangles his various uses, contexts, and meanings of analogy, arguing that his concerns reflect a deliberate effort to distinguish his views from issues in *Naturphilosophie*.

[11] The variorum text appears at Coleridge, 2001, 2: 1100–02.

Coleridge limited the scope of analogical reasoning while preserving the homogeneous root of powers and qualitative distinctions between them. This is best illustrated in his dual portrayal of the plant–animal relation as both antithetical and hierarchical. In Chap. 13, Dillon Struwig traces the development of Coleridge's theory of matter, examining his extensive critical commentary on Kant's seminal *Metaphysical Foundations of Natural Science*, the starting point for the dynamical account of matter favoured by the *Naturphilosophen*. Struwig focuses on Coleridge's search for a set of terms that would clarify and consistently maintain the ontological distinctions between (i) matter, material substance, and body, and (ii) the 'fundamental forces' or 'spiritual powers' whose divinely grounded causal activity is necessarily manifested through all such physical entities. In Chap. 14, Andrew Cooper explores Coleridge's *Theory of Life* (1816? Cooper suggests 1819 is more likely) amid debates in British medical institutions about the scientific status of comparative anatomy. Cooper argues that Coleridge proposed organizing comparative anatomy under the idea of life and highlights his method's influence on key figures like Joseph Henry Green, William Whewell, and Richard Owen. The chapter concludes that Coleridge's adaptation of German physiology significantly shaped British comparative anatomy, playing a more pivotal role in its development than previously acknowledged.

My Chap. 15 examines Coleridge's *punctum saliens*, inspired by Aristotle's and William Harvey's observations of the incipient heartbeat, as a metaphor for emergence across levels of being. I explore how Coleridge related embryological theory to idealist views of matter as arising from opposing forces, influenced by Cambridge Platonism and German idealism. The chapter traces the *punctum saliens* from Aristotle and Harvey to Coleridge's metaphysics, incorporating insights from Leibniz and Kant on the dynamical theory of matter, showing how Coleridge converted biological observations to metaphysical use. Allison Dushane (Chap. 16), then examines how Coleridge and Schopenhauer developed theories of the will influenced by romantic-era science and anti-dualist views of matter, yet with differing conclusions. Both integrated scientific findings while rejecting science as the sole arbiter of truth. Coleridge theorized dynamism in nature to define human will, agency, freedom, and God positively. Schopenhauer, in contrast, framed the will as a material foundation of life, proposing a negative definition of freedom. These contrasting views shape modern conceptions of subjectivity and selfhood.

In Chap. 17, Kimberley Page-Jones connects Coleridge's *Theory of Life*, Blumenbach's concept of formative force, and Coleridge's later writings on humans, animals, and race. While Blumenbach's theoretical formative nisus appealed to Coleridge, he rejected functional adaptation to nature. In his contributions to Green's 1828 lectures, Coleridge sought to reconcile a dynamic anthropological view of life with Christian interpretations of human origins. Interconnecting these writings, Page-Jones illuminates the 1820s debates on life and race, highlighting the paradox of new sciences fostering dynamic views of nature while asserting human superiority. In Chap. 18, Andrea Timár reads Coleridge on 'the riddle of humanity' as the co-existence in each individual of 'life and mind', considering our shared 'human nature' and the uniqueness of the person. Can life, he asked, shape itself into mind, and, if it can, how does human nature develop into human uniqueness? Timár argues that passions play a prime role in the passage from life to mind and from human nature to human uniqueness. Re-contextualising *Opus Maximum*, she also shows that maternal love guides the infant from being a general member of the human species to becoming a unique human individual, a person. Concluding the volume, Jim Mays (Chap. 19) gives a careful reading of Coleridge's late and often-neglected 'Human Life', a poem also discussed by Dushane, Page-Jones, and Timár in the immediately preceding chapters of this book. Mays's chapter takes off from what one might call its complicated double-rhyme scheme. It turns out that this poem is more carefully constructed than it is usually taken to be, even though it remains a poem that attempts—as Coleridge said of some of John Donne's—to say too much.

1.3 A Call to End Crude Idealism–Materialism Polarization

A hope of the current volume is that subtler and more comprehensively gleaning approaches to materialisms and idealisms will be encouraged to flourish in departments of history and philosophy of science, literature, and the other arts and humanities. Concomitant with that hope is a belief in the need for scholarly communities to abandon as unhelpful the kind of boo–hooray, or crudely emotivist, polarizing commentary that has often been published over the last forty years, and which has been quick to use an affiliation to idealism of one kind or another as a basis for taking some thinkers less seriously. That attitude, however, rests on ignorance of the preponderance in dynamical idealist philosophers, scientists, and literary writers, among others, of serious thinking about what matter is and how it emerges from something that is not matter but which is its sustaining foundation, be that energy, spirit, or cosmic ideas, laws, or principles.

The Marxist origin of the new materialism that took shape in the 1980s is detectable in blanket dismissals of 'romantic ideology' rather than engaging with these thinkers as possibly having found valuable insights into the energies of physical and living existence (McGann 1983, passim; Levinson 2018 [1989], 33–66). That early new-materialist perspective erred in using simplistic polarization as a method of intellectual history, as when, for instance, Jerome McGann (1983, 5) declared 'conceptual-idealist' metaphysics to be 'deplorable' from 'a Marxist view'. It is also worth recalling that while Karl Marx presented his materialism as a polemic against idealism, he retained many romantic elements in his thought. Relatedly, object-oriented ontology, represented by Graham Harman and others in that early twentieth-first-century group, generally opposes idealism of all stripes but also more prominently contends against materialism not in the ontological but in the consumerist sense, doing so from a greater ecological awareness. New materialism and object-oriented ontology are not the only adversaries to the revival of interest in the varieties of idealism. More moderate opposition exists in the qualified, partially reductive naturalism of analytic philosophy.

The prominent theme of the current volume is that idealism has its own materialism. More accurately, that among the highly divergent varieties of idealism, those which affirm the dynamical theory of matter offer a qualified theory of materiality that is contained within a broader energism where matter is an emergence. The development of dynamical theories of matter from various kinds of idealism, as this book examines, shows several advantages of idealism worth exploring. Emergence, in the dynamical theory of matter, is not a term used to conceal a mystery, rather, the concept works more basically here, much as a group emerges when a number of people find common cause or identity, or in phase changes in physics, such as the emergence of wetness when sufficient molecules of the right kind accumulate at the right temperature and pressure.

Methodological othering and suspicion as reactionary of literary and philosophical texts misses the radicality and instinct towards a deeper basis of reality that the current volume illuminates in the thinkers it identifies as dynamical idealists. Materialisms that do not recognize the dynamical idealists fail to see a genealogical branch of idealism that is intimately interconnected and at times identical with the history and theory of developing energic theories that considered matter as not ultimately foundational and that materiality itself has a dynamic, energic origin that has too often been overlooked throughout the history of new materialism.

Happily, some new (e.g. Marjorie Levinson, whose view has evolved into something more sophisticated and realistic than her earlier Marxist materialism) and some speculative materialists (e.g. Manuel DeLanda, who is better described as a speculative *physicalist*), have conveyed an understanding of ecology and systems of energy exchange that draws them nearer to aspects of dynamical idealism. For instance, Levinson complicated her earlier and simpler materialist 'embrace of the Marxian methodology' and 'Marxian borrowings' (2018 [1989], 37–8) with her return to Baruch Spinoza and her later references to post-classical physics, which can be read as moving from classical corpuscularian and Marxist materialism into a more informed and philosophical physicalism (Levinson 2018, chaps. 5, 9, and 11). DeLanda, often considered a leading speculative materialist, is nearer to energism. He observed in interview that the:

> problem with the term 'materialism' is that not only matter but also energy and physical information are needed to account for self-organizing phenomena and the processes which fabricate physical entities. Also, some forms of materialism may imply reductionism (of the mind to matter, for example) and that is not at all implied by the term 'realism'. (DeLanda 2005, 66.)

In another striking parallel with Coleridge's matter as 'coagulated spirit', though with a different ultimate contemplation of human life, denying immortality and affirming a persistence of ecology within the eternal circulation of energy, DeLanda argues that:

> Our organic bodies are, in this sense, nothing but temporary coagulations in these flows [of circulating energy]: we capture in our bodies a certain portion of the flow at birth, then release it again when we die and micro-organisms transform us into a new batch of raw materials. (1997, 103.)

The various authors of this volume—intellectual historians, philosophers, and literary scholars—each find in their different ways a fertility in working with the idea of the new genealogical category of 'dynamical idealists'. These chapters extend the history of energy-based (rather than foundationally materialist) philosophy and physics of matter to dynamical idealist counterparts who perceived energy and energic ecology as foundational in the biology, psychology, anthropology, literature, and views of culture of the period 1612–1843. With my fellow contributors to the current volume, I hope that our work will encourage further research into identifying and more accurately honing and complicating a cluster of dynamical idealist, energic worldviews.

References

Barrow, John D. 2009. *The book of nothing: Vacuums, voids, and the latest ideas about the origins of the universe*. New York: Knopf Doubleday.

Beiser, Frederick C. 2008. *German idealism: The struggle against subjectivism, 1781–1801*. Cambridge, Mass: Harvard University Press.

Berger, Benjamin. 2023. *Schelling, Hegel, and the philosophy of nature: From matter to spirit*. New York: Routledge.

Böhme, Jakob. 1764a–81. *The high and deep searching of the threefold life of man, through or according to the three principles* (1620). In *The works of Jacob Behmen, the Teutonic philosopher*. 4 vols. Vol. 2: 195 pp. Tr. John Sparrow, John Ellistone, and Henry Blunden. Ed. George Ward and Thomas Langcake. G. Robinson.

———. 1764b–81. *The mysterium magnum, or an explanation of the first book of Moses, called Genesis, in three parts* (1623). In *The works of Jacob Behmen, the Teutonic philosopher*. 4 vols. Vol. 3. Tr. John Sparrow, John Ellistone, and Henry Blunden. Ed. George Ward and Thomas Langcake. London: G. Robinson.

Carlyle, Thomas. 2008. In *Sartor resartus: The life and opinions of Herr Teufelsdröckh in three books (1831)*, ed. Kerry McSweeney and Peter Sabor. Oxford: Oxford University Press.

Coleridge, Samuel Taylor. 1956–71. In *The collected letters of Samuel Taylor Coleridge*, ed. E. L. Griggs, vol. 6. Oxford: The Clarendon Press.

———. 2001. In *Poetical works, part one: Poems (reading text)*, ed. J. C. C. Mays, vol. 2 vols. Princeton: Princeton University Press.

———. 1957–2002. *The notebooks of Samuel Taylor Coleridge*. Vols 1–3. Ed. Kathleen Coburn, 1957–73; vol. 4. Ed. Kathleen Coburn and Merton Christensen, 1990; vol. 5. Ed. Kathleen Coburn and Anthony Harding. Princeton: Princeton University Press.

Cousin, Victor. 2025. Preface to the 1826th edition. In *Philosophical fragments*. Tr. Daniel Whistler. Ed. Delphine Antoine-Mahut and Daniel Whistler. Oxford: Oxford University Press.

Dalton, John, William Hyde Wollaston, and Thomas Thomson. 1893a. *Foundations of the atomic theory: Comprising papers and extracts by John Dalton, William Hyde Wollaston, M. D., and Thomas Thomson, M. D. (*1802–1808*)*. Edinburgh: William F. Clay.

Dalton, John, Joseph-Louis Gay-Lussac, and Amedeo Avogadro. 1893b. *Foundations of the molecular theory: Comprising papers and extracts by John Dalton, Joseph-Louis Gay-Lussac, and Amedeo Avogadro (1808–1811)*. Edinburgh: William F. Clay.

DeLanda, Manuel. 1997. *A thousand years of nonlinear history*. Princeton: Princeton University Press.

DeLanda, Manuel, John Protevi, and Torkild Thanem. 2005. Deleuzian interrogations: A conversation with Manuel DeLanda and John Protevi. *Tamara: Journal of critical postmodern organization science* 3 (4): 65–88.

Dunham, Jeremy, Iain Hamilton Grant, and Sean Watson. 2011. *Idealism: The history of a philosophy*. London: Routledge.
Ellis, Brian. 2001. *Scientific essentialism*. Cambridge: Cambridge University Press.
Friedman, Michael. 2013. *Kant's construction of nature: A reading of the metaphysical foundations of natural science*. Cambridge: Cambridge University Press.
Grant, Iain Hamilton. 2006. *Philosophies of nature after Schelling*. London: Bloomsbury.
Greene, Brian. 30 September 2005. 'That famous equation and you'. New York times.
Hegel, G. W. 1975. *Hegel's logic, being part one of the encyclopaedia of the philosophical sciences (1830)*. Tr. WilliamWallace. Oxford: The Clarendon Press.
Holden, Thomas Anand. 2004. *The architecture of matter: Galileo to Kant*. Oxford: The Clarendon Press.
Kant, Immanuel. 2004. *Metaphysical foundations of natural science (1786)*. Tr. and ed. with an introduction by Michael Friedman. Cambridge: Cambridge University Press.
Ladyman, James, Don Ross, with David Spurrett and John Collier Ross (2007). Every thing must go: Metaphysics naturalized. Oxford: Oxford University Press.
Levinson, Marjorie. 2018. *Thinking through poetry: Field reports on romantic lyric*. Oxford: Oxford University Press.
McFadden, Johnjoe, and Jim Al-Khalili. 2014. *Life on the edge: The coming of age of quantum biology*. London: Bantam Press.
McGann, Jerome. 1983. *The romantic ideology: A critical investigation*. Illinois: University of Chicago Press.
Rovelli, Carlo. 2014. *Seven brief lessons on physics. Tr. Simon Carnell and Erica Segre*. London: Penguin Books.
Schelling, Friedrich, and Sämmtliche Werke. 1856–61. In *Friedrich Wilhelm Joseph von Schellings Sämmtliche Werke*, ed. Karl Schelling, vol. 10, 301–390. Stuttgart: J. G. Cotta'scher Verlag.
Schelling, Friedrich. 1978. *System of transcendental idealism (1800)*. Tr. Peter Heath. Charlottesville: University of Virginia Press.
———. 1997. *Ages of the world* (second draft, 1813). In Slavoj Žižek and Friedrich Schelling. *The abyss of freedom and ages of the world*. 105–82. Tr. Judith Norman. Ann Arbor: University of Michigan Press.
———. 2000. *The ages of the world (third draft, c.1815)*. Tr. Jason Wirth. Albany: State University of New York Press.
———. 2004. *First outline of a system of the philosophy of nature (1799)*. Tr. K. R. Peterson. Albany: State University of New York Press.

Peter Cheyne is professor in the School of Letters at Hiroshima University; life member at Clare Hall, Cambridge; and visiting fellow in philosophy at Durham University. He is the author of *Coleridge's Contemplative Philosophy* (Oxford UP, 2020); editor of *Imperfectionist Aesthetics in Art and Everyday Life* (Routledge, 2023) and *Coleridge and Contemplation* (Oxford UP, 2017); and co-editor of *The Philosophy of Rhythm: Aesthetics, Music, Poetics* (Oxford UP, 2020).

Part I

Chapter 2
'Wonderfully Did this Gifted Seer Fly Before his Age': Samuel Taylor Coleridge, Jacob Böhme, and *Naturphilosophie*

Cecilia Muratori and James Vigus

Abstract Samuel Taylor Coleridge (1772–1834) wrote copious marginalia in his copy of the English translation of the works of the early modern German mystical philosopher Jacob Böhme (1575–1624). This chapter outlines the development of Coleridge's reading of Böhme, which took place over a long period. It then shows that Coleridge did not treat Böhme purely as a theologian or visionary, but critically grasped and appropriated the scientific thought of the 'teutonic theosopher' using the language of contemporary *Naturphilosophen*, above all Henrik Steffens. Thus, deploying the natural-philosophical paradigm of polarity in his interpretation of Böhme's account of the generation of light in nature, Coleridge significantly imports the modern term 'gravitation'. The chapter concludes by examining the prominent place of Böhme's 'spiritual chemistry' in Coleridge's later thought. The ambivalence with which Coleridge continued to view the supposed struggle of the German thinker with pantheism is also addressed.

Samuel Taylor Coleridge's engagement with the works of the early modern mystical philosopher Jacob Böhme (1575–1624) surpassed that of all the latter's romantic-era readers, and it resulted in the largest quantity of marginalia that Coleridge wrote on any single author.[1] Fig. 2.1 Only very recently, however, have Coleridge's Böhme-marginalia begun to be studied thoroughly. Peter Cheyne has argued that

[1] *This essay is a fully collaborative production, but James Vigus was primarily responsible for Sect. 2.1 and the conclusion, and Cecilia Muratori for Sect. 2.2, with the introduction entirely shared. Translations are the authors' own, if not stated otherwise.

C. Muratori
Dipartimento di Studi Umanistici, University of Pavia, Pavia, Italy
e-mail: cecilia.muratori@unipv.it

J. Vigus (✉)
English Department, Queen Mary University of London, London, UK
e-mail: j.vigus@qmul.ac.uk

P. Cheyne (ed.), *Matter and Life in Coleridge, Schelling, and Other Dynamical Idealists*, International Archives of the History of Ideas Archives internationales d'histoire des idées 253, https://doi.org/10.1007/978-3-031-78315-9_2

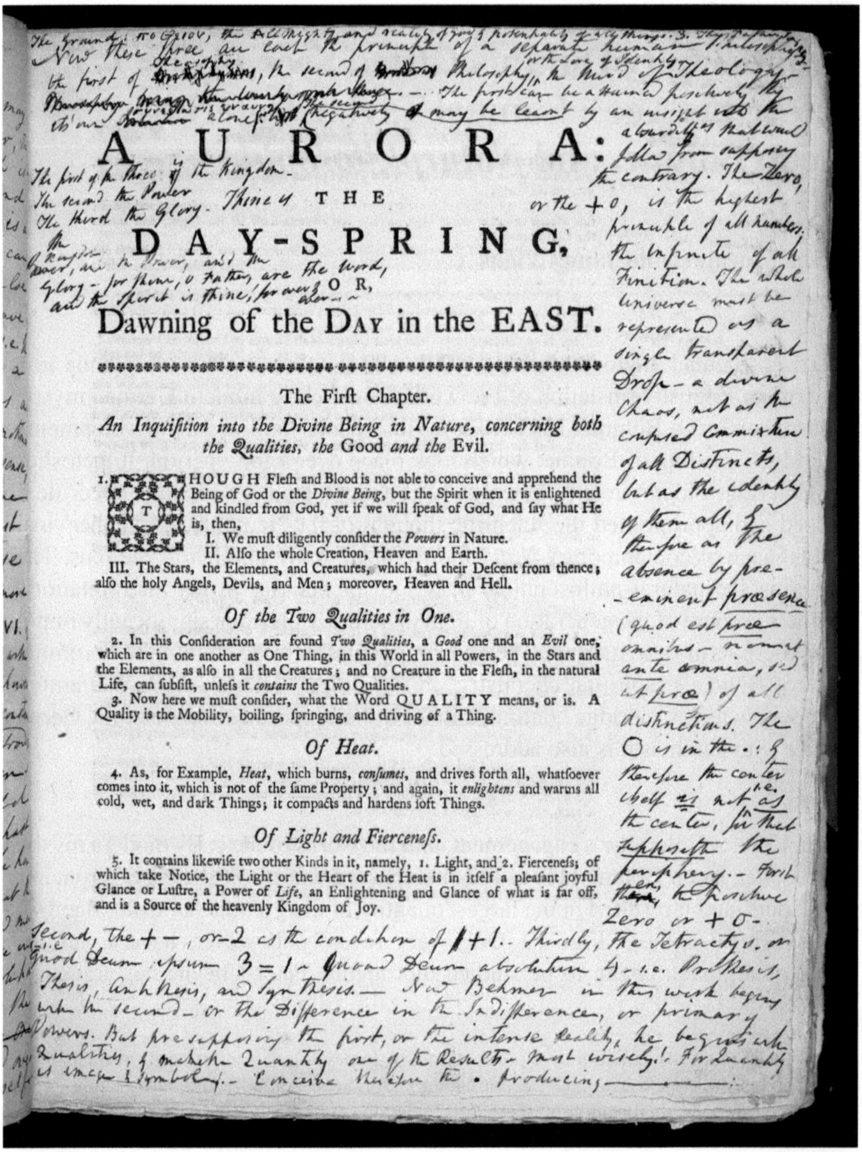

AURORA:

THE

DAY-SPRING,

OR,

Dawning of the DAY in the EAST.

The Firſt Chapter.

An Inquiſition into the Divine Being in Nature, concerning both the Qualities, the Good *and the* Evil.

1. THOUGH Fleſh and Blood is not able to conceive and apprehend the Being of God or the *Divine Being*, but the Spirit when it is enlightened and kindled from God, yet if we will ſpeak of God, and ſay what He is, then,

I. We muſt diligently conſider the *Powers* in Nature.

II. Alſo the whole Creation, Heaven and Earth.

III. The Stars, the Elements, and Creatures, which had their Deſcent from thence; alſo the holy Angels, Devils, and Men; moreover, Heaven and Hell.

Of the Two Qualities in One.

2. In this Conſideration are found *Two Qualities*, a *Good* one and an *Evil* one, which are in one another as One Thing, in this World in all Powers, in the Stars and the Elements, as alſo in all the Creatures; and no Creature in the Fleſh, in the natural Life, can ſubſiſt, unleſs it *contains* the Two Qualities.

3. Now here we muſt conſider, what the Word QUALITY means, or is. A Quality is the Mobility, boiling, ſpringing, and driving of a Thing.

Of Heat.

4. As, for Example, *Heat*, which burns, *conſumes*, and drives forth all, whatſoever comes into it, which is not of the ſame Property; and again, it *enlightens* and warms all cold, wet, and dark Things; it compacts and hardens ſoft Things.

Of Light and Fierceneſs.

5. It contains likewiſe two other Kinds in it, namely, 1. Light, and 2. Fierceneſs; of which take Notice, the Light or the Heart of the Heat is in itſelf a pleaſant joyful Glance or Luſtre, a Power of *Life*, an Enlightening and Glance of what is far off, and is a Source of the heavenly Kingdom of Joy.

Fig. 2.1 Coleridge's annotation to the title page of Böhme's *Aurora*. Compare Coleridge, *Marginalia*, 1: 560–3. *The Works of Jacob Behmen, the Teutonic Theosopher* (London: 1764, fourth edition 1781) I i 22, British Library, shelfmark C 126 k 1. Copyright: British Library Board 22/08/2023

'Coleridge derived some of the most crucial elements in his philosophy from Böhme' (Cheyne 2020, 125). According to Cheyne, Coleridge constructed his dynamic system of psychology, especially from 1818, in dialogue with Böhme. Specifically, Coleridge sought to fuse Böhme's immanentist 'schema of "The Seven Forms of Spirits" with the Christian neo-Platonic, hierarchical, linear ascent from sense to contemplation' (Cheyne 2020, 125).[2] However, the role of the Böhme marginalia in the natural-scientific aspect of Coleridge's thought has yet to be explored in detail. In particular, Coleridge's attempt 'to [rebuild] *Naturphilosophie* on the basis of a theistic metaphysics' involved recourse to Böhme, as we will see (Harding 1999, 218). Looking beyond scholarship on Coleridge, meanwhile, Böhme's capillary influence on eighteenth- and early nineteenth-century philosophies of science has also been only partially explored, so that much commentary tends to consider Böhme merely as a source of inspiration for divinatory rather than scientific discourse.[3] This chapter therefore follows a twofold path: by studying Böhme's role in Coleridge's *Naturphilosophie*, it traces the significance of Böhme's thought within the romantic period. It thus contributes a new case study to the scholarly debate on the border between science and theosophy.[4]

Indeed, Coleridge's natural-philosophical grasp of Böhme is a good example of the principle that 'rather than speculating how writers and artists of the romantic period did or did not manage to cross the boundary into the [incipient] natural sciences, we need to remind ourselves that that boundary was still very porous' (Berns 2017, 137). Thus, it was in the margin of Böhme's *Aurora* that Coleridge reflected on contemporary scientific progress, establishing a complex relationship between the newest discoveries from the laboratories and Böhme's early modern mindset regarding the study of nature: 'Humphry Davy in his Laboratory is probably doing more for the Science of Mind, than all the Metaphysicians have done' (Coleridge 1980–2001, 1: 566). More specifically, in Coleridge's time, interpreting Böhme

---

Space constraints preclude general introductory material on these two authors and their thought. For an introduction to Coleridge and science, see Hessel 2022. Böhme was long excluded from (the history of) philosophy due mainly to his dense prose and convoluted style; he has only recently become a more regular reference point in studies of early modern thought and its international reception. For a brief overview of Böhme's life and thought see Muratori 2022a. On the reception of Böhme in England, the classic sources were Lewis Bailey 1914 and Struck 1936, prior to a series of recent studies: Hessayon 2020 and a section of Martin and Muratori 2023 (including Hessayon 2023).

[2] Previously, the fullest discussion of Coleridge and Böhme was McFarland 1969, 325–32. Coburn wrote briefly on Böhme as one of Coleridge's 'earliest heroes' (1979, 30). Jessen's survey (2014) focuses more on William Blake than Coleridge.

[3] One of the most detailed studies on Böhme's presence in nineteenth-century philosophy is Brown 1977. Magee 2001 considers Böhme as a blueprint for the idealists' engagement with esoteric materials, building on Walsh 1978. In contrast, see Muratori 2016, 29–42, for an overview of Böhme's reception within eighteenth- and early nineteenth-century *Naturphilosophie*.

[4] On the emergence of this border, which was particularly blurred during early modernity, see Coudert 2011, xvi. See also Latour (1993, 13), on the danger of trying to evaluate history by drawing drastic separations between 'winners and losers'—in this case, between the realm of science and that of magic.

through *Naturphilosophie* and vice-versa was a plausible if unorthodox mode of scientific investigation. But which specific ideas from Böhme did Coleridge select for interaction with contemporary scientific endeavours? This is the key question which the present chapter seeks to answer.

Coleridge's approach to Böhme evolved over a considerable period. The trajectory of his reading was non-linear, his concern with specific aspects of Böhme's philosophy repeatedly resurfacing. By 1818, Coleridge had concluded that Böhme's work contains significant flaws, most seriously a tendency to pantheism or the identification of the divine with the creation. Nevertheless, Böhme long remained for Coleridge an example of how science and theosophy can create a synergy, pursuing the same aim. This is why Coleridge loved Böhme's 'spiritual Chemistry' (Coleridge 1980–2001, 1: 602). In Böhme's work, he found that theosophy provides the impulse for scientific development. This, we suggest, is the key aspect of Coleridge's reading of Böhme. In Sect. 2.1 we outline the main stages in the development of Coleridge's reading. Section 2.2 turns to Coleridge's appropriation of the language of the Baltic *Naturphilosophen*, especially Henrik Steffens, arguing that this framework enabled him to sketch out a unique interpretation of Böhme, one that treats the mystic's scientific endeavour not as antiquated, but as ahead of its time.[5] The concluding part considers Böhme's place in Coleridge's later thought.

2.1 'Conjuring over' Böhme: The Development of Coleridge's Engagement with Böhme

According to a letter he wrote to Ludwig Tieck, Coleridge first '*conjured over*' Böhme as a schoolboy at Christ's Hospital in the 1780s (Coleridge 1956–71, 4: 751, 4 July 1817). This was probably the period when, 'even before my fifteenth year, I had bewildered myself in metaphysicks'; and when Coleridge's fellow pupil, Charles Lamb, recalled him unfolding, in 'deep and sweet intonations, the mysteries of Jamblichus, or Plotinus' (Coleridge 1983, 1: 15 and n.). Coleridge's earliest awareness of Böhme is likely to have come via Coleridge's clergyman father: the broad-minded John Coleridge was acquainted with Samuel Badcock, who reviewed the final volume of the 'Law Edition' of Böhme in 1782 (Mays 2014, 25). In the 1790s, some imagery from Böhme's first book, *Aurora* (1612–13), surfaces in Coleridge's Gutch notebook. Coleridge here scribbled a brief phantasmagoria in fragmentary, poetic prose, including gothic vocabulary associated with the idea of hell, such as 'upboyling anguish' and 'wrathfires' (Coleridge 1957–2002, 1: §272–3; Beer 1963, 184). The passage reveals his fascination with Böhme's account of 'concupiscence', glossed by Coleridge as his 'assertion respecting the nature of the guilty Act' by which first Lucifer and then Adam and Eve fell (Beer 1959, 126).

[5] Levere 1981, the pioneering work in this field, discusses Böhme in only a limited context (156–8). The other major treatment of Coleridge and *Naturphilosophie* is Modiano 1985.

Related imagery was soon to emerge on a larger scale in Coleridge's poem 'The Rime of the Ancient Mariner'. He would long appreciate the 'poetical' quality of Böhme's writing: for example, he compares a passage of *Aurora* favourably with Milton's *Paradise Lost* (Coleridge 1980–2001, 1: 600; also 1: 592). Böhme's 'poetical' language influenced other major poems, too, especially 'Dejection: An Ode' (1802), where Coleridge's description of the wind as a 'lutanist' echoes Böhme's term for Lucifer in *De Signatura Rerum* (Mays 2019, especially 104).[6]

Around 1796–7 Böhme was crucial to Coleridge in a further respect. In *Biographia Literaria*, published in 1817, a couple of confessional passages recount a philosophical and religious crisis. 'The importance of this account can hardly be overestimated', as John Beer pointed out, 'for it lays out the terms of an intellectual conflict that dogged Coleridge for most of his life' (Beer 2003, 26). Coleridge had become a Unitarian at the University of Cambridge, and he absorbed the achievements of Joseph Priestley, whose work promised to unify chemistry and divinity (Wylie 1989). The empiricist worldview of Unitarianism, however, came to seem untenable to Coleridge: reflecting on the foundations of morality and personhood, he recounted, 'I found myself all afloat ... Doubts rushed in' (Coleridge 1983, 1: 200). This was in part due to the power of the sceptical objections of David Hume upon empiricist premises (Coleridge 1983, 1: 141–2). Further, Priestley's scheme failed to acknowledge the 'incomprehensibility' of the divine: 'How is it that Dr Priestley is not an atheist?', Coleridge had asked. 'He asserts in three different Places, that God not only does, but is, every thing.—But if God be every Thing, every Thing is God—: which is all the Atheists assert—. An eating, drinking, lustful God—with no *unity of Consciousness*—these appear to me the unavoidable Inference from his philosophy' (Coleridge 1956–71, 1: 192–3 [19 March 1796]). 'The pressure of that question grew more insistent in the following years' (Halmi 2012, paragraph 14). It was a question that Coleridge reframed as a fear of pantheism: the identification of God with the world.

Gradually, through the studies enabled by his trip to Germany in 1798, Coleridge rejected Unitarianism and became an ardent Trinitarian. In the process, Kant, with his Copernican revolution that prioritized mind over world, 'took possession' of Coleridge 'as with a giant's hand' (Coleridge 1983, 1: 153). Prior to his absorption of Kant, however, Coleridge found other sources of speculative hope. His testimony in *Biographia Literaria* indicates that reading certain 'mystics' prevented him from despairing once he began to link Priestleyan Unitarianism with atheism or pantheism. Coleridge names three 'mystics' in this context: George Fox the Quaker, Böhme, 'and even ... Behmen's commentator, the pious and fervid William Law' (Coleridge 1983, 1: 151). Coleridge thinks primarily of Law's later work, especially *A Serious Call to a Devout and Holy Life* (1728), which he elsewhere mentions specifically (Coleridge 1993, 384). The word 'commentator' may simply refer in general terms to Law's well-known fascination with Böhme (Rivers 2004). The

[6] Imagery from Böhme is also central to the meaning of Coleridge's dark poems 'Ne Plus Ultra' and 'Limbo' (both 1811; see Ridenour 1963), and important in the 1817 version of 'Ode to the Departing Year' (Cheyne 2020, 142).

other possibility is that Coleridge may have shared a common misconception about Law's involvement in the publication of *The Works of Jacob Behmen* (1764–81), the first edition of Böhme's collected works in English translation. This edition, frequently referred to as the 'Law edition', was published in honour of William Law after his death. It merely reproduces the seventeenth-century translations of Böhme's writings, many prepared by John Sparrow (on Sparrow as translator see Hessayon 2018), with the addition of some texts, and especially of images, 'left by the Reverend Mr. Law', as the editors state (Böhme 1764–81, 'Advertisement'), that were the work of the German émigré in London Dionysius Andreas Freher (1649–1728: see Muratori 2022b). Coleridge recounts further:

> the writings of these mystics acted in no slight degree to prevent my mind from being imprisoned within the outline of any single dogmatic system. They contributed to keep alive the *heart* in the *head*; gave me an indistinct, yet stirring and working presentiment, that all the products of the mere *reflective* faculty partook of DEATH, and were as the rattling twigs and sprays in winter, into which a sap was yet to be propelled … (Coleridge 1983, 1: 152)

Coleridge records a strong 'feeling of gratitude' to these men.[7] It would be easy to assume that this passage is camouflage for Coleridge's notoriously inadequate citation of contemporary German authors such as Friedrich Schelling, whose work constituted his main sources for the philosophical passages of *Biographia Literaria*. Nevertheless, supporting testimony for Coleridge's account is to be found in the marginalia to Böhme. There, Coleridge suggests that the kind of spiritual depression he experienced is nothing less than a 'pre-condition' for reading Böhme productively. A reader of Böhme requires not just a 'loving and humbled state of mind', asserts Coleridge, but also 'an experience of the shadowyness and death of all mere notional Learning—a feeling, a weary feeling, if not a clear sense, that notions are but words, and excellent as signs become mendacious when they would mean themselves' (Coleridge 1980–2001, 1: 561). Coleridge refers primarily to Böhme's notorious struggle with language, expressed in the abundant use of neologisms—an attempt to measure with words the boundaries of what can be said about the divine. For Coleridge, Böhme's linguistic creativity promised to revitalize the intellectual striving of the modern age.

Coleridge was deeply attracted by Böhme's combination of contemplating the divine in nature—to which Priestleyan Unitarianism had also aspired, but in Coleridge's view upon a flawed foundation—with a Trinitarian theology. Further, in a rivalrous collaboration with William Wordsworth, Coleridge developed a theory of the special meaningfulness of the language of 'ordinary', unschooled people. This common concern found expression in their volume of poems *Lyrical Ballads* (1798). It helps to explain Coleridge's interest in 'many a simple page' of Böhme, given that the latter was traditionally represented as an unlearned cobbler.[8] In

---

[7] Coleridge 1983, 1: 152; see also 1: 161. In 1803, Coleridge planned to include Böhme as well as Fox in a work on 'Revolutionary Minds' (Coleridge 1957–2002, 1: §1646), versions of which scheme occasionally resurface, including in his 1810 letter to Lady Beaumont, discussed below.

[8] Coleridge 1983, 1: 151. This is at least partly a myth, since Böhme probably received at least a basic education and was in fact a shoemaker, a skilled artisan, and not a cobbler, or mere repairer.

Biographia Literaria he exclaims: 'O! It requires deeper feeling, and a stronger imagination, than belong to most of those, to whom reasoning and fluent expression have been as a trade learnt in boyhood, to conceive with what *might*, with what inward *strivings* and *commotion*, the perception of a new and vital TRUTH takes possession of an uneducated man of genius'.[9] This passage echoes the inward struggle in Böhme's *Aurora*. For instance, in chapter 14 Böhme had remarked on the disproportion between the magnitude of the revelation he received and his limited means, resulting in an irresistible urge to write: 'But I marvel that God should reveal himself thus fully to such a simple Man, and that he thus impels him to set it down in Writing'.[10] Böhme also compared himself to the patriarch Jacob, '*who wrestled with him* [i.e. God] *all night, till the Dawning of the Day*, or Morning *Redness* broke forth'—where 'morning redness' is the literal translation of the original German title of Böhme's *Aurora*, namely *Morgenröte im Aufgang*, the morning redness arising.[11]

At the same time, the passage from *Biographia Literaria* reflects a larger, if tacit aim: to distance Coleridge's thought from his political radicalism of the early-to-mid-1790s. Thus, he focuses not on Böhme's controversial status and his appropriation by religious radicals of the English Civil War period, but rather on the 'uneducated' mystic's private struggle. Böhme exemplified one of Coleridge's favourite desynonymizations: he was an 'enthusiast', 'contra-distinguished, from a fanatic', meaning that this 'meek and shy quietist' valued internal conviction rather than external applause (Coleridge 1983 [1817], 1: 147; compare Coleridge 1993, 391–2).

It was not until 1808 that Coleridge began to read Böhme regularly with pen or pencil in hand, when he received a four-volume set of the so-called 'Law edition' as a present from Thomas De Quincey (Coleridge 1980–2001, 1: 553). This set of marginalia is unique in Coleridge's extensive reception of German texts in that he read Böhme in English. He imports his thoughts about several other German thinkers into the marginalia, but he read those authors almost entirely in German. Unusually among British recipients of Böhme, Coleridge is concerned about problems of translation, as when he fastidiously corrects 'spewed out' to 'evomited'. Regarding Böhme's theory of the seven qualities of nature in *Aurora*, he remarks: 'I have not the original German; but I cannot doubt, that the Translation is here erroneous. For the Sour we ought to read the astringent' (Coleridge 1980–2001, 1: 580,

On the topos of the mystical insight of uneducated cobblers, see Muratori 2012.

[9] Coleridge 1983, 1: 150. Compare Coleridge 1993, 392: 'the illuminated Teutonic Theosopher and Shoemaker, honest JACOB BEHMEN'.

[10] Böhme 1764–81, 1: 139; compare Böhme 2013 [1612], ch. 14.39, 426–7. Quotations from *Aurora* are usually from the Law edition, namely the one that Coleridge knew and annotated; chapter and paragraph number are as given in the version known as *Die Urschriften* (Böhme 1963–6), reproduced and translated in Böhme 2013 [1612]; see also Böhme 1730.

[11] Böhme 1764–81, 1: 54; Böhme 2013 [1612]; ch. 6.24, 206–7.

666; see also 588, 613).[12] He records a wish to have the expensive German volumes purchased for him (Coleridge 1980–2001, 1: 688).[13]

In 1810 he provided an interim report in a letter to Lady Beaumont. The letter responds to a pamphlet Lady Beaumont had provided, entitled 'An account of an English hermit, by a respectable Clergyman', which included a discussion of the so-called hermit's interest in Böhme.[14] Coleridge writes: 'Of Jacob Behmen I have myself been a commentator', a word that anticipates his above-quoted description of Law as Böhme's 'commentator' (Coleridge 1956–71, 3: 278, 21 January 1810). At this point, he strikes a relatively sceptical note. He suggests that Lady Beaumont will not 'derive any advantage' from reading Böhme, because the theosopher is insufficiently practical: rather than address the moral aspects of leading a Christian life, he elaborates theories that—even if true—merely 'delight the *curiosity*' (Coleridge 1956–71, 3: 278). This anxiety reappears in the marginalia, especially when Coleridge exhorts anyone who might be reading the volumes he annotated: 'Read then in meekness—lest to read him at all, which might be thy folly, should prove thy Sin' (Coleridge 1980–2001, 1: 558). In the 1810 letter, Coleridge conjectures that the obscurities in Böhme derive from a subjective bias: he associated his intuitions with certain accidental images and feelings, including the language of 'chemical experiments' that he would have learned about from the alchemist Robert Fludd. Nevertheless, Böhme was, he tells Lady Beaumont, 'an extraordinary man'. As this letter indicates, Böhme's difficult prose served as a paradigm case for Coleridge's long-held principle 'always *to suppose myself ignorant of a Writer's Understanding, until I understand his Ignorance*'.[15]

Coleridge's main engagements with Böhme probably took place in 1808–10, then June 1817–18; some of the marginalia are dated, the final date being 1826 (Coleridge 1980–2001, 1: 555–6). At least three main stages of Coleridge's reading may be detected, based both on his handwriting (some entries are in pencil, others in a very clear hand in ink, still others in ink but in a less clear hand) and on his occasional dating and cross-referencing (e.g. Coleridge 1980–2001, 1: 555). His marginalia testify to a growing appreciation of Böhme, as he read, re-read, and commented upon *Aurora*, the *Three Principles*, *Mysterium Magnum*, *The Threefold Life of Man*, *Forty Questions of the Soul*, *Signatura Rerum*, *Clavis*, and *On the Election of Grace*. As with much of his marginalia, Coleridge expected that others would read his notes: at one point, for example, he directly addresses his sons, Hartley and Derwent (Coleridge 1980–2001, 1: 615). Writing marginalia enabled him to try out

[12] Coleridge 1980–2001, 1: 580, 666; see also 588, 613.

[13] In 1817 (and again in 1819) Ludwig Tieck planned to send Coleridge a copy of Böhme in German ('Jac. Böhm in Quart.'), but there is no evidence that this intention materialized: Zeydel 1931, 93–4.

[14] The 'Account' was sent from Lady Beaumont to Dorothy Wordsworth in December 1809. The 'hermit' in question was the nonjuror Thomas Gardiner (Beaumont 2021, 105, 108, 211–14).

[15] Coleridge 1956–71, 3: 278. An earlier formulation of this principle appears in Coleridge 1957–2002, 1: §928 fol. 27; the best-known version is Coleridge 1983, 1: 232. On Böhme and Fludd see Saifullin 2023.

ideas, free from the anxiety of reception, anticipating a limited audience of sympathetic readers. He informed such readers that his understanding had developed radically: around 1808 he had been 'incapable of coming near Behmen' (Coleridge 1980–2001, 1: 648). Yet, by 1818 Coleridge was claiming that he had succeeded in disentangling Böhme's deep insights from his errors and obscurities. In a long note to *Aurora* he now identified two 'errors' in Böhme. First, less seriously, Böhme's theory of *Natursprache*, or language of nature, according to which words reveal their intrinsic significance through their sounds, is in Coleridge's opinion superficial—'a mere scum'.[16] Assonances between words, which Böhme understands as proof of a conceptual link, are (for Coleridge) nothing more than the basis for personal associations. In chapter 20 of *Aurora*, Böhme had claimed that the language of nature contains the knowledge of all things: when Adam named the animals in the Garden of Eden, he did so in the language of nature, pronouncing words that matched each creature's qualities. For Böhme, the language of nature does not correspond to any spoken language, but is rather the '*Root or Mother* of *all* languages, which are in this world'.[17] Yet, for Coleridge, the best proof that the theory of *Natursprache* is arbitrary is the fact that the assonances only work within a particular language, specifically German. The same link between sounds and concepts cannot be reproduced across languages, despite Böhme's claim that the language of nature, as the 'mother' of spoken languages, should remain visible, or rather audible, in all languages, not just German.

Second, and crucially, Böhme's cosmogony threatens to collapse into pantheism. According to Coleridge, the crux of the problem is Böhme's understanding of the Absolute as an all-encompassing Divinity, from which the Godhead 'evolves', and all creatures 'bud forth'.[18] For Coleridge, this is ultimately a kind of pantheism akin to Spinozism. But he does consider himself well placed to recognize the contours of the problem because it was the speculative danger that troubled him in the 1790s (and beyond) and that—as we have seen—he credited Böhme with having helped him keep at bay, albeit without fully exonerating the German thinker.

2.2 Böhme the Natural Philosopher

It was Coleridge's further study of *Naturphilosophie* that enabled him to 'come near' Böhme after 1810, as this section will show. Coleridge had studied in Göttingen with Johann Friedrich Blumenbach in 1799 (Levere 1981, 18–19; van Woudenberg

[16] Coleridge 1980–2001, 1: 602; also 607, 634, 652. See further Cheyne 2020, 128. Elsewhere Coleridge comments on the subjectivity of Böhme's imagery: 'J. Bœhmen's mind may be well illustrated from Dreams—there is meaning, important meaning, in both; both the exponents are almost accidental—such infinity of synonimes exist in the language of vision, considered as the language or representatives of Sensations & Notions' (Coleridge 1957–2002), 3: §3692.

[17] Böhme 1764–81, 1: 204; compare Böhme 2013 [1612], 604–5, ch. 20, 90–91.

[18] Coleridge 1957–2002, 1: §602.

2017, 98–100), and his readings in German *Naturphilosophie* very clearly accompany his readings of Böhme. In Coleridge's interpretations, Böhme almost becomes the forerunner of the scientific explorations of Coleridge's age. In view of the reception of Böhme in Germany from the end of the eighteenth to the first half of the nineteenth century, this might at first glance hardly seem original. For instance, philosophers who experimented with magnetism, including Franz von Baader and Schelling, were also readers of Böhme. Again, the physicist Johann Wilhelm Ritter, who experimented with magnetic cures, was in direct contact with the group of early German romantics with the Schlegel brothers in Jena at its centre, in which Böhme was read and appreciated (Muratori 2016, part 1, especially 29–30; Mayer 1999).

It is the natural-philosophical idea of the magnetic polarities which resonates with Böhme's account of the struggle between opposites, as discussed in the early chapters of *Aurora*, where Böhme describes how the 'qualities' in nature oppose each other and thus create life. In his marginalia to *Aurora*, Coleridge writes that 'this polar Unity, of which the Magnet is the earthly Symbol, Behmen calleth the two Qualities in one—namely, Expansion or the +, and the Negation or Counter-power, = – / and this later he calleth evil—i.e. contemplated in its separate action, and because he hasteth to the moral interest which predominates in him' (Coleridge 1980–2001, 1: 563).[19] The latter tendency in Böhme, the preoccupation with a 'moral interest' as opposed to regarding the natural world as ethically inert (as 'our gallicizing Laboratory Atomists' might: Coleridge 1980–2001, 596), was one that Coleridge especially valued.

Coleridge takes *Aurora* as the starting point for his interpretation of Böhme: the vast majority of the marginalia refer to this book, Böhme's first and most natural-philosophical work. Coleridge establishes some general parallels between Böhme and contemporary science, notably through the link we have just noted between Böhme's 'qualities' and the polarities of the magnets, just as the Schellingian scientists did. Coleridge does far more than that, however. He seeks to provide a sort of step-by-step guide to reading Böhme in scientific terms—for 'as few Philosophers better deserve, so none more require a running Comment'—and in so doing he elaborates a unique interpretation (Coleridge 1980–2001, 633). For Hegel, Böhme stood on the threshold of modernity, hinting at further philosophical developments which were only realized later, by Descartes, when philosophy became properly modern.[20] For Coleridge, on the other hand, Böhme appears even to have been ahead

[19] On the concept of 'quality' see Bohme 2013 [1612], ch. 1.3–5, 116–17; Böhme 1764–81, 1: 23: 'Now here we must consider, what the word QUALITY means, or is. A Quality is the Mobility, boiling, springing, and driving of a Thing. … As, for Example, *Heat*; which burns, *consumes*, and drives forth all, whatsoever comes into it, which is not of the same Property; and again, it *enlightens* and warms all cold, wet, and dark Things; it compacts and hardens soft Things.… It contains likewise two other Kinds in it, 1. Light, and 2. Fierceness'. For a summary of Böhme's main philosophical tenets, including the doctrine of the qualities, see Schmidt-Biggemann 2023.

[20] Hegel 2009, 93–103, especially 103: 'This path is profound and inward to the highest degree; it is moving and full of sensibility, and someone familiar with Boehme's own forms will discover this depth and inwardness; but of course we cannot completely reconcile ourselves to the form his

of his time. Böhme's theosophical insight—as Coleridge's marginalia seem to claim—may now be understood within the framework of certain modern approaches. Indeed, Coleridge superimposes his readings of *Naturphilosophie* on his interpretation of the 'German theosopher' (Coleridge 1980–2001, 1: 577).

With regard to certain theories presented in *Aurora*, Coleridge even surmises that the lack of a physical explanation for some of Böhme's descriptions is 'rather proof of the imperfection of our present Physiology than an Objection to Böemen's Position' (Coleridge 1980–2001, 1: 577). In this note, Coleridge discusses Böhme's notion of the 'counsellors', or 'Räte': the eyes, ears, nose, tongue, and sense of touch. This is a kind of theory of perception, which is couched simultaneously in physical and moral terms. Böhme talks of 'quel Adern' (translated by John Sparrow as 'Fountain-veins'), which are connected to the heart.[21] This provides Coleridge with the starting point to link Böhme to a theory that had been discussed since Aristotle, namely that there are nerves stemming from the heart. Aristotle, in the *Parts of Animals*, claims that the heart has sinews (*neura*), a view that was extensively debated from the Renaissance into modernity.[22] Coleridge evidently regards Böhme as standing in this tradition; yet this is far from being a sign that Böhme's knowledge of physiology was questionable or outdated. Rather, 'this View may be beyond the physiology and metaphysics of the Day', one that scientists were only catching up with in the early nineteenth century (Coleridge 1980–2001, 1: 577).

This comment is one of the crucial ones in the marginalia, because it also reveals Coleridge's *modus operandi*. This note, more clearly written than many later ones, bears a date, albeit one that appears mysterious: 1800. Since Coleridge acquired the volumes in 1808, this is either a slip, or a record of the date when he first conceived this idea (as noted by Haven in Coleridge 1980–2001, 1: 577n.). Coleridge also notes underneath the comment: 'approved on reperusal 27 June 1817' (Coleridge 1980–2001, 1: 577). This shows clearly that his reading of Böhme was a long-term work in progress, which led him constantly to reassess his views on Böhme's philosophy, albeit with certain themes that he invariably regards as crucial. One of these is Böhme's relation to science. It is an example of how reading the marginalia poses two challenges: how to establish the chronology of Coleridge's engagement with Böhme, as we have discussed above; and how to place the marginalia within the larger frame of Coleridge's studies of related authors and topics. The major intertexts include W. G. Tennemann's history of philosophy and the

thought takes—especially not in its details. With this comment we conclude our discussion of the preliminary period of modern philosophy and proceed to its first period proper, which we begin with Descartes'.

[21] Böhme 1764–81, 1: 48; compare Böhme 2013 [1612], ch. 5.23, 192–3.

[22] *Marginalia*, 1: 577. Böhme 1764–81, 1: 48–50 (compare Böhme 2013 [1612], ch. 22–33., 192–7). See Aristotle, *Parts of Animals* III.4 (666b14–17); id., *Generation of Animals* V.7 (787b11–19). On Aristotle's conception of neuron see Gregoric and Kuhar 2014. On the reception of Aristotle's anatomy in the Renaissance, see Cunningham 1997. On the reception of Aristotle's theory of the sinews of the heart see Ishizuka 2006.

natural-philosophical works by Steffens, Lorenz Oken and H. C. Oersted. Steffens's work is crucial to Coleridge's interpretation of Böhme.

One of Coleridge's marginal notes comments in detail on a passage from *Aurora*, chapter 9.23, where Böhme explains how light is born out of the interplay of the qualities:

> Wen das licht in der Bittern qualitet geboren wird / Das ist / wen das Bitter vnd trucken quellen / das sisse quel wasser des lebens fenged / vnd trincked des / So wird der Bitter geist lebendig / in dem Herben geist / vnd ist der Herbe geist nun wie ein Schwanger geist / der des lebens Schwanger ist / vnd mus das leben Immer gebären /. Den das Süsse wasser / vnd in dem Süssen wasser das licht / steiged nun Immer in der Herben qualitet auff … Den die Herbe qualitet libed das süsse wasser / Darumb / das in dem Sissen wasser der geist des lichtes geboren wird / vnd trencked die Herbe / Harte vnd kalte qualitet vnd erleuchtet sie. vnd wermed Sie / Den in dem wasser Hitze / vnd licht stehed das leben.

> When the Light is generated in the bitter Quality; that is, when the bitter and dry Fountain-sources *catch* the sweet Fountain-water of Life, and *drink* it, then the bitter Spirit becomes living in the astringent Spirit, and the astringent Spirit, which is as a Spirit impregnated with Child, is impregnated with Life, and must continually generate the Life. For the sweet Water, and the Light in the sweet Water, rise up continually in the astringent Quality, and the bitter Quality triumphs continually therein. … For the astringent Quality loves the sweet Water. And first, because in the sweet Water the Spirit of Light is generated, and *imbibes* or gives Drink to the astringent, hard and cold Qualities; also it enlightens them, and warms them; for in Water, Light, and Heat, the *Life consists*.[23]

Coleridge's discussion of this page contains the following conclusion: 'Water is the gravitation in its greatest proximity to Light—therefore, is Water the differencible Indifference, or Depth (i.e. Body, Base) in its most passive state of modifiability' (Coleridge 1980–2001, 1: 597). 'Gravitation' is a term evidently not derived from Böhme's vocabulary. In his letter to Tieck of 1817, Coleridge had stated that 'Sound was = Light under the præpotence of Gravitation, and Color = Gravitation under the præpotence of Light', a thought that he links to his schoolboy reading of Böhme's *Aurora*.[24] Obscure though this formulation appears, it is does have a basis in Böhme's text: in chapter 10 of *Aurora*, for instance, we read that 'das licht / oder der Plitz / macht den klang' ('the Light or Flash, makes the Ringing', where 'ringing' could also be rendered as 'sound').[25]

The comment in the marginalia and the statement in the Tieck letter together become more transparent if one considers both the term 'Gravitation', and the equally mysterious expressions 'under the præpotence of Gravitation' and 'under the præpotence of Light', as translations from German. In particular, Henrik Steffens's terminology in his *Grundzüge der philosophischen Naturwissenschaft*,

[23] Böhme 1764–81, 78–9; Böhme 2013 [1612], ch. 9. 22–3, 272–4, from which the German is taken.

[24] Haven (1966, 178) argues that Coleridge overstates his 'claim to have anticipated the *Naturphilosophen*' in having conceived such an elaborate theory before he had read their work.

[25] Böhme 2013 [1612], ch. 10.15, 288; Böhme 1764–81, 1: 84.

which Coleridge knew at least since 1815, was undoubtedly a source of inspiration.[26] Steffens writes:

> Das Wasser enthält den Keim aller Differenzen (als Abweichungen) in sich, die aber in ihm verschlossen ruhen. Es bezeichnet die Aktivität des Lichts unter der Potenz der Schwere, so wie die relativen Differenzen der festen Erdmasse die Passivität der Schwere unter der aktiven Potenz des Lichts. (Steffens 1806, 55)
>
> [Water contains the germ of all differences (as deviations) in itself, yet they rest enclosed in it. It designates the activity of light under the predominance of weight, just like the relative differences of the compact matter of the earth [designates] the passivity of weight under the active predominance of light.]

The expressions used by Coleridge ('under the præpotence ...') perfectly match Steffens's 'unter der Potenz der Schwere ... unter der ... Potenz des Lichts'.[27] Steffens even directly mentions the formation of colours, just as Coleridge does in the letter to Tieck: 'So deuten die Farben die Grade der Abweichung unter der Potenz des Lichts an' ('Colours represent the degree of deviation under the power of light').[28]

Steffens understands 'Schwere' and 'Licht' as polar opposites: their struggle is the most basic opposition in nature. 'Schwere' is the permanence, the being in space, the reality of nature. 'Licht' is the impermanence, the movement in time, the eternal ideal of nature. Thus, although Coleridge's rendition of 'Schwere' with 'gravitation' is not incorrect, what is meant in the natural-philosophical tradition that lies behind this concept is something more akin to the Latin 'pondus': the heaviness that characterizes matter.[29] In Böhme's explanation of the qualities, and especially of the generation of light from water, Coleridge thus reads the same interplay of opposites theorized by Steffens in the *Grundzüge*: 'Profoundly therefore doth Behmen affirm that in Water, Heat, and Light the *Life* consists!'[30] Furthermore, Coleridge regards light and 'Schwere' as opposites that are akin to Böhme's insistence on the struggle between light and darkness. In a detailed letter to his Swedenborgian friend, Charles Augustus Tulk, Coleridge thus interprets the division of light from darkness in *Genesis* as the origin of the 'Licht'/'Schwere' opposition: 'And God divided the *Light* from the *Darkness*—i.e. Light from Gravitation: ... thus Darkness and the Depth became relative, and the two Poles of the material

[26] On his first visit to Steffens, the Danish poet Adam Gottlob Oehlenschläger found Böhme's *Aurora* lying open on his desk (Oehlenschläger 1850, 205). Further relevant connections are drawn in Hansen 1999.

[27] A variation of this is 'Gravitation under the form and presidentcy of Light' (Coleridge 1980–2001, 1: 593–4). Abrams (1972, 462) expounded the context in *Naturphilosophie*, but unnecessarily dismissed 'Coleridge's reference of his views on light and sound' to Böhme as 'misleading'.

[28] Steffens (1806), 54. Likewise, annotating Oken's *Erste Ideen zur Theorie des Lichts, der Finsternis, der Farben und der Wärme* (1808), Coleridge notes: 'Thus, Sound is Light in Metal, Color Metal in Light' (Coleridge 1980–2001, 3: 1025).

[29] In another note: 'Sound, which is but Light under the paramouncy of Gravitation. It is the Mass-Light' (Coleridge 1980–2001, 1: 572), 'Mass' again connotes 'pondus'. See further Levere 1981, 156–8.

[30] *Marginalia*, 1: 597.

Universe are established, viz. Light and Gravitation' (Coleridge 1956–71, 4: 771, September 1817). This gravitation–light polarity henceforth became key to Coleridge's whole intellectual system.[31] In the letter to Tulk, terming that system '*Heraclitus redivivus*', he again asserts a pre-Schellingian tradition involving Bruno and Böhme (Coleridge 1956–71, 4: 775). Heraclitus was indeed on Coleridge's mind as he read Böhme (Coleridge 1980–2001, 1: 584).

Coleridge thus reads *Aurora* as a real (not metaphorical), prophetic (not outdated) account of how nature works, and of how it came into being. Coleridge pushes Böhme even further in the direction of modern chemistry, when he claims that Carbon and Nitrogen are the 'poles of gravitation', and 'oxygen and hydrogen' those of light (Coleridge 1980–2001, 1: 597). This gives rise to a scheme in the shape of a diamond that Coleridge repeats throughout the marginalia in different forms (Coleridge 1980–2001, 1: 633) (Figs. 2.2 and 2.3). Again, this aligns with Steffens' explanation that from the two principal forces of 'Licht' and 'Schwere' derives what he terms a 'quadruplicity'.[32] For Steffens, 'the north-south polarity stands under the power of weight. .... The east-west polarity stands under the power of light. The north-south polarity is determined by the magnetic opposition, the east-west polarity through the electric one.... Each material body contains the whole quadruplicity, because as matter is it the union of all dimensions' (Steffens 1806, 41).[33]

The relationship of magnetism and electricity is prominent in Coleridge's marginalia to Böhme, too. It is no coincidence that Coleridge also annotated works by Oersted, who discovered electromagnetism. For instance, Coleridge writes that Böhme 'supplies the fundamental position of [... the] Dynamodynamis of Magnetism, i.e. Animal Magnetism' (Coleridge 1980–2001, 1: 581). Here, too, Coleridge reads Böhme through the filter of German *Naturphilosophie*. Again, with Steffens in mind, Coleridge complains in one note that 'The reader must constantly bear in remembrance Behmen's inconsistency in naming the four powers—thus in §26 we must understand Attraction, Contraction, Repulsion and Dilation as the elementary Powers' (Coleridge 1980–2001, 1: 641).[34] This is because Böhme usually lists seven powers, or qualities (the exact list varies throughout his work), rather than four, as Coleridge would wish.[35]

---

[31] See Levere (1981, 133), who, however, omits Böhme from an otherwise cogent list of influences. Coleridge's *Opus Maximum*, which is beyond the scope of this chapter, is a key text.

[32] On Steffens' quadruplicity see Schmidt 2013, especially 108–9.

[33] 'Die Nordsüd-Polarität steht unter der Potenz der Schwere, und verliert sich in der Masse des Planeten, als das Seyende, Gebildete. Die Ostwest-Polarität steht unter der Potenz des Lichts, verliert sich in das Werden des in Bildung Begriffenen. Die Nordsüd-Polarität wird durch den magnetischen Gegensatz, die Ostwest Polarität durch den elektrischen bezeichnet.... Eine jede Materie enthält die ganze Quadruplicität, weil sie als Materie die Einheit aller Dimensionen ist.'

[34] Coleridge 1980–2001, 1: 641.

[35] The number seven has of course a Biblical echo, which Böhme exploits directly when he talks of the spirits of God in *Aurora*, ch. 12.3 (Böhme 2013 [1612], 338).

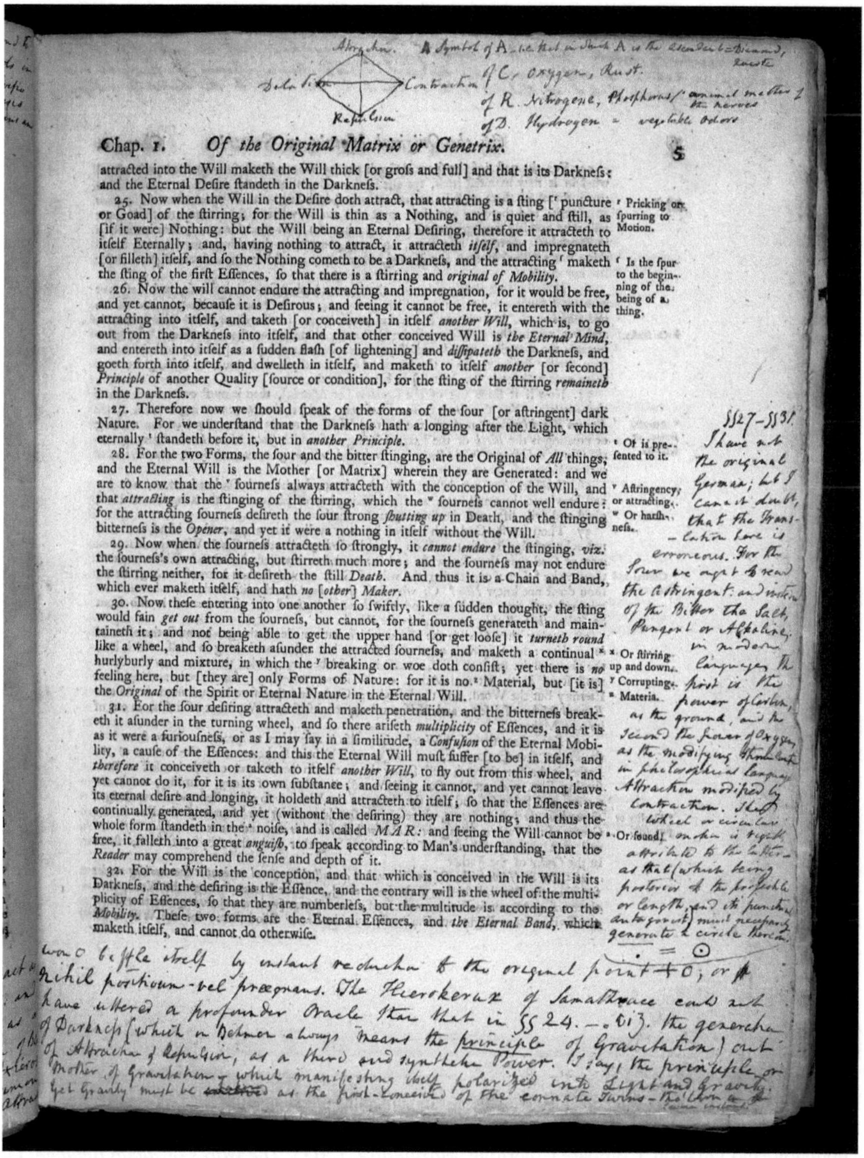

Fig. 2.2 Coleridge's diamond, within Böhme's *The Threefold Life of Man*, displaying Attraction, Contraction, Repulsion, and Dilation. Compare Coleridge, *Marginalia*, 1: 666. *The Works of Jacob Behmen*, II i 5, British Library. Copyright: British Library Board 22/08/2023

Steffens had indeed posited that the entire world is characterized by quadruplicities. In space, there are four main areas (north, south, east, west); in time, there are four seasons and also four parts of the day (morning, midday, afternoon, night); matter is made of four elements (fire, water, air and earth); and four substances

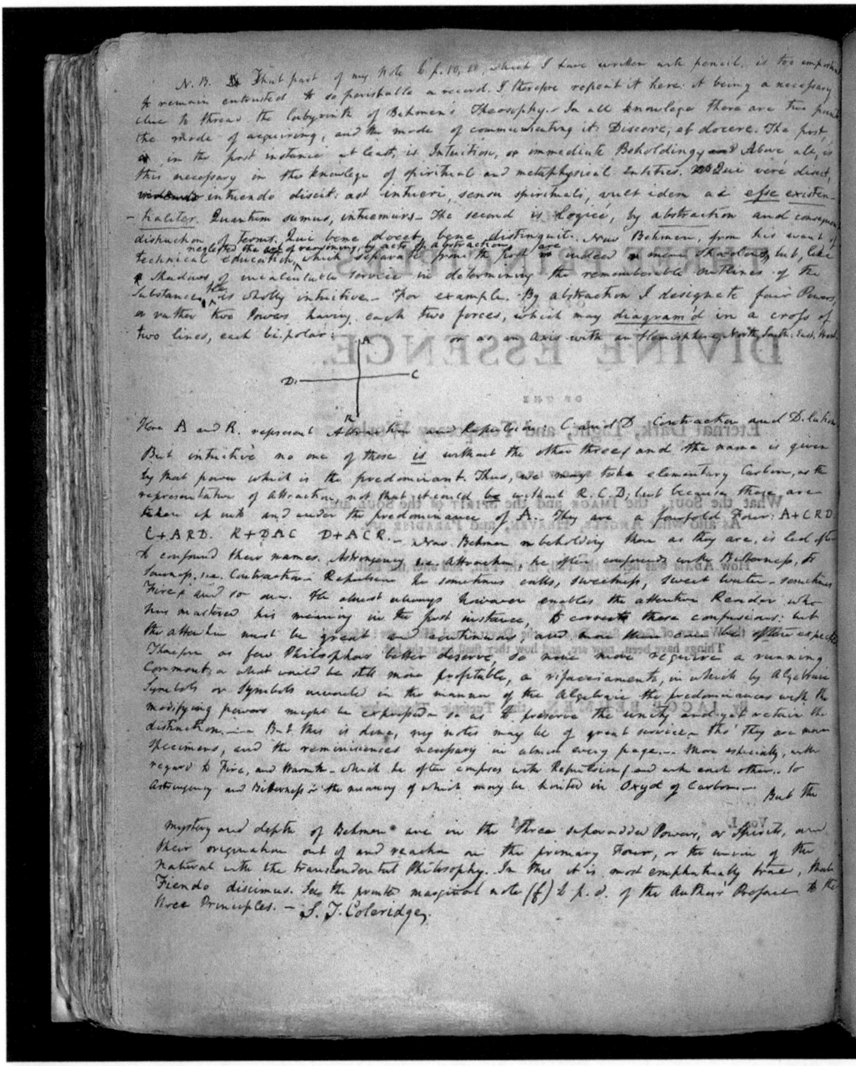

Fig. 2.3 Another version of Coleridge's diamond, on blank page facing 'Preface to the Reader' in Böhme's *The Three Principles of the Divine Essence*, displaying Attraction, Contraction, Repulsion, and Dilation. Compare Coleridge, *Marginalia*, 1: 633. *The Works of Jacob Behmen*, I ii [2], British Library. Copyright: British Library Board 22/08/2023.

(oxygen, hydrogen, nitrogen, carbon) are responsible for the changes and oscillations in nature (for instance through oxidation and hydrogenation: Steffens 1806, 43–4, 60, 148–9). Steffens thus organizes the whole of nature according to the idea of quadruplicity. For instance, he writes that 'precious metals contain the whole quadruplicity in themselves': gold appears to him to be closest to east, silver to west, thus connecting the areas of the globe, and also the idea of magnetic polarities, to the different characteristics of the metals (Steffens 1806, 89–90). For Steffens, 'in individual parts, just as in the whole, one therefore finds the same inner organization'—an organization always divided into four.[36]

Coleridge's opinion of Steffens's *Naturphilosophie* fluctuated over time, just as did his appreciation of Böhme. In one of the 1818 marginalia, he directly links Steffens, along with Newton, Kant and Schelling, to Böhme: here he states that he had by then 'outgrown' the 'mistake[n]' conviction that there are 'four primary forces' (structured along two axes as magnetism and electricity), and had instead reached the conclusion that there are only two opposing '*primary* forces', 'Attraction and Contraction'.[37] This latter view would appear more compatible than the 'four forces' version with Böhme's theory of the fight between opposites, so that in this respect Coleridge moved closer to Böhme rather than further away. In his 1818–19 *Lectures on the History of Philosophy*, indeed, Coleridge warmly endorsed not only the 'spirit of love which runs through' Böhme, but specifically his vision of polarity: 'I trace in him the love of action and that constant sense of the truth that all nature is in a perpetual evolution, that two great powers are for ever working, manifesting themselves alike in the apparently inadequate and inanimated, and in intellectual nature, namely, the powers by which each particular endeavours to detach itself from nature and the counteracting powers by which nature is still bringing back each of her creatures into itself' (Coleridge 2000, 2: 484). What Coleridge values in Böhme is thus the dynamism of his vision of nature. The 'two great powers' are the two qualities introduced by Böhme in chapter 1 of *Aurora*, in which they are presented as the powerhouse of life itself. This dynamism is guaranteed by the fact that the qualities—the good and the evil one—are in fact 'One Thing', and therefore must fight each other 'in this World in all Powers, in the Stars and the Elements, as also in all the Creatures'.[38] Coleridge is therefore right in pointing out that for Böhme nothing is truly inanimate. The mechanism of the two qualities (or powers, in Coleridge's terminology), is thus the point of articulation between Böhme's natural philosophy and his understanding of a moral struggle taking place between Good and Evil. For Coleridge, this becomes a struggle that enlivens nature, and at the same time points beyond nature, as he underlines the directionality of these powers: one that seeks detachment from nature, and the other which instead causes full embedment in nature itself. This is why Coleridge is interested not simply in Böhme's general idea of a combination of natural philosophy and theology, but in the details of this combination: namely in the mechanism of polarity, which, he believed, linked Böhme directly to the new research of Steffens.

[36] Steffens 1806, 45: 'Im Einzelnen wie im Ganzen findet man daher immer dieselbe innere Organisation'.

[37] Coleridge 1980–2001, 1: 660. Compare Coleridge 1983, 1: 297: 'the transcendental philosopher says; grant me a nature having two contrary forces, the one of which tends to expand infinitely, while the other strives to apprehend or *find* itself in this infinity, and I will cause the world of intelligences with the whole system of their representations to rise up before you.'

[38] Böhme 1764–81, 23; compare Böhme 2013 [1612], ch. 1.2, 116–7.

2.3 Conclusion: Böhme's 'Spiritual Chemistry' in Coleridge's Later Thought

Despite the various changes in Coleridge's views, what remained stable was his appreciation of Böhme's philosophy as a system that is both theologically and natural-philosophically valuable. Böhme is for Coleridge the herald of a different, vitalistic, natural philosophy, opposed to the 'Corpuscular or mechanic philosophy' (1956–71, 4: 956 [October 1819?], to the editor of *Blackwood's Magazine*): thus, as he says annotating Böhme, 'wonderfully did this gifted Seer fly before his Age' (1980–2001, 1: 641). Yet, historically, it was corpuscularianism that ultimately became established at the expense of the kind of metaphysical, natural-philosophical approach that Coleridge values in authors as diverse as Böhme and Steffens. The fertile ambiguity of some of Böhme's and of Steffens' modes of expression, which can and should be applied both to the study of nature and to the philosophical understanding of the whole of reality, could no longer co-exist with mechanical views of nature.[39] Indeed, Coleridge grew increasingly dissatisfied with the mainstream scientific developments of his age: having early endorsed the experimental method of Humphrey Davy, he gradually decided, still commenting in the margin of Böhme, that the English chemist's 'atomism' was inadequate (1980–2001, 1: 566, 572). In this context, Böhme embodies the promise of a theological and natural-philosophical explanation of the universe, of the kind that Coleridge himself sought to provide.

But why did Coleridge persist in thinking through 'the labyrinth of Böhme's theosophy', when apparently more satisfying material was to be found in the *Naturphilosophie* through which he interpreted Böhme (1980–2001, 1: 632)? The main reason is that Coleridge found in Böhme at least the potential for a dynamic philosophy of nature that could still preserve divine transcendence and so avoid falling back into pantheism. Schelling, whose 'System' Coleridge polemically labelled 'little more than Behmenism' in a letter to Tulk in 1818, lacked the merit of resisting pantheism.[40] In contrast, Coleridge sometimes found the Trinitarian theology of Böhme in the early sections of *Aurora* 'masterly and orthodox' (Coleridge 1980–2001, 1: 565; also 588). This is despite the fact that, as noted before, Coleridge's principal reservation about Böhme's thought is that 'he approaches so perilously near to Pantheism' (Coleridge 1980–2001, 1: 600). Coleridge requires a philosophical-religious system to begin with a personal God: a supreme intelligence, neither an impersonal force nor an 'eating, drinking, lustful God'. The crux of the critique articulated by Coleridge in 1818 is twofold. First, Böhme had spoken

[39] Paul 1973 emphasises the 'Vieldeutigkeit' (polyvalence) of Steffens' thought, which combines natural-philosophical with aesthetical ideas. According to Paul, the hybridity of Steffens is also reflected in his legacy, which is negligible in terms of pure scientific advancement, yet is found at the crossroads of natural philosophy, philosophy, literature and theology (16).

[40] Coleridge 1956–71, 4: 883, 24 November 1818. Compare Coleridge 1980–2001, 1: 619: 'Schelling and his Followers are in the same case as Behmen—only that Behmen oscillates or rather leaps from Error to the Truth or to the Neighborhood of the Truth, while Schelling *settles* on the *Error*.'

of a 'birth of God'. Exemplarily, in chapter 12 of *Aurora* we read that 'the Spirits of God rise up continually in one another, and have a Sport or Game of Love in their Birth or Geniture, and yet every Spirit keeps his natural *Seat* or Place, in the Birth or Geniture of God'.[41] This implies a dynamism within God, due to the interplay of the seven spirits. It also implies, in line with a longstanding mystical tradition, differentiating this living, dynamic God, and a Divine Abyss that must somehow remain untouched by these enlivening mechanisms, and must therefore be thought as prior to God as Creator, and thus as impersonal.[42] In Coleridge's terms, this is tantamount to 'irreligious Pantheism' (1983, 1: 152). Böhme's error in this respect is one of 'Pars pro Toto', mistaking the part for the whole: in Coleridge's view, what he is in fact describing at such moments, triggered by the dynamics of the seven spirits, is the birth of *nature*, not of God (Coleridge 1980–2001, 1: 601).

Second, again consistently with his desire to assert a biblically-based trinitarian Christianity, Coleridge is anxious about the logical precedence that Böhme accords to Lucifer in the order of creation. The central role of Lucifer in Böhme is a direct derivation of his linking the natural and the theological levels, on the basis that everything that lives must contain in itself an interplay of good and evil (and indeed Böhme uses the adjective for moral evil, 'böse', when discussing the dynamics of the two qualities, the good and the evil one). The preface to *Aurora* is exemplary in this respect, where Böhme describes how the opposition in nature corresponds to a struggle between moral good and moral evil within the human being, too—and Lucifer is here the representative of the evil force.[43] Paradoxically, therefore, it is Böhme's combination of natural philosophy and theology, which Coleridge appreciated, which leads him to assign to Lucifer such a central role. Coleridge's explanation, which at first sight appears cryptic, combines these reservations about Böhme's worldview:

> [Böhme] preposterizes [makes an after-event come first] the Consequent of the Fall into the absolute First, and ignorant of the only sense, in which it is other than an Abstraction, he makes it a mere abstraction, and then by a second lawless abstraction, that of Subject and Object from the idea of the Absolute, he identifies the former with God as Omneity [allness], out of which the Living Godhead evolves, as an eternal Birth, as the Creatures bud forth ⟨&⟩ individualize, each out of its pre-impregnated Nature. In the first instance his error is radically the same as that of Spinoza, and in both instances the same as that of Schelling and his followers.[44]

---

[41] Böhme 1764–81, 1: 111; compare Böhme 2013 [1612], ch. 12.57, 354–5.

[42] From around 1620, Böhme will employ the term 'Ungrund' (the Un-Grounded) to define this Divine before God. In *Aurora*, this concept is already present when Böhme talks of God the Father as a dark valley ('fünster Thal'). Böhme 2013 [1612], ch. 3.22, 154–5.

[43] Böhme 2013 [1612], preface, paragraph 22, 82–3.

[44] Coleridge 1980–2001, 1: 602. Compare 1: 645–6 on the possible metaphysical schemes, and 1: 643: 'It is vexatious, that from the want of one distinction, of the second Birth from the first, Behmen should have so confusedly reflected truths of which, in their unity and co-inherence, he had so clear an intuition'.

Coleridge once again locates Böhme within a tradition; but this time it is the tradition of Giordano Bruno, Spinoza, and Schelling (pantheistically interpreted), not the tradition of *Naturphilosophie* (as philosophical speculation intertwined with laboratory practice).[45] Indeed, Coleridge uses Schelling's terminology in referring to Böhme's idea of an abyssal Divine, called in this note the 'Absolute' (a term which does not belong to Böhme's vocabulary). Coleridge fears that by using the same explanation for the dynamism in God and in nature, Böhme ultimately identifies the one with the other. This attitude seems to Coleridge particularly strong in *Aurora,* where Böhme repeatedly calls nature the body of God ('der Leib Gottes').[46] Crucially, in the marginalia to *Mysterium magnum* (which are much less copious than those to *Aurora*) Coleridge detects an evolution in Böhme's philosophy, as the 'pantheism' of *Aurora* seemed to be softened in this later work: 'With great delight I find, that Behmen guided by the light of a sincere love of truth, worked himself out of the Pantheism (God = Chaos) of his earlier writings: and seems in this Tract to have emerged into the full Day' (Coleridge 1980–2001, 1: 677). Yet, this favourable impression did not last long—literally the space of a page, and a day—for Coleridge soon writes with disappointment: 'As I read on, I have found that this first Chapter is a deceptive Promise: that Behmen soon deviates into his original error, πρωτοψευδος, and places the polarities *in* the Deity (Coleridge 1980–2001, 1: 678).

Coleridge never resolved this ambivalence in assessing Böhme's thought, any more than he could more generally achieve an ultimate synthesis of a dynamic view of nature with Christian belief. In the conclusion to his 1825 *Aids to Reflection* he re-states his two reservations about Böhme's work, the arbitrariness of the *Natursprache* and the tendency to pantheism (Coleridge 1993, 384). Nevertheless, Böhme survived Coleridge's disillusion with Schelling and even with Steffens, retaining a central place in his mind in the 1820s. This was possible because he detected in Böhme a 'more glorious Antecedent', a true conception of a transcendent 'Deity in the eternal Plenitude', even if this is visible to the reader only in flashes (Coleridge 1980–2001, 1: 600). In Cheyne's words, Coleridge 'surmises that Böhme's pantheist tendency contains the seeds of its own correction through intimating a higher, unseen presence' (Cheyne 2020, 137). Linked to this, Coleridge saw greater authenticity in the 'unlearned'[47] Böhme's struggle for expression than in the academic style of the 'Grand Siegnoir', Schelling (Coleridge 1956–71, 4: 792, 13 December 1817 to J. H. Green). Coleridge continued to accord Böhme

[45] On Coleridge and the (German) pantheism controversy: McFarland 1969, Berkeley 2007.

[46] Böhme 2013 [1612], ch. 2.16 (p. 132), ch. 23.48 (p. 688) and 92 (p. 700), ch. 24.15 (p. 708).

[47] Coleridge 1980–2001, 1: 616. The metaphor of the flash recurs in this passage: 'being a poor unlearned Man he contemplated the Truth and the forms of Nature thro' a luminous Mist, the vaporous darkness rising from his Ignorance and accidental peculiarities of fancy and sensation, but the Light streaming into it from his inmost Soul. What wonder then, if in some places the Mist condenses into a thick smoke with a few wandering rays darting across it & sometimes overpowers the eye with a confused Dazzle? The true wonder is, that in so many places it thins away almost into a transparent Medium, and Jacob Behmen, *the Philosopher*, surprizes us in proportion as Behmen, the Visionary, had astounded or perplexed us' (Coleridge 1980–2001, 1: 558).

special status as a religious thinker, echoing the epithet 'divinely illuminated' (from the 'Life' that prefaced the Law Edition) in the conclusion to his 1825 *Aids to Reflection* (Coleridge 1993, 392). Whether or not Böhme was inspired in the same sense as the biblical authors, when he claimed to have had experienced 'a thorough view of the Universe' compressed into fifteen minutes, depends, as Coleridge cautiously notes, on how inspiration is defined (Coleridge 1980–2001, 1: 560). Coleridge merely emphasises Böhme's own claim to have written down '*unaided* and *partial* recollections of inspired Truths' (Coleridge 1980–2001, 1: 560; compare Coleridge 1957–2002, 4: §5118). How seriously Coleridge took this question emerged when he attempted to explicate the Mosaic creation narrative in the terms of *Naturphilosophie*. In this project, he even treated Böhme as an authority on what was and was not authentic in the opening of Genesis.[48]

Coleridge's extensive reading of Böhme, then, was driven by a strong agenda, which at times led him to override Böhme with the German *Naturphilosophen*, a combination that frequently makes the marginalia difficult to interpret. Yet, this method also enabled Coleridge to establish a unique approach to Böhme, one that went beyond declaring a generic similarity between the 'mystic' and new scientific theories, such as magnetism, as others had already done, and instead strove for a precise system of references, making of Böhme a theologian and scientist ahead of his time, and even ahead of Coleridge's own age. Coleridge's many comments frequently reveal how 'Jacob Behmen, the *Philosopher*, surprizes us in proportion as Behmen, the Visionary, had astounded or perplexed us' (1980–2001, 1: 558) (Fig. 2.4). For the reader in the twenty-first century, is the task of tracking Coleridge's interpretation of 'Behmen, the *Philosopher*' as he laboured at the '*Heraclitus redividus*' system a worthwhile one? Coleridge himself made a case for it when he told the politician Lord Liverpool that under the predominant corpuscularian system, 'a few brilliant discoveries have been purchased at the loss of all communion with the life and the spirit of Nature' (1956–71, 4: 761). Our present-day predicament underlines that the need for a different path, in search of living ideas, has hardly diminished since then. In Coleridge's interpretation of Böhme's thought, the 'living ideas' are those intuitions that permit an anachronistic approach, as it can be properly appreciated only within an intellectual framework that was not available in Böhme's time: it is through the later approach known as *Naturphilosophie* that, for Coleridge, the work of Böhme can be properly understood.

[48] 'On Genesis 1.1–31 and 2.1–3', in Coleridge 1995, 2: 790–4. Compare Coleridge 1980–2001, 1: 612: Böhme's view is 'more Correspondent to the Mosaic account' than that of the *Naturphilosophen*.

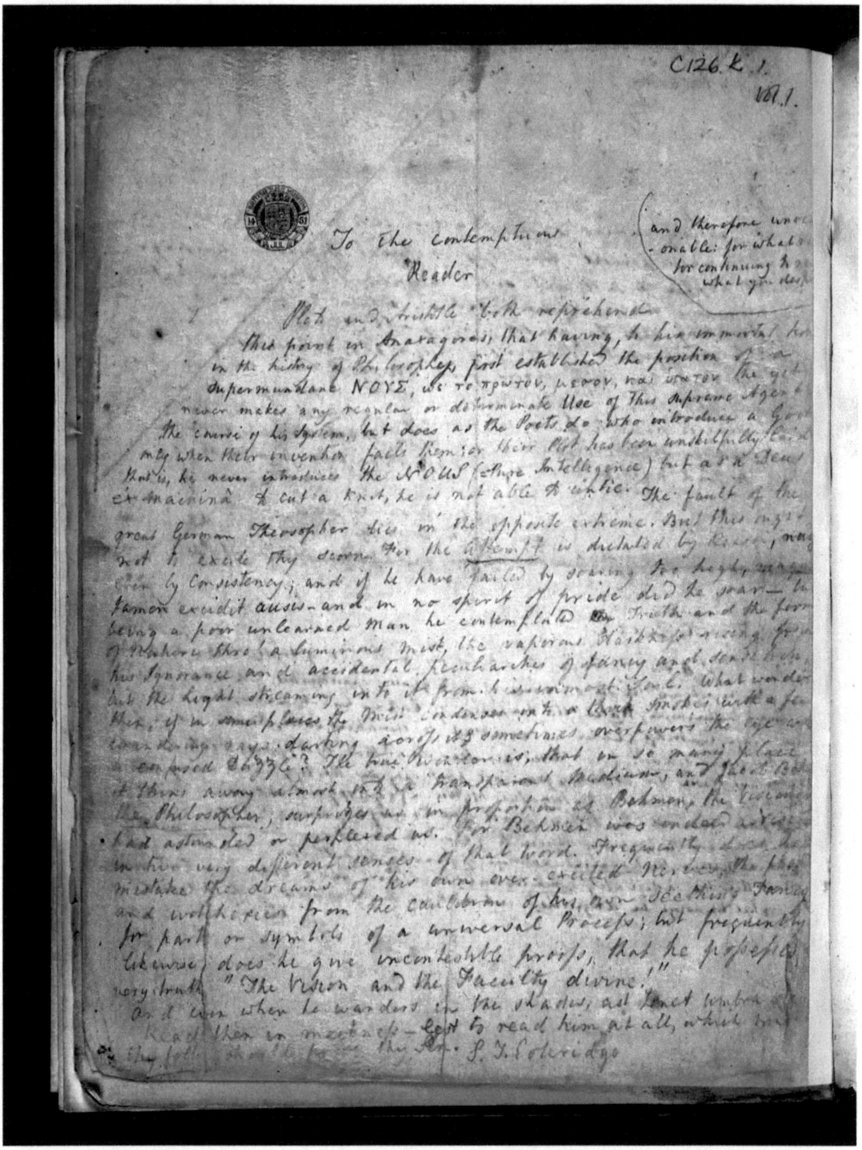

Fig. 2.4 Coleridge's annotation on an inserted flyleaf prior to Böhme's *Aurora*. Compare Coleridge, *Marginalia*, 1: 557-8. *The Works of Jacob Behmen* I (n.p.: inserted leaf), British Library. Copyright: British Library Board 22/08/2023

References

Abrams, Meyer Howard. 1972. Coleridge's 'a light in sound': Science, metascience, and poetic imagination'. *Proceedings of the American Philosophical Society* 116 (6): 458–476.

Beaumont, Sir George, and Lady Beaumont. 2021. *Collected letters of sir George and Lady Beaumont to the Wordsworth family*, ed. Jessica Fay. Liverpool: Liverpool University Press.

Beer, John. 1963. Coleridge and Böhme's 'Aurora'. *Notes and Queries* 208: 183–187.

———. 1959. *Coleridge the visionary*. London: Chatto and Windus.

———. 2003. *Romantic consciousness: Blake to Mary Shelley*. New York: Palgrave Macmillan.

Berkeley, Richard. 2007. *Coleridge and the crisis of reason*. New York: Palgrave Macmillan.

Berns, Ute. 2017. Science and knowledge. In *Handbook of British romanticism*, ed. Ralf Haekel, 135–163. Berlin: De Gruyter.

Böhme, Jacob. 2013. *Aurora (Morgen Röte Im Auffgang, 1612) and Fundamental report (Gründlicher Bericht, Mysterium Pansophicum, 1620)*. Ed. Andrew Weeks and Günther Bonheim, with Michael Spang. Leiden and Boston: Brill.

———. 1963–6. *Die Urschriften*. Ed. Werner Buddecke. 2 vols. Stuttgart-Bad Cannstatt: Frommann.

———. 1730. *Aurora, oder Morgenröthe im Aufgang ...*, in Böhme, *Theosophia revelata*, s.l. [Holland]: s.n., vol. 1.

———. 1764–81. *The works of Jacob Behmen*. Ed. George Ward and Thomas Langcake [so-called Law edition]. 4 vols. London: M. Richardson.

Brown, Robert F. 1977. *The later philosophy of Schelling: The influence of Boehme on the works of 1809–1815*. Lewisburg: Bucknell University Press.

Cheyne, Peter. 2020. *Coleridge's contemplative philosophy*. Oxford: Oxford University Press.

Coburn, Kathleen. 1979. *Experience into thought: Perspectives in the Coleridge notebooks*. Toronto, Buffalo and London: University of Toronto Press.

Coleridge, Samuel Taylor. 1956–71. *The letters of Samuel Taylor Coleridge*. Ed. Earl Leslie Griggs, 6 vols. Oxford: Clarendon Press.

———. 1957–2002. *The notebooks of Samuel Taylor Coleridge*. Ed. Kathleen Coburn, Merton Christensen and Anthony Harding. 5 vols in 10 parts. New York, London, and Princeton: Routledge and Kegan Paul.

———. 1980–2001. *The marginalia of Samuel Taylor Coleridge*. Ed. George Whalley and H.J. Jackson, 6 vols. Princeton: Princeton University Press.

———. 1983. *Biographia literaria* (1817). Ed. James Engell and W. Jackson Bate. 2 vols. Princeton: Princeton University Press.

———. 1993. *Aids to reflection* (1825). Ed. John Beer. Princeton: Princeton University Press.

———. 1995. *Shorter works and fragments*, ed. H. J. Jackson and J. R. de Jackson, 2 vols. Princeton: Princeton University Press.

———. 2000. *Lectures 1818–1819 on the history of philosophy*. Ed. J. R. de J. Jackson, 2 vols. Princeton: Princeton University Press.

Coudert, Allison P. 2011. *Religion, magic, and science in early modern Europe and America*. Santa Barbara, Denver, and Oxford: Praeger.

Cunningham, Andrew. 1997. *The anatomical renaissance: The resurrection of the anatomical projects of the ancients*. London: Routledge.

Gregoric, Pavel, and Martin Kuhar. 2014. Aristotle's physiology of animal motion: On neura and muscles. *Apeiron* 47 (1): 94–115.

Halmi, Nicholas. 2012. Coleridge's ecumenical Spinoza. In *Romanticism and Victorianism on the Net* 61 (April 2012): http://www.erudit.org/revue/ravon/2012/v/n61/1018604ar.html

Hansen, Jan-Erik Ebbestad. 1999. Böhme—Schelling—Steffens. Eine deutsche Filiationslinie. In *Henrik Steffens—Vermittler zwischen Natur und Geist*, ed. Otto Lorenz and Bernd Henningsen, 27–42. Berlin: Berlin-Verlag Spitz.

Harding, Anthony. 1999. Coleridge, the afterlife, and the meaning of 'hades'. *Studies in Philology* 96 (2): 204–223.

Haven, Richard. 1966. Coleridge and Jacob Boehme: A further comment. *Notes and Queries* 386: 176–178.

Hegel, G. W. F. 2009. *Lectures on the history of philosophy*. Tr. R. F. Brown and J. M. Stewart, with H. S. Harris. Ed. Robert F. Brown. Vol. 3. Oxford: Clarendon Press.

Hessayon, Ariel. 2018. Jacob Böhme's foremost seventeenth-century English translator: John Sparrow (1615–1670) of Essex. In *Jacob Böhme and his world*, ed. Bo Andersson, Lucinda Martin, Leigh Penman, and Andrew Weeks. Leiden: Brill.

———. 2020. 'Teutonicus': Knowledge of Boehme among English speakers before the English civil war. *Daphnis* 48:247–269.

———. 2023. 'To make such things knowne in my native language': Acquiring, translating and disseminating Böhme's writings in England and Wales. In *Jacob Böhme in three worlds: The reception in Central-Eastern Europe, The Netherlands, and Britain*, ed. Lucinda Martin and Cecilia Muratori, 301–322. New York and Berlin: De Gruyter.

Hessel, Kurtis. 2022. Coleridge and science. In *The new Cambridge companion to Coleridge*, ed. Tim Fulford, 161–177. Cambridge: Cambridge University Press.

Ishizuka, Hisao. 2006. The elasticity of the animal fibre: Movement and life in enlightenment medicine. *History of Science* 44 (4): 435–468.

Jessen, Elisabeth Engell. 2014. Boehme and the English romantics. In *An introduction to Jacob Boehme: Four centuries of thought and reception*, ed. Ariel Hessayon and Sarah Apetrei, 180–195. London: Routledge.

Latour, Bruno. 1993. *We have never been modern*. Tr. Catherine Porter. Cambridge, Mass: Harvard University Press.

Levere, Trevor. 1981. *Poetry realized in nature: Samuel Taylor Coleridge and early nineteenth-century science*. Cambridge: Cambridge University Press.

Lewis Bailey Margaret. 1914. *Milton and Jakob Boehme. A study of German mysticism in seventeenth-century England*. New York: Oxford University Press.

Magee, Glenn Alexander. 2001. *Hegel and the hermetic tradition*. Cornell: Cornell University Press.

Martin, Lucinda, and Cecilia Muratori, eds. 2023. *Jacob Böhme in three worlds: The reception in Central-Eastern Europe, The Netherlands, and Britain*. Berlin: De Gruyter.

Mayer, Paola. 1999. *Jena romanticism and its appropriation of Jakob Böhme: Theosophy, hagiography, literature*. Montreal: McGill-Queen's University Press.

Mays, J. C. C. 2014. *Coleridge's father: Absent man, guardian spirit*. Bristol: The Friends of Coleridge.

———. 2019. *Coleridge's dejection ode*. New York: Palgrave Macmillan.

McFarland, Thomas. 1969. *Coleridge and the pantheist tradition*. Oxford: Clarendon Press.

Modiano, Raimonda. 1985. *Coleridge and the concept of nature*. London and Basingstoke: Macmillan.

Muratori, Cecilia. 2016. *The first German philosopher: The mysticism of Jakob Böhme as interpreted by Hegel*. Dordrecht: Springer (international archives of the history of ideas 217).

———. 2022a. Böhme, Jacob. In *Encyclopedia of renaissance philosophy*, ed. Marco Sgarbi, 443–449. Berlin: Springer.

———. 2022b. Freher, Dionysius Andreas, in *Literaturwissenschaftliches Verfasserlexikon* [VL 17], ed. Friedrich Vollhardt et al., 158–172. Berlin: De Gruyter.

———. 2012. 'Sutores ultra crepidam.' Die Debatte über Schusterei und Schwärmerei im Anschluss an Jakob Böhme. In *Dichtung—Gelehrsamkeit—Disputationskultur. Festschrift für Hanspeter Marti zum 65. Geburtstag*. Ed. Robert Seidel, Reimund B. Sdzuj, and Bernd Zegowitz, 121–146. Köln and Weimar: Böhlau.

Oehlenschläger, Adam Gottlob. 1850. *Meine Lebens-Erinnerungen*. Vol. 1. Leipzig: Carl B. Lorch.

Paul, Fritz. 1973. *Henrich Steffens: Naturphilosophie und Universalromantik*. Munich: Fink.

Ridenour, George M. 1963. Source and allusion in some poems of Coleridge. *Studies in Philology* 60 (1): 73–95.

Rivers, Isabel. 2004. William Law, in *Oxford dictionary of national biography* (published online 23 September 2004, updated 4 October 2007), https://doi.org/10.1093/ref:odnb/16154.

Saifullin, Amir. 2023. Thought imagery: Vision and knowledge in the cosmogony of Jacob Böhme and Robert Fludd. In *Jacob Böhme in three worlds: The reception in Central-Eastern Europe, The Netherlands, and Britain*, ed. Lucinda Martin and Cecilia Muratori, 363–383. New York and Berlin: De Gruyter.

Schmidt, Sarah. 2013. Analogie versus Wechselwirkung–Zur 'Symphilosophie' zwischen Schleiermacher und Steffens. In *Friedrich Schleiermacher in Halle 1804–1807*, ed. Andreas Arndt, 91–114. New York and Berlin: De Gruyter.

Schmidt-Biggemann, Wilhelm. 2023. Angelus Silesius—ein Böhmist? In *Jacob Böhme in three worlds: The reception in Central-Eastern Europe, The Netherlands, and Britain*, ed. Lucinda Martin and Cecilia Muratori, 41–62. Berlin: De Gruyter.

Steffens, Henrik. 1806. *Grundzüge der philosophischen Naturwissenschaft*. Berlin: Realschulbuchhandlung.

Struck, Wilhelm. 1936. *Der Einfluss Jakob Böhmes auf die englische Literatur des 17. Jahrhunderts*. Berlin: Junker und Dünnhaupt.

Walsh, David. 1978. *The esoteric origins of modern ideological thought: Boehme and Hegel*. Ph.D. thesis University of Virginia.

van Woudenberg, Maximiliaan. 2017. *Coleridge and cosmopolitan intellectualism 1794–1804: The legacy of Göttingen university*. New York: Routledge.

Wylie, Ian. 1989. *Young Coleridge and the philosophers of nature*. Oxford: Clarendon Press.

Zeydel, E. H. 1931. *Ludwig Tieck and England*. Princeton: Princeton University Press.

Cecilia Muratori is assistant professor of the history of philosophy at the University of Pavia. She is the author of *The First German Philosopher: The Mysticism of Jakob Böhme as Interpreted by Hegel* (Springer, 2016), and *Renaissance Vegetarianism: The Philosophical Afterlives of Porphyry's* On Abstinence (Legenda, 2020). She is also associate editor of *Renaissance Studies* and co-edits *Brill's Series in Philosophical Historiographies*.

James Vigus is a senior lecturer in English at Queen Mary University of London. His publications include *Platonic Coleridge* (2009), a critical edition of Henry Crabb Robinson's *Essays on Kant, Schelling, and German Aesthetics* (2010) and several (co-)edited collections, including *Symbol and Intuition: Comparative Studies in Kantian and Romantic-Period Aesthetics* (2013).

Chapter 3
Coleridge on Leibniz, Continuity, and the Hidden Life of the Self

Charity Ketz

Abstract This chapter scrutinizes both thinkers' broad interpretations of the crisis of the composition of matter. Both viewed it as arising from the superimposition upon actuality of the conditions of possibility. Familiarly, Leibniz developed his concept of force in order to discover the individual at the point of action, where Coleridge analyzed consciousness in order to discover the dynamic relations between parts and wholes. Yet, in detailing the double role played by the divine influence upon an alienated nature, and in treating the relation between opposition and incorporation, Coleridge employs something like a Leibnizian concept of force.

'Know thyself' (Γνωθι σεαυτον) is one of Coleridge's most frequent injunctions.[1] But exactly what place can the Delphic command have in his dynamic philosophy? On one view, the command makes most sense for a philosophy committed to real powers, and dynamism (from the Greek root *dúnamis*, power) is the theory and study of these: being, mind, and the principle of their unity expressed on one side through polar forces and laws in nature and on the other through ideals empowering our rational, aesthetic and moral judgments. On this view too, since we live by natural and cultivated participation in the moral world, to know ourselves is to relate ourselves to some value or higher-order power. More precisely, we know ourselves by an inner quickening, by the intimation of 'an IDEA, that most glorious birth of the God-like within us' (Coleridge 1972 [1816], 50).[2] As we turn inward and

[1] e.g. Coleridge 1969 [1818], 514–21, 509; 1983 [1817], 1: 252, 243–4, 283, 291 and 2: 240; 1993 [1825], 10, 11–25, 178 and passim; and 1972 [1816], 79.

[2] Coleridge 1969 [1818], 1: 508–9 echoes this image, describing 'the obscure quickening of an inward birth' in seeking 'to discover the originals of the forms presented … in [one's] … own intellect'.

C. Ketz (✉)
English Department, Pennsylvania State University, University Park, State College, PA, USA
e-mail: charityketz@berkeley.edu

P. Cheyne (ed.), *Matter and Life in Coleridge, Schelling, and Other Dynamical Idealists*, International Archives of the History of Ideas Archives internationales d'histoire des idées 253, https://doi.org/10.1007/978-3-031-78315-9_3

contemplate its presence there, its image appears more clearly before us and conforms us to it.

But then, if it is the transcendent idea we regard and not any peculiarity of its being incarnate in us, the individual has minimal significance in his or her individuality. Indeed, the whole ethical project of self-knowing seems based on just this turn from the particular to the transpersonal and transtemporal. Yet to adapt from Coleridge's historical consideration of the question, there is still what, drawing on Pythagorean tenets, he calls 'determinate number' and 'the numbers themselves, inasmuch as these are … the representatives of the divine power, not contemplated in its own perfection but as a self Will so influenced' (2002b [1819–23], 367). At a lower level, physical movement is from somewhere determinate to somewhere else determinate, and the initial coordinate not only allows but partly determines the movement. Or else the particular individual is the mysterious ground by which the variegated splendor of 'that great living Idea' manifests itself as through air or water, and then, as Coleridge says, individuality is not only inconceivable but unreal (2002b [1819–23], 211; 1969 [1818], 520).[3]

That conclusion, framed in Plotinian terms and built around the sense that, intellectually considered, there is no transition possible between the universal and the particular, strangely closed Coleridge's 'Essays on the Principles of Method' (in *The Friend*, 1818 edition), which had foregrounded the term's etymology: 'μεθοδος, literally *a way* or *path of Transit*' (1969 [1818], 521, 457). And it did not sit well. Later that year, in a paragraph he repeatedly redrafted, Coleridge asks 'Are *we* then indeed NOTHINGS?…How and whence did this sterile and pertinacious Nothing acquire its *plural* Number?' (1969 [1818], 522–3 n. 1).[4]

In an 1826 notebook entry, Coleridge begins to liken individuality to the distinction and unity of the persons of the Trinity but then, shifting the comparison, says that the form of individual consciousness is 'distinct but even in thought inseparable, and co-inherent' with the personal will, requiring not only the universal reason but its peculiar presence as one of 'the *co.efficients* of *actual* personality' (1990 [1826], 4: §5377). Using a favorite figure of his, that of the sun's light which is given and withdrawn in determinate periods, he explains that the personal will disappears with the disappearance of this '*dispensed*' reason. The problem of the transition between universal and particular introduced at the end of the 'Essays on … Method' cannot be solved through the individual will without this presupposed apportioning. Not a something, this apportioning determines a potential relation, allowing the will to relate itself uniquely and retain its stream of material expressions.

This chapter will do three things. First, it reviews what sort of answer can satisfy the question of self-knowing. Then it reconsiders some complexities of Coleridge's account of perception in relation to his engagement with Leibnizian continuity. Finally, it considers the value of the individual's glimpse of him- or herself in

---

[3] For the description of the material principle as air that Coleridge's text seems to recall, see Plotinus 2018 [220–70?], 3.6.14, 324.

[4] Copy L qtd here.

transit, alternately appearing—mattering—and reshaping circumstance. The first section, 'Orienting self-knowledge', shows that while Coleridge frames self-knowing in broadly Platonic and Augustinian terms (as the identification of particulars with a higher-order power), he also points to another potentiality co-present with the particular will. The second section, 'Harmony', explores Coleridge's claim that consciousness requires a mental element that can harmonize with universal mind. The third section, 'Continuity', explains the splitting up of self-knowing into universal truths or else a tally of facts through the Leibnizian formulation of continuity. The fourth section, 'Hidden life', inspects Coleridge's claim that nature, as an ascending scale of relative powers, depends upon a subject. The final section, 'How to cross a space', considers Coleridge's solution to the problem of particular self-knowing as the elicited and sustained realization of a potential relation rather than as something's intermittent incorporation of the whole as 'living force'.

3.1 Orienting Self-Knowledge

In *Biographia Literaria*, Coleridge depicts self-knowing as the dawning awareness of a mysterious power at work in us, one that is more nearly us than anything that would mark us out as distinct from it or one another. Quoting Plotinus on the nature of intuitive knowledge, Coleridge speaks of 'the sacred power of self-intuition' by which we might 'understand the symbol, that the wings of the air-sylph are forming within the skin of the caterpillar' (1983 [1817], 1: 241–2). The power by which these successive states are of the same being, and by which we know that this is so, is as much above any and all of them as it is above any and all of our states, except insofar as we intuit the unity. To know that is obliquely to realize the unity we are.

Practically speaking, self-knowing is supposed prevent us from well-meaning but misspent activity or, as Plato's Socrates would have it, mistaking localized beauty for its paradigm or hurrying into care of the body or the city without a sense of what it would mean to care or without insight into that central idea around which everything must and does turn. At the same time, it transmutes self-knowing's central question—what should be done?—for that reduces to the perfective knowing of finding what is at the essence of one's being.

For a long Christian-Platonist tradition, the task of self-knowing could well appear as one of disentangling what must be continuously present in the seeker from the accidental features of embodied experience that attest it.[5] For Augustine, when the mind seeks to know itself, 'it already knows that it is mind', but it is commanded to know itself in order to 'live according to its nature', which is to say, to be subject to some things and to control others according to its place in the cosmos (Augustine 2015 [420?], 292–3, 296–7). Self-knowing realizes substance. For without being relationally enacted, this mind is not substantially present to itself. The trouble, says

[5] For a defense and explanation of 'Christian Platonism', see Hedley 2000, 10–12 and passim.

Augustine, is that the deluded mind cannot distinguish itself from the sensible objects without which it could not think itself; and it cannot do so because it does not wish to do so and, hence, loses its proper life—or as we might say, its centre. Identifying with momentary states of the world, it reifies itself.

Coleridge often speaks this way to powerful social and psychological effect.[6] But he also presses the point of community to its absolute limit. In the last of his 'Essays on … Method, he addresses his readers:

> Hast thou ever raised thy mind to the consideration of EXISTENCE, in and by itself, as the mere act of existing? Has thou ever said to thyself thoughtfully, IT IS! heedless in that moment, whether it was a man before thee, or a flower, or a grain of sand? (1969 [1818], 514).

Shortly after, he writes:

> The finite form can neither be laid hold of, nor is it any thing of itself real, but merely an apprehension, a frame-work which the human imagination forms by its own limits, as the foot measures itself on the snow; and the whole truth of which we must again refer to the divine imagination, in virtue of its omniformity (1969 [1818], 520).

This is one of the paragraphs Coleridge later faulted for its susceptibility of pantheistic interpretation. In the years that followed, his solution to the threat of pantheism, and the unintelligibility of the individual, would partly, but only partly, be solved through individual will.

In his *Opus Maximum* fragments, he argues that though every being besides God only shares in the divine unity, the derivative nature of all created being does not imply imperfection. Even when we turn to plant life, we find 'the positive power, of the plant' where no negation or privation is implied. For any form and any individual, the question should still return on us, 'how was the finite possible?' (2002b [1819–23], 218–9). Yet, through the whole series of individuals, none now has more than an 'imperfect unity' and 'no form of its own … no being for its form' except such as is borrowed from 'its own opposite and from the opposite of being, from negation or limit, and from multeity' (2002b [1819–23], 211). The explanation is the apostasy, the fall; what has been lost through it is access to one's own being and, thereby, to one's will.

In *Aids to Reflection*, observing that an antecedent unity must be in every seed in nature as it contains 'means and instruments' that cannot be otherwise explained, and urging readers to reflect on 'the countless millions of Seeds of the same Name, each more than numerically differenced from every other' that require a certain harmony of things, a unified and supporting context, and an 'individual Law or principle of Union in each', Coleridge asks whether it is 'rational to infer One universal Presence, a One present to all and in all' and 'not most irrational to suppose that a finite Will can exclude it'. If someone is determined to act in harmony with others 'must not something be attributed to this allpresent power as acting *in* the Will?' (1993 [1825], 76–7).

---

[6] Coleridge 2002b [1819–23] 2: 120–7 for his link of familial disconnect and self-reification.

But if this is the case, that (1) a principle of unity is in each physical seed as its law, and in the individual will to determine it to harmonious activity, and yet (2) individual will is irreducible to reason, and if (3) the whole series of individuals is somehow without being of its own, the problem is not the gap between universality and particularity. Rather, this will stands apart from its own particularity, which is to say, its current potential for expression. This solution of Coleridge's, which the second half of this chapter takes up, illuminates the other side of the question of self-knowledge, where good exists in a particular form within the individual.[7]

For Coleridge, we locate ourselves in motion. Before turning to that argument of his, let us glance at two instances in which he ostensibly treats the question of particular self-knowing as futile or perverse. Opening with the Delphic injunction, the poem 'Self-Knowledge' (1834?) begins with the unintelligibility of our individual selves, 'What is there in thee, Man, that can be known?', 'canst thou make thyself?'. It concludes, 'ignore thyself and strive to know thy God!' (2001, 1: 1154). If the point is initially equivalent to one Coleridge made through Giambattista Vico in the *Theory of Life*, namely, that we know geometric constructions because we make them, and if knowing our individual lives means the ability to lay their groundwork and demonstrate why we exist, then life is unknowable. But we are not seeking to know by positing ourselves and gleaning something in logical analysis. Nor are we seeking only the fact of our standing 'in the Idea'. Such standing would not tell us anything in particular even if it is the ground of our particularity. The poem 'Self-Knowledge' leaves us in Platonic suspense. Only if we could know God, that ineffable, could we know this ineffable, our individual selves.

Sometimes, Coleridge portrays the attempt to comprehend one's individual self as morally guilty. This is the perspective of an 1825 notebook entry that works out the difference between Lucifer's fall and Adam's. Whereas Lucifer 'strove to soar above the Light':

> the Adamic Man sought to have his Will as his Life—to *enjoy* his Will, and to ~~understand~~ know his Life. *He* was *tempted*—i.e. his fall was not purely & totally the birth (τοκος αμητρῶρ [motherless birth]) of his Will (1990 [1825], 4: §5249).

If Adam's fall had been the 'motherless birth' of his will, he would have constructed himself. That he cannot actualize but only align his will with some external vision, divine or demonic, is what keeps him from being like Lucifer and irredeemably falling, though it also proves the source of his unease and apostasy. He agrees to demand his will as his being and seeks his life in that impossible coincidence rather than knowing his will as something knowable because it is aligned with universal reason and rather than enjoying a life he can receive but not make and, therefore, not know as his construction.

---

[7] Coleridge scholarship that has taken up the question of self-knowing has often explored the individual's assimilation to an idea but not the apportioning of the idea to the particular. See Muirhead 1930, 263–4 and Cutsinger 1987, 34–45. See Hedley 2000 for a scholarly account of the difference between physical and spiritual unity that does not share my concern for the particular. Hedley, 2000, 65–6, 169–80 and passim. See Cheyne 2014 for an exception to the pattern worked out through the particularity of art.

Neither of these instances rules out particular self-knowing. Such knowing, Coleridge rather suggests, does not occur apart from knowing God, nor does it occur as understanding. What, then, can be said of it?

3.2 Harmony

Chapters 5–7 of *Biographia* are often described in terms of the critique of (Hartleian) associationism. Correspondingly, Coleridge is understood to supply 'the willful and poetic imagination' in response to the insufficiency of mechanical explanations of perception (Engell and Bate in Coleridge 1983, xc-xci). But if his solution is to insist on the activity of something like the Kantian necessary imagination, whose combining activities some individuals can stylize through secondary imagination, it is one he will soon criticize. By 1819, he argues that while affirming this primary construction 'reconciles Leibnitz with Kant', it cannot explain perception nor does it yet belong to mind (1990 [1819], 4: §4545). Affirming it explains only a general possibility, not that of someone.

Instead, he describes a two-level, chiastic process by which the primary construction, which 'is all, or all but all compulsion', is raised to perception.[8] At the lower level, what he calls objective memory flows forward into the perceptual object and gives it its unity. But this unity is only phenomenal, working in but not for the subject. It is continually and fractionally constructed in relation to 'indistinguishable Multeity' and so indistinct, like eddying water in relation to a stream or an ever-changing hue in relation to white as the totality of colours (1990 [1825–6], 4: §5292). It is, as he says in another place, as though this 'Monad or Indistinction' would distinguish itself but cannot this way be 'opponible' to the common identity (1990 [1819], 4: §4513). At a higher level, subjective perception is already distinct in that it requires the subject to complete it. This unity, achieved through particular (material) expression, is then integrated with the plurality of subjective memory. Because we can find no 'differencing difference of position in reference to the *Product*', Coleridge says, what had been an unperceived sensation is elevated into what we commonly call perception and there is 'Internal Vision in Clair Voyance' (1990 [1819], 4: §4545). Something will-ish operates at each level, a spontaneous appetence and momentary aggregation of some portion of multeity at the lower level and a subjective combining 'by laws of association' at the higher. But what raises potential into actual perception is the power of mind for the subject. Whereas the actuality of universal mind corresponded with potential perception, mind as a new subjective possibility enables actual perception, which then appears with the subject's memory, with being someone in particular.

[8] Cheyne 2020 characterizes Coleridge's philosophy as offering a two-level, interpenetrating view of reality.

If Coleridge's model still looks Kantian here, he soon casts it in terms of an enduring individual. While the 'elementary Nature' of the first synthesis has nothing but pure position, the second, incipient in some animals, is not completed in apprehension of universal truths. Animals bear the look of the first synthesis—sempiternal type reflected from vanishing will—and so cannot act. Being immediately what they are, they display 'the conquest of Essence over Form' and have no history. Finality is superinduced, and 'the Eternal' has not yet appeared as 'the Power of the Temporal' (1990 [1820], 4: §4648). When mind works through 'laws of association', it discovers someone's individuality. Insofar as human knowing is special, it includes an oppositional act—to redispose matter—for which we are peculiarly qualified. Any self-knowing could not be perfective without exercising this aspect of individual history. In stressing the priority of essence over personal form, we would seem to know ourselves as animals, rejecting a creative potential in us and in the whole.

What is new in the second synthesis—'subjective perception'—is the subject, which requires '*Fidelity* to our own Being as far as such Being is not and cannot become an object of the sense' (1995 [1820], 1: 834). But that requirement is also unique. Not an explanation of how the change came about, the object of fidelity presents the subject with a peculiar task. Leibniz frequently invoked the scholastic dictum, '*actiones sunt suppositorum* [actions belong to individual substances]' (1989a [1686], 40 §8).[9] Coleridge would concur this much: only individuals act. His insight, albeit inchoate, is that individuality is not secured by distinguishing the ontological from the conceptual. Rather, a law never implies an individual, but a task always does.

To put the problem this way is to return to Coleridge's historical approach as he distinguishes it from that of the 'best Schellingians' who would take 'the very resistance to the redemptive Power' as part of the redemptive process so that even the elementary nature is not fallen from its being but always already minimally beatified, known. For Coleridge, though some potential for subjectivity must be presupposed wherever there is mind—throughout intelligible nature—it is not the case that the point-of-viewish self is still bound up in the divine mind, still duple or capable of moving into its larger life and self-identity. Historically considered, this elementary nature is objectless appetence, without time or relation, the unaccountable beneath reason, 'subsisting *in Will* … and *potentially* only' (1990 [1820], 4: §4648). It must receive some quality to be, some task to be a subject.

In the first case (where being means being determined), we might recall Leibniz's argument that a mental principle invests any portion of mobile matter; movement arises from distinctity not the reverse. Taking issue with his own juvenile explanation of qualities emerging through continuity's (matter's) transition into contiguity (through motion), Leibniz imaged the interchange of homogeneous portions of matter as a condition of perfect indiscernibility. Were a sphere to rotate within another, not even an angel would be able to determine whether it rotated or was at rest

[9] See also, Leibniz 1989b [1668], 117 §31; 1989a [1698], 160; 1989a [1712], 200.

without a criterion of distinction; nor could one ever arise (1989a [1698], 163–5).[10] In this application of the identity of indiscernibles, all created being must be absolutely and uniquely determinate, not partly self-created.

For Coleridge, too, will, as 'the communicated Ground of all communicated Being', lacks any quality by which it could relate to the laws of nature (2002a [1828], 5: §5813). It must be

> rendered intelligible by the energy, which combining therewith makes it no longer Hades: even as Number is brute Multëity raised into … intelligibility, and made the possible Object of Mind, by Unity (2002a [1828], 5: §5931).[11]

In being so raised, will is neither its potentiality nor primarily reflective of 'Unity' but has the relationality of 'Number', which depends on unity without reducing to it. Anything that might be called a particle of nature is so invested, is virtually duple in nature, either without 'apparent *Mind*: as in the Plants' or with it, as rational beings that can employ something in this mental principle as their way to mind someone or something and thus be present to the source of mind and themselves somewhat as thoughts grow distinct in activation (1995 [1833]; 2: 1510).

In 1827, Coleridge notes to himself 'Q^{y}—*Reason* = mind exclusively objective?', observing that while mind is defined as 'a Subject that is it's own Object', 'the "*I*"' is defined as 'a Subject recognizing itself as a Subject' and so 'cannot be the WHOLE *Mind*'. He adds, though the 'I'

> is indeed the mind itself, of which the Mind is conscious … it is not of necessity conscious of it *as* itself. Something more, therefore, seems to be required for the existence of an 'I', than mind in the universal idea! — What is this? (Coleridge 2002a [1827], 5: §5670).

His provisional answer begins by distinguishing true consciousness, which requires will, from mental reflexivity, which does not, so that conscience is 'the *real* antitheton of the I' and the '*Dreamatis* Personnæ, in which the Subject is recognized as an *Object*', the parodic form.

As represented by the syntheses, and etymologically implied by 'consciousness', apperception requires a self-gathering beyond the 'I'. But its correlate is not simply given, is not 'mind in the universal idea'. Here Coleridge's notion of animals that are '*scious*, but not conscious' may help (2002b [1819–23], 73). What is lacking in them is not will but another aspect of mind that could harmonize with universal mind. Coleridge, following Schelling's view on 'animals, those incessant somnambulists' Schelling (1983 [1805], 255), often represents animals as somnambulants (1995 [1820], 1: 837; 1969 [1818], 524; 2002b [1819–23], 75). It seems that if the Personëity determining laws of nature is (subjectively) withdrawn, the particular will reflects the impersonality of law; the point of view provides only the barest disturbance. If the Personëity is subjectively manifest, and the transforming nature is shadowed, anticipated, carried along and then woken, what would otherwise be

[10] For Leibniz's juvenile view, compare Leibniz 1989b [1669], 96 but, contrastingly, 1989b [1669], 110–2.

[11] For more on Coleridge's linking of fallen or 'Natural Man', potentiality and 'Hades', see Harding 1999, 73–9.

dreaming nature finds itself a subject confronting reason considered as 'the representative of the Will of God' (2002b [1819–23], 84).

'*Conscience*', Coleridge elsewhere claims, 'is not a Result or Modification of Self-Consciousness; but its Ground and Antecedent Condition', arising through recognition of the '*I*' and the '*Thou*' as 'harmonies or correlatives', 'equal … but yet not the same' (1990 [1824–6] 4: §5167; 2002b [1819–23], 75, 78). The 'I' has to take something as itself. Yet it cannot take this other reflexively '*as* itself' without irrationally losing the distinction necessary for consciousness and falling into gross pantheism. Subjectively wandering then, it casts any number of object-like images. Coleridge completes the solution by differentiating the 'empirical' from the 'substantial I'. The first looks too much like animal scious-ness, the latter enables peculiar selfhood.

> Of the Substantial I I only anticipate the solution in the immanent Life in the communicative Logos … which being communicated … subsists subjectively by it's Unity with, and correlation to, the Holy Spirit objectively (2002a [1827], 5: §5673).

Because reflexivity is secured for the 'I' by the communicated Logos and Holy Spirit, the 'I' can be drawn from its dreaming inert-ness and newly take on the look of mind, 'perpetually construe[ing]' its objectivity 'as a Subject by the act of self-renunciation, constituting it's antithesis to the *Objectivity* of the Holy Spirit' and 'in this blessed sterility, it is *free*'.

Although somewhat ambiguous, the suggestion contained in 'sterility' seems to be that the 'I' will not, as nature does, simply proliferate. Instead, it will recast the law of mind (reflexivity) as its own seeking. In common parlance, to number is to place something distinct among others of its kind; it is also to understand a qualitative progress through a sequence. If 'Number' provides what Coleridge seeks in this note, it is this power to instantiate (Personëity is objective) or else individuate (Personëity is subjective).

For Coleridge, 'dynamic ratios' are 'not of powers, but … are Powers … αριθμοι Πυθαγοραιοι, Numeri numerantes [Pythagorean numbers, numbering/creative numbers]' (1990 [1825–6], 4: §5294). They are the eternal outpouring of the divine mind, structuring relations, giving individuals a means of relating to others and to mind in the universal idea. As representative of the divine influence, they might be glimpsed as goads toward this (2002b [1819–23], 361). But they are also what enables the incompleteness of the will to be an actual individual; they are '*the condition*, the *sine quâ non* of a *free*—i.e. of *actual*…Will', offering something the 'I' can progressively realize as itself and a means by which the 'I' is recognized by and with the divine mind (1990 [1826], 4: §5377). In brief, I read this waystation—will-and-communicated-Logos—as peculiar subjectivity, which may be apprehensible on the way to universal mind—or hidden in the new 'I'.

Let us take stock. Harmony requires the concord of distinct parts. In rejecting Hartleyan associationism with its dualistic hypothesis, in which the soul is passive before sense impressions, and the materialist variants of his followers (viewing consciousness as produced by the mechanical relation of 'harp', or brain, and 'breeze', or sensory stimulation), Coleridge points to harmony as a 'mode of relation, the very *esse* of which is *percipi*' (the very being of which is to be perceived; 1983 [1817], 1: 117–18). The problem with these accounts of association is not only that an impression has to generate a perception and the subject receiving it or that, as a vanishing effect, it can do neither. The problem is that the passive soul/brain has fallen from time and cannot hold this impression without recovering some meaningful place in time.

But now, as Leibniz treats harmony and continuity as synonymous, and as Coleridge focuses the *Biographia* critique on 'contemporaneity', which he calls 'Leibnitz's *Lex Continui*', is it time to scrutinize the principle Leibniz sometimes called a law and sometimes referred to a more general principle of order,[12] or described as belonging to a set of clear hypotheses 'consistent with themselves and the phenomena', among which was the general axiom '*in changes there are no leaps*' (Coleridge 1983 [1817], 1: 123; Leibniz 2013 [1699], 69–71; 2013 [1698], 11).

3.3 Continuity

In 1686, Leibniz published his 'Brief Demonstration' arguing that mv^2 (similar to what we now call kinetic energy) rather than mv (now called momentum) is the correct measure of motive force and that this rather than the latter, Cartesian measure was conserved in the universe. In making his argument, Leibniz assumed a principle of equipollence. That is, bracketing factors like friction, a heavy body falling from any height '*will have…precisely the power necessary for it to rise back to the same height*'; or as he summarized elsewhere, power rather than action is conserved in the universe (1989b [1686], 299).[13]

A year later, responding to Malebranche and the Abbé Catelan, who had published a defense of Cartesian physics, Leibniz argued that Cartesian physics violated 'a principle of general order'. He then formulated continuity as a principle of harmony:

> *when the difference between two instances in a given series…can be diminished until it becomes smaller than any given quantity whatever, the corresponding difference in…their results must…also be diminished or become less than any given quantity whatever* (1989b [1687], 351).

In the same paper, Leibniz argued that geometric principles relating to the infinite could be applied to physics, and when they were, the results were more orderly. Just as theorems proved for an ellipse could be applied to a parabola considered as an ellipse one of whose foci is infinitely far removed, so equality could be considered an infinitely small inequality, and rest an infinitely small velocity. Without this

[12] cf. Leibniz 1989a [1695], 132; 1996 [1705/1765], § 60.

[13] See Leibniz 1989a [1691], 106–10 for further statements on the principle as a relation between power and action.

application to physical entities of a principle related to the infinite, a large body with zero velocity should be moveable by any body of non-zero velocity—a drifting feather (1989a [1678–82], 249).

Adverting to this paper in a letter to mathematician Johann Bernoulli (and implicitly to Burcher de Volder), Leibniz refers to '*the law of continuity*' and adds 'there is no assignable change in an instant'. Just as moving bodies pass from place to place through intermediate places, so with changes in degree. That bodies avoid leaps in their changes 'is due to an elastic force existing in them'. Yet he cautions that this elastic force is not essential to bodies but can be assumed to arise through 'the structure of the system of the whole universe' (2013 [1698], 11). A year later, he repeats the point to De Volder, adding that motion is 'destroyed by the collision and restored by an elastic force' but this is to be understood as a successive process 'in accordance with the law of equilibrium through the intervention of dead forces', where dead forces are two or more infinitesimal velocities (2013 [1699], 71). Colliding bodies gradually compress, lose forward motion, come to rest, and restore themselves through a retrograde motion until they recover their initial size and velocity. They can do so because primitive force can be directed outward, become elastic force on collision, and return inward, so that each body regains its own motion (1989a [1695], 132). Force appears a quality of this orderly cosmos that can become incorporated in things.

What is relevant for Coleridge's criticism is less the mathematical, physical, or biological applications of the law of continuity and more their common metaphysical underpinning. For Leibniz, the structure of the cosmos and of each created being can be explained through the correlation of power and action. In an essay defending the inherent force of nature, he writes:

> there can be no action [*actio*] without a force for acting, and conversely, a power [*potentia*] which can never be exercised is empty....nevertheless, action and power are different things, the former successive, the latter persisting (1989a [1698], 160).

In an early explanation of creation consistent with his subsequent views, Leibniz distinguishes between the divine mind, which consists in the ideas, and the concurrent divine mind, one idea of which concurs with A, another with B, and so on (1989b [1669], 118). Rather than being regarded in the aspect of unity, through concurrence, the ideas are extended from the divine mind and given a kind of alterity by being severally regarded. Like the Plotinian soul whose sequential regard for the ideas becomes the principle for the cosmos, in Leibniz's account, this act of divine self-regard is paradigmatic for the activity of all nature. The divine mind is to concurrence as power is to action.

Complementing this account of creation is another figure Leibniz often employed, that of a city comprised of its living points of view. God, he writes:

> views all the faces of the world in all ways possible...The result of each view of the universe, as seen from a certain position, is a substance which expresses the universe in conformity with this view, should God see fit to render his thought actual and to produce this substance (1989a [1686], 46–7).

And:

> Every substance is like a complete world and like a mirror of God or of the whole universe, which each one expresses in its own way, somewhat as the same city is variously represented depending upon the different positions from which it is viewed (1989a [1686], 42).

What is happening here can be glimpsed theologically through the power-action dyad. It is essential to God that his ideas are of something—'there are no ideas in God except as there are things outside of him'—and equally essential that these things be efficacious—'the ideas of God are the substances … of things' and 'actions belong to *supposita* [individual substances]' (1989b [1668], 117–8).[14] It can also be glimpsed through Leibniz's commitment to the doctrine of ideas in which no reason can be given 'for the ratio of 2 to 4 being the same as that of 4 to 8'. These possibilities or ideas of things 'are numbers' coinciding with God, who, as a mind, cannot but be pleased with the most perfect harmony and necessitated to do the best (2002b [1671], 146). As God moves around the city, there is a gain as he accords with himself. His self-limited self-regard necessarily opens up perspectives with power to continue as themselves. Were it not so, these perspectives would not exist and God would have no power. In one sense, they are objective to him, but only insofar as he regards their possibility in himself.

Let us return to the law of continuity. As formulated, a differential at one level (the instance) is a differential at another (its result). A principle of polarity is at work, so that even God is put in relation with creatures through an idea, 'idea is the union of God and creatures, so that the act of agent and patient is one' (1989b [1668], 118). What is essentially true of one pole is also true of another; hence, rest can be a special case of motion, equality of inequality, and geometric principles relating to the infinite can be applied to finite figures. Something that appears at an ineffable remove, 'the ideas of God', reappears on the plane of existence as the correlative of the 'substances of things', 'the same in fact, different in relation…as action and passion' (1989b [1668], 119).

How far the difference between action and power, contingency and necessity guards the difference between creatures and God cannot be addressed here.[15] Unlike what Coleridge calls 'dispensed reason', concurrence is not for us but for divine self-realization. But there is a crucial difference in divine and creaturely knowledge. When created beings appetitively reach back to the city centre (as divine concurrence does), they always obtain something of 'the whole perception toward which [they] tend'. Yet, impeded by an inertial principle, what they capture of 'the passing state' just is their endeavor: their momentary relation to the infinite lower-level monads surrounding them, which constitute their bodies (1989a [1714], 214–15; 1989a [1714], 207). As a dominant monad becomes less active, surrounding monads may become more active and vice versa (1989a [1686], 48). Again we find what, at a lower level, elastic collisions analogized when their primitive force was briefly redirected outward and converted into elastic force. As monads are blocked in their

[14] For Leibniz's distinction between complete beings and complete concepts, see Di Bella 2005, 1–54.

[15] For a related answer, albeit in terms of divine freedom, see Adams 1994, 23–52.

reach toward the whole, their limitation stirs 'parts asleep in the abyss of things' (Leibniz 1989a [1697], 155). The whole of creation slowly advances.

Let us pause here. For what Leibniz calls perception is continual but not conscious; apperception is conscious but not continual. He himself treats perception as phenomenal, as the moment by moment responsiveness of beings to the whole through their bodies (which are 'nothing other than *to move*'), a responsiveness that, in some, is like being in a stupor, and in others is only more distinct through memory, which provides a sequence (Leibniz 2001 [1672–6], 17; 1989a [1714], 216–7). Further, he claims that rational souls are such because they know necessary truths by '*an inborn light within*', which allows us to say 'I', to be capable of demonstrative knowledge, and so on (1989a [1702], 191; 1989a [1714], 209, 217). This 'inborn light' enables us not only to mirror the universe as all monads do but also to reflect divinity, to concentrate the whole in ourselves and produce our thoughts as God produces beings 'by a kind of emanation' (1989a [1714], 211; 1989a [1697], 154; 1989a [1686], 46).

But though demonstrative knowledge offers something like a view of the ideas, and though we are sensible of bodies, 'where there must be something numbered, repeated, and continued' and likewise of certain diffused qualities 'such as whiteness … in milk, color, ductility, and weight … in gold, and resistance … in matter', we have no distinct knowledge of individuals (2013 [1699], 73). It is easy to think here of Leibniz's doctrine of petite perceptions, that to hear the roaring of the sea is confusedly to hear each part. Each part, though, has to be understood as the momentary interrelation of the multitude's states that change not instantaneously but slowly and in equilibrium with one another.

3.4 Hidden Life

Let us now return to Coleridge's interesting claim that contemporaneity

> is the *limit and condition* of the laws of mind, itself being rather a law of matter, at least of phænomena considered as material…it is to *thought* the same, as the law of gravitation is to loco-motion. In every voluntary movement we first counteract gravitation, in order to avail ourselves of it (1983 [1817], 1: 124).

Then imaging thought as a leap, in which someone resists gravity by a voluntary act and, by another partly voluntary act, yields to it in order to land on a proposed spot, he adds, 'now let a man watch his mind while he is composing; or…while he is trying to recollect a name' (1983 [1817], 1: 124). Finally, he images the same two-part process by way of an insect that propels itself across a watery current by pulsing forward and releasing its body back into the current to gain 'a momentary fulcrum for a further propulsion' (1983 [1817], 1: 124).

Let us take the resisted gravitation as a key term here. In *Opus Maximum*, Coleridge describes 'two species of spirit' where 'spirit' names a power or 'that which cannot appear in itself, either not at all or as manifested in another' (2002b

[1819–23], 304, 301). These species of spirit are polar oppositions, one distinguished from and comprehending the other by transcending it, as universal to particular. The other is considered as 'matter', whose internal opposition is transformed and organized through its opposition to the comprehending power, which then allows the comprehending power to be thinkable. By converting a presupposed opposition of centrifugal and centripetal powers, gravitation works specifically upon bodies that limit each other's direction. In this relation, it is 'spirit', though in relation to locomotion, it is 'matter'. So too, Coleridge will say, 'when water is aerated, the drops partake of the gas not as drops, but as drops of that aerated water' and when drops enter a stream, they lose their form as drops; mechanical laws describe flows (2002b [1819–23], 305). But 'remove from any machine its particular figure, and it ceases to be a mechanical and becomes a statical power' (2002b [1819–23], 305, 306).

There are three key elements to this argument. Again, there are stadial expressions of the opposition between power and action. The statical power is differently expressed at each level of nature's ascent: homogenously at the level of striving masses in equilibrium, positively and 'energic[ly]' at the level of organic form. Nature moves from stasis to stasis; 'throughout all Nature there is seen an evolution from within' (2002b [1819–23], 120). But thus sketched, every stage is only relatively unifying and specifying, the whole an arrangement of potentialities. As far as explanation goes, there is an infinite regress, something 'still parabolic instead of an ellipse'. In terms of existence, nothing is yet one, for there is no subject, no 'dative of manifestation' before which nature can be objective and in whom 'the specific and the universal' can be simultaneously realized (2002b [1819–23], 311, 315).

As far as the *Biographia* criticism goes, the suggestion is that contemporaneity (action) and continuity (power) invoke the scala natura without offering a presiding subject by which the ascent could be made. 'Contemporaneity', the momentary standing of things, must be guarded and transcended—its elements selectively arranged—for there to be thought. Only a subject can do both. The progressive, hologram-like quality of nature needs something like Kant's transcendental unity of apperception, the 'I think', 'I perceive', which can always be said before any experience.

Yet, significantly, Coleridge's image of the water insect that loses itself in the current in order, then, to forge its way in opposition, figures consciousness's power to be reduced into objectivity and then recover itself as an individual subject. It emphasizes the individuality rather than the universality of thought. For thought, after all, needs to alight on a recollected name, a particular composition.

Let us say that consciousness is not only that by which the universalizing and specifying powers in nature appear but a higher-order version of nature's rung-by-rung ascent, one that moves through the universal for the sake of the individual rather than the reverse. We might still want to know how thought is moving from step to step. In the *Opus Maximum* fragments, Coleridge places the will as difference-maker between an 'I' and another so that they appear as 'harmonies' (2002b [1819–23], 75). But insofar as the individual will by itself is unrelated to reason or is 'no more than a potential entity, the dimidium sui [the half of oneself]', it is unclear how will is supposed to propel itself (1990 [1826], 4: §5377).

We seem, then, to have landed at an odd spot for an inquiry into particular self-knowing. Will must act at the point where it is. As Coleridge often says, it must be able to originate a state or the word has no meaning (2002b [1819–23], 18, 32, 194, 161). And it must be extended or it is not even will but merely a material principle. That is, it must appear through another, as the water at the base of a fountain rises and blossoms in the watery column's form (1995 [1826], 2: 1335–7; 1990 [1825–6], 4: §5298). In the confrontation with another or with an unknown, will as the difference-maker must be subordinated so that the similitude, the rhyme of an answer, appears. The figure of the leap shows as much: the thinker cannot remain in pure space. However this intermediate moment is conceived, as a formal suspense or as evidence of the re-potentiating will that must appear through another, the problem of particularity looks recalcitrant.

In the 'Essays on the Principles Method' and the *Opus Maximum* fragments, Coleridge depicts a scientific inquirer hanging over nature's clear streams to find the source of both nature's products and our reflective life. In the 'Essays on … Method', this inquirer lengthens out the stay as there is no representative of our free agency there, yet he cannot resist the sensation that what he contemplates through forms of his own abstracting and comparing intellect has a reality outside and independent of him (1969 [1818], 509). The later *Opus Maximum* version upbraids this 'Mad Narcissus'—not for seeking the source of the subjective among the sensible forms of nature but for the reverse, seeking the will made objective in nature's stream (2002b [1819–23], 104). The duration and frustration of this moment, then, is nothing other than the un-intermitted will that seeks the whole through its own formalization, potentiation. And in the *Opus Maximum* version, Coleridge ascribes the result of pantheism—loss of particularity—to this overactive will, 'Mad Narcissus, that in barren Self-love transformist thyself to Form without Substance … Whence did this Nothing acquire its plural number?'

The same result is achieved through two extremes. In the *Opus Maximum* fragments, the observer hangs over a longed-for image of his will that he cannot substantiate on the stream. In the 'Essays on … Method', 'the finite form' is not 'of itself real', and individuality as such, like 'the transparent air', can neither be beheld

> with nor without the substance. Not without, for then the forms cease to be, and are lost in night. Not with it, for it is the light, the substance shining through it, which thou canst alone really see (1969 [1818], 520).

In both the recantation to this portion of the 'Essays on … Method' and in *Opus Maximum*, the forms of feeling rise up within the inquirer demanding an account of plurality, saying, 'the forms are numberless: & whence the number and the diversity of the Senses?' and 'what is that inward Mirror, in and for which these Nothings have at least a relative reality' (2002b [1819–23], 104). The will that would find the world in itself is obviously wrong. But so is the high-minded turn to unity—if it devalues the many.

3.5 How to Cross a Space

Earlier we found the energy of unity raising the 'Multeity' or the 'purely potential...ground of individuation' into intelligibility and number (2002b [1819–23], 379). In so doing, 'Unity' did not give the natural will a reduced form of its own oneness—'the subternumeral confusion of "each" infinitely repeatable'—but, rather, particular place (2002b [1819–23], 314). We found that consciousness can arise when this unity assumes a subjective aspect: the Logos 'being communicated' and already one with the Holy Spirit. From at least 1820, Coleridge teases out both consciousness and human individuality in these terms, so that the subjective way the Logos faces us corresponds to the transforming of perception.

His interest is not just the possibility of conscientious individuality but its training. 'Music in the un*self*conscious Muscles of the Musician's Fingers' must be potentiated and also formed before it is the musician's stirring or sleeping will; the question of freedom's exercise 'would still leave us in the same inconclusive state of mind as the new-born infant without the idea of the parent' (1990 [1820], 4: §4677; 2002b [1819–23], 14). His construction of nature really details a process of education.

As Coleridge has it, at the apostacy, the Divine Will restrains the fall to chaos and, in consequence, it separates into light and dark states in response to light, that is, to the divine as a distinguishing power that fits itself to the chaos. Several times he considers what he calls 'matter of light', a midpoint between light as such and 'aether' or darkness, characterizing it in terms of two opposed tendencies. Where beings had enjoyed unity and distinctity in the godhead, following the apostacy, these qualities come apart for them. On one hand, there is 'particularity devoid of Distinctiveness': 'a nisus to division without that distinction which necessarily implies a community, or the existence of another in the self, as another and yet equal to self'. On the other, there is a homogenizing tendency, 'the positive indistinction, or absence, of particularity' (2002b [1819–23], 325). Neither correlative is fully actual; each becomes real by the realization of the other and contains the quality of the other potentially. Each develops alternately in opposition to the other or else gradatively 'when both are conceived either as equal or under the predominance of some one [primary or composite pole]' (2002b [1819–23], 366–7; 372–3). Likewise, all material objects display a 'relatively interchangeable and amphoteric nature', so that we find any number of analogues in the physical realm: as forces, elements of chemical reaction, and so on.

Even this much reality has been called forth by the divine influence, yet because of this influence, there is still a nascent form of the appetitive gathering we found in objective perception as an apostatic tendency toward the center that is in all but can never be fully realized since, Coleridge says, 'in the place itself the power altogether ceases' (2002b [1819–23], 371). Presumably, being 'in the place' means without the whole toward which it reaches and by which it is real. Gravity, specific and general, manifests the same essentially relative character. Because this tendency is futile, he says, it is realized through its opposite: an unresisting plenum, which does not exclude light that enters in as a '*potence*' or quality, giving 'outwardness and distance' to this potentiality, and as the potentiality, which Coleridge calls 'aether', is not itself space, it is negatively realized as 'amassment' (2002b [1819–23], 371).

Coleridge's construction of nature progresses in slow, zig-zagging increments. For present purposes, we need only focus on the fact that the divine influence is playing both parts, acting 'in the Self-Will as Self and for self', then compelling the self to work in it. As the reality of both acts is God, 'in the one the God incarnate, in the other the God επφανης, or super-apparent', neither can be neutralized and the result is a motion cycloidal, that is, the path or progress made by each revolution (2002b [1819–23], 358).

Here, comparison with Leibniz's concept of 'living force' will help. In one thought experiment, Leibniz images a tube that has been rotating at uniform speed around an immobile center. In it, a ball is suddenly released from a hindrance. At first, the numerical value of the motion by which the ball moves away from the center is infinitely small in comparison to the combined centripetal and centrifugal forces. But after it has been rotating for some time, the 'complete centrifugal impetus...comparable to the rotational impetus...must arise in the ball' (1989a [1695], 121). To the extent possible (here represented by the relative mass of the ball, as impetus is velocity times mass), the whole is in each but as the divergent direction of each.

When Coleridge says Leibniz 'did not comprehend the origin and true character of the merely potential', that he 'ever speaks of continuity as actual, whereas it is no case predicable but of and in proportion to the potential of the Deity or the plenitude', he seems to mean that the world does not become itself through itself, does not complete divine decree through the infinity of its relations. For Coleridge, harmony is not continuity—nature reintegrated from the whole or moment-by-moment repotentiated because infinitesimally divided. Rather, our nature is 'actualized by number or the power of number, ... the continuity thus counteracted' (2002b [1819–23], 367).

We might conclude with two central principles in new guise. 'For each individual the spirit itself is Time' (2002b [1819–23], 191–2, 187). And, regarding Holy Spirit, the 'primary, absolute, co-eternal intercirculation of Deity', Coleridge writes, 'as an humble aid to the lower faculties of the mind, it is permitted...to recollect the third sound discovered and produced by Tartini' (1995 [1826], 2: 1334).

For Coleridge, time, like consciousness, is a spiritual reality that is internal to us but thinkable only in relation to something engulfing us that is actually time and which we cannot positively think. Only negatively can we conceive 'the truth of the eternity as coexistent with the perception of the eternal in Time and Space'. For every time we can conceive is measured with respect to a relatively unsuccessive time. On our planet

> The relatively immoveable sun is at once dimly awakening the dark into twilight, dawning, rising, risen, climbing ... even so must we conceive the agency of the Logos as parcelled out ... in the logos itself the succession could not be ... but neither in the materia could succession be utterly alienated from the logos, for succession presumes distinction, and the logos is essential distinctness, the number of all numbers (2002b [1819–23], 355).

As in consciousness, the Logos apportioned gives times and seasons individual beauty and character. Although this manifestation is only negatively of the eternal itself, the eternal in time is a positive power. Neither beings nor times are not built from the relating of inexpressible reality and instantaneity. Rightly understood, the relation to the cosmic center is individual, the influence of the Logos making that relation apprehensible there.

In his *Principles and Power of Harmony*, Giuseppe Tartini had described how two sounds made by any musical instrument, and held, will generate a third. If, as Coleridge has it, this sound images the Holy Spirit, who asserts the absolute identity of 'that which affirms' and 'that which is affirmed', this third sound also figures apperception (1969 [1818], 521). For there, the individual must continually relate potential individuality (will) to actual individuality (the communicated Logos) so that by the identity of communicated Logos and Holy Spirit we can apperceive identities in the world. The inner harmony we achieve—in our spirit—pulls us out of potentiality, just as apperception pulls us out into relation, but we never just are the Holy Spirit any more than we own consciousness. For Coleridge, we seem able to say that just as time reveals consciousness's character, so apperception yields some, at least indirect consciousness of individual character. For we are actual through particular self-integration, not before it. Yet through the reflexive unity of the communicated Logos and Holy Spirit, all knowing is potentially self-situating, self-knowing that way.

References

Adams, Robert Merrihew. 1994. *Leibniz: Determinist, theist, idealist*. New York: Oxford University Press.

Augustine. 2015. *De trinitate (c.420). Tr. Edmund Hill*. New York: New City Press.

Cheyne, Peter. 2014. The art of poetic life-writing. *The Coleridge bulletin* 44:77–89.

———. 2020. *Coleridge's contemplative philosophy*. Oxford: Oxford University Press.

Coleridge, Samuel Taylor. 1969. The friend. Vol. 1 (1818). Ed. Barbara E. Rooke.

———. 1972. *The statesman's manual* (1816). Ed. R. J. White.

———. 1983. *Biographia literaria* (1817). Ed. James Engell and W. Jackson Bate.

———. 1990. The notebooks of Samuel Taylor Coleridge. Vol. 4 (1819–26). Ed. Kathleen Coburn and Merton Christensen.

———. 1993. *Aids to reflection* (1825). Ed. John Beer.

———. 1995. Shorter works and *fragments*. ed. H. J. Jackson and J. R. de J. Jackson.

———. 2001. In *Poetical works, part one: Poems (reading text)*, ed. J. C. C. Mays, vol. 2. Princeton: Princeton University Press.

———. 2002a. The notebooks of Samuel Taylor Coleridge. Vol. 5 (1827–34). Ed. Kathleen Coburn and Anthony Harding.

———. 2002b. *Opus maximum* (*c*.1819–23). Ed. Thomas McFarland, with the assistance of Nicholas Halmi.

Cutsinger, James. 1987. *The form of transformed vision*. Mason, GA: Mercer University Press.

Di Bella, Stefano. 2005. *The science of the individual: Leibniz's ontology of individual substance*. Dordrecht: Springer.

Hedley, Douglas. 2000. *Coleridge, philosophy and religion*. Cambridge: Cambridge University Press.

Harding, Anthony John. 1999. Coleridge, the afterlife, and the meaning of 'Hades'. *The Coleridge bulletin 14*: 73–79.

Leibniz, G. W. 2001. *The labyrinth of the continuum: Writings on the continuum problem, 1672–1686*. Tr. Richard T. W. Arthur. Yale: Yale University Press.

———. 2013. *The Leibniz–De Volder correspondence with selections from the correspondence between Leibniz and Johann Bernoulli (1698–1706)*. Tr. Paul Lodge. New Haven: The Yale University Press.

———. 1989a. *Philosophical essays (1675–1715)*. Hackett: Tr. Roger Ariew and Daniel Garber. Indianapolis.

———. 1989b. *Philosophical papers and letters (1666–1716)*. Tr. and ed. Leroy E. Loemker. Dordrecht: Klumer Academic Publishers.

Muirhead, John. 1930. *Coleridge as philosopher*. New York: Macmillan.

Plotinus. 2018. The enneads (c.270). Ed. Lloyd P. Gerson. Tr. George Boys-Stones, John M. Dillon, Lloyd P. Gerson, R. A. H. King, Andrew Smith and James Wilberding. Cambridge: Cambridge University Press.

Schelling, Friedrich. 1984. Aphorisms as an introduction to Naturphilosophie (1805). *Tr. Fritz Marti. Idealistic studies* 14 (3): 244–248.

Charity Ketz is assistant teaching professor of English at Penn State University. Besides her research in literature and philosophy, she also teaches and writes poetry.

Chapter 4
Plastic Nature from Cudworth to Shaftesbury and Germany

Thomas Wormald

Abstract Plasticity, elaborated by French philosopher Catherine Malabou, is one of the most interesting and rich concepts in contemporary materialist thought. Malabou developed this concept in outlining how Hegel exports the plastic arts—its movement of activity and passivity, such as we can characterize both the artist and their material itself as plastic—outside of their traditional aesthetic realm to encompass the movement and nature of the self-forming subjectivity, thought, and being itself. This paper traces the conceptual epigenesis of plasticity from its notional kernel in Ralph Cudworth's concept of plastic nature through Shaftesbury to eighteenth- and nineteenth-century thinkers including Herder and Hegel in Germany and Coleridge in Britain to reveal the paradoxical presence of the idealist concept plastic 'life' at the heart of contemporary materialist thought.

Famous thinkers such as Spinoza and Leibniz have long been recognized as integral figures in the creative metamorphoses that occur in mid-to-late century German aesthetics and *Naturphilosophie,* and subsequently, as decisive influences on and viable resources for post-Kantian continental thought. The legacy of Cambridge Platonism, however, is relatively neglected if not occluded as a serious intellectual inheritance in contemporary theory. Indeed, the names of Spinoza, Leibniz, Kant, Hegel, and Schelling preponderate in modern scholarship as credible and generative philosophical resources, yet, you will find scant evidence of cutting-edge theorists in contemporary continental thought opining on the hidden radicality of Ralph Cudworth or the Third Earl of Shaftesbury.[1] This chapter aims to remediate this

[1] While historians of philosophy, such as the ones adduced throughout this essay like A. O. Lovejoy, John Zammito, Frederick Beiser and M. H. Abrams, among others, recognize the role of Cambridge Platonism in shaping German Romanticism and Idealism, I mean to suggest that in surveying contemporary Continental critical theory, one will not find Cambridge Platonism engaged with in

T. Wormald (✉)
Centre for the Study of Theory and Criticism, University of Western Ontario, London, ON, Canada
e-mail: twormald@uwo.ca

P. Cheyne (ed.), *Matter and Life in Coleridge, Schelling, and Other Dynamical Idealists*, International Archives of the History of Ideas Archives internationales d'histoire des idées 253, https://doi.org/10.1007/978-3-031-78315-9_4

neglect by relating Ralph Cudworth's concept of plastic nature to twenty-first-century French philosopher Catherine Malabou's notion of plasticity. According to her own account, Malabou extrapolates the concept of plasticity from Hegel, and attributes to him its development. While Malabou indicates that Hegel ultimately borrows and expands this concept from Goethe, she nonetheless does not pursue any substantial further genealogical elaboration. Here, I outline the development of Cudworth's idea of plastic nature and how it ends up as a defining feature of thinking in the late German eighteenth century, establishing Cudworth's plastic nature as the conceptual 'embryo' of a principle of plasticity in other thinkers, including Malabou. The genealogical line runs from Cudworth to Shaftesbury and then through mid-to-late eighteenth century Germany, principally through Herder. In connecting plastic nature to plasticity, and in filling out the genealogy which in Malabou's account only starts with Hegel, we hope to establish 'plastic nature' as the conceptual kernel of plasticity, and that Cambridge Platonism consequently exerts an insufficiently acknowledged influence on contemporary thought, and, more specifically and paradoxically, contemporary materialism. This shows the presence of an idealist concept, that of plastic 'life', at the heart of twentieth-century materialist thought. I shall outline Malabou's idea of plasticity and how it operates as the basis of her materialism, before returning to examine Cudworth's idea of plastic nature and to reconstruct its development through Shaftesbury and its subsequent transmission to Germany and its influence on Herder and Goethe.

4.1 Malabou's Plastic Materialism

The idea of plasticity has ascended to philosophical prominence through the work of contemporary French philosopher Catherine Malabou. Beginning with her *The Future of Hegel: Plasticity, Temporality, Dialectic* (1996; 2004), Malabou has successively developed plasticity as the conceptual and imaginative metabolic center of her ongoing attempt to elaborate a new materialist philosophy that is at once philosophical, biological, and political. Expressed simply, plasticity is the capacity to both give and receive form. Plasticity expresses at once something active or formative (e.g. like a plastic artist, or sculptor) and passive or formable (e.g. like clay), the pliable or formable material on which the plastic artist works. From this seemingly modest kernel, Malabou has generated profoundly transformative re-interpretations of canonical thinkers including Hegel, Heidegger, and Kant, opened stimulating dialogues within diverse fields such as gender studies, psychoanalysis, and neuroscience, and repeatedly petitioned for renewed engagements between the continental tradition and the life sciences in works such as *What Should We Do with Our*

the same manner that other philosophical resources from the same period *are*. For example, in Deleuzian studies, broadly construed, Spinoza and Leibniz are routinely utilized and discussed as profoundly generative thinkers for contemporary thought. It is simply not the case with Cambridge Platonists.

Brain?—an entreaty to take seriously the philosophical and political possibilities revealed by neuroplasticity—and *Before Tomorrow: Epigenesis and Rationality*, an exploration on the role of the biological theory of epigenesis in Kant's conception of reason and the transcendental.

While Malabou originally discovers this idea in the margins of Hegelian thought, she amplifies it further, making it the basis of a wider ontology that integrates the insights of contemporary biology—such as neuroscience and epigenetics—for a materialist philosophy grounded in a cognizance of our material shaping of one another and the imperative to see all forms of life as engaged in practices and experiments of self-formation that should be preserved or guarded. Malabou's philosophy of plasticity represents a wider trend in contemporary theory: new materialism, a philosophical and theoretical denomination describing perspectives that share the basic commitment to refute what are purportedly long-held reductive conceptions of matter as mechanistic, passive, and inert in favor of reconceptualizing matter as vital, active, and dynamic. It is precisely this 'focus on activity' and 'notions of dynamism, self-transformation, plasticity' (Wolfe 2017, 51) which these new materialisms take, rightly or not, to be new. Through her concept of plasticity, Malabou articulates a philosophical vision motivated by the conviction that any material or symbolic form—whether the 'form' is a brain or an organism, a self or a collective, a work or a corpus—is ultimately self-forming and not subject to or determined by a source transcendent or external to it. Perhaps we could say that the animating core of her thought is rigorous dismissal of any and all (pre)determinisms.[2] Against the idea that forms of life can be reduced to any kind of (pre)determinism—whether this is material or symbolic, genetic or cultural—Malabou finds that contemporary biology fruitfully offers conceptual resources to think a notion of plastic materiality that sees 'life as possessing its own modes of self-transformation, self-organization, and self-directedness' (Malabou 2015b, 45) and to think of life as being immanently sufficient to give shape or form to itself without the aid of or recourse to any exterior, 'transcendental instance' which would organize, direct, and, ultimately, determine it (2015a, 48). We do not, Malabou asserts, 'need to affirm the existence of a beyond or an outside of the real to confer meaning to reality, as if a prior structure, necessarily nonmaterial, was requested to give sense to materiality itself' (2015b, 40–1). Life is sufficient to shape or form itself; there is a transformative tendency internal to materiality, a 'self-transformative tendency of life', that thereby precludes the need of any external or transcendental source to guide or command it (2015b, 44–5). Unconstrained by any predeterminism or preformation that would 'necessarily [orient] and [determine] every' development through means of 'an internal tension toward a *telos*', (2015a, 49). Malabou wishes to conceive matter as

[2] A cursory glance at the titles of her principal works reveals this thread of her theoretical preoccupations: *The Future of Hegel, The Heidegger Change, What Should We Do with Our Brain? Before Tomorrow: Epigenesis and Rationality.* There is an emphasis on the *future* because it is not determined, and because it is not determined, the various forms of our life are things that we can *change*; there is something for us to *do*—forms are not invented 'before tomorrow', and therefore, they are not preformed but must be created.

its own self-forming architect, capable of *giving form to itself, a form which it itself produces* that is not determined by something over and above it. One of Malabou's organizing metaphors is that of sculpture, and she envisages self-sculpting sculptures or 'statues' that are 'alive', dynamic plastic forms shaping themselves in constant mutual engagement with the world and others.

4.2 'Plastic' Prior to Hegel?

A pillar of Malabou's account of plasticity is her extrapolating this concept from where she finds it Hegel. 'The concept of plasticity', she writes, was 'discovered for the first time in the preface to Hegel's *Phenomenology of Spirit* ... discovered in the heart of Hegelian philosophy' (2010, 8–9).[3] Her second major claim is that Hegel novelly enlarges the application of plasticity by displacing it from its original domain and amplifying and expanding it from a previous strictly aesthetic meaning to having a more comprehensive ontological or metaphysical operation and sense. Hegel is the first philosopher to authorize, elevate, and substantialize plasticity, to 'accord it the value of a concept', (Malabou 2011, 79) the alchemization into a proper concept meaning that it operates philosophically or does some kind of ontological work. Malabou's etymological argument goes that the substantive 'plasticity' enters the English, French (*plasticité*), and German (*Plastizität*) languages in the eighteenth-century, developing closely from two existing related words that shared the same root of the Greek *plassein*, meaning to model or to mold: the substantive 'plastics' (*die Plastik*), and the adjective 'plastic' (*plastisch*) (Malabou 2004, 8). The extant use of 'plastics' and 'plastic' were deployed, Malabou attests, in the 'native land' or 'original domain' of plasticity, 'the field of the art', with plastic arts characterizing 'the art of 'modelling', and, in the first instance, the art of sculpture' while generally referring to or being inclusive of those arts 'whose central aim is the articulation and development of forms' such as 'architecture, drawing and painting' (2004, 8) According to Malabou, Hegel 'rips [plasticity] away from' these traditional 'strict aesthetic ties (or sculptural ties, to be precise), definitively conferring the metaphysical dignity of an essential characteristic of subjectivity upon it' (2010, 13). Specifically, Malabou claims that Hegel *ontologizes* the formative process of the plastic ideal of Greek sculpture as the comprehensive modality of formation as such in the system, operating across registers that are aesthetic, ethical, political, and biological. A subject is not merely a passive recipient of predicates—like clay receiving shape from a sculptor—but gives form or shape to itself.

While Malabou's idea of plasticity has been widely circulated and engaged with, there has been scant interest to scrutinize her claim that Hegel invents this concept or examine if there is any relevant conceptual prehistory. Is it *true* that prior to Hegel

[3] Malabou reiterates this, *inter alia*, in a 2010 interview when asked to clarify the development of the concept of plasticity: 'I found it for the first time in Hegel' (Malabou and Vahanian 2008, 6).

'plastics' and 'plasticity' never had an ontological or philosophical meaning or operated in a conceptual register that exceeded aesthetics? And, if this is not true, if plastics *has* operated in a philosophically or ontologically significant way prior to Hegel, what is the relation, if any, between this previous use and plasticity in Hegel? Malabou herself provides a partial answer to the first question, revealing that, while Hegel is the first to substantially invest plasticity with a philosophical meaning by applying the aesthetic to the formation of subjectivity, Hegel himself borrows this term from Goethe (Malabou and Lawless 2016, 103–4).[4] While this is a philosophically creative reading of Hegel, the claim that Hegel has the exclusive merit or propriety over the ontological application of the aesthetic idea of plastic arts, requires some qualification and revision. The fact of the matter is that 'plastic' has a long history, one which has, since its inception, evinced the intertwining of the aesthetic, biological, and ontological. From Greek antiquity in Galen, through to the Latin-Arabic world of Avicenna and Averroes, to Ficino and Leoniceno and the Italian Renaissance, through figures like Jacob Schegk, Daniel Sennert, and William Harvey, the idea of a plastic force in nature has itself changed shape or form many times being called, variously, *dunamis diaplastike*, *virtus formativa*, *vis conformans,* and *vis plastica*. However, this idea gains a unique kind of conceptual consistency in the seventeenth century in the work of Ralph Cudworth and the idea of plastic nature. The question is what does the concept of plastic nature have to do with plasticity, if anything at all, and second, if there is a meaningful connection between plastic nature and plasticity, how do we chart a path from Cudworth to Hegel? Let us first assess the concept of plastic nature and then determine how it arrives to or becomes *available* for Hegel, becoming the plasticity that Malabou uses as the key notion in the elaboration of her contemporary materialist philosophy.

4.3 Cudworth's 'Plastic Nature'

Cudworth is one of the earliest and most insightful modern theorists of materialism and one of the staunchest critics of determinism in both material and theological forms—he is also one of the most marginalized or neglected of the early modern period by contemporary continental theory. While mostly neglected or not taken seriously by all but scholars and specialists of early modern British thought, Cudworth's theoretical analytics in *True Intellectual System of the Universe* (1678), though shrouded by unwieldy presentation, arguably merit the attention of and deserve to be read as closely as much more canonical modern philosophical works such as the *Meditations,* the *Ethics*, and the *Leviathan*. Cudworth develops the

[4] Hegel confirms the importance of Goethe in a letter, writing in reflection that 'When I survey my intellectual development, I find you everywhere interwoven therein and might well call myself one of your sons; from you my inner life acquired nourishment and the strength to resist abstraction and found its just course by keeping your constructs [Gebilden] in sight as if they were beacons' (Hegel 1954, 84).

concept of plastic nature in *True Intellectual System of the Universe* (1678) in his now infamous 'Digression on Plastic Nature', a digression which has paradoxically become one of his central and definitive philosophical contributions. Stemming from a dissatisfaction with what he considered to be overly mechanical and deterministic explanations of the natural world that excised vitality from nature, Cudworth seeks to forge a middle way between ontological conceptions of nature that either ascribe the formation of the world entirely to God or entirely to matter—whether this matter is 'passive' (Wolfe, 2017, 49, 51) and fortuitously moved as in Epicurean physics, represented in Cudworth's time by Hobbes, or whether this matter is conceived hylozoically, a term Cudworth coins (1678, 105), being 'active' in the sense that it is self-moving or self-forming.[5] The problem each materialism shares is that it omits a transcendent creator—Epicurean materialism eschews a creator and any active or formative principle altogether, ascribing everything to chance or 'fortuitous mechanism', and thereby denying any 'Final and Intending Causality in Nature' or a divine order and purpose governing the world (Cudworth 1678, 147). On the other hand, the active materialism of hylozoism rightly admits a vital principle but eschews a transcendent creator by ascribing this formative or '*Self-active power*' (Cudworth 1678, 105) to matter itself. According to Cudworth, this materialism slightly improves on the former atomistic materialism insofar as it admits the existence of life and vitality in the world, a 'self-active' or 'plastic power', (106, 107) but it errs in its omission of a transcendent creator, 'adulterat[ing] the notion of matter' by '…blending and confounding it with life', their 'Fundamental Error' being a belief in the '*Life of Matter*' (Cudworth 1678, 839). While Cudworth's opposition to atheistic materialisms is hardly unsurprising, Cudworth found equally objectionable predominant 'theistic' explanations for natural generation, whether this was of the variety that saw God designing a mechanical universe which he monitored at a distance (say, in Descartes) or a stronger version of voluntarism that 'sees God everything every thing in Nature should be done Immediately by God himself' (149). This idea is as untenable as the notion that the world is the result of pure chance or fortuitous mechanism as it ultimately compromises God's omnipotence. First, if God is so powerful, so much exertion, effort, and attention to assure the functioning of mundane and minute operations should be unnecessary (149). Second, that God 'should…set his own Hand, as it were to every Work, and immediately do all the Meanest and Triflingest things himself Drudgingly, without making use of any Inferior and Subordinate Instruments' (149) is indecorous and demeaning; it debases God to imply that God should 'condescend to do all the meanest Offices himself' (149–50). Further, the formation of living beings 'would seem to be but a Vain and Idle Pomp, or a Trifling Formality, if the Agent were omnipotent', and this leaves both the gradual development of organic beings and the '*Errors* and *Bungles*' that populate nature without adequate explanation (149–50).[6]

[5] One scholar attests Cudworth's conception of hylozoism refers to Spinoza's substance monism (Hampton 2017, 419), with another averring the theories of Glisson and Cavendish (Begley 2017).

[6] An '*Omnipotent Agent*' would presumably overcome any 'Indisposition of Matter' and 'dispatch its work in a Moment … [and] do it *Infallibly* and *Irresistibly*' and leave no '*Bungle* or *Fumble* in

While one would consider Epicurean materialism and theological voluntarism to be antithetical, Cudworth shows that they are actually speculatively identical in that, whether the world is determined immediately by God or by chance, both deprive nature of its vitality. Similar to Schelling's claim that thinkers like Descartes, Fichte, and Kant 'do away with or murder nature, depriving it of a living ground', (Schelling [1809] 2006, 26).[7] Cudworth also accuses modern philosophy of reducing the 'whole World' to '…nothing else, but a mere Heap of Dust, Fortuitously agitated… a Dead Cadaverous thing, that hath no…Vitality acting in it…' (Cudworth 1678, 149–50). To construe life in this fashion is to think 'a kind of Dead and Wooden World, as it were a Carved Statue, that hath nothing…Vital' (148).

Cudworth then rejected that nature was fortuitously formed by chance (Epicureanism, Hobbes), that it was formed by self-forming matter (Hylozoism), or that it was a mechanical creation of God, whether preestablished and superintended from a distance or formed and maintained directly by God's hand (variations of voluntarism such as Descartes, Malebranche, and Leibniz). To ameliorate this impasse, Cudworth appeals to the idea of plastic nature, or the idea of a nature that is plastic. It is principally through the aesthetic metaphors of the architect and artisan that Cudworth makes his concept of plastic nature legible—God is the head, and plastic nature the 'hands'. Cudworth writes: 'Nature may be called the … *Manuary Opificer* that *Acts* subserviently under the *Architectonical Art* and Wisdom of the Divine Understanding' (Ibid., 156). God is '*Supreme Architect* and Master-builder of the World' and plastic nature is the '*Immediate* Workman and Operator' (Cudworth 1678, 153.) This 'workman' or 'operator' is to be conceived as '*Art itself, acting immediately on the Matter, as an Inward Principle*', as if the '*Art of the Shipwright, were in the Timber it self* … the Mind of the Architect…transfused into the Stones, Bricks and Mortar'. Plastic nature is not '*Art Acting* … from without and at a distance but *Immediately* upon the thing it self which is Formed by it' (Cudworth 1678, 153).[8] Cudworth then articulates an ontological economy of production: God provides an ideal design and blueprint, plastic nature works to materialize it. However, imperatively, plastic nature only follows directives and orders and never supplies its own. Plastic nature '*does Do without Knowing the Reason of what it Doth*' (Cudworth 1678, 156). This, Cudworth writes, explains the possible '*Imperfections* and *Defects*' of the art of plastic nature: that 'though it Act *Artificially*

anything' (Cudworth 1678, 149–50).

[7] See Schelling ([1809] 2006, 26): 'nature is not present' to modernity for it 'lacks a living ground'.

[8] Kant's wrestling with the analogues of art and life in the section 'Critique of the Teleological Power of Judgment' seems to uncannily echo Cudworth's own attempts to articulate what is specific to nature's self-forming capacity while circumventing the problem of hylozoism. Kant seems to implicitly reject Cudworth's solution, characterizing it as a 'an alien principle standing in communion' with nature where the 'soul is made into an artificer' (246). Kant rebuts Cudworth's proposal of a plastic nature as artificer because it is necessarily *immaterial* and, consequently, the resultant 'product must be withdrawn from (corporeal) nature' (246). Because plastic nature is immaterial, nature cannot be said to have organized or produced itself—yet, this objection dissipates if we enlarge our conception of matter to be capable of self-organization and self-formation; that is, if we admit the existence of *material plastic nature.* Kant 2000 [1790], 246.

for the *Sake of Ends*…it self doth neither *Intend those Ends*, nor *Understand the Reason of that it doth*'. Nature, Cudworth observes, 'is not *Master* of that *Consummate Art* and Wisdom according to which it acts, but only a *Servant to it*, and a *Drudging Executioner* of the Dictates of it'. In this manner, plastic nature is 'alive' or active enough that it can shape and form, but it is nonetheless passive in that it is merely executing the design or plan of a supervising architect, God. Plastic nature 'act[s] for the sake of those Ends, that are not intended by it self, but some Higher Being', like a 'Saw or Hatchet in the hand of the Architect or Mechanick doth' (Cudworth 1678, 162). Essentially, plastic nature operates as an intermediary force between matter and God. It is more than matter because it is active—it shapes or forms the world—but it is passive, so it is *like* matter, in that it is subordinate to God, but it only shapes or forms according to God's plan. By conceptualizing plastic nature as a formative or shaping power both *distinct from God and matter* but *immanent to matter* that forms or shapes the world according to and on God's behalf, Cudworth believes he has sufficiently accounted for the fact that nature is granted some kind of life—that it exceeds mechanism or dead materialism in that it is formative, plastic, self-shaping—while avoiding the pitfalls of hylozoism by stipulating that while it is formative it is nonetheless passive or subordinate and teleologically oriented in that it follows dictates or orders supplied by a transcendent origin or mind. Plastic nature solve Cudworth's theological problem because God obtains an alibi in that nature is conceived as being minimally autonomous, being able to act independently of God enough to shoulder responsibility for 'errors and bungles' that occur in nature and the world.[9] Plastic nature is conceived ideally as a mimetic or reproductive formative power—formative but formed, plastic nature exercises its creative power to follow the instructions and design of the superintending architect who is God.

In plastic nature, Cudworth develops ideas that prove historically significant, if not paradigmatic, for future thought: nature as aesthetic, independent but teleologically oriented, and dynamically self-forming. Further, in rebutting mechanistic or reductive materialism evacuated of any living or vital principle, Cudworth is

[9] It should be noted here that the development of plastic nature is conceptually pivotal to Cudworth's overarching philosophical project, both to provide an intelligible natural philosophy that properly represents the relationship of God and nature, but, also, for the work that this ontological account of the world accomplishes for Cudworth's ethical thought as well: by providing an answer as to how God can both be present in the world but not directly responsible in the shaping or forming of his creation, plastic nature helps establish a conceptual space that preserves the ability for individuals to shape or form themselves, providing an answer as to how human beings are still sufficiently autonomous or free to determine their own actions while God is still considered pervasive in the world. The invocation of the conceptual space opened by plastic nature serves not to repudiate just material and divine determinist natural philosophies but to equally rebut combat their moral philosophical equivalents in Epicureanism and Calvinism. We structurally occupy the same place as plastic nature—between matter and God, we are the *manual labourers* or *plastic* artisans on the ground, so to speak, materially executing or fulfilling providence. The *ethical* project of Cudworth and Shaftesbury consists in petitioning us to become *conscious* of our plasticity, that is, our existence as *plastic* beings who can shape or form ourselves for ends or purposes which are not *predetermined* but which we ourselves determine.

uncannily resonant with recent theoretical preoccupations and successive developments in German thought. The animating thrust of Cudworth's contribution to natural philosophy appears profoundly prescient: that is, the contention that any account of the natural world cannot entirely deprive nature of vitality or life, that is, be *entirely mechanistic* and founded on mere chance, and, secondarily, that, for any account of the world to be intelligible, we must grant nature a minimal amount of autonomy. Cudworth's insistence that there must be *more* to the natural or material world than mere matter, that there must be life or something *spiritual* to nature, resonates through contemporary petitions to re-enchant nature or to see nature as vital and agentive, with its own life. Beyond the notions of nature conceived as aesthetic, vital, and as dynamically self-forming, Cudworth arguably formulates in rudiment the conceptual infrastructure, in ways that are fundamentally unaltered in the next century, of the problems of teleology and biological generation in eighteenth-century debates of preformationism, spontaneous generation, and epigenesis. Cudworth's expression of the need to think a formative or plastic force in nature that does but does not know', that acts for the sake of ends that it itself does not understand, that is not merely mechanical or divine of nature that unconsciously pursues 'conscious' or 'intelligent' ends, these contorted formulations of the specific *kind* of nature that are necessary to intelligibly think life in ways that escape the pitfalls of both materialist and divine positions are uncannily similar to the ones Kant employs later on teleology, art, and the living being in *Critique of the Power of Judgment*. And, in Cudworth's critique of both mechanical and divine natural generation, we have prefigurations of what would later become critiques of preformationism and spontaneous generation: natural forms cannot be randomly produced by material causation, unfold a divinely predetermined seed, or be shaped directly or miraculously by a divine hand. Instead, we have to think of life as having its own formative force—it is more than mere matter, but not God—that is, it is not predetermined but *is* teleologically oriented. In this manner, Cudworth arguably sketches the conceptual parameters of what will become a defense of the epigenesis of natural forms.

4.4 From Cudworth to Shaftesbury to Germany

Cudworth's notion of plastic nature continues through Shaftesbury, obtaining a more refined and aesthetically pleasing form.[10] In his canonical interpretation of Cudworth, J.A. Passmore contends that 'This much at least we can assert without

[10] The connection between Shaftesbury and Cudworth, and Cambridge Platonism more broadly, is suggested throughout scholarship. The key interpretive issue is not whether Shaftesbury had read or was familiar with Cudworth's work and ideas, but rather pertains to the extent of influence, what modifications, and other questions of this ilk. Shaftesbury was acquaintances with Damaris Masham, Cudworth's daughter, through their mutual friend, and Shaftesbury's tutor, John Locke, while Shaftesbury's first publication was a preface to a collected edition of Benjamin Whichcote's

qualification, that Shaftesbury was fundamentally a Cambridge Platonist' (Passmore 1951, 99). While not quite as forcefully as Passmore, Michael Gill similarly observes that Cudworth and the Cambridge Platonists 'so molded the general shape of Shaftesbury's thought that it is fitting to think of his philosophy as a continuous outgrowth of Cambridge Platonism itself' (Gill 2006, 78). It is 'Shaftesbury principally', Cassirer writes, 'who saves the Cambridge School from the fate of a learned curiosity and makes it a philosophic force in the centuries to come' (Cassirer 1953, 160).

Shaftesbury further develops Cudworth's plastic nature, that is, nature as a plastic power or capacity that should be imagined as '*Art itself, acting immediately on the Matter, as an Inward Principle*' (Cudworth 1678, 150). Two interrelated features of plastic nature particularly inform and are emphasized in Shaftesbury's own development of this idea: first, the aesthetic notion of nature as plastic, as a self-forming 'artist' or agent, and, second, the connected aesthetic notion of plastic arts as consisting in the harmonious relation of parts and whole. The most influential aspect of Shaftesbury's thought 'lay in Shaftesbury's concept of nature under the standpoint of artistic capability" and his vision of an 'artistically productive force' as immanent to every form of nature.[11] While Shaftesbury 'looks upon the world as a work of art', Shaftesbury's paradigm-shifting insight resides in the claim that the essence of aesthetic activity, and hence, nature, consists not in the *product* or the *work*—it 'is not exhausted in any particular work, nor in the infinite profusion of its works'—but rather that 'it manifests itself solely in the creating and forming process' (Cassirer 1951, 84–5). In Shaftesbury's own powerfully felicitous phrase, 'the beautifying, not the beautified, is the really beautiful' (Shaftesbury 1711b, c [1709], 322). Nature is not a collection of *things*, a product or a 'dead form'—nature is not a 'Dead Statue'—but, in Shaftesbury's words, nature is a 'form which forms', or a *forming form* (322–3). To play off Cudworth's phrase, nature must be understood as a self-forming sculpture, not as a 'Dead' but as a *living* statue that is seen as self-shaping and self-forming. Through his own theory of forms in the *Moralists*, Shaftesbury then supplies his own modulation of Spinoza's distinction between *natura naturata* and *natura natarans*, or, what Schelling will later characterize as nature as *object* (or product) and nature as *subject* (or productivity). As opposed to forms which are incapable of self-formation and must be shaped externally—what Shaftesbury calls dead forms, pure objects or products—nature itself has 'inner form': the capacity to shape or form is *immanent* or internal to nature itself. Thus, Nature itself 'in its deeper sense is not the sum total of created things but the creative power from which the form and order of the universe are derived' (Cassirer 1951, 326). The essence of nature is its capacity to *give form*, and moreover, give

Sermons. Through his friendship with Damaris Masham, Shaftesbury would have had access to Cudworth's private manuscripts. Scholars note that there are decisive changes in both Locke's *Essay* and Shaftesbury's thought after Locke had taken up residence with Masham in 1691.

[11] See Dilthey, qtd. in Matysik, 2012, 115–16. It is the idea that 'Nature is animated by an inherent divine power', that 'She is *one* and uniform throughout', appearing as 'the most perfect artist, working with unique technique' (Dilthey 1985 [1910], 287–8).

form to itself. In other words, the essence of nature is its plasticity. Thus, when Shaftesbury petitions us to follow nature or to look at nature as a model, he is enjoining us to see ourselves as self-forming artists, and to aspire to emulate the plasticity of nature itself.[12]

One of the most powerful ways that Shaftesbury expresses his vision of nature as plastic is through the infamous declaration that we should each become a 'Second Maker, a just Prometheus under Jove', allying the figure of Prometheus with that of the 'sovereign artist or universal plastic nature' (Shaftesbury 1711d [1710], 93). Nature is analogized as a Promethean figure, a self-forming artist who exerts its creative capacities in a manner that is 'just to Jove'. We are ourselves 'notable architects' (Shaftesbury 1711b, c [1709], 323) who, like Prometheus or a sovereign artist, are capable of self-formation—we then should be *like* plastic nature and shape and form ourselves so that our individual self, like a microcosm of nature, 'forms a whole, coherent and proportioned in itself, with due subjection and subordinacy of constituent parts' (Shaftesbury 1711d [1710], 93). Here, the critical modification Shaftesbury makes with respect to Cudworth's plastic nature is that Shaftesbury explicitly solicits us to emulate it. The paradigmatic shift between Cudworth and Shaftesbury with respect to plastic nature is that Shaftesbury claims something that one could never imagine Cudworth deigning to: plastic nature is something not only to be celebrated, but actively emulated. One could scarcely imagine Cudworth rhapsodically encouraging his readers to become like that 'Drudging Executioner', hailing us to assume our work as servants and manual labourers mindlessly following the orders of God. This differs palpably from Shaftesbury's many breathless exultations: 'O glorious nature! Supremely fair and sovereignly good! … O mighty nature! Wise substitute of Providence! Empowered creatress!' (Shaftesbury 1711b, c [1709], 298) that he petitions us to emulate. Whereas Cudworth consistently asserts plastic nature's as a lowly subordinate and subservient unconscious formative agent, Shaftesbury petitions us to *be like* that 'universal plastic nature', (Shaftesbury 1711d [1710], 93) that is, to become conscious of ourselves as formative agents. Shaftesbury enjoins us to become a conscious plastic nature, to deliberately emulate the formative and productive plastic powers of nature to fashion ourselves and to recognize our role in shaping or forming the world as a formative agent contributing to its ongoing development. Whereas Cudworth's plastic nature is a 'manual labourer', a 'drudging executioner', Shaftesbury's plastic nature is much more venerably rendered as a poet and artist. Thus, 'the conception of Nature as a model for the creative artist of genius', as a Promethean figure who shapes themselves as a plastic artist, that becomes so transformative for thinkers like Herder and Goethe, and major currents flowing through Germany in latter stages of the eighteenth-century 'is traceable to Shaftesbury' (Kistler 1954, 389).

The second important feature that Shaftesbury inaugurates for the modern period is making explicit what was implicit in Cudworth's turn to the aesthetic metaphor of

[12] Shaftesbury's distinction here significantly inspires Moritz's 'On the Artistic Imitation of the Beautiful' (1788).

plastic nature: an incipient organicism developed from plastic aesthetics. Like the plastic artwork, Shaftesbury reads nature aesthetically, as a system of coordinating parts that work together to form a larger harmonious totality. The interrelation of microcosm and microcosm is, again, an integral part of Cudworth's natural theology, but Shaftesbury renders it even more emphatic. All art should express what Shaftesbury calls 'plastic truth' which consists in 'the proper or due subjection or subordination of parts to a whole', (Shaftesbury 1711b [1709a], 67) where 'particulars…yield to the general design and all things [are] subservient to that which is principal' (66). This is what the genuine artist achieves, and why 'that sovereign artist or universal plastic nature' is our best aesthetic model because nature rightly 'forms a whole, coherent and proportioned in itself, with due subjection and subordinacy of constituent parts' (Shaftesbury 1711d [1710], 93). For Shaftesbury, the virtues of proportion, harmony, and order are just as much natural and biological as they are aesthetic. The organic system of nature is tantamount to a living, self-forming plastic artwork. The natural world is a 'mighty union' (Shaftesbury 1711c [1709b], 275) of which there are a numerous nested parts, a multitude of 'private systems' which ultimately conspire to form 'one universal system' (Shaftesbury 1711a [1699], 169). Just as individual or particular facets of a plastic work must be conceptualized not as isolated but as fundamentally related for the intelligibility of the total work, so too in nature we must constantly understand the interrelation of the individual and the whole; there is no creature that 'could be understood to be absolute and complete in [themselves], without any real relation to anything in the universe besides' (168). In this way, 'Shaftesbury prophetically anticipated the later recognition that the cosmos and every particular organism is a system, in which the parts are coordinated into a whole by the unity of purpose".[13] Because of this unique character of Shaftesbury's conceptualization of nature—as one natural system or whole comprised of interdependent and subserving parts—Shaftesbury can be considered 'among the first important upholders of an organicist approach to the world' (Rogers 1972, 251).[14] Indeed, an 'organic doctrine appears fully explicitly in Shaftesbury', providing the guiding principle of 'whole that precedes its parts and makes them possible' (Beiser 2011, 82). The portrait of being that Shaftesbury offers is of a creative, organic nature, a nature conceived as an interdependent, living totality of forming forms which mutually shape and form one another. All forms are formed but forming, drawing their plastic possibility from a 'single One', that is a 'mighty Genius', a 'Sole animating and inspiring power', (Shaftesbury 1711c [1709b], 307) an 'original soul, diffusive, vital in all, inspiriting the whole' in which they participate (303). Shaftesbury' universe is a 'living thing' comprised of coordinating and mutually constitutive self-forming or plastic beings; not a random

[13] O. Walzel, qtd. in Zammito 1992, 390, n. 26.

[14] See also Beiser 2011, 82. Here, Beiser argues that Shaftesbury's organic holism is a central influence on the thought of Justus Möser, a thinker whose admirers include Goethe, Hamann, Herder, and Schlegel (65). Beiser again rebuffs the conventional ascription of Leibniz as the primary influence and instead centers Shaftesbury as the decisive agent in the development of organic thought in mid-to-late century eighteenth century German thought.

ricocheting of atoms, nor that of a 'machine run by an absentee machinic' (Abrams 1953, 185).

4.5 From Cudworth and Shaftesbury to Germany

This conception of nature as aesthetic and self-forming, as dynamic, organic, and vital, in short, as plastic, is transmitted to Germany through both Cudworth and Shaftesbury's work. Cudworth's primary work migrates to Germany through the Latin translation of *True Intellectual System of the Universe* in 1733 by J. L. Mosheim, eventually finding its way onto Kant's bookshelf (Warda 1922)[15] and into the curriculum at Tubingen where Hegel, Schelling, and Hölderlin were students (Hedley 2000, 40–2). While there have been some attempts to affirm a direct connection between Cudworth and Kant,[16] most suggestions are somewhat reticent or diffidently made. Yet, while readers have long expressed perplexity at the intrication of art and biological life in Kant's *Critique of Judgment,* this purported enigma is arguably put into relief if Cudworth, Shaftesbury, and the philosophical legacy of Cambridge Platonist are properly brought into the frame as integral influences on thought during this time.[17] While efforts can be made to track the direct influence of Cudworth on developments in German thought in the mid-to-late century,[18]

[15] Kant discusses, and dismisses, plastic natures as early as 1754, writing in the essay 'The question, whether the Earth is ageing, considered from a physical point of view', that 'Those who assume the existence of a general world spirit in this sense do not understand by it some non-material power, a world soul, or plastic natures, the creations of a bold imagination, but a subtle though universally active matter which, in the products of nature, constitutes the active principle and, as a true Proteus, is able to assume all shapes and forms' (Kant 2012 [1754], 180). The reproach of plastic natures as non-material resembles Leibniz's critique in 'Considerations on Vital Principles and Plastic Natures, By the Author of Pre-Established Harmony' (1705) where Leibniz comments that while he and Cudworth agree in many important respects, they diverge on the point of requiring 'certain immaterial plastic natures' (Leibniz 1989b [1705], 589).

[16] Such as Lovejoy (1908).

[17] Another source of evidence of Cudworth's influence on German thought more generally is cited to be the adoption of the motto *hen kai pan*, or, 'One-and-All', by Lessing in his confession of Spinozism to Jacobi (Assmann 1987, 139). While traditionally understood to be an expression of Spinozism, the phrase at the heart of the Pantheism Controversy purportedly derives not from Spinoza but can be attributed to Mosheim's translation of Cudworth's *True Intellectual System*—in which the phrase appears verbatim amidst discussions of pantheism—from which the phrase 'entered the philosophical vocabulary and become[s] a fundamental metaphysical stance'. (Mihai 2020). However, while the primary dissemination of Cudworth's work certainly contributed to the diffusion of his thought, the text over which Lessing and Jacobi meet to discuss is potentially more instructive: Goethe's Prometheus. While it has long been recognized that Goethe's works on Prometheus were immensely influential, it is hardly emphasized that it is from Shaftesbury that this figure enters into German thought. Herder also avers, writing to Jacobi, that Shaftesbury has his 'own *hen kai pan*' (Herder 1989 [1784], 123).

[18] An example is Alexander Hampton's article that connects Cudworth and Herder. The connection could be strengthened by the inclusion of an intermediary which would more plausibly explain

Shaftesbury presents a potentially more viable path as the main conduit of transmission for plastic nature, Cudworth, and Cambridge Platonism in Germany. The evidence for Shaftesbury's importance to mid-to-late century German thought is ample and overwhelming, with 'hardly any philosophy or poet of any standing in the movements of Enlightenment, *Sturm and Drang* and early Romanticism who did not react in one way or another to Shaftesbury's work'. (Klemme 2001, xix). Shaftesbury was heralded as a 'prophet' of Romanticism (Tuveson 1965, 405), there is an 'unmistakable line' that runs straight through the eighteenth-century' from 'Shaftesbury to Romanticism as a whole' (Boyer Jr. 2003, 183.) Indeed, Shaftesbury's injunction that we each become our own 'Second Makers', (Shaftesbury 1711d [1710], 93), 'self-improving Artists', and 'the architect[s]' of our own 'life and fortune', (Shaftesbury 1711c [1709b], 322) epitomizes the 'romantic imperative' that one should approach one's life and the world as a work of art.[19] In fact, this might be said to be Shaftesbury's cardinal contribution to modern thought.[20] Shaftesbury's major works were circulating in Germany already by the mid-century, J. J. Spalding having translated Shaftesbury's central pieces the *Moralists* in 1745 and the *Inquiry* in 1747, while Hamann finished a translation of the *Characteristics* in its entirety in 1755. As early as 1754, Lessing and Mendelssohn, in their anonymous pamphlet *Pope, A Metaphysician!: An Anonymous Pamphlet in Defense of Leibniz*, were celebrating Shaftesbury as a philosophical equal of Leibniz.[21] He is an integral figure for both Christoph Wieland (Elson 1913, 6) and Karl Phillip Moritz's aesthetics,[22]and historians such as Dilthey, Cassirer, Berlin[23] and Beiser each single out Shaftesbury's priority as an influence on a luminous coterie of names such as Hamann, Herder, Moser, and Jacobi, Lessing, Goethe, Kant, Moritz, Schiller, and Schelling. Shaftesbury was recognized and celebrated by Leibniz, and almost 'universally revered' by all significant figures of the time. (Boyer Jr. 2003, 184.) Calling him the

how Cudworth gets to Herder, i.e. through the work of Shaftesbury. So, when Hampton writes that Herder uses 'Cudworth's philosophy to create a revised version of Spinoza's metaphysics' (Hampton 2017, 418), we would qualify this that Herder does use Cudworth's philosophy but receives it through Shaftesbury. See Hampton (2017, 417–31).

[19] This is a phrase from Schlegel that Beiser takes to name his book the *Romantic Imperative* in where he provides an account of early German romanticism, suggesting that the titular imperative is to turn nature, science, and life into an 'art' and vice versa. See Beiser 2003, *Romantic Imperative*, 19.

[20] Shaftesbury's might be the first person in modern thought to issue the 'exhortation to…to live our lives as though they are works of art' (Gill 2018, 3).

[21] 'Overall, I must confess, that *Shaftesbury* very often appears to one to agree with *Leibniz* so happily, that I wonder why one did not long since compare these two philosophers. I wonder even why the Academy itself did not prefer to give out the task of investigating the system of *Shaftesbury*, and holding it against the Leibnizian, rather than the system of *Pope*. They would, in that case at least, have placed philosopher against philosopher, and profundity against profundity, instead of enveloping in an unequal battle poet with philosopher' (Lessing and Mendelssohn 1999 [1754]).

[22] See Todorov (1984), 153; Halmi (2007, 72).

[23] Berlin cites Shaftesbury as one of the critical figures of the 'Counter-Enlightenment'. See Berlin (1979, 9).

'beloved Plato of Europe', Herder himself writes that Shaftesbury, 'this virtuoso of humanity', had a profound influence on the best minds of his century. (Herder 1881 [1793], 158). Indeed, as one commentator observes, 'The roster of those who were influenced by Shaftesbury's philosophy in word and in spirit reads like a list of the literary and philosophic greats of the Enlightenment' (Grean 1967, ix–x). The most important figure for tracking Shaftesbury's impact, however, is Herder, as it is Herder who is perhaps one of Shaftesbury's most vocal advocates and it is Herder who introduces Shaftesbury to Goethe.

Shaftesbury's influence on Herder was pervasive and persistent. Hamann introduces Herder to Shaftesbury in 1767 (Kistler 1954, 386), with Shaftesbury appearing that year in the 'Fragments on Recent German Literature'[24] and being defended by Herder in 1770 correspondences.[25] Herder included a poetic translation of the *Moralists* in the second edition of *God: Some Conversations*, specifically the section known as the 'Hymn to Nature',[26] as well as appending another in the third collection of his *Letters for the Advancement of Humanity* (1794).[27] Herder hails Shaftesbury as one of his central influences, citing him as his 'favourite companion' (Zammito 2002, 213) and naming Shaftesbury as a philosophical peer of Spinoza and Leibniz.[28] Herder's yearning for a philosophical framework adequate to this unique vision of nature alchemized in an inspired synthesis of Spinoza, Leibniz, and Shaftesbury (Zammito 1992, 183). When Herder's son approached him for counsel as to how he should best learn the philosophy of Spinoza and Leibniz, Herder responded that Shaftesbury was 'the author who contained the Spinoza-Leibniz philosophy in the most beautiful and select form' (Herder 1949 [1787], 221).[29] For Herder, Shaftesbury's thought distills and represents in best summary form what Herder saw as the core truths of Spinoza and Leibniz's philosophy: Spinoza's famed stipulation of the identity of God and being—the idea of pantheistic immanence—

---

[24] For discussion of Shaftesbury and the concept of *kalos k'agathos* and the virtuoso see Herder (2004 [1767], 45–7).

[25] Herder protects Shaftesbury against the charge of atheism: 'An atheist who preaches nothing so much as order, harmony and highest wisdom in the structure of the whole world … that atheist, with his great World-Spirit (for me the sublimest name for God) is more to me than ten such 'pettifoggers' in philosophy' (Herder, qtd in Burkhardt 1940, 8).

[26] See Herder (1949, 71).

[27] According to Clark, Herder includes this translation of Shaftesbury to corroborate the idea that human beings are developed and formed, that art is the very nature of man, and thus that we should pursue everything with a feeling for this: that every activity, philosophy, science, art, is not at a remove from 'humanity' but is a particular process or modulation in the shaping *of* humanity (Clark 1955, 371.)

[28] In the first edition preface to *God: Some Conversations* (1787), Herder writes that '[f]or some 10 or 12 years, I have had in mind a little essay which was to be called, 'Spinoza-Shaftesbury-Leibniz'' (Herder 1949 [1787], 67).

[29] 'Appendix B', *God, Some Conversations*, 221. This letter is also referenced in Hampton 2019, 89 fn. Elsewhere, Herder avers, 'Spinoza is not for you. His works are a precious stone buried deep in poor rock, and you cannot possibly master them. On the other hand, Shaftesbury's writings are the best that I can recommend to you. His *Rhapsody* and *Theocles* contains the Spinozist-Leibnizian philosophy in the most beautiful and exquisite summary form'. (Herder, qtd in Bell 1984, 38).

and Leibniz's notions of dynamic force, individuality, and plurality. This synthesizing of Leibniz and Spinoza represents one of the most generative and profound hallmarks of this moment in German thought: the suffocating determinism of Spinoza's immanent monism is alleviated by the introduction of Leibnizian principles of dynamism, individuality, and plurality. However, this traditional account seemingly elides the fact that while Leibniz certainly provides a resource to dynamically infuse Spinoza, the two thinkers still nonetheless both espouse at bottom two kinds of geometrically indebted rationalisms. The pathos or affect that perfuses suffuses the work of late eighteenth-century German thought is inexplicable if the sole reference points are the work of Spinoza and Leibniz[30]: how do vital and aesthetic philosophies of nature emerge from the affective aridity of rationalism? Herder himself admits as much, stating that Leibniz ultimately spoke to his head, proclaiming 'Leibniz, Leibniz! Where is thy spirit?'[31] So while integrating Leibniz into Spinoza's mechanical system dynamized it, the system remained to be vivified as such. This is the critical role that Shaftesbury, plastic nature, and the legacy of Cambridge Platonism play.

As one commentator pithily renders it, 'If Leibniz spoke to Herder's head, Shaftesbury spoke to his heart' and instilled in him 'an appreciation of the importance of the aesthetic' (Spitz 1955, 458). While appealing to different aspects of Herder's sensibilities, Leibniz's and Shaftesbury's thought dovetails in significant respects, with Leibniz himself saying that Shaftesbury's *Moralists*, composed in 1709, contained almost 'all' of his *Theodicy* (1710; Leibniz 1989a [1711], 633). Leibniz reflects on their shared thematics, writing of the *Moralists*:

> The universe all of one piece, its beauty, its universal harmony, the disappearance of real evil, especially in relation to the whole, the unity of true substances, and the great unity of the supreme substance of which all other things are merely emanations and imitations are here put in the most beautiful daylight (Leibniz 1989a [1711], 633).

Moreover, while Shaftesbury and Spinoza mutually affirm a one-and-all, it is, as Dilthey argues, not Spinoza's 'rationalist-mechanist' pantheism but Shaftesbury's uniquely 'aesthetic-vitalist'[32] vision that provides the critical affective stimulus that energizes the imagination of the late eighteenth century and provides it with its aesthetic, creative, and vital inflection.[33] John Zammito corroborates, observing that it is this "new conception' of nature articulated by Shaftesbury that proves to be 'the decisive stimulus to the later pantheism of Goethe and Herder' (Zammito 1992,

[30] The aesthetics of Wolff and Baumgarten are similarly bereft of the kind of emotional, creative charge offered by Shaftesbury. Winckelmann should be noted, but he was influenced by Shaftesbury and drew upon Shaftesbury's more overt classicist elements.

[31] Herder, qtd in Burkhardt 1949, 9. See also Spitz 1955, 454.

[32] Dilthey distinguishes two modern strands of pantheism: 'one aesthetic-vitalist', deriving from Bruno and Shaftesbury, going through Herder and Goethe, and one 'rationalist-mechanistic'. See Dilthey 'Shaftesbury und Spinoza', 167–77, qtd. in Matysik (2012, 115–6).

[33] If we follow Beiser's insistence to remember the critical role of Platonism in the developments of both German Romanticism and Idealism, Leibniz and Shaftesbury perhaps best represent the two sides of Platonic thought: rational and affective, mathematical and poetic.

183).[34] Zammito concurs with Cassirer, who earlier stated that Shaftesbury is central to a 'powerful current of new feeling for nature', and it is 'Shaftesbury's apostrophes to nature' which 'exert a decisive influence', giving expression to those fundamental forces which shaped the philosophy of nature of Herder and the young Goethe' (Cassirer 1951, 84–5). This organic, aesthetic conception of nature is the lens through which Herder ultimately re-interprets and re-shapes the philosophies of Spinoza and Leibniz. Zammito articulates perfectly how Herder uses Spinoza and Leibniz to mutually re-shape one another:

> Herder read Leibniz through Spinoza and Spinoza through Leibniz to find a philosophical mode for articulating his consistently naturalist insight. Force and dynamism were essential for the new natural philosophy, but it was equally essential that these be seen as immanent in nature. Leibniz was important for Herder because he dynamized the natural order … Herder sought to revise Leibnizian dynamism from a transcendent to an immanent monadology. Just this inspired his reinterpretation of Spinoza to balance Leibniz (Zammito 2018, 181).

To this superlative encapsulation we only add the decisive role that Shaftesbury and the idea of a plastic nature plays as the third member of Herder's formative triumvirate. Herder imagines not just an immanent monadology where forms do not interact—whether, respectively, by being parallel or in a preestablished harmony—but an immanent monadology of plastic forms which give shape and form to one another in a way that is not preestablished or determined in advance but, as Herder was to argue, in the uniquely organic way of developing or becoming.[35] The 'mathematical reasoning of Spinoza finds an aesthetic form in Shaftesbury' (Pascal 1952, 213), as substance is re-imagined as a fundament of creativity and mutability and nature a theatre of transformations, and the Leibnizian monad is no longer imagined as 'windowless' but fundamentally 'windowed',[36] open to, affected by and affecting, the world. To the monism and monad of Spinoza and Leibniz—to the notion of an immanent world consisting of a dynamic plurality of individual actants—Herder creatively incorporates a historicized and biologized interpretation of Shaftesbury's aesthetic-vital notion of a plastic nature, meaning, one that both gives and receives form and one that is conceived organically, in which plurality and unity co-exist. It is then through Shaftesbury that Herder develops the profound importance of the

[34] Zammito notes that this connection is also observed by Ernest Tuveson.

[35] 'For [Herder] perfection is neither rational nor affective, but a disposition or habitus realized in activity. Consequently, his view of the mediation of providence differs from both the British and the German traditions. It is not mediated through a sense, nor through reason, but is realized in activity. The hand of God is neither invisible nor personified in the philosopher ruler, but embodied in the hands of the child which, through testing and moulding its surroundings, cultivates itself' (Caygill 1989, 186).

[36] 'Herder speaks not in terms of a single organism, but rather in terms of a complex of organisms, each of which is an active whole in itself comparable to the Leibnizian monad. Unlike the Leibnizian monad, however, the Herderian monad has windows. This is an important qualification. For though the individual units of the whole are distinct entities, deriving their source of activity from within, they are nonetheless considered by Herder as being not wholly self-contained but interconnected' (Barnard 1969, 31).

plastic idea of the giving and receiving of form, that is, of a fundamental plasticity that is the essence of both *symbolic* (the self-development and cultural self-formation expressed in aesthetic activity and notions such as tradition and *Bildung*)[37] and *biological* (epigenesis, organic forces or *Kräfte*)[38] life[39] that manifest powerfully in key texts such as the *Outlines* and *God: Some Conversations*.

It is through Herder that Shaftesbury's aesthetic conception of nature reaches Goethe, with it being Herder who recommended Shaftesbury's work to Goethe (Kistler 1954, 388). Through Herder, Shaftesbury's Prometheus 'began to take possession' of Goethe's imagination, with Herder employing the term 'a second Prometheus' already in 1769 (Price 1953, 105.) As a figure who famously gives shape or form, Prometheus is a model of plasticity and the plastic artist *par excellence*, shaping human beings out of clay in his image. The titan is typically invoked as a symbol or image of creative rebellion: Prometheus represents the formative power to create against and independently of the authority of the Gods. Following Shaftesbury's allying of 'plastic nature' and Prometheus, Prometheus should be read as a figure of the creativity of both the artist *and* nature: a Promethean nature, or a Promethean artist, is one which would not be determined or constrained by limits, models, or designs, but capable of genuine self-formation. Shaftesbury's impact on Goethe appears immediately after Goethe's meeting Herder in 1770, manifesting in the early work on Shakespeare and genius, 'On German Architecture' (1772) and the Prometheus pieces over the years 1773–5. Texts from this period all bear the unique hallmarks of Shaftesbury's influence in their themes of self-formation, creativity, and nature, and more broadly evince Shaftesbury's formative role in the so-called Storm and Stress period. Classic studies by Dilthey, Walzel, and Weiser all strongly contend for the impact of Shaftesbury on Goethe arguing especially for the resonances between Shaftesbury's 'Hymn to Nature' in the *Moralists* and the Goethe-Tobler fragment 'Die Natur' where 'The contemplation of Nature as a dynamic, all-pervading force occurs repeatedly in almost literal agreement' (383). It is Shaftesbury's 'conception of nature as an artist', and the allying of nature and art that particularly struck Goethe, remaining constantly with him (Price 1932, 106). After Goethe's study of Shaftesbury begins in the early 1770's, with 'serious study' commencing in 1776 (Ibid.) his thinking of nature was 'less naïve and assumed more aesthetic and pantheistic implications', (Kistler 1954, 388) continuing on through to the years of his Italian journey (1786–88) and the development of

[37] For more, see Horlacher (2004, 409–26) and Horlacher (2016). Horlacher connects the development of the concept of *Bildung* in Herder and, German thought more broadly, through the reception and translation of Shaftesbury's idea of 'inner form' or the self as plastic.

[38] 'Herder's employment of the term *Kraft*, rather than using the word *Plastik*, can be traced to his use of Mosheim's Latin translation, whose often liberal rendering of Cudworth's work sometimes omits the word entirely, but most often renders it as *vis genetrix*, or 'generative force'. (Hampton 2017, 428).

[39] The aesthetic pantheism and idealized nature of Shaftesbury forms and shapes Herder, while Herder shapes and forms it in turn, *materializing* it through the biological science of his day.

his morphology.[40] The effect of Shaftesbury's thinking on Goethe appears to be long-lasting, as Goethe gives prominent place to Shaftesbury in commemorating Wieland in 1813 (Bruford 1962, 34), and writes in 1817 that: 'What has been formed is immediately transformed again, and if we would succeed, to some degree, to a living view of Nature, we must attempt to remain as active and as plastic as the example she sets for us'.[41] 'To the end of his life', Goethe's philosophy was 'fundamentally, though not exclusively, Shaftesburian' (Price 1932, 108).

4.6 Conclusion

In tracing out the development of plastic nature in Cudworth, through to its influence on Shaftesbury, and, subsequently, its formative role in shaping the thought of Herder and Goethe, we have hoped to have compellingly accounted for the decisive contributions of Cambridge Platonist thinking on the profoundly generative metamorphoses of thought regarding aesthetic, nature, and philosophy in the late-eighteenth German intellectual world. This chapter's contention has been that any account or understanding of German thought in late-eighteenth century that does not center the idea of plastic nature and the influence of Cudworth and Shaftesbury is incomplete and deficient, if not distorted and illegible. To the already well-recognized figures such as Leibniz and Spinoza, we should add Shaftesbury, and the philosophical lineage of Cambridge Platonism, as the third dominant influence which leads to subsequent conceptions of nature as dynamic, immanent, organic, and self-forming, that is, *plastic*. It is only by incorporating the formative role of the idea of plastic nature that the decisively aesthetic orientation of the radical creative transformations in *Naturphilosophie* are ultimately legible. The reintroduction of its philosophical ancestry serves to demonstrate the, again, unacknowledged influence that Cambridge Platonist thought exerts on contemporary thought and, moreover, the enigmatic life that Cudworth's plastic nature has had. Malabou's contemporary plastic materialism is precisely what Cudworth rejected the most: the idea of self-forming or plastic matter that shaped itself independently of the guide of any transcendent architect. There is, then, paradoxically, an idealist conception of life at the heart of contemporary materialist thought. In this way, Pierre Bayle was eminently correct in his debate with Le Clerc in 1704: Cudworth's articulation of atheistic hylozoism ultimately undermined the protective, apologetic edifice Cudworth wanted to construct.[42] Against all Cudworth's most fervent protestations, it turns out that what he feared and abjured most came to be true: plastic nature came to have a life of its own, escaping the hands of its creator.

[40] See Price (1932, 103–8).

[41] Goethe's 'Unbildung and Bildung' (1817) as cited in Brady (1987, 274).

[42] Sarah Hutton (2020, 160) explains, 'Bayle charged that Cudworth's conception of Plastic Nature opened the door to atheism by rendering God superfluous since Bayle took Plastic Nature to be an entity operating independently of God'.

References

Abrams, M. H. 1953. *The mirror and the lamp: Romantic theory and the critical tradition*. Oxford: Oxford University Press.

Assmann. 1987. *Moses the Egyptian*. Cambridge, Mass: Harvard University Press.

Barnard, F. M. 1969. *J. G. Herder on social and political culture*. Cambridge: Cambridge University Press.

Begley, Justin. 2017. 'The minde is matter moved': Nehemiah grew on Margaret Cavendish. *Intellectual History Review* 4 (27): 493–514.

Beiser, Frederick C. 2003. *The romantic imperative*. Cambridge, Mass: Harvard University Press.

———. 2011. *The German historicist tradition*. New York: Oxford University Press.

Bell, David. 1984. *Spinoza in Germany from 1670 to the age of Goethe*. London: University of London Press.

Berlin, Isaiah. 1979. *Against the current: Essays in the history of ideas*. Princeton: Princeton University Press.

Boyer, Ernest, Jr. 2003. Schleiermacher, Shaftesbury, and the German enlightenment. *The Harvard theological review* 96 (2): 181–204.

Bruford, W. H. 1962. *Culture and society in classical Weimar 1775–1806*. Cambridge: Cambridge University Press.

Burkhardt, F. H. 1940. *Introduction. God, some conversations*. Tr. F. H. Burkhardt. New York: Veritas.

Cassirer, Ernst. 1951. *The philosophy of enlightenment*. Tr. by Fritz C. A. Koellen and James P. Pettegrove. Princeton: Princeton University Press.

———. 1953. *The platonic renaissance in England*. Ohio: Nelson.

Caygill, Howard. 1989. *The art of judgement*. Oxford: Blackwell.

Clark, Robert T. 1955. *Herder: His life and thought*. Berkeley: University of California Press.

Cudworth, Ralph. 1678. *The true intellectual system of the universe*. London: Richard Royston.

Dilthey, Wilhelm. 1985. *Poetry and experience* (1910). Ed. Rudolf A. Makkreel. Tr. Christopher Rodie. Princeton: Princeton University Press.

Elson, Charles. 1913. *Wieland and Shaftesbury*. New York: Columbia University Press.

Gill, Michael. 2006. *The British moralists on human nature and the birth of secular ethics*. Cambridge: Cambridge University Press.

Gill, Michael B. 2018. Shaftesbury on life as a work of art. *British Journal for the History of Philosophy* 26 (6): 1–21.

Grean, Stanley. 1967. *Shaftesbury's philosophy of religion and ethics: A study in enthusiasm*. Ohio: Ohio State University Press.

Halmi, Nicholas. 2007. *The genealogy of the romantic symbol*. Oxford: Oxford University Press.

Hampton, Alexander J. B. 2017. An English source of German romanticism: Herder's Cudworth inspired revision of Spinoza from 'Plastik' to 'Kraft. *The Heythrop journal* 58 (3): 417–431.

———. 2019. *Romanticism and the reinvention of modern religion*. Cambridge: University Press.

Hedley, Douglas. 2000. *Coleridge, philosophy, and religion: Aids to reflection and the mirror of spirit*. Cambridge: Cambridge University Press.

Hegel, Wilhelm Friedrich. 1954. In *Briefe von und an Hegel*, ed. Johannes Hoffmeister, vol. 3, 1821–1831. Hamburg: Felix Meiner.

Herder, Johann Gottfried. 1881. Briefe zu Beförderung der Humanität (1793). In *Herders sämmtliche werke*. Vol. 7. Ed. B. Suphan. Berlin: Weidmannsche Buchhandlung.

———. 1949. *Preface to the second edition. God: Some conversations*. Tr. Frederick H. Burkhardt. New York: Hafner.

———. 1989. Letters to Jacobi (February 6, 1784). In *Against pure reason: Writings on religion, language, and history*, ed. Marcia Bunge, 120–125. Minneapolis: Fortress Press.

———. 2004. Fragments on recent German literature (1767). In *Philosophical writings*, ed. Michael Forster. Cambridge: Cambridge University Press.

Horlacher, Rebekka. 2004. *Bildung*: A construction of a history of philosophy of education. *Studies in Philosophy and Education* 23:409–426.

———. 2016. *The educated subject and the German concept of Bildung*. New York: Routledge.

Hutton, Sarah. 2020. Philosophy, religion, and heterodoxy in the philosophy of Henry more, Ralph Cudworth, and Anne Conway. *Church history and religious culture* 100:157–171.

Kant, Immanuel. 2012. The question, whether the earth is ageing, considered from a physical point of view. In *Kant: Natural science*, ed. Eric Watkins, 165–181. Cambridge: Cambridge University Press.

———. 2000. *Critique of the power of judgment* (1790). Tr. Paul Guyer and Eric Matthews. Ed. Paul Guyer. Cambridge: Cambridge University Press.

Kistler, Miark O. 1954. Sources of the Goethe-Tobler fragment. *Monatshefte* 46 (7): 383–389.

Klemme, Heiner F. 2001. General introduction. In *The reception of British aesthetics in Germany: Seven significant translations, 1745–1776*, ed. Heiner F. Klemme and Manfred Kuehn. Bristol: Thoemmes Press.

Leibniz, Wilhelm Gottfried. 1989a. Remarks on the three volumes Entitled Characteristics of Men, Manners, Opinions, Times (1711). Philosophical papers and letters: A selection. Boston: Kluwer Academic Publishers.

———. 1989b. Considerations on vital principles and plastic natures, by the author of pre-established harmony (1705). In *Philosophical papers and letters*, ed. Leroy E. Loemker, 586–591. Boston: Kluwer Academic Publishers.

Lessing, Gotthold, and Moses Mendelssohn. 1999. Pope, a metaphysician!: An anonymous pamphlet in defense of Leibniz (1754). *Fidelio* 8:4.

Lovejoy, A. J. 1908. Kant and the English Platonists. In *essays philosophical and psychological in honor of William James*. Ed. members of the Columbia University Department of Philosophy. New York: Longmans, Green.

Malabou, Catherine. 2004. *The future of Hegel: Plasticity, temporality, dialectic*. Tr. Lisbeth During. London: Routledge.

———. 2010. *Plasticity at the dusk of writing: Dialectic, destruction, deconstruction. Tr. Carolyn Shread.* Cambridge: Polity.

———. 2011. *Changing difference: The question of the feminine in philosophy. Tr. Carolyn Shread.* Cambridge: Polity.

———. 2015a. Whither materialism? Althusser/Darwin. In *Plastic materialities: Politics, legality and metamorphosis in the work of Catherine Malabou*, ed. Brenda Bhandar and Jonathan Goldberg-Hiller, 35–47. Durham, North Carolina: Duke University Press.

———. 2015b. Will sovereignty ever be deconstructed? In *Plastic materialities: Politics, legality and metamorphosis in the work of Catherine Malabou*, ed. Brenda Bhandar and Jonathan Goldberg-Hiller, 47–61. Durham, North Carolina: Duke University Press.

Malabou, Catherine, and Kate Lawless. 2016. The future of plasticity. *Chiasma: A site for thought* 3:99–108.

Malabou, Catherine, and Noelle Vahanian. 2008. A conversation with Catherine Malabou. *Journal for cultural and religious theory* 9 (1): 1–13.

Matysik, Tracie. 2012. Spinozist monism. In *Monism: Science, philosophy, religion, and the history of a worldview*, ed. Todd H. Weir, 107–134. New York: Palgrave Macmillan.

Mihai, Adrian. Accessed 2020. The reception of Ralph Cudworth (1617–1688). Online. The Cambridge Platonism sourcebook.

Pascal, Roy. 1952. *The German Sturm und Drang*. Manchester: Manchester University Press.

Passmore, J. A. 1951. *Ralph Cudworth*. London: Cambridge University Press.

Price, Lawrence Marsden. 1932. *The reception of English literature in Germany*. Berkeley: University of California Press.

Rogers, Pat. 1972. Shaftesbury and the rhapsody of aesthetics. *The British journal of aesthetics* 12 (3): 244–257.

Schelling, Friedrich. 2006. *Philosophical investigations into the essence of human freedom (1809)*. Tr. Jeff Love and Johannes Schmidt. New York: State University of New York Press.

Shaftesbury, Anthony Ashley Cooper. 1711a. An inquiry concerning virtue or merit (1699). In *Characteristics of men, manners, opinions, times*, ed. Lawrence E. Klein, 163–231. Cambridge: Cambridge University Press.

———. 1711b. *Sensus communis*, an essay on the freedom of wit and humour in a letter to a friend (1709a). In *Characteristic of men, manners, opinions, times*, ed. Lawrence E. Klein, 29–70. Cambridge: Cambridge University Press.

———. 1711c. The moralists: A philosophical rhapsody, being a certain recital of certain conversations on natural and moral subjects (1709b). In *Characteristics of men, manners, opinions, times*, ed. Lawrence E. Klein, 231–339. Cambridge: Cambridge University Press.

———. 1711d. Soliloquy, or advice to an author (1710). In *Characteristics of men, manners, opinions, times*, ed. Lawrence E. Klein, 70–163. Cambridge: Cambridge University Press.

Spitz, Lewis W. 1955. Natural law and the theory of history in Herder. *Journal of the History of Ideas* 16 (4): 453–475.

Todorov, Tzvetan. 1984. *Theories of the symbol*. Ithaca: Cornell University Press.

Tuveson, Ernest. 1965. Shaftesbury on the not so simple plan of human nature. *Studies in English literature, 1500–1900. Restoration and the eighteenth century* 5 (3): 403–434.

Warda, Arthur. 1922. *Immanuel Kant's books*. Berlin: Martin Breslauer.

Wolfe, Charles. 2017. Varieties of vital materialism. In *The new politics of materialism*, ed. Sarah Ellenzweig and John H. Zammito, 44–65. Londom: Routledge.

Zammito, John H. 1992. *The genesis of Kant's critique of judgment*. Chicago: University of Chicago Press.

———. 2002. *Kant, Herder and the birth of anthropology*. Chicago: Chicago University Press.

———. 2018. *On the gestation of German biology*. Chicago: Chicago University Press.

Thomas Wormald is a Ph.D. candidate at the Centre for the Study of Theory and Criticism at the University of Western Ontario. He is the co-editor of *Thinking Catherine Malabou: Passionate Detachments* (Rowman and Littlefield, 2018) and his work can be found in journals such as *Symposium: Canadian Journal of Continental Philosophy, Cosmos and History: Journal of Natural and Social Philosophy* and Interdisciplinary Literary Studies A Journal of Criticism and Theory.

Chapter 5
Novalis's Poetic Understanding of Nature in the Age of Romanticism

Alexander J. B. Hampton

Abstract Like his contemporary Coleridge in Britain, Novalis on the continent sought out a means to understand the logos of nature and express it through poetics. This chapter explicates Novalis' contention that poetry is such a language by examining his philosophical, linguistic, and religious writings. Through an examination of key works including *Fichte Studien*, *Monolog*, and *Die Christenheit oder Europa*, poetry emerges not as a subjective expression of one's impression of nature, but as an external metaphysical force, the constitutive logos of nature, and the ideal structuring principle of society. The attentive individual can become attuned to this poetry of nature, and in so doing speak the language of nature itself according to Novalis, as expressed in Novalis' enigmatic novella *The Novices of Sais*.

Like his contemporary Coleridge in Britain, Novalis, in the context of German romanticism, sought to understand and express the language of nature through poetics. It is not uncommon to find statements such as the following throughout the writings of the *Frühromantiker*:

> Poesie is the absolutely real. This is the core of my philosophy. (Novalis 1960–2006 [*Über Goethe*, 1798], 2: 647, §473)

> The Poet understands nature better than the scientific mind. (Novalis 1960–2006 [*Blüthenstaub*, 1798], 3: 468, §1093)

Through statements like these, Novalis sets out how poesie (poetry) is a key concept in his overall philosophy, particularly for his understanding of nature. This chapter explores what his intended poetical understanding of nature might be. Perhaps surprisingly, Novalis is not asserting the power of the subjective self to re-conceptualize nature according to philosophically idealist categories. Rather, he is asserting a philosophical realism in the view that all of nature is constituted through real,

A. J. B. Hampton (✉)
Department for the Study of Religion, University of Toronto, Toronto, ON, Canada
e-mail: a.hampton@utoronto.ca

P. Cheyne (ed.), *Matter and Life in Coleridge, Schelling, and Other Dynamical Idealists*, International Archives of the History of Ideas Archives internationales d'histoire des idées 253, https://doi.org/10.1007/978-3-031-78315-9_5

intelligible ideals, and that these metaphysical forms are key to understanding the natural world and creating a harmonious relationship with it.

Novalis addressed these themes striving to conceptualize and re-articulate the human relationship to the natural world in the light of recent philosophical developments. Among these were the development of Kantian and post-Kantian idealism; the continued growth of natural philosophy and disciplines such as chemistry, minerology, botany; and the changing human relationship to the natural environment occasioned by these intellectual developments, made increasingly manifest in anthropogenic damage to the natural environment. In light of his approach to the challenges of his age, and the intellectual resources he would deploy to address them, Novalis may be described as a post-Kantian Christian neo-Platonist, with all of the fascinating complexities that such a characterization connotes (Cheyne 2020, 183, 222–5; Hampton 2019, 125–32; 187–212).

The complex intertwining of idealism and realism in the thought of Novalis can be explicated by examining his philosophical, linguistic, and religious thought through major works in his oeuvre, revealing the character of the age of German romanticism. First, this examination considers Novalis's philosophical aim of finding a middle path between Spinozistic monism and Fichtean idealism in the *Fichte Studien*. It then proceeds to a consideration of his linguistic thought, focusing on the *Monolog*. Next it progresses to the cultural embodiment and expression of philosophy and linguistics in the political and religious ideas expressed in *Die Christenheit oder Europa*. In each of these, Novalis develops a concept of poesie which is transcendent, autonomous, and constitutive of reality as a metaphysical force. In the final section of this examination, Novalis's novella *Die Lehrlinge zu Sais* (*The Novices of Sais*) demonstrates how the attentive individual can become attuned to this poetry of nature, and in so doing learn to speak the language of nature itself.

5.1 Philosophy

Novalis's philosophy was developed in relation to the philosophical landscape of his own day. This was defined by two opposing, yet deeply influential philosophical positions, represented by Fichte and Spinoza. Novalis's reflections on both are recorded in his *Fichte Studien*. For Novalis, Spinoza and Fichte represented opposing manifestations of the same problem: philosophical foundationalism, the basis of modern philosophy. The goal of modern philosophy was to establish an immanent absolute foundation, that is, a fundamental first principle that did not require a transcendent ground. Transcendent first principles, whether the Good of Platonism, or the God of monotheism, were untidy in a modern context. They forever receded from circumscriptive definition, revealing themselves in myriad particulars rather than in simple concrete concepts, and they left themselves open to endless interpretability and approximation. Over a long period, the view of the world as grounded in a transcendent first principle withdrew from ascendency through a process of conceptual evolution and reform. Within this increasingly immanent framework, the conception of reality shifted from a theocentrism where self and nature participated in a hierarchical, vertically orientated cosmos wherein meaning ultimately resided in the transcendent, to an anthropocentrism where self and nature existed within a horizontally orientated cosmos conceptualized in immanent terms alone (Hampton 2019, 1–19).[1]

In this context, modern philosophy aimed to conceptualize an immanent first principle. Spinoza adopted an *ab extra* position and made substance monism, 'God or nature' as he described it, the foundation for philosophy. Alternately, Fichte adopted the *ab intra* foundation of the self for his metaphysical bedrock. We can observe Novalis comparing and inverting both positions in a note in the *Fichte Studien*:

> Transcendent nature is at the same time immanent—thus also the immanent person is at the same time transcendent—and also the reverse. Nature, however, as we have seen, is part immediately, part mediately conditioned. Similarly with the person—both in different ways. What is here immediately conditioned is there mediately conditioned and also vice versa.
>
> They are one and the same being—only reversed. They correspond down to the finest detail.
>
> Pictorially they are like two pyramids that have a single tip. They are like a single line. In one direction it is the image of nature—in the other [it is] the image of the I. (Novalis 2003 [1795/96], 2: 157, §153)

For Novalis, both philosophical foundations collapse transcendence into immanence, and render either self or nature exclusively active or passive, when in reality both condition each other in mutual interplay. Consequently, while these two philosophical positions addressed the need for an immanent first principle, they also created new problems.

This was particularly the case for the human–nature relationship. Fichte turned external reality into a reflection of the self, threatening a collapse into solipsism. Spinoza made the self into a mere mode of the substance of God, portending determinism. Although these critiques did not necessarily appreciate the nuances of either philosophy, the threat of turning nature into a reflection of the self, or of making the self a passive part of nature, was real. It was seen in the spectre of determinism that arose in the *Pantheismusstreit* between Jacobi and Mendelssohn, something which Goethe referred to as "an explosion" (Goethe 1988, X, 49). Equally it was seen in the increasing domination and destruction of the natural environment observed by the early Romantics, which evinced an understanding of nature devoid of its inherent meaning and value, and turned it into a raw resource for experimentation and exploitation. Novalis referred to this as a process of transcribing 'the creative music of the universe into the uniform clattering of a monstrous mill', as we shall come to see (Novalis 1997 [*Christenheit oder Europa*, 1799], 3: 515).

To this state of affairs, Novalis responds that not just an 'I' and a 'not-I' are necessary for philosophy, but a third principle, a theistic absolute, is also required as the ground of both. He writes: 'The act by which the I posits itself as I must be

[1] Also Taylor (2007), Blumenberg (1966), Funkenstein (2018).

connected with the antithesis of an independent Not-I and of the relationship to a sphere that encompasses them—this sphere can be called God, and I' (2003 [*Fichte Studien*, 1795/96], 2: 107–08, §8). Only this third element provides an absolute that transcends both of the foundational principles of self and nature. Summarizing his position, he writes, 'Spinoza ascended as far as nature—Fichte to the I, or the person. I to the thesis God' (2003 [1795/96], 2: 157, §151). The theistic absolute that Novalis describes changes our relationship to nature as well as to ourselves. When the transcendent absolute takes the place of the immanent foundations of self and nature, the self is no longer capable of determining nature (Fichte), nor nature the self (Spinoza). Rather, both are determined in relation to this third transcending principle.

For Novalis, to think in terms of a transcendent foundation challenges this increasing domination of nature by humans, and this has the result of changing the status of nature from passive material: 'Instead of Not-I, You' (Novalis 2007 [*Allgemeine Brouillon*, 1798/99], 3: 429, §820). That is, nature changes from the status of a 'Not-I' on which the self is projected, to an interlocutor, a 'You'. It gains this status, not because it becomes a conversant speaking person, but because its ground is the same as the self. Philosophical criticism, what Novalis referred to as his 'inversion method [Umkehrungsmethode]' demonstrated in his consideration of Spinoza and Fichte, is essential for recovering nature as a dialogue partner:

> It lets us divine nature, or the external world, as a human being—it proves that we can and should only understand everything as we understand ourselves and our loved ones, as we understand us and you. (Novalis 2007 [1798/99], 3: 429, §820)

Novalis made it his goal to find a form of speech that reflects this reality, where nature and the self speak a common language. He alludes to this language in a fragment that reflects on the common etymology of 'Stimme', 'Stimmung', 'Stimmen', 'Bestimmen', and 'einstimmen':

> Voice [Stimme]—tune/mood [Stimmung]—to tune/put in a certain mood [stimmen]—determine [bestimmen]—agree with [einstimmen]. 'Voice' expresses a self-constituting thing. 'Mood' arises out of two things acting and two things acted upon. (Novalis 2003 [*Fichte Studien*, 1795/96], 2: 283, §634)

The interrelationship between self and nature is something that is determined (bestimmen) by the mood or attunement that arises out of the relationship between both, through which one's self-constituting is expressed.

5.2 Language

Novalis thought that language could express this third, transcending principle. In his own work, Novalis seeks the correct form of speech, or *Stimme*, that will atune human language to the language inherent in nature itself. This leads my chapter to consider language, and Novalis's attempt to find the correct *Stimme*, which as it turns out is *Poesie*. In his short essay *Monolog*, Novalis considers the independent and autonomous quality of language: 'One can only marvel at the ridiculous mistake that people make when they think—that they speak for the sake of things. The particular quality of language, the fact that it is concerned only with itself, is known to no one' (Novalis 1997 [*Monolog*, 1798/99], 2: 672). According to Novalis, we misunderstand language when we think of it as something that is ours. We mistakenly consider ourselves to be its inventor and misconstrue its meaning to be our own. Then, with this misunderstood language we go on to commit a second error by applying it to nature. As a result, language becomes a tool for us to impose our own meaning on nature, rather than holding a dialogue with the meaning that exists outside of ourselves, inherent in nature. Furthermore, we are alienated from the capacity of language to form this dialogue between ourselves and the world.

In the context of this mistaken approach, it is language that gives meaning to the world outside of ourselves, and nature itself is considered devoid of meaning. This was the modern mindset which Novalis saw exemplified in the work of one of his teachers at the Freiburg mining academy, Abraham Gottlob Werner, a lecturer in geology and mineralogy. As a student, Novalis made detailed notes on Werner's system of mineral classification (Novalis 1984 [*Werner-Studien*, 1798/99], 3: 135–6). These record his objection to the 'one-sided' use of external characteristics alone to classify minerals (Novalis 1984 [1798/99], 3: 144). Despite being based on objective empirical observations, Werner's system of classification was symptomatic of the narrow instrumental use of language in the natural sciences, divorced from the realist ideas that participated in and constituted phenomenal reality. This had the effect of limiting our understanding, giving words the status of little more than semantic abstractions, arbitrarily assigned to things. This essentially rendered humans deaf to the language nature already spoke. However, Novalis observes that 'It is not man alone that speaks—the universe also *speaks*—everything speaks—infinite languages' (Novalis 2007 [*Allgemeine Brouillon*, 1798/99], 3: 267–8, §143). The ultimate outcome of the monological imposition of meaning upon nature would be to further fracture human understanding into disparate disciplines, undermining the unity of knowledge (Novalis 1984 [*Werner-Studien*, 1798/99], 3: 135–6). When language is misused according to Novalis this manifests itself as foolishness, the inability to articulate what we really wish to say, particularly because our misused words can never capture their intended objects. The foolishness of language, its evading our desired ends, leads people to hate language, Novalis contends. However, it is precisely this foolishness and frustration that shows language's 'infinitely serious side'. The resistance of language to our desires shows that it has its own meaning unto itself, and this draws Novalis to make a comparison with mathematics and music. Mathematics has its own logic. If we wish to do mathematics, we must play by its rules, rather than making mathematical formulae conform to our arbitrary will. Extending this analogy to music, Novalis writes: 'whomever has a fine feeling for its fingering [Applictur], its rhythm, its musical spirit, who perceives within himself the delicate working of its inner nature, and moves his tongue and hand accordingly, will be a prophet' (Novalis 1997 [*Monolog*, 1798/99], 2: 672). With the skill of a musician, by listening and attuning one's own speech with the nature of true language itself, one is able both to understand and communicate by it. The prophetic quality Novalis describes opens up the wider philosophical claim being made about language. The language that Novalis describes here precedes all things, human and nature, and constitutes their 'inner nature'. It is language that gives things their form and consequently their meaning. As such, language is far more than an intra-human convention for communication, an arbitrary system of signs and signifiers; rather, it is a metaphysical force constitutive of physical being. In attuning oneself to language therefore, one is attuned to the transcendent reality beyond both the self and nature, the theistic absolute that was the outcome of Novalis's consideration of the need to overcome the immanent foundationalism of Fichte and Spinoza. It is in the context of this prophetic attunement that Novalis finally introduces the concept of poesie: 'in saying this I believe I have described the essence and function of poetry in the clearest possible way' (Novalis 1997 [*Monolog*, 1798/99], 2: 672).

5.3 The Metaphysics of Poesie

It is worthwhile to unpack the use and meaning of Novalis's term 'Poesie'. The term itself was brought into German use in the sixteenth century. It came to be superseded by words such as 'Dichtung' and 'Dichtkunst' by Novalis's time, which offered the more limited understanding of poetry as literary. As a result, his use of 'Poesie' is intentionally evocative. Its source is the Greek term πόίησις (póiēsis), meaning production, in particular, original and creative making that brings reality into being. It is defined by Plato in the *Symposium* (205b): 'Everything that is responsible for creating something out of nothing is a kind of poetry.' Plato is careful to differentiate between what he considers to be the good and bad uses of poetry, most famously in Book X of the *Republic*. Plato (1997 [*c*.375 BC]) censures both Homer and Hesiod for creating 'false and contemptible representations of the gods', representations that cause them to appear more human than divine (*Republic* 377a–378c, 596a–600e). According to Plato, this interested poetics produces art thrice removed from reality (*Republic* 597b–e). Written only for dramatic effect it serves only to reflect our own interests back to ourselves. Interested, or misused poetry, it constituted a kind of instrumentalization of language and denial of true disinterested poesis; a critique that reflects Novalis's own position on the misuse of language.

True poesie is possessed of a divine quality. It is not a language whose meaning is subjective and private to the mind of a particular author. Rather, it is subjective to one mind only, the mind of God, whose subjectivity constitutes objectivity for all creation. Poetry is a 'divine gift', a glimpse into the intelligible reality that constitutes all phenomenal existence, including both the self and nature (Plato 1997, *Phaedrus* [*c*.370 BC] 245a; *Ion* [*c*.398 BC] 235b–c). This relationship between the divine mind and creation receives elaboration in the *Timaeus*, where the demiurge looks to the intelligible forms, the models whereby it shapes the cosmos (*Timaeus* [*c*.360 BC] 28a–b, 29e–30b). When adopted into the Hellenized Judaism of the first century BC by Philo, the forms become the divine ideas in the mind of the one God, who is both architect and builder (Philo, *On the Creation*, V.20). The Hellenized reading of Genesis then came to deeply influence early Christianity's doctrine of creation, with the ex nihilo act of creation understood as a divine act of poesis, as can be observed in further Platonizing readings of Genesis, as well as the Johannine concept of the Logos.

Novalis makes a similar claim to Plato's about the disinterested use of language when he writes: 'When someone speaks merely for the sake of speaking, he utters the most splendid, original truths. But if he wants to speak about something definite capricious language causes him to say the most ridiculous and confused things' (Novalis 1997 [*Monolog*, 1798/99], 2: 672). The understanding of language as being disinterested, as something which humans could never fully possess because of its divine quality, is present throughout the Christian Platonic tradition, from Clement, Aquinas, and Hildegard, right up to Jakob Böhme, Coleridge and Moritz. The other side of this elusiveness however, meant that if one could become attuned to poesie, then one could become a prophet, as Novalis says.

Novalis's friend Friedrich Schlegel described Romantic poetry as that which can 'hover at the midpoint between portrayed and portrayer, free of all real and ideal self-interest' (Schlegel 1958–2002 [1798], 2: 315). This reflects the anticipatory verse where God's creative spirit or breath (רוּחַ [*ruach*]) hovers over the waters in the pregnant moment preceding the 6 days. Poesie occupies a middle position, between ideal and real, infinite and finite, transcendent and immanent. In its attunement to the generative Logos or language of nature, poesie is the constitutive language of being, and the poet who captures it becomes its prophet.

For Novalis, poesie provides the means to overcome the twofold error that characterizes the misuse of language. As a force continuously constitutive of reality, it shows how divine language antecedes and constitutes both the self and nature. Returning to the poetic use of language makes it possible to overcome the alienation that the misuse of language causes. Poesie, rather than constituting a means to impose our own meaning upon nature, returns us to a dialogical participation in it. The main task that faced Novalis, following from the observations summarized in the *Monolog*, was how to put his insights into practice. The irony that he articulated these insights in discursive, non-poetic prose was not lost to him: 'anyone who knows how to write truths like these but does not have ear and sense enough for it will be outwitted by language itself and mocked by people as Cassandra was by the Trojans' (Novalis 1997 [*Monolog*, 1798/99], 2: 672). Poesie, as Novalis concludes, is not so much a theory one learns as a praxis one must perform.

5.4 Religion

The example of Cassandra raises the question as to how this philosophy and language of poesie was to be put into practice. Thus far we have explored Novalis's position that both philosophy and language must be grounded in the transcendent absolute. The same must be true for the civilizational organization of society as a whole in which both are embodied. This is where religion, according to Novalis, is required:

> In the beginning poets and priests were one, and only later times have separated them. The genuine poet, however, is always a priest, just as the genuine priest has always been a poet. And should not the future again re-establish the old state of things? (Novalis 1960–2006 [*Blüthenstaub*, 1798], 2: 441, §71)

Before proceeding further, however, what Novalis means by religion here must be qualified. Rather than an institution with its beliefs and doctrines, what Novalis foremost means is the civilizational orientation towards the transcendent basis of all immanent reality.

This is at the core of his essay *Die Christenheit oder Europa* (*Christianity or Europe*, 1799), in which Novalis situates his own age within a broader history of Europe. The essay itself has always been contentious, even when it first circulated, and it is easily misinterpreted on first reading (Novalis 1960–2006 [1799], 4: 646, letter to Schleiermacher). It begins with the opening lines 'Those were beautiful, sparkling times, when Europe was a Christian land, where one Christendom inhabited this civilized part of the world' (1997 [1799], 3: 507). Yet, as becomes apparent, Novalis does not argue for a return to mediaeval Christendom, nor even necessarily a return to Christianity. Instead, his aim is to call his reader's attention to the form beyond the particular. In Novalis's mediaeval ideal, all finite actions are orientated towards infinite ends, and all creatures are to be understood in relation to their creator. This was reflected in the religious practices of mediaeval Christendom—the cult of saints, the veneration of relics, and pilgrimages to holy sites—but it was also reflected more widely, in artistic, scientific, technical and mercantile practices, 'by the powerful drive upward of all other human forces, and by the harmonious development of all institutions' (Novalis 1997 [1799], 3: 509).

This view leads Novalis to defend as just the control exercised by the church over the development of knowledge. Theoretical (or scientific and philosophical) knowledge needed to be contained within the carapace of holy knowledge:

> With justice, the wise supreme head of the church set himself against the impudent products which arose from human propensities at the expense of the holy message, and against dangerous untimely discoveries in the realm of knowledge. Thus he prevented bold thinkers from publicly asserting that the earth was an insignificant planet, for he well knew that if people lost respect for their own dwelling place and their earthly fatherland, they would also lose respect for their heavenly home and its kind, would give precedence to limited knowledge over eternal faith and become accustomed to scorn everything great and deserving of wonder, regarding it as the lifeless effect of laws. (1997 [1799], 3: 508–9)

For Novalis, valuing the Earth is uniquely bound up with valuing the heavens. To view creation through the lens of the holy was to view it through the context of transcendent realism. All of creation participated in the divine and was a finite expression of infinite divine being. Accordingly, all creatures were infinitely more than their particularity. As expressions of the divine, they were deserving of human wonder, respect, and consequently needed to be conserved. This is the foundation of a dialogical human–nature relationship, culturally embodied in the practices and beliefs of the mediaeval church. Alternately, Novalis describes a human propensity to value limited knowledge over infinite faith, the former being complete and the latter approximate. The devaluation of the holy leads to the devaluation of creation. Modernity gives precedence to reductionist knowledge that is able to conceptualize nature as the 'lifeless effect of laws' (Novalis 1997 [*Christenheit oder Europa*, 1799], 3: 508–9). Once circumscribable by finite human thought, he argues nature loses its uniqueness and its wonder. Earth becomes, in Novalis's diagnosis, 'an insignificant planet' (Novalis 1997 [1799], 3: 508–9).

In Novalis's historical narrative the mediaeval ideal could not sustain itself, 'humanity was not mature enough' (Novalis 1997 [*Christenheit oder Europa*, 1799], 3: 515). Consequently, what comes to dominate the modern thought of Novalis's day was the privileging of the finite reductivism of the anthropocentric standpoint, and consequently a philosophy which seeks out and destroys any form of thought that fails to conform to its finite logic, whether religion, imagination, or feeling. Novalis continues:

> The product of the modern way of thinking was called philosophy, and it was held to embrace everything that was opposed to the old way, principally therefore every objection to religion…. Still more hatred of religion extended itself very naturally and consequentially to all the objects of enthusiasm, made heretics of imagination and feeling, rectitude and love of art, the past and the future, it just managed to place mankind at the head of the series of natural beings, and turned the infinite, creative music of the universe into the uniform clattering of a monstrous mill, driven by the stream of chance and floating on it, a mill of itself without builder or miller and really a true *perpetuum mobile,* a mill grinding itself. (1997 [1799], 3: 515)

Here, the target of Novalis's criticism is the naturalistic and mechanistic thought of Enlightenment philosophy. Though its initial object of criticism was the church, in his mind it also turned against the sources and manifestations of religion, viz. feeling and art. In their stead, the terrifying image of nature as an autophagic mill appears, a mechanistic organism that in consuming and reproducing itself has arbitrarily produced humans as a product, and presumably will eventually devour them like Cronus did his children, to produce another equally unintended product (1997 [1799], 3: 515).

However, it is in this image that Novalis first identifies the poetic impulse that is the means of salvation from this nightmare. Though rationalized religion 'tirelessly engaged in cleansing nature, the earth, human souls, and learning of poetry, rooting out every trace of the sacred', the poetic abides precisely because it is not human-dependent (1997 [*Christenheit oder Europa*, 1799], 3: 516). Ironically, he writes: 'What a shame that nature remained so marvelous and incomprehensible, so poetic and infinite, in spite of all efforts to modernize it' (1997 [1799], 3: 516). 'Immortal meaning', he goes on to explain in *Die Christenheit*, 'cannot be destroyed, but it can be confused, paralysed, supplanted by other meanings', such as in the immanent philosophy of Spinoza or Fichte, or the instrumentalized language of Werner's modern natural science (1997 [1799], 3: 509).

As the language and expression of this immortal meaning, poetry is therefore the means to re-invent religion anew and restore our relationship with nature: 'Poetry elevates each single thing through a particular combination with the rest of the whole .... [It] is ... the key to philosophy, its purpose and meaning; for poetry shapes the beautiful society—the world family—the beautiful household of the universe' (Novalis 1997 [*Logologische Fragmente* 2, 1798], 2: 533). Mediaeval Christendom, or at least the ideal Novalis sets before his readers, constituted a poetic form of understanding. The same poetics that constituted nature, bringing forth an endless multiplicity of life, constituted human activities—art, science, technology and trade.

According to Novalis the events that undermined the mediaeval ideal and alienated humans from nature were part of a cycle of growth, decay and renewal. The benefit of his present age, writes Novalis, was its historical perspective, which was able 'to recognize in those wonderful follies remarkable crystallizations of historical matter' (Novalis 1997 [*Christenheit oder Europa*, 1799], 3: 520). Both mediaeval Christianity and the scientific Enlightenment proved their legitimacy, but also exhausted themselves in the process of history. One of the tasks of Romanticism, according to Novalis, was 'to return the sacredness of nature', and this required the reunification of poetry and religion, the subject of one of Novalis last works, *The Novices of Sais* (1802) (Novalis 1997 [1799], 3: 520).

5.5 Poetics of Nature

Novalis worked on *Die Lehrlinge zu Sais* during the Freiberg period, probably from late in 1797 to sometime in 1800. Central to the work is the relationship between knowledge of outer nature and inner self, and the role language plays in the connection or estrangement of both. The work remained unfinished, leaving it consisting of two chapters of uneven length: The first on the novices (*die Lehrling*) introduces the characters and the idea of nature as a cipher, while the second, on nature, outlines their education through a series of shifting voices offering reflections on the human relationship with nature. *Die Lehrlinge* is not a traditional narrative work, as it lacks conventional characterization and plot. Instead, it largely comprises inner monologues or dialogues whose speakers are often difficult to identify. Though the location does not play a distinctive role, the setting of Sais, the antique city in the Nile delta, evokes a classical topos. In the first century AD, Plutarch famously recorded that the shrine of Athena at Sais, which he identified with Isis, carried the inscription 'I am all that has been, and is, and shall be; and my veil no mortal has hitherto raised' (Plutarch 1969 [100], V, 25 IX.354).

The first lines of *Die Lehrlinge* describe nature as a script of ciphers (*Chiffernschrift*) that contain a hidden grammar (*Sprachlehre*) (Novalis 1960–2006 [1802], 1: 79). Novalis suggests that our exploration of the outer natural world is equally an exploration of the inner self, because they are unified by the absolute ground which transcends them both. Knowledge of the self and knowledge of nature both have this same source and end, with the inner and the outer connected by a doctrine of signatures and correspondences that momentarily reveals itself:

> Manifold are the ways of man. He who follows and compares them will see wonderful figures arise; figures that seem to belong to that great written cipher that one sees everywhere: on wings, eggshells, in clouds, in the snow, in crystals and rock formations, on freezing waters, in the inside and outside of the mountains, the plants, the animals, the people, in the lights of heaven, on scored disks of pitch or glass, in iron filings around a magnet, and in the strange conjunctures of chance. (Novalis 1960–2006 [1802], 1: 79)

In all these figures, Novalis continues, we may glimpse a key to this wondrous script (*Wunderschrift*), perhaps even a glint of its grammar (*Sprachlehre*), and yet these intimations of a higher language refuse to accommodate themselves to any fixed form—precisely because their source of meaning is not the human mind but the divine mind. Even if we come to be attendant to the language that constitutes nature, this only occurs fleetingly providing a momentary glimpse of a unified language, which once again dissolves and swims before our eyes.

This expresses the tension between the freedom of this script of ciphers and our own often instrumental, interested use of language as described in the *Monolog*: '[Language] is such a wonderful and fruitful secret—because when one merely speaks for the sake of speaking, he pronounces the most splendid, most original truths' (Novalis 1997 [1798/99], 2: 672). If we allow the script of ciphers to speak of its own accord, however, we are able to gain a glimpse of the transcendent reality beyond immanence: not an escape from immanence but a more true view of reality, of the constitutive meaning given to us and things, rather than the meaning we give to things. 'No mortal', says the novice, 'can lift the veil, so we must seek to become immortal; he who does not desire to lift it is no true novice of Sais' (Novalis 1960–2006 [1802], 1: 82). Speaking this original prophetic language of poetics is the end of the novice's education in *Die Lehrlinge*.

The second part of *Die Lehrlinge*, entitled 'Nature', begins with a comparison of the way language operates in its original form and the way it operates when it falls away from this form: 'It may have taken a long time before people thought to designate the manifold objects of their senses with a common name, and to place themselves in opposition to them' (Novalis 1960–2006 [1802], 1: 82). As this development was furthered, it led to the increasing division of understanding, like light split into an ever-wider spectrum. The 'sickly predisposition' of later forms of language lacked the power to mix these colours back together in either their original form or to bring about a new unity (Novalis 1960–2006 [1802], 1: 82). Prior to this separation of self and nature, the ideal and real were one: 'To those earlier people, everything seemed human, familiar, and social, there was freshness and originality in all their perceptions, each one of their utterances was a true natural breath, and their impressions corresponded to the world around them' (Novalis 1960–2006 [1802], 1: 83). This shift in language and epistemology which Novalis details is given no historical specificity, yet it corresponds to the gradual decline of realist transcendence, and with the establishment of immanent foundationalism by modern philosophers such as Spinoza and Fichte. As such, it corresponds with the decline of the mediaeval worldview outlined in *Die Christenheit oder Europa*.

One consequence of this development, Novalis explains, is the loss of the linguistic unity natural to poetry and science: 'Natural scientists and poets have, through speaking one language, always shown themselves to be one people' (Novalis 1960–2006 [*Die Lehrlinge zu Sais*, 1802], 1: 84). In this relationship, science gathered and arranged, whilst poets moulded the vastness of nature into smaller pleasing natures (Novalis 1960–2006 [1802], 1: 84). Separated, however, both instrumentalize language to their own ends, losing contact with the whole and become destructive forces: 'Under their hands friendly nature died, leaving behind only dead, quivering remnants', while the poets 'soar to heaven, engaging in god-like fancies, alienating themselves from the everyday, whilst squandering the fruits of their cheerfulness' (Novalis 1960–2006 [1802], 1: 84). In this example, rather than transcending itself, language becomes an expression of the ego, revealing its own abstraction in natural science, or its own fancy in poetics.

Alternatively, the novice who is attuned to the ciphers of nature expresses the whole, the original 'one language' of nature, even if only momentarily. This poetics of nature mends the fracturing that Novalis expresses with a catalogue in multiple voices, revealing the many but disconnected ways in which nature is conceptualized. Overall, according to Novalis, there are those who seek friendship with nature and those who confront it with the aim of domination. This opposition, he explains, results from the loss of original language and its instrumentalization to finite immanent ends. However, amidst this clamour of voices that characterize this part of *Die Lehrlinge*, that of a youth intercedes to explain the unifying role of poesie:

> Only the poets have felt what nature can be to mankind … They find everything in nature. To them alone its soul remains no stranger, and not in vain do they seek all the ecstasies of the golden age in its presence. For them nature has all the variety of an infinite soul, and more than the cleverest, most alive of men, it astounds us with ingenious turns and fancies, with correspondences and deviations, with grandiose ideas and trifling whimsies. So inexhaustible is nature's fantasy, that no one will seek its company in vain. It has power to beautify, animate, confirm, and even though an unconscious, unmeaning mechanism seems to govern the part, the eye that looks deeper discerns a wonderful sympathy with the human heart in concurrences and in the sequence of isolated accidents. (Novalis 1960–2006 [1802], 1: 85–7)
>
> What Novalis describes here is the emergence of a new dialogical relationship with nature, achieved through the medium of Poesie: 'Does the cliff not become a unique Thou, whenever I speak to it? And what am I but the stream, when I look sadly down into its waters and lose my thoughts in its flow?' (Novalis 1960–2006 [1802], 1: 89).

Attentive to a unifying language between the inner self and outer nature, the generative poesie of the poet gives momentary and approximate expression to a new sense of unity that transcends yet upholds the multiplicity of nature. This is the voice of the self, only to be heard in an age before recording as something echoed distinctly back as the voice of the cliff. It is the image of oneself, utterly dependent on the water's reflection-bearing quality. Because of this unified multiplicity, it is necessarily poetics that can be attuned to and express the language of nature. For Novalis, this language is not determined by us but by the theistic absolute and, as such, reveals to us the poetic language in which we participate but can never fully or discursively capture. Poetics is the lens through which to view the holy of nature.

References

Blumenberg, Hans. 1966. *Die Legitimität der Neuzeit*. Frankfurt am Main: Suhrkamp.

Cheyne, Peter. 2020. *Coleridge's contemplative philosophy*. Oxford: Oxford University Press.

Friedrich von Hardenberg Novalis. 1960–2006. Historische-kritische Ausgabe: Novalis Schriften. Ed. Richard Samuel, Hans-Joachim Mähl & Gerhard Schulz. 6. Stuttgart: Verlag W. Kohlhammer.

von Hardenberg Novalis, Friedrich. 2003. *Fichte studies*. Tr. and ed. Jane Kneller. Cambridge: Cambridge University Press.

———. 2007. *Notes for a romantic encyclopaedia: Das Allgemeine Brouillon*. Tr. and ed. David W. Wood. Albany, NY: State University of New York Press.

———. 1997. *Philosophical writings*. Tr. and ed. Margaret Mahony Stoljar. Albany: State University of New York Press.

Funkenstein, Amos. 2018. *Theology and the scientific imagination from the middle ages to the seventeenth century*. Princeton: Princeton University Press.

von Goethe, Johann Wolfgang. 1988. In *Johann Wolfgang von Goethe. Hamburger Ausgabe*, ed. D. Kuhn and R. Wankmuller, vol. 14. Munchen: Deutscher Taschenbuch Verlag.

Hampton, Alexander J. B. 2019. *Romanticism and the re-invention of modern religion: The reconciliation of German idealism and platonic realism*. Cambridge: Cambridge University Press.

Plato. 1997. In *Complete works (399–56 BC)*, ed. John M. Cooper and D. S. Hutchinson. Indianapolis: Hackett.

Plutarch. 1969. *Moralia (100). Tr. Frank Cole Babbit*. Cambridge, Mass: Harvard University Press.

Schlegel, Friedrich. 1958–2002. Kritische Friedrich-Schlegel-Ausgabe. Ed. Ernst Behler et al. 35. Munich: Schöningh.

Taylor, Charles. 2007. *A secular age*. Cambridge, Mass: Belknap Press of Harvard University Press.

Alexander J. B. Hampton is associate professor at the University of Toronto, specialising in metaphysics, poetics, and nature. His publications include *Romanticism and the Re-Invention of Modern Religion* (Cambridge UP, 2019), *Christian Platonism: A History* (co-ed., Cambridge UP, 2021), the *Cambridge Companion to Christianity and the Environment* (co-ed., Cambridge UP, 2022), and *The Metaphysics of Divine Participation* (Cambridge UP, 2025).

Part II

Chapter 6
Mind and Matter in Hegel and Schelling

Tilottama Rajan

Abstract This chapter explores the relation between mind and matter, spirit and nature, in Schelling's and Hegel's *Naturphilosophie*, framed through and against J. H. Green's concept of 'physiogony' and Coleridge's complication of this idea. Physiogony escalates Schelling's hypothesis of the *Stufenfolge*, or graduated stages of nature, in the *First Outline*, into an aspirational discipline, and aptly describes Hegel's desire in the *Philosophy of Nature* to trace an *a priori* logic of nature wherein each stage (following Kant) is cause and effect of itself, resolving the contradictions of the previous stage. However, a newly dynamical sense of matter renders this logic impossible and makes this impossibility generative. For Both Hegel's text (despite itself) and Schelling's radically experimental *First Outline* (more deliberately) break open the idealist binding of the volatile, organismic forces of nature, and disclose a dynamic interfolding of mind and matter, in which nature, rather than following the dictates of mind, generates new forms of thought.

This chapter explores the relation between mind and matter, spirit and nature, in Friedrich Schelling's and G. W. Hegel's *Naturphilosophie*, framed through and against J. H. Green's concept of 'physiogony' and S. T. Coleridge's complication of this idea. Physiogony escalates Schelling's hypothesis of the *Stufenfolge*, or graduated stages of nature, in the *First Outline of a System of the Philosophy of Nature* (1799), into an aspirational discipline, and aptly describes Hegel's desire in the *Philosophy of Nature* (part 2 of his *Encyclopaedia of the Philosophical Sciences*) to trace an *a priori* logic of nature wherein each stage (following Immanuel Kant) is cause and effect of itself, resolving the contradictions of the previous stage. However, a newly dynamical sense of matter renders this logic impossible and makes this impossibility generative. For both Hegel's text (despite itself) and

T. Rajan (✉)
Department of English and Writing Studies, University of Western Ontario,
London, ON, Canada
e-mail: trajan@uwo.ca

P. Cheyne (ed.), *Matter and Life in Coleridge, Schelling, and Other Dynamical Idealists*, International Archives of the History of Ideas Archives internationales d'histoire des idées 253, https://doi.org/10.1007/978-3-031-78315-9_6

Schelling's radically experimental *First Outline* (more deliberately) break open the idealist binding of the volatile, organismic forces of nature, and disclose a dynamic interfolding of mind and matter, in which nature, rather than following the dictates of mind, generates new forms of thought.

6.1 Introduction

In his introduction to the *First Outline of a System of the Philosophy of Nature*, which he wrote after he completed the *First Outline* and printed immediately after it, but with a separate title page, Schelling says it would be impossible to get 'a glimpse of the internal construction of nature' except by an 'invasion of Nature … through freedom', which he calls an 'experiment'. An 'experiment' is a 'question put to Nature, to which it is compelled to give a reply', and is itself 'a production of phenomena [*Hervorbringen der Erscheinungen*]' (Schelling 2004a [1799], 196–7; 1856–61, 1/3: 276).[1] Schelling's statement can be read alongside J. W. von Goethe's reflection that a science matures when it reaches a 'crisis' that confronts the 'universalist' with the 'singularist' who *wants* to but cannot find one type [*Typus*] everywhere (Goethe 1998 [1833], §419: 52), putting both deductive and inductive systems at risk. But it follows logically from newly dynamic theories of matter that complicate, even if they do not erase, the boundaries between the inorganic and organic. For if there is no absolute boundary between organic and inorganic, there is no clear dualism of mind and matter, opening the possibility that mind is a potency of matter, and that material singularities can also reconfigure mind. In other words, a subject-object dualism in knowledge is also impossible as mind and matter continuously unfold and enfold each other.

This reflexivity of mind and matter can be seen if we contrast two apparently similar passages from 1797 and 1799. In the more conventionally idealistic 1797 'Treatise Explicatory of the Idealism in the Science of Knowledge', Schelling writes that spirit only 'gradually… approximate[s] its own nature' by 'appear[ing] for itself in an *external* form, namely, *as organized, animated*, matter'. Though spirit must 'intuit *itself*' outside itself in matter in 'the succession of its representations', matter is the 'visible analogue' of an immanently developing *spirit* with ontological if not temporal priority (1994b [1797], 93). In 1799, in the *First Outline*, it is again a question of a relation between inside and outside, but this time it is a dynamic relation rather than one of adequation. In the *First Outline*, Schelling writes that 'the *empirically infinite* is only the external intuition of an *absolute* (*intellectual*) *infinity* whose intuition is originally in us, but which could never come to consciousness without external, empirical exhibition' (2004a [1799], 15). But in this complex back and forth, even as the empirical world is 'only' the intuition of the 'intellectual',

[1] Where I quote the German, the second reference is to the relevant volume and page number in Schelling's *Sämmtliche Werke*.

what comes to consciousness through the outside is not an ontologically prior inside, or spirit, made manifest in a 'gradual succession of organisms' (Schelling 1994a [1797], 93), nor even one epigenetically developed within the preformed concept of a gradative evolution. Rather what comes to light is a productivity in nature that requires intellectual intuition to grasp what lies within the mathematical infinity of the empirical, while this productivity in turn simulates or serves as an intuition of the absolute infinity of intellectual possibilities that mind can encounter only by grasping itself through nature.

With such absolute knowledge as absolute infinity in mind, I shall take up the *First Outline* as one of Schelling's most radical and formally experimental texts, standing to the *Naturphilosophie* as the unpublished *Ages of the World* (1815) stands to the philosophy of spirit, as an archive of ideas.[2] Absolute knowledge, as Schelling elaborates it in 'On the Nature of Philosophy as Science' (1821), leaves '*everything*' behind, all presuppositions, all '*definable* science[s]', since to define is to confine (1997 [1821], 210–41, 217). Indeed, this much later essay, synthesizing the third to the eleventh of Schelling's *Erlangen* lectures (1820–1) and published by Schelling's son in the *Sämmtliche Werke* he edited, is a useful gloss on the *First Outline*. For it too poses the desire for 'one system' to 'become absolute master over the others' against the 'asystasy' of knowledge: the fact that the very desire for system presupposes that knowledge 'does not exist in a system' but 'in inner conflict' (1856–61, 168–9). In keeping with this asystasy, which requires the 'human spirit [to] have searched in every possible direction' (1856–61,168), the *First Outline* does not unfold as an argument like the *System of Transcendental Idealism* (1800) a year later, nor does it have the narrative and de-cisionary structure of the *Freedom* essay. I will therefore put it in dialogue with Hegel's *Philosophy of Nature* (1842), which wrestles differently with a post-Newtonian nature that is dynamic rather than a Cartesian one of pure mathematics. For although Hegel's argument suffers multiple diremptions on the way to a goal it never attains, his *Philosophy of Nature* tries to narrativize the life sciences through what Schelling calls the *Stufenfolge*, which Schelling describes as a dynamic process in which individual types of natural phenomena develop as one production 'inhibited at various stages' ([1799] 2004a, 49). To consider Schelling's *First Outline* (*Erster Entwurf*) together with Hegel's *Philosophy of Nature* (the expanded second part of the latter's *Encyclopaedia of the Philosophical Sciences in Basic Outline* [*im Grundriss*]),[3] lets us both highlight the role of Schelling's text as an encyclopedia of the natural sciences, and consider both

[2] I discuss this notion of texts as archives further in Rajan (forthcoming 2025).

[3] Note that the *Philosophy of Nature*, as printed both in modern German versions of the *Encyclopaedia* (1830) and English translations of the text, is not the version that appeared in the actual 1830 *Encyclopedia* but the enriched and expanded version edited as part of Hegel's posthumous *Werke* (1832–47) by Karl Ludwig Michelet (1842, 1847). Michelet's edition, like editions of several other texts for the *Werke*, combines all three versions of the *Encyclopaedia* with further notes Hegel made for his lectures in Berlin, material from Hegel's Jena period in the first years of the nineteenth century, and notes taken by Hegel's students on his Berlin lectures. Whereas Michelet's edition comprises 696 pages, the relevant part of the 1830 Encyclopaedia is only 160 pages.

texts from the perspective of their dynamic interaction with and impact on philosophy. In other words, this chapter is also specifically concerned with the disciplines or fields through which both thinkers focalize nature, with the ontological personalities of disciplines, and with a certain in-discipline in Schelling's text.

For as Schelling writes, even while publishing it, the *First Outline* is a 'treatise… written … as a guide for lectures' and cannot be subjected to 'the same demands' as a book 'intended for the public' (2004a [1799], 3). As such, it is not even a set of lectures printed as chapters on recognizable disciplines, like the 1803 lectures *On University Studies*, which Schelling describes as an 'outline [*Grundriss*]' that can stand in for an 'encyclopaedia of the sciences' (1966 [1803], 41; 1856–61, 1/5: 247). Rather, the *First Outline* (*Entwurf* or draft) is a chaos of (un)divided parts without a clear architecture. Mark Wigley has critiqued the profoundly conservative role of 'architectural discourse' (Wigley 1997, 25) in philosophy, evident in Kant's use of the word architectonic to describe a system conceived as the coinherence of parts in a whole (Kant 1998 [1781], 691–2). As Wigley writes, architecture 'represents building to itself as complete' and 'secure', and so by analogy the 'edifice' or 'grounded structure' represents philosophy as 'the construction of propositions that stand up' (1997, 25, 9). But rather than a 'work' with an architecture, the *First Outline* is truly a 'text' in Roland Barthes' sense of not 'clos[ing] on a signified' but being a weave of possibilities produced and shifted by its writing. Barthes elaborates this distinction between text and work in terms strikingly appropriate to the *First Outline*. It is not that the work is 'classic' and the text is its 'avant-garde …. decomposition'; rather, the text is a 'methodological field' and is '*experienced only in an activity of production*': 'it follows that the Text cannot stop (for example on a library shelf)' (1977, 155–8).

Schelling himself sees nature in the *First Outline* as a 'productivity' that cannot be arrested in a 'product' ([1799] 2004a, 15n, 16n), using the image of a whirlpool or 'vortex' to capture the product formed within this resistance to hypostasis (2004a [1799], 18 n.). But in the midst of this whirlpool I will argue that there *is* one recurrent idea: an idea that indexes the ambitions of Idealism for a newly dynamic sense of matter by channelling the volatile, organismic forces of nature into a form of purposiveness. This idea, though it is constantly at risk, is the Idealist projection of the *Stufenfolge* or 'graduated stages' by which nature, or more precisely 'nature as subject' (2004a [1799], 17), organizes itself in increasingly complex forms, as '*one* production captured at different stages' (2004a [1799], 39). The notion of a 'history of nature' that also grounds reason in history sounds Hegelian, and this 'anthropological idea', as we might call it, is indeed the structuring prototype for Hegel's philosophies of nature, history and aesthetics. But Schelling proposed it well before Hegel in the *Treatise Explicatory*. He returned to it in the *System of Transcendental Idealism* (Schelling 1978 [1800], 122, 125, 199), and the hypothesis—though unsecured from the anthropology that guides it in the *System*—also distinguishes the *First Outline* from the earlier *Ideas for a Philosophy of Nature* (1797) and *On the World Soul* (1798). To elaborate, in the *First Outline* Schelling was disappointed with the existing concept of 'Naturgeschichte' or 'natural history' as a merely taxonomical 'Naturbeschreibung' that frustrates the purposiveness implicit in the word 'history'. For as Michel Foucault points out, natural history, as the collection and arrangement of nature's contents, contains very little that is historical, seeking to place 'the proliferation of beings occupying the surface of the globe' in space rather than time, in 'the field of a mathesis' where 'development' is no more than the traversal of a 'preordained table of variations' (1970 [1966], 235–6, 276). Schelling therefore wanted to give the term natural history a 'much higher meaning' that would yield 'eine Geschichte der Natur selbst', replacing a static account of nature's contents with a dynamic 'history of nature' itself, in which nature 'brings forth the whole multiplicity of its products through continuous deviations from a common ideal', which it realizes 'not indeed in the individual, but in the whole' (2004a [1799], 53; 1856–61, 1/3: 68).

Responding to Schelling, Coleridge's friend Joseph Henry Green gave this history of nature the name 'physiogony', in lectures that he delivered at the Royal College of Surgeons between 1824 and 1828, and from which he drew the 'Recapitulatory Lecture' (1828) that concludes *Vital Dynamics* (1840). *Vital Dynamics* was followed by *Mental Dynamics* (1847) to provide a titular simulacrum that would re-contain the disciplinary ferment of the life sciences in a progress from nature to spirit. Green distinguishes three approaches to nature, increasingly hypostatizing the last. Physiography is natural history in its conventional sense of *natura naturata* (nature natured) or the enumeration of the products of nature. Physiology (which is broader than animal physiology) studies *natura naturans* (nature naturing) or the laws and powers behind nature, and is a dynamic rather than mathematical science. Finally, physiogony or the history of nature traces a development 'from the *polypi* to the *mammalia*, as so many embryonic states, of an organism, to which nature from the beginning had tended, but which Nature alone could not realize' (Green 1840, 101–2). By claiming that nature 'labour[s] in birth with man' 'to complete the evolution of the organic realm', physiogony makes nature a 'preface and portion of the history of man' and part of his 'self-knowledge' (1840, 38, 73). It not only moves from the static to the dynamic, but it gives the dynamic a clearly evolutionary direction.

6.2 Hegel: Spirit Seeking Itself in Nature

Green's term physiogony captures what Arthur Lovejoy calls the temporalizing of the Chain of Being, whereby the '*plenum formarum*' is no longer just an 'inventory of nature' but the gradual carrying out of a 'program of nature' (1936, 242–4), however, as an 'Idea' that 'lies in Reason' rather than a literal evolution (Schelling 2004a [1799], 49; trans. modified). Yet as an environment for this 'Idea',[4] the *First Outline* is unique because its lack of architectonic boundaries makes it a resource

[4] A lower case 'idea' is loosely interchangeable with a concept or notion, and has no metaphysical force. I have followed translators of Hegel in capitalizing 'the Idea' in the singular, when it has the force of a totalizing, albeit aspirational, drive.

for thought rather than a grounding of concepts. This gives experimental status to the *Stufenfolge*, which is less determinately anthropological in the *First Outline* than in Schelling's *System of Transcendental Idealism* or Green's *Vital Dynamics*. I will return to the unique epistemology of the *First Outline,* but turn now to Hegel, who provides a logic of disciplines absent from the *First Outline*, while involving this logic in a deeply reflexive relation with the nature it tries to regulate. Hegel's *Philosophy of Nature* has been read as an account of the graduated stages of nature that knits philosophy and nature together through a series of increasingly tight 'matter/thought' or 'concept/matter relations' (Stone 2005, 54, 57). This *Stufenfolge* has its goal in the Kantian organism conceived, in the third *Critique*, as natural purpose (Kant 2000 [1790], 244); namely, as a whole whose parts, or rather 'members', are both cause and effect of each other, being both 'product and also productive', and 'reciprocally end and means' (Hegel 1970b [1842], 279, 377).

Hegel's claim for the progression from matter to organic life is generally seen as a 'logical' rather than 'natural' evolution (Ferrini 2011, 203; Gambarotto 2018, 119), though I would add that it is an 'Idea' he struggles to realize.[5] For as Coleridge says, in a comment that strikingly approximates the Hegelian *agon*, 'life' and 'mind' are distinct. Mind is complete in itself as 'a Subject possessing its object in itself', whereas life is 'a Subject with an inherent tendency to produce an Object, wherein and whereby to *find* itself', and is incomplete and lacking (Coleridge 1995c [1828], 2: 1426–7, 1437). In the *Philosophy of Nature*, Hegel likewise produces nature as an object in which 'philosophy' struggles to recognize itself, an 'alien existence, in which Spirit does not find itself' (1970b [1842], 3). This formulation, like Coleridge's, lacks the confidence of Green's account of nature as 'preface and portion of the history of man' and thus part of his 'self-knowledge and the outwardly realized history of our own consciousness and conscious being' (Green 1840, 43). Coleridge nevertheless wants to see life as having an 'ascension towards mind' and thus as intelligent (1995c [1828], 2: 1427). Life is a term that wavers in being attached to either subject or object, and it can accordingly be defined as the milieu in which the interchange between them occurs, as the *cogito* finds itself in life, so that mind unfolds as a life. Paul Ricoeur describes the resulting 'phenomenology' (rather than logic) in terms similar to Coleridge's. Whereas 'consciousness' is 'directed toward another' that 'is lacking to it', 'spirit' 'is entirely complete within itself', yet falls short of this completeness until the final moment (Ricoeur 1991 [1986], 230–1). Hegel's phenomenology is therefore 'a phenomenology of mind that remains … in consciousness', or 'a phenomenology of spirit *in* the element of consciousness' (231). The final moment serves as a 'hermeneutic tribunal', 'a criterion of meaning' or 'measure of truth' with respect to which 'all the modalities that precede it … [are] lacking' (231). Yet despite terms such as spirit and mind, Hegelian

[5] The word 'logical' may not state the case properly, as Hegel does give this evolution an ontological force, describing 'a system of grades of which one arises necessarily from the other but not in such a way that one is generated by the other naturally but rather *in the inner Idea lying at the base of Nature*' (Hegel 1986 [1840], 143; emphasis mine). See also Hegel (1970b [1842], 20).

consciousness is by no means 'transcendental', as the 'very thing that, in Kant, had been set up as a tribunal is born suffering the pain of uprooting' (233).

The *Philosophy of Nature* unfolds as an attempt to reach 'Mind' through the biological analogue of the Kantian organism as a unity that holds its members together. But Hegel goes beyond organizing nature through strategies of dialectic and teleology that forward this Kantian Idea. He also puts forth what Schelling calls a *Stufenfolge der Wissenschaften*,[6] both in the *Encyclopaedia* as a whole and in its individual modules, including the *Philosophy of Nature*. The philosophy of nature is thus organized as a scale of disciplines, which is meant to forward the programme of an evolution from matter to spirit; as such, it comprises one 'circle' in the larger circle of philosophy where each 'member [*Glied*] has an antecedent and a successor' (Hegel 1970b [1842], 2). The phrasing associates Kant's theory of organisms as means-end purposes with knowledge itself, which Hegel says (again echoing Kant's definition of architectonic) 'must not be an aggregate' but 'must present itself as an organism' (1970b [1842], 6). This focalizing of material nature through a system of sciences may partly account for Hegel's insistence on the greater rigour attending his philosophy of nature over Schellingian 'fantasy' on the one hand, and on the other hand over a British natural philosophy based in physics, which remains empirical and has not raised itself to the level of a philosophical science (1970b [1842], 2, 6). Yet this does not mean that Hegel claimed to have grasped the essence of 'this Proteus' he called nature (1970b [1842], 3). For while he referred to Logic (the first part of the Encyclopaedia) as 'Wissenschaft' (science or systematized knowledge) and while he also attaches this honorific to the study of nature in the *Philosophical Propaedeutic*, thereafter Hegel entitled the second part of the Encyclopaedia 'Philosophie der Natur'.

An emphasis on disciplines is part of the eighteenth-century encyclopaedias against which Hegel, in 1817, positioned his own *Encyclopaedia of the Philosophical Sciences in Basic Outline* (1990 [1817], 53). That this disciplinary emphasis motivated his project is clearest in the first, bare-bones version of his *enkýklos paideía* (Greek: circle of learning or all-round education) in the *Philosophical Propaedeutic* lectures Hegel gave at the Gymnasium in Nuremberg (1808–11), which were published by Karl Rosenkranz in 1840. Across this text and the three versions of the *Encyclopaedia* (1817, 1827, 1830), the scale begins with 'Mathematics', which is subsumed into 'Mechanics' after 1817. It then moves to 'Physics', including meteorology and chemistry, and then to the third division, 'Organics'. But three points are worth making about Hegel's restless modifications of his headings and divisions.

[6] Schelling refers—although critically—to a 'Stufenfolge' of 'Wissenschaften' in the German original (1856–61, 1/5: 61) of *On University Studies* (1966 [1802], 55). The English translation is inaccurate. While Hegel does not specifically use this phrase, in the *Aesthetics* he does speak of the aesthetic spirit as going through a 'Stufenfolge der bestimmter *Weltanschauungen*' and a 'Stufengange besonderen Gestaltungsformen' (1970a, 72, 73; 1970c, 103, 104). The terms 'Stufengrade', 'Stufenleiter', and 'Stufengang' are also throughout the *Philosophy of Nature*, describing a logic of nature 'regarded as a *system of stages*, one arising necessarily from the other, and being the proximate truth of the stage from which it results' (Hegel 1970b [1842], 20).

First, in the *Propaedeutic* and the 1817 *Encyclopaedia*, 'Mechanics' was part of the second division on 'Physics', but thereafter it became the first division, absorbing 'Mathematics' as a metonym. The result of removing 'Mechanics' from 'Physics' was to make 'Physics' more purely dynamic, since inert matter and quantification are put behind us as ways of grasping nature.[7] Here Hegel also broke with Kant, who had said that in 'any special doctrine of nature there can be only as much *proper* science as there is mathematics therein' (Kant 2004 [1786], 6). Second, with the demotion of 'Mechanics' to the first division after 1817, the second division on 'Physics' was also trisected into the 'Physics' of the 'universal', 'particular', and 'total individuality', thus making the dialectic of individuation a key theme of the lectures. Finally, in the *Propaedeutic* and 1817 *Encyclopaedia*, the third division is entitled 'Physik des Organischen', mirroring 'Physik des Unorganischen' in the second division on 'Physics'. The organic is thus still part of 'Physics'. But in later versions of the *Encyclopaedia* (1827, 1830) and in Karl Ludwig Michelet's enriched, posthumous editions of the *Philosophy of Nature* (1842, 1847) the third division becomes just 'Organik' to mark a sharper, if troubled, division between physico-chemistry and organics.[8] The organic, in turn, is subdivided into terrestrial nature (the material of geology and mineralogy), vegetable nature (botany and plant physiology), and the animal organism (physiology and medicine), taking Hegel to the threshold of spirit. Without going into detail, the ascent of disciplines is by no means smooth. The organic is profoundly complicated by its subdivisions, as the very logic of the *Aufhebung* as cancellation and preservation allows mechanical and vegetable structures to return in the animal (for instance in the bones and ganglia respectively). Moreover, since the human is included under the animal, the transition to the human sciences in the larger *Encyclopaedia* is also highly problematic, as animal somatology returns in the convolutes of anthropology and psychology, preventing them from becoming pure sciences of spirit.[9]

Three further points are worth making of the *Philosophy of Nature* itself in its more evolved form. The movement from its first to its second division narrates a transition in knowledge from the static Newtonian disciplines of mathematics and mechanics to dynamic physics. Hegel set out this goal as early as his *Habilitationsschrift*, *De Orbitis Planetarum* (1801), which wanted to emancipate physics from the mechanical notion of an 'inertial matter' moved from outside to 'forces, internal and immanent' to matter (2006 [1801], 9). But secondly, Hegel's *Stufenfolge* is doubly removed from any claim to epistemological certainty, since he organizes nature not just through concepts but through the sciences as the imperfect

[7] In the *Propaedeutic* Hegel previews the second division as 'Inorganic Nature' and then as 'Physics of the Inorganic' (1986 [1840], 143), and then divides the actual section, which is just called 'Physics', into 'Mechanics' and 'Physics of the Inorganic' (1986 [1840], 145–6). In 1817 the Second Division is entitled 'Physics of the Inorganic' and subdivided into 'Mechanics', 'Elementary Physics', and 'Physics of Individuality'.

[8] I emphasize the word 'troubled', since although 'Organik' is used in the table of contents, Hegel still uses 'organisches Physik' in the actual subdivisions.

[9] I discuss this in Rajan (2021, 122, 124–5).

shapes or *Gestalten* by which we grasp nature. This double remove concedes the idealism, even solipsism, inhabiting the philosophy of nature, rather than simply offering what Alison Stone calls a 'strong *a priori*' reading of nature itself (2005, 21). And third, *The Philosophy of Nature* is beset by an 'ever-increasing wealth' of empirical 'detail' avoided by Kant's third *Critique*. This detail not only proves 'refractory to the unity of the Notion' (1970b [1842], 444). It also means that Hegel does not fully observe the Kantian precaution of proposing his program as an Idea of Reason. The *Philosophy of Nature* is not purely about *mind* self-organizing its understanding of nature. For Hegel does not hold the Idea, a word charged with affect, firmly separate from nature. In that sense Schelling is wrong to accuse him of a withdrawal into 'pure thought' that 'hid[es] the lack of true life' (1994 [1833–4], 136, 143). Rather Hegel posits the Idea as an Idea *in* nature, not an Idea *of* nature. This is not equivalent to converting a regulative idea into a constitutive Idea; rather Hegel attempts an invasion of nature through freedom in which 'nature' talks back, binding him even more to nature as the immanent medium of the development and disappointment of the Idea. Hegel's graduated stages are not only posited as unfolding dynamically; they involve a dynamic but ruinous interfolding of mind and matter, as 'spirit' tries to find itself in nature and cannot do so.

It is here that Hegel's *Philosophy of Nature* and Schelling's *First Outline* converge because both take the dynamic theory of matter beyond just matter and into our current concern: a dynamic relation of *mind* and matter in the context of 'life'. But it is here that they also diverge. For in the labour of the negative that constitutes his lifelong struggle with nature, Hegel is tormentedly Fichtean in the resistance he feels between the 'I Am' and the 'It Is'. Hegel holds subject and object apart, unable to withdraw mind back into itself, as Kant does in his theory of the sublime, but unable to embrace mind and matter as folds of each other, as Schelling does in the *First Outline*. This is not to critique Hegel, since the *First Outline* is a product of *Frühromantik*, while the 'tragic tone' of Hegel's work, as Ricoeur calls it (1967, 206), had its own impact on Schelling after their traumatic break in 1807. In repeating the familiar argument that the post-Kantians make reflective judgment determinant and render the regulative constitutive, Maurizio Esposito writes that they 'ontologized' and 'naturalized' a 'strategy of reason' (2013, 31). The key word here is 'ontologized'. By positing consciousness *in phenomena*, Hegel brings out the ontological stakes of what Kant poses as an epistemological strategy. But contrary to Esposito, Hegel cannot hypostatize purpose in nature. On the one hand consciousness is posited outside itself in nature. On the other hand, this nature in its unassimilable materiality is also inside consciousness as the extimacy of mind to itself. For as Hegel writes, the 'sciences', which should prove the emergence of spirit from nature, are themselves man's 'non-organic nature', which resists him and which he struggles 'to make his own' (1970b [1842], 276).

The ontological task of the *Philosophy of Nature* unfolds in the climate of disciplines that are no longer static and thus self-certain. As John Zammito says, science in the late eighteenth century shifted from 'mathematical kinematics' to 'experimental physics' and 'natural history—respectively, the problems of '"imponderable fluids", such as electricity, magnetism, chemical bonding, light and heat'", and that 'of "organized form" or life' (2007, 52).[10] Life in all its complexity is the milieu of Hegel's thought, as distinct from Kant's, and life is the milieu in which the *Philosophy of Nature* actually develops itself.[11] In Coleridge's terms, Kant comprehends life within mind, whereas for Hegel mind is thrown outside itself into life. Idealism, nevertheless, wants to see life as intelligent, and Kant's theory of the organism as natural purpose gave Hegel a *télos* that would bind the volatile forces unloosed by a dynamic rather than mathematically based physics and chemistry within this ascension towards Mind. But the point was to get from experimental physics and chemistry to the organism through the newly developmental *Geschichte der Natur* Schelling describes, in which nature proceeds through 'continuous deviations from a common ideal' that it realizes not 'in the individual, but in the whole' or genus process (2004a [1799], 53).

Accordingly, we can follow Alison Stone in seeing the desired logic of the *Philosophy of Nature* as one where nature's progress from the inorganic to the organic, and through planets, plants, etc. to the animal organism, gradually approximates a body made up of interdependent 'members' (*Glieder*) rather than 'parts' (*Teile*) (Hegel 2010 [1812], 681). An organism made up of parts (like the plant) does not fully possess those parts, whereas in one composed of 'members' the universal is revealed within the part(icular) so that the body is 'intelligibly structured' throughout (Stone 2005, 46–50, 75; Hegel 1970b [1842], 277, 291). The movement towards an integrated body is effected through a series of *sciences*, including crystallography, meteorology, geology, botany, animal physiology and finally medicine, which Hegel tries to put in an intensifying scale of what Green and Coleridge will call individuation and integration. In his argument with British natural philosophy (1970b [1842], 6), it is individuation that Hegel sees as crucial to recovering physics for philosophy, which he presumably does not see Schelling as having accomplished in his 'speculative physics', the experimental discipline that provides the subtitle to his Introduction to the *First Outline*.

Yet the ascent of nature is not as simple as Stone suggests, even if we reconceive it as a dialectic of individuation and integration rather than as straightforwardly linear. Thus a crystal is 'articulated through and through' and is not 'mechanically compounded', but because it is 'rigid', 'mechanism is resumed' or returns in organism (Hegel 1970b [1842], 160); a comet is not rigid, but is an 'unstable mass of vapour', lacking 'a nucleus' and centre of individuation (Hegel 1970b [1842], 99). The difference between these deviations from the ideal raises the problem of a disunion between force and form that is supposedly resolved in the organism: the fact that form without force is rigid but forces without form annihilate their matter rather than reveal themselves in their parts. Or as Hegel elaborates, 'the *body of rigidity* has only a formal being-for-self', whereas 'the body of dissolution ... behaves

[10] Zammito is presumably using the term 'natural history' more broadly than in its normal usage, as an enumeration of nature's products or what Green calls 'physiography'.

[11] As theorized by Canguilhem (2008 [1952], 98–120) 'milieu' is not the environment over against the organism, but 'mi-lieu' or being in the middle of; life as mi-lieu is thus the dynamic interaction between the organism and its environment.

aberrantly' (1970b [1842], 99). These deviations from the ideal also raise issues having to do with individuation, integration, and what Coleridge calls 'specification'. That a crystal is a 'specific total' means that because of its molecular structure its parts are uniform rather than differentiated and individualized, and exist in 'apposition' rather than being integrated.[12] Such discrepancies between individuation and integration persist higher up the scale in planets (covered by Hegel under Physics), and remain even in Organics. For planets, the terrestrial organism and plants all deviate from the ideal of the animal organism as a whole that holds its 'members' together rather than falling apart into subsystems (Hegel 1970b [1842], 276). But the point is that throughout the space of nature, entities deviate *differently* from this ideal, falling short of the philosophical norms in various ways that expose aporias and tensions between these norms. Indeed, this may be why Schelling, as we will see, cannot construct a scale of being or a *Stufenfolge der Wissenschaften* culminating in *Organics*.

Individuation and integration are the contrary 'tendencies' that Green wants to see reconciled in the organism (1840, 105). But Green manages to synchronize them by conceiving individuation on the Kantian model of organisms as natural purposes that are both means and ends. In this model individuation as an increasing complexity and specialization of parts is kept in check because it also entails the interdependence of parts and the 'subordination of the lower to the higher' (Green 1840, 41). The 'tendency to integration in the parts' makes individuation simply a microcosm of the other 'great tendency': 'integration' itself, whereby nature creates a 'comprehensive whole' through 'advancing integration', within a 'physiogony or genetic process' of 'gradative evolution' (Green 1840, 37, 39). Interestingly, for Green, the scientific terms have moral extensions: 'Individuality has its acme in integrity' (Green 1840, 43). But for Coleridge, who in his *Theory of Life* similarly wants to see individuation in terms of organisms as natural purposes whose parts are all 'reciprocally means and end' (1995a [1816?], 1: 510, 512–13), the tendencies are also 'counteracting' pulls. Paralleling Goethe's distinction between the universalist and the singularist, Coleridge describes a force of 'attachment or reduction' into 'the universal life' and of 'detachment' from it (1995a [1816?], 1: 515, 557), a tension already taken up by Schelling (2004a [1799], 53–4). Indeed, when Coleridge distinguishes among the tendencies to 'totalize', 'specificate', and 'individualize', and when he offers the pebble and crystal as examples of the first two tendencies and the daisy and fly as examples of the third, his sense of individualization comes closer to what we might call 'singularity' (1995b, 2: 1029). For Coleridge does not describe the daisy and fly, unlike the pebble and crystal, in terms of the part-whole relationship; nor does he approach them as representatives of a class, as Green does in discussing individuation. The term individual, when thought within the

[12] In his essay 'Life' (1822) Coleridge distinguishes among the terms 'individualize', 'totalize', and 'specificate'. A daisy or fly are 'individualized', which connotes some kind of singularity. A 'pebble is a *mere* total', internally undifferentiated. A crystal, as a 'specific total', has uniform and not differentiated parts and so is only 'an aggregate by apposition', in which there is no 'evolution (*ab intra*)' (Coleridge 1995b [1822], 1029).

part-whole relationship is still normative, in that even deviations are considered in relation to the whole or an ideal of wholeness. When individuation is approached in relation to a class of entities (birds or fish, for example), the term is likewise being thought in relation to the genus-process, as when Jean-Luc Nancy refers to a 'higher form of substance or subject taking charge … of separate individualities' (1991 [1985], 27). Singularity, by contrast, better describes a poetic sensitivity to difference, at the level of Nancy's 'infra-individual' and 'singular differences' (2000 [1996], 13–15).

A detachment or individuation—let alone singularization—at odds with the whole explains why the various deviations from the ideal along the Chain of Being cannot really be 'continuous', as Schelling hopes in his experiment with nature (2004a [1799], 53). Thus, Coleridge in his *Theory of Life* describes how the 'class of fishes' marks a 'higher step' over that of insects yet 'sinks a step below' the latter in its mode of impregnation (1995a [1816?], 1: 546–7). Similarly, as Hegel moves from Physics to Organics, plants have the advantage over planets in being organisms that develop, as Coleridge would say, '*ab intra*' (1995b, 1029). The parts of a planetary system remain external to each other, while the plant grows from within and 'differentiates itself into distinct parts'. But although the plant thus enters the 'first… stage of subjective vitality', it 'falls apart into a number of individuals' rather than being 'a subjective unity of members' (Hegel 1970b [1842], 303). Planets, which are lower on the scale than plants, thus advance beyond them in forming a system, which in turn is associated with volition. For whereas the plant is 'impotent to hold its members in its power' (Hegel 1970b [1842], 276), there are a 'number of earths or planets which together form an organic unity', whose 'arrangement … in space is the act of the planets themselves' (1970b [1842], 104).[13]

Coleridge rationalizes this uneven development in his *Theory of Life* when he writes that every 'grade of ascension' is accompanied by 'an apparent retrograde movement', but when nature 'drops a faculty' she never fails to pick it up again' at a higher level (1995a [1816?], 1: 546–8). And Green similarly streamlines his logic of nature by arguing that in the 'evolution of the organic realm' there is 'an apparent sinking back' necessary to concentrate 'the organific energies' for a 'higher and more complete ascent' (1840, 36, 114). The logic is that of Hegel's *Aufhebung* as a dialectical spiral rather than linear progress, and as I argue elsewhere, Green tacitly supplements Schelling with Hegel to rein in the volatility of *Naturphilosophie* (Rajan 2018, 56–7). That task was made easier by the fact that the work by Hegel Green knew was the *Logic* rather than the still unpublished *Philosophy of Nature* in the more detailed eclectic text that was not edited by Michelet until 1842. But it is another question whether this logic can contain the singularities that are throughout the *Zusatze* that Hegel added to the propositions in his *Encyclopaedia* (*Grundriss*), the minutely granular details that Schelling added to his *First Outline* (*Entwurf*), or the entries in Coleridge's Notebooks wherein a poet's eye for detail is transferred to

[13] It is worth noting that Hegel considers planets both under 'Mechanics' and 'Physics'. In the former section he accuses the planets, though they have their 'centre within [themselves]' and 'possess determinateness', of being 'just' a 'plurality' (1970b [1842], 77, 80).

the way he engages with the sciences. Coleridge cordons off these details in his notebooks and fragments. But as Novalis says in his *Romantic Encyclopaedia*, every 'molecule of science' can be philosophically potentiated (2007 [1798–99], §489), so that the empirical, instead of being determined by the transcendental, retroacts upon it in a dynamic interfolding of mind and matter.

A gradative evolution of natural phenomena is also Hegel's aim, as he strives to attain the maximum of 'individuation and integration' in the animal organism, narrating this progress through the relevant sciences. But this is not the end of the story, since the *Stufenfolge* undergoes a profound diremption in the section on the animal, which ends with pathology. Throughout the *Philosophy of Nature*, the Kantian organism as an integrated whole rather than aggregate has been the light at the end of the tunnel. But when Hegel reaches this climax, he writes that the animal can 'as well not conform to its genus', which resistance he calls 'disease'. In disease one of the organism's 'systems or organs … establishes itself in isolation … against the activity of the whole' (Hegel 1970b [1842], 428). Here the very unity and integrity of the animal body, from which the plant is exempt, means that all of its systems—nervous, digestive, circulatory etc.—'interpenetrat[e]' in this dis-memberment (1970b [1842], 372). Or as Hegel puts it, hovering on the border between physiology and psychology, only an animal can become ill and 'turned against' its very 'structure', whereas a stone is 'destroyed in the negative of itself [and] chemically decomposed' (1970b [1842], 429).

Thus the very superiority of biology to chemistry as a way of grasping organic vitality, which is meant to distinguish Idealist biology from British natural philosophy, becomes a *pharmakon.* Likewise the integration of the animal body, which is made up of 'individual' and 'specialized' subsystems (Schelling 2004a [1799], 127), proves its undoing. For the 'isolated activity' that occurs when a member separates itself from the whole is still a 'moment of the whole', as the 'entire organism … is involved and its activity deranged because one wheel *(Rad)* in it has made itself the centre' (Hegel 1970b [1842], 428, 433). Moreover, Hegel seems unable to approach the animal through normal physiology, as he does the plant. The result is that the 'biological normal', as Georges Canguilhem writes, is 'revealed only through infractions of the norm and that concrete … awareness of life exists only through disease'. As Canguilhem argues in *The Normal and the Pathological*, disease thus preempts health, since the norm can be established only *a posteriori* through deviations from it, making any 'common ideal' purely ideal (1991 [1943], 118). This understanding exemplifies how philosophical paradigms are shifted by the dynamic retro-action of nature on mind within the 'milieu' of life.[14]

---

[14] It will be evident that the above reading departs from the tendency, in Hegel commentary, to read the *Philosophy of Nature* in terms of the concept of 'life' in the *Science of Logic*, or indeed simply to assume that it fits into the *Logic*. Hegel, as is well known, understands logic as a form of life rather than a set of 'empty, dead forms of thought' (Hegel 2010 [1812], 676). But he himself differentiates 'logical life as idea from natural life as treated in the *philosophy of nature*' (2010 [1812], 666–7). Commentators like Robert Pippin in Hegel's realm of shadows (Pippin 2019) and Karen Ng take up 'life' only in the *Logic*, thus approaching it not through 'matter' but as generat-

6.3 Hegel and Schelling: The Dilemma of 'Life'

In his Jena lectures, especially in 1805/6, Hegel was already working out the *Stufenfolge der Wissenschaften* that forwards this problematic *télos* of the organism. In terms of the two regions of dynamic nature Zammito describes, dynamic matter and organic vitality, Hegel wants to chart a progression from physics and chemistry as sciences of matter, to biology which, as Canguilhem says, is concerned with 'life'. The irreducibility of life to mere matter is due to the 'originality' and singularity 'of the biological phenomenon'. But in valorizing biology, Canguilhem does not mean to dismiss physics and chemistry, but to '"*comprehend*" matter within life, and the science of matter—which is science itself—within the activity of the living' (2008 [1952], 69–70). His essays in Part 3 of *Knowledge and Life*—particularly those on vitalism, the machine as organism, and the milieu—are concerned with the complex dialogue among mechanics, physics, organics, and finally medicine that is also central to Hegel and Schelling. It is in medicine that biological originality produces the most profound diremption from the regularity of the Idea. What Iain Hamilton Grant calls the 'eliminative idealism' of transcendental philosophy (2006, 108, 202) is designed, in Hegel's case as he moves up the scale of sciences, to protect biological originality, particularly from chemistry. But it also tries to repress this originality within the notion of autopoietic 'Kantian wholes' (Kaufman 2016, 67). Yet the Kantian regulation is finally defeated by the 'instability and irregularity' of the organism (Canguilhem 2008 [1952], 70), as biological originality turns out to include illness. For a stone cannot fall ill, as Hegel tells us. Or as Canguilhem puts it, 'there is no physical, or chemical, or mechanical pathology' (1991 [1943], 127).

The sciences in the *Philosophy of Nature* are also in Schelling's *Naturphilosophie*, but they exist in a consciously experimental (dis)solution, given how many of these texts bear titles such as 'ideas for', 'hypothesis' or 'first outline'. We could say, then, that at least in his *Naturphilosophie* Schelling embraces a 'non-eliminative idealism' (Grant 2006, viii) that is more open to 'life', including the understanding of matter within life. In terms of Zammito's two regions of dynamic nature, Schelling never quite resolves their (dis)connection and the related question of whether a transcendental direction can emerge from, or be synchronized with, natural

---

ing 'a theory of conceptual form and activity' (Ng 2020, 63). While crediting logic with an organic form, their conception of the organic, being immanent to logic, therefore remains logical, as in Ng's emphasis on inner purposiveness. Arguably even those who do take up the *Philosophy of Nature* (e.g. Stephen Houlgate) are oriented by the *Logic* and concerned with a *concept* of nature. This concept involves 'positing as negative' any 'objectivity indifferent to [the life-process]' and finally as '*genus-process* … sublating [life's] singularization' (Hegel 2010 [1812], 679). In the process Hegel commentary largely sidesteps what Canguilhem calls the 'irregularity' of life (2008 [1952], 70), so that even when it recognizes contingencies in the *Philosophy of Nature*, as Houlgate does, these are not paradigm-changing for philosophy, because we know and expect that nature will be the Idea itself in the form of externality and otherness (Houlgate 2005, 107–15), this externality already having been accounted for and de-fanged in the *Logic* (Hegel 2010 [1812], 679). The Hegel commentary, in short, gives logic a preformationist precedence over nature that curtails the latter's epigenetic power in relation to philosophy in the *Philosophy of Nature*.

philosophy. The 1797 version of *Ideas for a Philosophy of Nature* unfolds purely in 'physics and chemistry', though in revising its preface for the 1803 edition, Schelling promises a continuation that 'include[s] the principles of a theory of organic nature, or so-called physiology' (1988 [1797/1803], 5). But in the preface to *On the World Soul* (*Von der Welstseele*, 1798) Schelling puts this promise more negatively, writing that he will *not* continue *Ideas* until he can formulate a 'scientific [*wissenschaftlichen*] physiology' that 'rounds off the whole'. The *World Soul,* he insists, does *not* provide this physiology (1856–61, 1/2: 351). Yet the *World Soul* is subtitled a 'Hypothesis of the Higher Physics for the Explanation of the Universal Organism', describes Blumenbach's *Bildungstrieb* as having produced a 'total revolution' in 'philosophical thinking' (1856–61, 1/2: 522), and it evokes the Kantian notion of organisms as natural purposes (1856–61, 1/2: 520).

So why does the *World Soul* not round off the whole? The hope for a progression from physics/chemistry to organics marks a desired advance of German Idealism over British natural philosophy. I use the term Idealism rather than *Naturphilosophie* deliberately, to suggest an episteme that seeks to gather a more individuated and (dis)integrated body of knowledge in *Naturphilosophie* into a system itself conceived on the model of Kant's *Naturzweck*, in which each science would be both means and end, or as M. J. Petry puts it, a sphere in its own right and a level in an ascending scale of knowledge (1970, 31). But despite projecting a theory of organic nature as his desired capstone in the 1803 revision of the 1797 preface to *Ideas for a Philosophy of Nature*, in the actual preface to the second edition of *Ideas*,[15] Schelling describes his promised further part as 'organic physics' (1988 [1797/1803], 7), which still puts the text short of 'organics' as the culmination of the Hegelian *schema.* Schema: because, as Schelling himself says of the schematic, we are dealing with a construct in which the universal and particular have not been synthesized (Schelling 1989 [1804/5], 46), or in which the particulars appropriate to the universal have not been conceived.

So perhaps Schelling never filled out his schema and never wrote his third part because organic physics cannot really bring the 'real' series of disciplines to an end, allowing us to move on to the 'ideal' series in the transition Hegel schematized. In his middle work, especially *Ages of the World* (1815), Schelling becomes preoccupied with the difficulty of beginning and ending, or of moving from the first to the third through a linear or dialectical progression, without again going 'back to the first' or second (2000 [1815], 13, 36). Despite dwelling much more on the organism than *Ideas* (1797), we can argue that the *World Soul* does not decisively proceed beyond *Ideas* because it remains too entangled in chemistry. In *Ideas* a chemistry of forces rather than elements, an organic rather than inorganic chemistry, had been the result of '*dynamics* as the basic science of a theory of Nature', which provided the text's '*philosophical* part' (Schelling 1988 [1797/1803], 5). It seemed, therefore, that progress was being made. But in the *World Soul* Schelling expresses frustration

[15]The 1803 *Ideas* contains both a revised preface to the 1797 edition and a preface to the 1803 edition.

with even this chemistry, because it discloses only '*incomplete organizational processes*', furnishing only 'analogies' (1856–61, 1/2: 189), and so does not mark a real advance. And in the *First Outline*, he complains that chemistry provides only 'effects instead of causes' (2004a [1799], 110), and hence, as he says in *On University Studies*, is not yet a science (1966 [1802], 132).

At the heart of this dissatisfaction is the irreducibility of life to physico-chemical causes characteristic of what Maurizio Esposito calls 'Romantic biology': a tradition he traces from the post-Kantians, via British Idealism, to twentieth-century scientists such as J. S. Haldane and D'Arcy Thompson. In Hegelian terms, the *World Soul* does not quite cross the threshold between a 'physics of the total individuality' with which the second division of the *Philosophy of Nature* ends, into Hegel's third division on Organics. Instead, it approaches the organism through what Grant calls a 'field-theoretical [view] of nature' in which 'individuation and organization' occur through forces rather than forms (2010, 87). There are of course reasons for this drawing back. Briefly, Schelling often approaches differences in nature quantitatively. But for Hegel 'merely quantitative' differences are 'indifferent', and domains such as Physics and Organics, or even more so nature and spirit, must be separated by qualitative differences (1970b [1842], 26–7). The quantitative/qualitative distinction means that the early Schelling unfolds nature in space, while Hegel develops it in time, where time contains a teleological promise but also entails finitude and fatality. Thus, we need not necessarily read Schelling's difficulty in proceeding to a 'third' part through Hegel. For after all in *Ages* (1815) Schelling will cast doubt on any linear or even dialectical progression from first to third, noting that the 'third is incapable of continuance' because the 'unity [cannot] elevate itself and be outside the antithesis', and because 'each of the three has an equal right to be that which has being' (2000 [1815], 19, 36). So perhaps Schelling did write an (in)complete physiology or biology, not in the *World Soul* but in parts of the *First Outline,* but he could not give it the synthesizing authority of 'the third' because he questioned the very possibility of synthesis.

Yet if I evoke Hegel, it is also because Schelling cannot be as purely aligned with a quantitative and spatial understanding of nature as has now become fashionable.[16] It is because the schema of a temporal development that is most clearly formulated by Hegel also retreats and returns in Schelling's work, thus providing a scaffolding that Schelling's texts (dis)assemble. This temporal schema emerges powerfully in the middle work, especially the *Freedom* essay and the 1811 *Ages*, which participate in both the philosophy of freedom and the philosophy of nature. But as we have seen, it is present, albeit more abstractly, as early as the *Treatise Explicatory*, where Schelling writes of a 'gradual succession [*Stufenfolge*] of organisms' and a 'transition from an inanimate [*unbelebte*] to an animate nature [that] clearly reveals a productive force' developing 'towards complete freedom' (1994b [1797], 93; 1856–61, 1/1: 387).

[16] See Woodward (2016, 23–41), Heuser (2016, 43–47), Furlotte (2018, 6).

Put differently, we cannot just oppose Schelling as a dynamic materialist or Deleuzean transcendental empiricist to Hegel as an Idealist, whether a complacent pan-logician or a 'tormented' idealist for whom nature is a series of 'botched' attempts at the 'absolute' (Krell 1999, 25, 96). For much of Schelling's work is also deeply Idealist. But nor can we follow Hegel in dismissing Schelling as an Idealist without a sense of dialectic, without the labour of the positive that occurs in the differences between his texts. Rather than moving from *Naturphilosophie*, to the Identity philosophy, to the philosophy of freedom, as Manfred Shröter's edition represents Schelling's corpus,[17] Schelling works on several tracks at once, seeing natural and transcendental philosophy as 'one science' with 'opposite orientation[s]' (Schelling 2004b [1799], 194), and yet constantly probing their divergences. Even his idealisms are different: evolutionary and/or historical at times, and at other times involving a synchronicity wherein particulars are modes or potencies of the universal. It is notable that when Schelling revised *Ideas* in 1803 by refocalizing the speculative realism of 1797 through a Spinozistic neo-Platonism, he did so through a series of 'Supplements' that let the original version stand, as in a palimpsest.

6.4 Schelling's *First Outline*: Mind Intergenerated with Nature

It is the *First Outline* which, in a disciplinary sense, comes closest to the engram of what will become the Hegelian schema, beginning with experimental physics in the discussions of dynamic atomism and the universal fluid, and ending (if it can be said to end) with physiology, which segment includes, in the middle, an appendix on disease, if not on medicine per se. In between, Schelling floats the hypothesis of 'natural history' in the 'higher' sense of a 'history of nature', which could then be seen as a bridge from experimental physics to the organism, but a bridge that the actual physiology may well collapse. Notably, the *First Outline* has no clear division between sections that is not overrun by its proliferating details. There is no paginated table of contents, as in Hegel's *Encyclopaedia*, where divisions and subdivisions are clearly schematised in advance. While the English translation has three divisions extrapolated from the *Hauptabschnitten* in the *Grundriss*, or opening outline of the *First Outline*, these are not marked in the 1799 original. And we cannot simply treat the *Grundriss* as a table of contents, since the First Division has four subdivisons in the *Grundriss* and five in the text itself. The text is studded with upper and lower case Roman and Arabic numerals, and Latin and Greek letters,

[17] Shröter presents Schelling's work in six volumes (Schelling 1965-9), consisting of Early Writings (1793–98), Naturphilosophie (1799–1801), Identity Philosophy (1801–6), Philosophy of Freedom (1804–15), Historical Philosophy (1821–54), and Religion (1841–54); there are also six supplementary volumes following a different chronology. Among other anomalies, this mixed thematic-chronological format results in *Von der Weltseele* being in the *Early Writings*, and the *System* being under *Naturphilosophie*.

confusing any subordination of subdivisions to divisions, or any hierarchy of thoughts and afterthoughts. And in contrast to the complicating but graspable structure of *Satz* and *Zusatz* used in Michelet's enriched text of Hegel's *Philosophy of Nature*, Schelling uses a multiplicity of terms: *Satz, Lehrsatz, Zusatz, Folgesatz, Beweis, Anmerkung, Erlauterung, Problem, Auflösung*. Finally, there are also long footnotes by Schelling in the manuscript, inserted five decades later by his son (probably with Schelling's permission) when he republished the *First Outline* in the *Sämmtliche Werke* begun during Schelling's lifetime.

This method that decomposes system into asystasy makes the *First Outline* what Deleuze, using Joyce as an example, calls a 'great work which contains all the complicated series', and into which these 'divergent series … lead out and back in'. Within this great work, 'each series', which could be an idea, category, paradigm, discipline or pre-discipline, 'explicates or develops itself, but in its difference from the other series which it implicates and which implicate it' (Deleuze 1994 [1968], 123). Within the *First Outline*, lines of enquiry not quite stabilized as sciences develop and are implicated in other lines or folds, and the history of nature is only one of these series. Indeed, Schelling's larger corpus might also be a great work in which divergent series lead out and back in. These series, as Deleuze says, are not 'different points of view on the same story', to be absorbed into a Leibnizian pre-established harmony (Deleuze 1994 [1968], 123).[18] This synchronization is one of Schelling's experiments, when he argues that the temporal is a mode of the eternal (1988 [1797/1803], 272), or that natural and transcendental philosophy are different perspectives on one science (2004b [1799], 194). But the series or paradigms are 'divergent' series that 'unfold simultaneously', such that the 'point … of convergence lies in a chaos' or asystasy (Deleuze1994 [1968], 123). For as noted, throughout his life Schelling worked experimentally on several tracks at once, constructing multiple outlines, ideas and systems that involved natural and transcendental philosophy, religion and mythology. Until 1809, for example, the transcendental and natural series lead in and out of his corpus, reconfiguring each other in the pivotal *Freedom* essay of 1809, and then being again unbound in the *Ages* (1815) from the de-cision reached in the *Freedom* essay, to leave speculation behind and narrate a leap of faith from nature to freedom.

The *First Outline* is the great work that contains Schelling's *Naturphilosophie* up to this point; hence it must be (un)published as an archive rather than a finished work. As such the text is in restless transition between fields of knowledge—physiology, physics, geology etc.—as even hard sciences become life sciences subject to the excitability Schelling discusses, whereby, as soon as a balance among forces is found in an (intellectual) product, the 'organic activity' that prevents life from being 'exhausted … in its product' results in this balance disclosing a disproportion that requires a further product (Schelling 2004a [1799], 159–60). Yet the *Outline,* though presenting an unbound conceptuality that unfolds on what Deleuze and Guattari call a smooth, undivided space, or 'plane of immanence' rather than development

[18] See also Leibniz [1714] 1998, §57: 267–81.

(Deleuze and Guatarri 1987 [1980], 35–42), *is* a form of transcendental thinking. For it exemplifies how mind thinks as a fold in nature, since mind is part of the same continuum as matter. But as we suggested earlier, this is not a transcendental idealism that proceeds from inside to outside. When Schelling says that to 'philosophize about nature means… to create it' and that 'everything that exists is a construction of the spirit' (2004a [1799], 5, 13), he means that everything is in construction, or that there are no *a priori* categories of mind. To experiment with thought does require an 'a priori judgment' (Schelling 2004b [1799], 197), but the forms that mind takes are disclosed by the new forms in which nature causes it to think. Just as there are no *a priori* categories of mind, so too the 'state of *configuration*' is 'the most original in which nature is viewed' (Schelling 2004a [1799], 6). It is thus and not in the sense of any self-grounding correlationism[19] that Schelling can describe philosophy as 'a *natural history* of our mind' and say that 'the system of nature is … the system of our mind' (1988 [1797/1803], 30).

If both mind and nature are in constant configuration, the text itself becomes a dynamic field to be experimentally de- and re-composed. We will attempt this in two ways, the first being to construct it along vertical and horizontal axes, where the vertical is an experiment with idealism, while the horizontal is an 'unconditioned empiricism' (Schelling 2004a [1799], 24). The vertical axis of the *Stufenfolge* is transcendental not just in a Kantian and formal sense but also in the more common sense where 'transcendental' signifies a content: the possible emergence of spirit from matter. Yet, as we have seen in Hegel, such an emergence is an Idea of Reason, which Hegel simulates through a progression in the sciences that never happens in Schelling's work. This difficulty, which we have seen in Schelling's inability to gather his earlier texts into a system of sciences, is reflected in the *First Outline* in the way the vertical thrust of the *Stufenfolge* is constantly interrupted by a horizontal axis that distributes the text into a network of fields in which it is enveloped instead of being able to put them in sequence. In the *System of Transcendental Idealism*, when Schelling writes that 'nature is the unconscious poetry of the spirit' (1978 [1800], 12), nature, history and aesthetics are bound into an architectonic, if not sequentially, then through anticipation, analogy and homology. But in the *First*

[19] Schelling's relation to 'correlationism' is complex. Quentin Meillassoux's charge of correlationism against any grasping of the object or outside in relation to a subject or even a thinking sets up an impossible demand; it is hard to see how we can 'think the "in itself" without it becoming', in some way, '"for us"' (Meillassoux 2008, 4). As Bruce Morrisette has compellingly shown with regard to *chosisme*, the unacknowledged literary precursor of object-oriented-ontology, Alain Robbe-Grillet's meticulously mathematical descriptions of things void of human presence inevitably slides back into an anti-humanism that is 'for us', 'pour-soi' rather than 'en soi', so that the 'chosiste' is always covertly affected by the 'humaniste' (Morrisette 1963). In the *First Outline*, Schelling comes as close as possible to thinking 'substance', as Meillassoux argues philosophy did before the Kantian turn put thinking the 'correlation' at the centre (Meillassoux 2008, 6). The result *is* an opening to the 'absolute' and the 'great outdoors' (Meillassoux 2008, 1, 7), as an 'absolute (intellectual) infinity' (Schelling 2004a [1799], 15). Yet thinking the correlation cannot but remain a by-product of this experiment, and Schelling's text is better described as transcendental empiricism (Schelling 2004a [1799], 22, 24) than as object-oriented ontology.

Outline, through a series of events or collisions, scientific fields as snapshots of mind being refolded by matter generate competing energetics of thought rather than forwarding a single idea.

These fields, as figures of thought-matter, of mind thinking as matter, include physics as a theory of dynamic atomism that recasts thought as a finding of differences within differences; the earth sciences which open up processes of composition and decomposition in soils and other solids; and chemistry, as a space of volatile exchanges which allows us to know only 'effects instead of causes', and which here, unlike in *Ideas,* is mixed up with physiology. Physiology, in turn, is both the body's chemistry in John Brown's *Elements of Medicine* (1788), which Schelling frequently evokes, and physiology in Green's broader sense of *natura naturans* or the forces in nature, which is to say that the individual and universal organism, nature and the body, are thought and rethought through each other. But importantly these fields are rarely named as sciences. Rather an intensity or potency hypostatizes in an organ of knowledge through which nature is focalized, and the text then moves on, creating a different organ of knowledge. Thus, the text is less an organization of knowledge 'defined by determinate organs' than a body without organs which, as Deleuze says, moves through 'provisional organ[s]' of knowledge that endure only temporarily and which are then 'displaced' and 'posited elsewhere' (Deleuze 2003 [1981], 47). Indeed, Schelling indicates that organs form epigenetically in relation to situations and needs rather than being pre-formed (2004a [1799], 38n, 113), which we can extrapolate to sciences as organs of understanding that are not restricted to disciplines named a priori.

These fields then retroact on knowledge itself, in the way that Novalis describes in his *Notes for A Romantic Encyclopaedia*, written in the same year as the *First Outline*. For Novalis 'encyclopedistics' consists in applying 'the system to the parts—and the parts to the *system* and the parts to the parts' (2007 [1798/99], §460). In *On University Studies*, in a more conservative logic of emanation, if not hierarchies, Schelling describes 'the organic body of knowledge' as 'flow[ing]' from its 'central organs' to sciences located in its 'outermost parts' (1966 [1802], 42). Using this image of fluidity far less idealistically than in its slightly later counterpart in Schelling's transcendental philosophy, the *First Outline* begins by evoking an *unbedingt* or universal fluid which is determined in 'figures' that occur when actants (*Aktionen*) of thought-matter are bound, only to be unbound by the pressure of other actants within the fig. (2004a [1799], 22–31). These actants or 'original productivities' do not 'actually *exist*' but are the '*ideal grounds* of the explanation of quality' (2004a [1799], 21). In Schelling's theory of dynamic atomism, adapted from Leibniz, all products contain 'an infinite multiplicity of unified tendencies' or actants (2004a [1799], 19). Because monads too are not physical but ideal or ideational entities, Leibniz's notion of the monad as indivisible is strictly heuristic and, as Kant argues, is not in contradiction to its infinite divisibility if we consider it as a 'sphere' or 'orbit of activity' rather than an atom of matter (Kant 2002, 57–8). In the *First Outline* sciences and fields can thus be seen as momentary bindings of multiple forces, figures deconstructed by nature's complexity. So as Nature organizes 'to infinity', the 'sphere' or field to which it is 'limited must again contain an infinity', so that within it 'other spheres are again formed and within these spheres others' (Schelling 2004a [1799], 44).

A different experiment for constructing this text is to read it in series, following Novalis, who organizes his entries neither alphabetically nor in a Hegelian *enkyklos paideia* or course of learning, but in series. In the *Romantic Encyclopedia* terms like encyclopedistics, physics, or natural history, are repeated at intervals in what Deleuze calls the eternal return of difference (1994 [1968], 41, 53). In Schelling's encyclopedistics of nature, which is richer in content than Novalis' numbered Schlegelian aphorisms, terms are similarly repeated to give the idea a renewed or different potency. As a topic returns, another field makes an incursion into it, 'derang[ing]' it, in Schelling's words (2004a [1799], 39). Take the *Stufenfolge.* The term first appears in footnotes and as the idea of a single 'organism inhibited at various stages of development' (Schelling 2004a [1799], 43), a theory of evolution, albeit not Darwinian evolution grounded in actual species change. This 'absolute organism', Schelling explains, must be thought as 'a graduated series of productivity', since the series of 'products' is necessarily 'interrupted' as creation fails to move smoothly up the Chain of Being (2004a [1799], 43). Ten pages later the *Stufenfolge* is emphatically if hypothetically linked to a potential new discipline: 'the *history* of nature' (2004a [1799], 53), which Green will call physiogony. But in between Schelling speaks of 'increasingly graduated divergences' between entities, as he allows 'comparative anatomy' to temporarily hypostatize as an interruption of this history of nature (2004a [1799], 50). Interestingly in the transcendental series of texts which includes the lectures on academic study, comparative anatomy promises the 'inner unity and affinity of all organisms' (1966 [1802], 142). But in the natural series it is the site of divergences that potentially derange the analogical leap from nature to history.

This is to say that comparative anatomy raises the problem of *Naturbeschreibung* as diversifying a history of nature and not just failing to achieve it. For Schelling also asks about different products that detach themselves from the absolute process to assert their autonomy, as individuation struggles against universalization. Since products are formed by inhibition (*Hemmung*), the arresting of productivity is a limitation. Thus Schelling tries to valorize 'nature as subject' over 'nature as object', productivity over products (2004a [1799], 17; 2004b [1799], 202). But this *natura naturans* that 'continues irresistibly' (2004a [1799], 17), consuming what is in its way, risks being the Hegelian *Aufhebung* that Schopenhauer, in his bitter critique of Idealism, later calls the will in nature. For Schelling too, 'Life comes into existence in opposition to Nature' (2004a [1799], 68). Raising the problem of biodiversity, he thus asks how 'inhibition could be *permanent*', how these 'natures which have torn themselves away from universal Nature … can maintain an individual existence' (2004a [1799], 53). While he says they cannot do so, clearly they do: fish do not sacrifice themselves to the genus process and become birds. In short, the physiology (in Green's sense) that underpins physiogony is interrupted by physiography, as a return within the biological series of that dynamic atomism of understanding by which 'every organ organizes again into infinity' (Schelling 2004a [1799], 58–9).

After this segment the *Stufenfolge* disappears for a hundred pages. Schelling then adds a further dimension in the form of a *Stufenfolge* of functions: the schema sensibility> irritability> reproduction (or the reverse) common in physiology of the time, which he constructs as a graduated series, so as to posit 'a *unity* of FORCE of production throughout … organic nature' (2004a [1799], 149). Such a series, both in the individual organism and the absolute organism as it ascends from plants and polypi to mammals, could ideally synchronize with the graduated stages of nature, bringing the 'graduated series of production' and 'products' into convergence (2004a [1799], 140 n.). But the problem is that functions are not aligned with separate organisms, allowing us to move beyond irritability to sensibility in the chain of being itself. Rather, in complex organisms each function becomes its own sphere as the organism breaks down into 'individual systems of *specialized excitability*' (Schelling 2004a [1799], 127, 143). Not only does the *Stufenfolge* of functions thus become a kind of ourobouros that cycles back into itself. In terms of a *Stufenfolge* of disciplines, like the one Hegel maps at least as a schema, physiology also fails to provide the desired unification for Schelling. For as he first projected it, John Brown's physiology—his theory of excitability—was supposed to unify the functions and to synthesize the systems of chemistry and vital force (Schelling 2004a [1799], 68). But physiology proves to be the site at which vital force, or the *Bildungstrieb* that pushes the graduated stages of nature forward, is disrupted by the chemistry of organisms.

This diremption of vital force by its chemical intensity raises another line of flight, the problem of illness, which Schelling brackets in an Appendix but which extends throughout the *First Outline*. For if the *Stufenfolge* provides 'a perspective … for *the whole of organic nature* … another is furnished' for the individual by disease as a 'deviation' from 'rule' or 'proportion' (Schelling 2004a [1799], 159). The Appendix thus takes us back to the graduated series of divergences, with the added recognition that disease in one organism may be health in another: thus the polyp would 'be quite healthy' 'with the degree of irritability' that sickens the plant (Schelling 2004a [1799], 159). The tension between universal and individual perspectives further raises issues about the projection of normativity onto nature. For it seems that disease 'has the same factors as life', just in different proportions, and so is 'just as natural' as life. But this raises a question, lethal to the *Stufenfolge*, of what life is, as Schelling, reflecting from a different perspective on the anthropological Idea, observes ironically that life is 'extorted from Nature … a perduring sickness' (2004a [1799], 159, 160n).

In conclusion, the *Outline* is the 'chaos of a cyclic poem', to evoke Percy Shelley who anticipates Deleuze's notion of the great work (Shelley 2002 [1820], 512). It is one of those primal texts in Schelling's corpus into which many series lead in and out, another being the *Ages* (1815), which is similarly unsequenced and without an architecture. The *First Outline* is a kind of encyclopaedia of the sciences, but on the model of Novalis, for whom a science 'is applicable to everything' and 'everything is applicable to it' (Novalis 2007 [1798/99], §176). In Novalis' part-whole logic, we do not just read the parts within a whole as in Kant's architectonic, but apply 'the parts to the parts' (2007 [1798/99], §460). In Novalis' words, this 'encyclopedization' of a science is the genuine 'self-criticism' of that science, and the 'stimulus' for the 'self-(post)-development of another science' (2007 [1798/99], §487). Thus, the various sciences in the *First Outline* intersect and involve each other, creating feedback loops in which a science discussed earlier in the text, like dynamic atomism, is reinvolved as a way of un-thinking sciences that come up later, since, as Schelling says, 'every science' has 'its unconditioned' (2004a [1799], 13). As ideas are passed through this weave of sciences, which themselves exist in both real and ideal, or natural and transcendental series, these ideas, such as 'nature' or 'life', enter a dynamic process that involves both self-criticism and concept-creation.[20] In a longer essay, I would also suggest that such feedback loops connect what Schelling calls 'the ideal part' of philosophy in the *Freedom* essay (2006 [1806], 4) back to the *Naturphilosophie*, through what Deleuze calls the trinity of complication/explication/implication. For instance, the pair evolution/involution occurs several times in the *First Outline* (2004a [1799], 37 n., 77, 112, 187–8). Following his own sense that the ideal is a higher potency of the real (1966 [1802], 79, 103), which means that a material proposition can be philosophically 'potentialized' in Novalis' terms (2007 [1798/99], §487), so as to raise physics to speculation, Schelling repeats these terms in the 1815 *Ages* (2000, 83). The *Ages* explicates the idea that Nature is 'an infinite evolution from one original involution' (Schelling 2004a [1799], 77) with reference to history, which is complicated by nature, as what has been expanded in the later text is contracted back into or re-involved in the great whole of the *First Outline.* Through this process of explication and implication, or expansion and contraction, the ideal series itself as it had existed in the *System of Transcendental Idealism* is subject to criticism. In this sense the early text is also a first outline for thinking Schelling's corpus as a whole.

References

Barthes, Roland. 1977. From work to text. In *Image-music-text*. Translated by Stephen Heath, 155–164. London: Fontana Press.

Canguilhem, Georges. 1991. *The normal and the pathological* (1943, rev. 1966). Translated by Carolyn R. Fawcett. New York: Zone Books.

———. 2008. *Knowledge of life* (1952). Translated by Stefanos Geroulanos and Daniela Ginsburg. New York: Fordham University Press.

[20] I take the term from Deleuze and Guattari's *What is Philosophy?* (1994 [1991], 5, 8, 11). 'Philosophy', as they understand it, is what the book's translators call a 'conceptual vitalism' (1994 [1991], viii) in which concepts are continuously fabricated and displaced on 'a plane of immanence' where they unfold from and enfold each other without any one concept becoming a transcendental signified (1994 [1991], 35–60). This constant folding is due to the 'concept' being non-simple and composed of a multiplicity of sub-concepts (1994 [1991], 15–23). That idea resonates with Schelling's dynamic atomism of products in the *First Outline*, where a product is composed of multiple, bound actants which continually push against this closure.

Coleridge, Samuel Taylor. 1995a. *Hints towards the formation of a more comprehensive theory of life* (1816? First pubd 1848). In *Shorter works and fragments*, ed. H. J. and J. R. de J. Jackson. Vol. 1: 481–557. Princeton: Princeton University Press.

———. 1995b. Life. In *Shorter works and fragments*, ed. H. J. and J. R. de J. Jackson, vol. 2: 1027–32. Princeton: Princeton University Press.

———. 1995c. On the passions (1828). In *Shorter works and fragments*, ed. H. J. and J. R. de J. Jackson, vol. 2: 1419–1453. Princeton: Princeton University Press.

Deleuze, Gilles. 1994. *Difference and repetition* (1968). Translated by Paul Patton. New York: Columbia University Press.

———. 2003. *Francis Bacon: The logic of sensation* (1981). Translated by Daniel W. Smith. London: Continuum.

Deleuze, Gilles, and Felix Guattari. 1987. *A thousand plateaus: Capitalism and schizophrenia* (1980). Translated by Brian Massumi. Minneapolis: University of Minnesota Press.

———. 1994. *What is philosophy?* Translated by Hugh Tomlinson and Graham Burchell. New York: Columbia University Press.

Esposito, Maurizio. 2013. *Romantic biology, 1890–1945*. London: Routledge.

Ferrini, Cinzia. 2011. The transition to organics: Hegel's idea of life. In *A companion to Hegel*, ed. Stephen Houlgate and Michael Baur, 203–223. Oxford: Blackwell.

Foucault, Michel. 1970. *The order of things: An archaeology of the human sciences* (1966). Translated by anon. New York: Vintage.

Furlotte, Wes. 2018. *The problem of nature in Hegel's final system*. Edinburgh: Edinburgh University Press.

Gambarotto, Andrea. 2018. *Vital forces, teleology and organization: Philosophy of nature and the rise of biology in Germany*. Cham: Springer.

Goethe, Johann Wolfgang von. 1998. *Maxims and reflections* (1833). Translated by Elizabeth Stopp. Harmondsworth: Penguin.

Grant, Iain Hamilton. 2006. *Philosophies of nature after Schelling*. London: Continuum.

———. 2010. Introduction to Schelling's *On the world soul*. *Collapse* 6: 80–87.

Green, Joseph Henry. 1840. *Vital dynamics: The Hunterian oration before the Royal college of surgeons*. London: W. Pickering.

Hegel, G. W. F. 1970a. *Aesthetics: Lectures in fine art*. Translated by T. M. Knox, 2 vols. Oxford: Oxford University Press.

———. 1970b. *Philosophy of nature* (1842). Translated by A. V. Miller. Oxford: Oxford University Press.

———. 1970c. *Vorlesungen über die Ästhetik* (1835). In *Werke*, vol. 13, ed. Eva Moldenhauer and Karl Markus, vols. 13–15. Frankfurt: Suhrkamp Verlag.

———. 1986. *The philosophical propaedeutic* (1840). Translated by A. V. Miller, ed. Michael George and Andrew Vincent. Oxford: Oxford University Press.

———. 1990. *Encyclopaedia of the philosophical sciences in outline* (1817). In *Encyclopedia of the philosophical sciences in outline and critical writings*. Translated by Stephen A. Taubeneck, ed. Ernst Behler, 45–263. New York: Bloomsbury.

———. 2006. *De orbitis planetarum* (1801). Translated by David Healan. Berlin. http://hegel.net/en/v2123healan.htm.

———. 2010. *The science of logic* (1812). Translated by and ed. George Di Giovanni. Cambridge: Cambridge University Press.

Heuser, Marie-Luise. 2016. Space philosophy: Schelling and the mathematicians of the nineteenth century. *Angelaki* 21 (4): 43–57.

Houlgate, Stephen. 2005. *An introduction to Hegel: Freedom, truth and history*. 2nd ed. Malden, MA: Blackwell.

Kant, Immanuel. 1998. *Critique of pure reason* (1781). Translated by and ed. Paul Guyer and Allen W. Wood. Cambridge: Cambridge University Press.

———. 2000. *Critique of the power of judgment* (1790). Translated by Paul Guyer and Eric Matthews, ed. Paul Guyer. Cambridge: Cambridge University Press.

———. 2002. The employment in natural philosophy of metaphysics combined with geometry, of which sample I contains the physical monadology (1756). In *Theoretical philosophy 1755–1770*. Translated by and ed. David Walford, 47–66. Cambridge: Cambridge University Press.

———. 2004. *Metaphysical foundations of natural science* (1786). Translated by and ed. Michael Friedman. Cambridge: Cambridge University Press.

Kaufmann, Stuart. 2016. *Humanity in a creative universe*. Oxford: Oxford University Press.

Krell, David Farrell. 1999. *Contagion: Sexuality, disease, and death in German idealism and romanticism*. Bloomington: Indiana University Press.

Leibniz, G. W. 1998. Monadology (1714). In *Philosophical texts*. Translated by Richard Francs and R. S. Woolhouse, 267–281. Oxford: Oxford University Press.

Lovejoy, Arthur. 1936. *The great chain of being: A study of the history of an idea*. Cambridge, MA: Harvard University Press.

Meillassoux, Quentin. 2008. *After finitude: An essay on the necessity of contingency* (2006). Translated by Ray Brassier. London: Bloomsbury.

Morrisette, Bruce. 1963. *Les romans de Robbe-Grillet*. Paris: Les Éditions de Minuit.

Nancy, Jean-Luc. 1991. *The inoperative community* (1985). Translated by Peter Connor et al., ed. Peter Connor. Minneapolis: University of Minnesota Press.

———. 2000. *Being singular, plural* (1996). Translated by Robert D. Richardson and Anne E. O'Byrne. Stanford: Stanford University Press.

Ng, Karen. 2020. *Hegel's concept of life: Self-consciousness, freedom, logic*. Oxford: Oxford University Press.

Novalis. 2007. *Notes for a romantic encyclopaedia: Das allgemeine brouillon* (1798/99). Translated by and ed. David W. Wood. Albany: State University of New York Press.

Petry, Michael John. 1970. Introduction to G. W. F. Hegel, *The philosophy of nature*, ed. Michael John Petry, 3 vols. Vol. 1: 11–178. London: George Allen and Unwin.

Pippin, Robert. 2019. *Hegel's realm of shadows: Logic as metaphysics in the science of logic*. Chicago: Chicago University Press.

Rajan, Tilottama. 2018. The asystasy of the life sciences: Schelling, Hunter, and British idealism. *Kabiri: The Official Journal of the North American Schelling Society* 1:47–68. https://journals.library.mun.ca/ojs/index.php/kabiri/article/view/1944

———. 2021. Openings: Biology and philosophy in Esposito, Bichat, and Hegel. In *Roberto Esposito: New directions in biophilosophy*, ed. Tilottama Rajan and Antonio Calcagno. Edinburgh: Edinburgh University Press.

———. 2025 (forthcoming). The archival text: Schelling's *First outline* (1799) and *Ages of the world* (1815). In *Media and mediation*, ed. Ralf Haeckel. Trier: Wissenschaftlicher Verlag.

Ricoeur, Paul. 1967. *Husserl: An analysis of his phenomenology*. Translated by Edward G. Ballard and Lester E. Embree. Evanston: Northwestern University Press.

———. 1991. Hegel and Husserl on intersubjectivity (1986). In *From text to action: Essays in hermeneutics II*. Translated by Kathleen Blamey and John Thomson, 227–245. Evanston: Northwestern University Press.

Schelling, F. W. J. 1856–61. *Friedrich Wilhelm Joseph von Schellings sämmtliche werke*, ed. K.F.A. Schelling. Stuttgart.

———. 1965–9. *Schelling's werke nach der originalausgabe in neuer anordnung*, ed. Manfred Shröter. Munchen.

———. 1966. *On university studies* (1802). Translated by E. S. Morgan, ed. Norbert Gutterman. Athens, Ohio: Ohio University Press.

———. 1978. *System of transcendental idealism* (1800). Translated by Peter Heath. Charlottesville: University of Virginia Press.

———. 1988. *Ideas for a philosophy of nature, as introduction to this* science (1797, second edn 1803). Translated by Errol E. Harris and Peter Heath. Cambridge: Cambridge University Press.

———. 1989. *The philosophy of art (1804–5)*. Translated by Douglas W. Stott. Minneapolis: University of Minnesota Press.

———. 1994a. *On the history of modern philosophy* (1833–4). Translated by Andrew Bowie. Cambridge: Cambridge University Press.

———. 1994b. Treatise explicatory of the idealism in the science of knowledge (1797). In *Idealism and the endgame of theory: Three essays by F. W. J. Schelling*. Translated by and ed. Thomas Pfau, 61–138. Albany: State University of New York Press.

———. 1997. On the nature of philosophy as science (1821). Translated by Marcus Weigelt. In *German Idealist Philosophy*, ed. Rüdiger Bubner, 210–243. Harmondsworth: Penguin.

———. 2000. *Ages of the world* (1815). Translated by Jason Wirth. Albany: State University of New York Press.

———. 2004a. *First outline of a system of the philosophy of nature* (1799). Translated by Keith R. Peterson. Albany: State University of New York Press.

———. 2004b. Introduction to the Outline of a system of the philosophy or nature, or, On the concept of speculative physics and the internal organization of a system of this science (1799). In *First outline*. Translated by Keith Peterson, 193–232.

———. 2006. *Philosophical Investigations into the essence of human freedom* (1809). Translated by Jeff Love and Johannes Schmidt. Albany: State University of New York Press.

Shelley, Percy Bysshe. 2002. A defence of poetry (1820). In *Shelley's poetry and prose*, ed. Donald H. Reiman and Neil Fraistat, 509–535. New York: Norton.

Stone, Alison. 2005. *Petrified intelligence: Nature in Hegel's philosophy*. Albany: State University of New York Press.

Wigley, Mark. 1997. *The architecture of deconstruction: Derrida's haunt*. Cambridge, MA: MIT Press.

Woodard, Ben. 2016. Lamps, rainbows and horizons: Spatializing knowledge in naturphilosophical epistemology. *Angelaki* 21 (4): 23–41.

Zammito, John. 2007. Kant's persistent ambivalence toward epigenesis, 1764–90. In *Understanding purpose: Kant and the philosophy of biology*, ed. Philippe Huneman, 51–74. Rochester, New York: University of Rochester Press.

Tilottama Rajan is distinguished university professor at the University of Western Ontario. Her four monographs include *Romantic Narrative* (Johns Hopkins UP, 2010) and *Deconstruction and The Remainders of Phenomenology* (Stanford UP, 2002). Her eight co-edited essay collections include, with Daniel Whistler, *The Palgrave Handbook of German Idealism and Poststructuralism* (2023). She is currently working on organizations of knowledge and the mutual impact of philosophy and science in Germany and Britain, 1780–1850.

Chapter 7
Grasping a Living Nature: Schelling, Goethe and the Roots of Process Philosophy

Philipp Höfele

Abstract This chapter argues that the concern Schelling shares with Goethe is, on the one hand, to present a concept of nature and life that takes better scientific account of the phenomena of dynamics, 'self-creation' and purposefulness. On the other hand, this ontological question is also linked to the project of establishing an appropriate methodology that pays attention to the phenomena themselves and their holistic significance for the whole of nature and the environment. Schelling and Goethe are united by a common front against a transcendental philosophy of nature oriented towards Newton, which they sought to correct both *methodologically* and *ontologically* with regard to the concepts of nature and life. Their approaches can therefore be seen as precursors of a process-philosophical understanding of reality, which is increasingly being taken into account today in environmental ethics in view of the challenges of the present.

Contrary to assumptions that the modern natural sciences are the result of one-dimensional historical progress, the concept of nature espoused by the *Naturphilosophie* around 1800 can undoubtedly claim a certain topicality.[1] This becomes more apparent against the background of the problems of the natural sciences of the twentieth and twenty-first centuries. On the one hand, new technologies, which seem to be an almost 'natural' part of our environment (Karafyllis 2019;

[1]The work leading to this publication was supported by the Deutsche Forschungsgemeinschaft (DFG, German Research Foundation) under Germany's Excellence Strategy—EXC-2193/1—390951807 as well as by the Young Academy for Sustainability Research at the University of Freiburg, which is funded by the Eva Mayr-Stihl Foundation. I thank James Fisher very much for correcting the English text.

P. Höfele (✉)
Freiburg Institute for Advanced Studies, University of Freiburg,
Freiburg, Baden-Württemberg, Germany
e-mail: philipp.hoefele@philosophie.uni-freiburg.de

P. Cheyne (ed.), *Matter and Life in Coleridge, Schelling, and Other Dynamical Idealists*, International Archives of the History of Ideas Archives internationales d'histoire des idées 253, https://doi.org/10.1007/978-3-031-78315-9_7

Schneider 2019), are increasingly dissolving the traditional conceptual distinction between nature and man-made technology, so that the technological seems to gain a foothold on the natural (Guzzoni 1995, 81–115). In the face of this dominance of human influence over animate and inanimate nature, in which man and the man-made become all-pervading components of nature, the natural scientists Paul J. Crutzen and Eugene Stoermer even spoke in 2000 of the 'Anthropocene' as a new geological age.[2] This is of philosophical significance not only from the perspective of environmental ethics, but also for the very reason that this anthropogenic reshaping of the natural poses epistemological and methodological questions. To what extent can one still speak of nature as such? Can nature still function as a space for self-understanding and orientation in its contrasting relationship to the human-technical? On the other hand, in view of the withdrawal of the traditional distinction between nature and man-made technology—as well as the heightened dynamics of the environment resulting thereof[3]—the problem of formulating an ontology that does not operate with static categories but does justice to the change and processual character of the 'natural' environment is also opened up.

In order to answer these methodological and ontological questions, which are aimed at the 'how' and 'what' of experiencing nature, it is advantageous to return to traditions of the philosophy of nature that have been largely forgotten in the present. Certain traditions of the philosophy of nature, starting with Galileo and Newton, have proved to be influential up to the present day. These traditions are to be regarded as insufficient for contemporary problems because they are mostly based on a static nature-mind dichotomy. Other traditions, on the other hand, have been put on the back burner and even disavowed as too 'metaphysical'.[4] In the twentieth century, Alfred N. Whitehead was one of the few to point this out in his work *Process and Reality,* published in 1929, with a view to the natural philosophical traditions that derive from Plato and Newton:

> The *Scholium* [of Newton on the first eight definitions of his *Principia*] betrays its abstractness by affording no hint of that aspect of self-production, of generation, of φύσις, of *natura naturans,* which is so prominent in nature. For the *Scholium,* nature is merely, and completely, *there,* externally designed and obedient. The full sweep of the modern doctrine of evolution would have confused the Newton of the *Scholium,* but would have enlightened the Plato of the *Timaeus.*[5] (Whitehead 1985 [1929], 93)

The historical connection emphasised by Whitehead is astonishing not so much because he considers Newton's *Philosophiae naturalis principia mathematica* (1687) to be insufficient, insofar as 'Biology is … reduced to a mystery' in his work and 'physics itself has now reached a stage of experimental knowledge inexplicable

[2] See fundamentally on this Crutzen and Stoermer (2000), Crutzen (2002). See also Höfele (2020).

[3] See the discussion on the so-called 'Great Acceleration' in the 'Anthropocene' in Steffen et al. (2011, 849–853), Steffen et al. (2015).

[4] Falkenburg (2017), Köchy (2017). See also Breidbach (2006, 141 f.), who states a break in the natural science tradition in Goethe's succession *c.*1840, about 10 years after his death.

[5] Whitehead refers to the *Scholium* on the first eight definitions, which he also quotes at length (1985 [1929], 70 f.). See Newton (1713 [1687], 5–11).

in terms of the categories of the *Scholium.*' (Whitehead 1985 [1929], 94) The perspective shown is rather astonishing in that he regards a tradition that is even older than Newton's as more suitable for understanding modern problems.

It is hardly a coincidence that Whitehead's 'process philosophy', in contrast to Newton's mathematization and mechanisation of natural processes, refers back to a philosophy of nature exemplified in Plato's *Timaeus,* to which the central theorist of *Naturphilosophie* around 1800, namely F. W. J. Schelling,[6] has been linked since his early philosophy. He did this in reaction to Kant's transcendental philosophy in particular, which is oriented towards Newton's mathematical physics, as well as in the course of an awakening interest in the life sciences (Steigerwald 2019). The concern Schelling shares with the *Naturphilosophie* of his time is, on the one hand, to present a concept of nature and life that takes better scientific account of the phenomena of dynamics, 'self-creation', productive *natura naturans* and purposefulness. On the other hand, this ontological question is also linked to the project of establishing an appropriate methodology that does not proceed in a purely rationalistic or transcendental-philosophical manner primarily from the subject as the all-integrating 'fixed point', but also pays attention to the phenomena themselves and their holistic significance for the whole of nature and the environment.[7] In the 'saddle period [*Sattelzeit*]' around 1800, Schelling and Goethe in particular were united by this common front against a transcendental philosophy of nature oriented towards Newton, which—according to the thesis of this article—they sought to correct both *methodologically* and *ontologically* with regard to the concepts of nature and life. Just as Whitehead diagnosed this for Plato, these authors would also have more to say to contemporary developments than the tradition they rejected. The approaches of *Naturphilosophie* around 1800 can therefore be seen as precursors of a process-philosophical understanding of reality, which is increasingly being taken into account today both in the theory of science and in environmental ethics in view of the challenges of the present.[8]

I intend to demonstrate this paradigmatically with two representatives of the philosophy of nature around 1800: firstly, with reference to Goethe as the inaugurator, above all in terms of science policy, of the 'culture of science' that established itself around 1800 in Jena and Weimar with its focus on *Naturphilosophie* (Sect. 7.1); secondly, with regard to Schelling, whose appointment to Jena was largely due to Goethe. For Goethe was very interested in Schelling's *Naturphilosophie* (Sect. 7.2). The starting point of each of the two sections will be (a) a natural philosophical debate in the context of the theory of colours, which first leads Goethe with Newton and then in 1798 Schelling together with Goethe. Despite the *methodological* similarities between Goethe and Schelling, which consist primarily in a common movement of demarcation from Newton, both draw different conclusions from this in terms of content. (b) Each of the two sections then goes beyond the discussion

[6] See Schelling's early commentary on *Timaeus* in Schelling (1994 [1794]).

[7] See concerning the notion 'holism' Köchy (1997, 62–64), Dierks (2016).

[8] See paradigmatically Seibt (2003), Muraca (2010).

within the framework of the theory of colours. Despite the common ground between Goethe and Schelling in that they distance themselves from mechanistic conceptions of nature, both pursue different *ontologies of the natural*. On the one hand, Goethe always cites the individual only as a sensual example of the (primordial) phenomenon, *(Ur-)Phänomen*, and sees this as an element of a development, or of a process of metamorphoses that is always already ordered towards a whole and is represented in the individual. Schelling, on the other hand, argues to a certain extent for a negative orientation towards phenomena or for a 'higher' empiricism that understands the imaginatively given only as a 'non-being'.[9] Even though Whitehead's process philosophy refers only marginally to Schelling and rather negatively to Goethe, he shares this movement of demarcation with both of them as well as the endeavour to implement a holistic and dynamic view of nature (Sect. 7.3).

7.1 The Plurality and Dynamics of Nature in Goethe's 'Primordial Phenomenon [*Urphänomen*]'

(a) Goethe's *Theory of colours*, published in 1810, is known to have grown primarily out of an inadequacy he diagnosed as regards the approach taken in Newton's work *Opticks: Or, a treatise of the reflections, refractions, inflections and colours of light* of 1704 (Goethe 1887–1919, 2.1: xii-xv). In this work, Goethe accuses Newton of not having lived up to his claim 'Hypotheses non fingo',[10] which was paradigmatically formulated in the second edition of the *Principia mathematica* of 1713. Nonetheless, Newton explicitly states at the beginning of *Opticks* that he does not want to explain the properties of light by hypotheses, but only by experiment and reason.[11] In the collection *Über Naturwissenschaft im Allgemeinen* (*On natural science in general*), first published posthumously in 1833, Goethe explains that 'hypotheses … are scaffolding which is erected before the building and which is taken down when the building is finished; they are indispensable to the worker; only he need not regard the scaffolding as the building'.[12] But this is precisely what Goethe reproaches Newton for in his *Theory of colours*, when he remarks with regard to the rays of light penetrating the *camera obscura*:

[9] On the relationship between Goethe and Schelling, see also Höfele (2021, 168–82).

[10] Newton (1713 [1687], 484, 'Scholium generale'). On Goethe's theory of colours, see Rehbock (1995).

[11] Newton (1704, 1): 'My Design in this Book is not to explain the Properties of Light by Hypotheses, but to propose and prove them by Reason and Experiments'.

[12] Goethe (1887–1919, 2.11): 132: 'Hypothesen sind Gerüste, die man vor dem Gebäude aufführt, und die man abträgt, wenn das Gebäude fertig ist; sie sind dem Arbeiter unentbehrlich; nur muß er das Gerüste nicht für das Gebäude ansehn.' (Unless otherwise indicated, all translations from German are mine.)

> It is striking how little there can be said here [in the *camera obscura*] of individual rays of sunlight, or bundles and clusters of rays, of ray cylinders, ray rods, and however one may imagine all this. For the convenience of certain linear representations, sunlight is assumed to be incident in parallel; but let it be known that this is only a fiction, which one can well allow oneself, where the break falling between the fiction and the true appearance is insignificant. But beware of making this fiction into a phenomenon again, and of continuing to operate with such a fabricated phenomenon.[13]

Goethe criticises Newton in this respect, saying that one does not see rays, but only images of light or the sun. According to Goethe, one may assume such rays for the purpose of representation, but one must not, as Goethe reproaches Newton in his *Beiträge zur Optik* (*Contributions to optics*) as early as 1791/2, regard 'these abstract mental products [*diese abstrackten Geistesproduckte*] as physical beings that actually exist [*als wircklich existirende physische Wesen*]',[14] which would lead to the construction of unjustified chains of causality with false claims to existence and consequently to 'fictitious phenomena'. Goethe, as a reader of Kant (Ho 2012), is nevertheless anything but a naïve realist. Thus he remarks in the preface to the *Theory of Colours:*

> It is, after all, a most strange demand, which is sometimes made, but which is not fulfilled even by those who make it: Experiences should be presented without any theoretical band, and the reader, the pupil, should be left to form his own conviction at will. For the mere sight of a thing cannot advance us. Every look at something passes into observing it, every observation into reflecting about it, every reflection into connecting it, and so one can say that we are already theorising with every attentive glance at the world. But to do this with consciousness, with self-knowledge, with freedom, and, to use a bold word, with irony, such a dexterity is necessary if the abstraction of which we are afraid is to become harmless, and the result of experience that we hope for is to become quite lively and useful. (Goethe 1887–1919, 2.1: xii)

Contrary to first appearances, Goethe is by no means concerned with a movement of abstraction that detaches itself step by step from experience. Rather, a 'quite lively experiential result' is to follow from this movement, which precisely avoids that abstraction 'consciously' and 'with irony'. In *Dichtung und Wahrheit* Goethe defines the 'ironic attitude' in such a way that it 'rises above objects, above happiness and unhappiness, good and evil, death and life, and thus attains the possession

[13]Goethe (1887–1919, 2.1): 128: 'Wie wenig also hier [bei der *Camera obscura*] von einzelnen Sonnenstrahlen, oder Strahlenbündeln und -Büscheln, von Strahlencylindern,—Stäben und wie man sich das alles vorstellen mag, die Rede sein kann, ist auffallend. Zu Bequemlichkeit gewisser Lineardarstellungen nehme man das Sonnenlicht als parallel einfallend an; aber man wisse, daß dieses nur eine Fiction ist, welche man sich gar wohl erlauben kann, da wo der zwischen die Fiction und die wahre Erscheinung fallende Bruch unbedeutend ist. Man hüte sich aber, diese Fiction wieder zum Phänomen zu machen, und mit einem solchen fingirten Phänomen weiter fort zu operiren.'

[14]Goethe (1887–1919, 2.5.2): 79 (Paralipomena, xviii): 'Nachdem die Wirckungen des Lichts, zur Bequemlichkeit der Demonstration, auf ideale Linien zurückgeführt, unter Linien vorgestellt und solche angenommene Licht Linien Strahlen genannt worden; so ist in der Lehre vom Licht und den Farben dadurch eine große Verwirrung entstanden, daß man diese abstrackten Geistesproduckte als wircklich existirende physische Wesen ansah.'

of a truly poetic world [*sich über die Gegenstände, über Glück und Unglück, Gutes und Böses, Tod und Leben erhebt, und so zum Besitz einer wahrhaft poetischen Welt gelangt*]' (Goethe 1887–1919, 1.27: 346), without evaporating like Romantic irony in an infinitely reflective hovering.[15] If we consider this, we see here that the 'chastened' Goethean irony mediates between polarities or, to be more precise, lets them stand together. 'Looking [*Ansehen*]', 'observing [*Betrachten*]', 'reflecting [*Sinnen*]', 'connecting [*Verknüpfen*]' and 'theorising' are not to be distinguished from each other as different levels of abstraction, but are to be understood as a unity.

This is precisely what Goethe's central methodological concept of the 'primordial phenomenon [*Urphänomen*]' describes, as he develops it, for example, in relation to the theory of colour.[16] When Goethe reproaches contemporary natural science, and Newton in particular, for allowing colour to be regarded as a 'compound for a simple thing, the simple [light] for a compound [*Zusammengesetzte für ein Einfaches, das Einfache für ein Zusammengesetztes gelten ließ*]', he criticises it for having placed 'a derived phenomenon in the upper place, the primordial phenomenon in the lower place [*ein abgeleitetes Phänomen an die obere Stelle, das Urphänomen an die niedere Stelle*]'; for

> a primordial phenomenon is the one we have presented so far. On the one side, we see the light, the brightness, on the other the darkness, the gloom, we bring the turbidity between the two, and from these opposites, with the help of imaginary mediation, the colours develop, likewise in an opposition …[17]

According to Goethe, the 'primordial phenomenon' and thus the structure of colours is revealed precisely in the interplay of light, darkness and 'turbidity'; it is thus still a real phenomenon, even if it also marks the limit of what can be experienced. According to him, 'primordial phenomena' are found in phenomena whose comprehension does not require taking refuge in unjustifiably hypostatised hypotheses. 'Primordial phenomena' as the essence of phenomena thus do not represent purely conceptual abstractions that are detached from experience and would only exist in a non-empirical world. Even if they are 'borderline experiences', they still remain part of the phenomena and their experience.

(b) As is well known, Goethe pursued this approach in the various fields of his extensive natural research and, on his trip to Italy in Palermo in 1787, he also advocated for it in an almost naïve sense with regard to botany on the basis of the concept of the 'original plant [*Urpflanze*]', which he later abandoned:[18]

---

[15] On 'Romantic irony', as articulated by F. Schlegel in the Athenaeum Fragment 116, see Behler (1992, 242–55). On irony in Goethe see Bahr (1998).

[16] See on this concept Wenzel and Schweizer (2012).

[17] Goethe (1887–1919, 2.1: 72 f.): 'Urphänomen ist dasjenige, das wir bisher dargestellt haben. Wir sehen auf der einen Seite das Licht, das Helle, auf der andern die Finsterniß, das Dunkle, wir bringen die Trübe zwischen beide, und aus diesen Gegensätzen, mit Hülfe gedachter Vermittlung, entwickeln sich, gleichfalls in einem Gegensatz, die Farben'.

[18] According to Goethe's own comment, this disappearance of the term 'original plant' takes place in Sicily at the same time as the rejection of a 'catalogue of characteristics' or a scheme of possible

> In the face of so many new and renewed creations, I remembered the old cricket: could I not discover the original plant among this crowd? There must be such a thing! How else would I recognise that this or that structure was a plant if they were not all formed according to one pattern?[19]

That which may seem naïve in these reflections on the 'original plant', 23 years before the remarks cited above on the 'primordial phenomenon', appears in a different light if one adds another consideration, also based on an observation in Sicily. For Goethe, the basic phenomenon to be found in the 'original plant' is not something static that has been deposited once and for all. In his report *Der Verfasser theilt die Geschichte seiner botanischen Studien mit* (*The author shares the story of his botanical studies*), Goethe describes how he became aware of the 'successive separation [*successive Trennung*]' of the 'simple, lance-shaped first leaves [*einfachen, lanzenförmigen ersten Blätter*]' of a fan palm, which increased 'until at last the fan-like was to be seen in perfect formation [*bis endlich das Fächerartige in vollkommener Ausbildung zu sehen war*]'. He was impressed by the 'succession of stages of these changes [*Stufenfolge dieser Veränderungen*]', of which a gardener had given him clippings, and revered them, as he writes, 'as fetishes, which seemed to excite and captivate my attention, and to promise me a prosperous result of my efforts.'[20] What these 'fetishes' reveal to him is precisely the 'changeability of plant forms [*Wechselhafte der Pflanzengestalten*]', their 'happy mobility and flexibility [*glückliche Mobilität und Biegsamkeit*]', which they 'are endowed with in order to adapt to so many conditions that act upon them above the earth's surface and to be able to form and transform themselves accordingly [*um in so viele Bedingungen, die über dem Erdkreis auf sie einwirken, sich zu fügen und darnach bilden und umbilden zu können*]' (Goethe 1887–1919, 2.6: 120). This almost limitless adaptability in the plant kingdom does not, however, prevent Goethe from searching for a 'sensual form of a supersensual original plant [*sinnlichen Form einer übersinnlichen Urpflanze*]' and from understanding 'the original identity of all plant parts' as the goal of his Sicilian journey: 'and I now sought to pursue this everywhere and to become aware of it again' (Goethe 1887–1919, 2.6: 121).

He generally views the individual in relation to the whole. The point of Goethe's doctrine of metamorphosis is that the individual is only ever named as a sensual example of the actual phenomenon of a plant, for instance. We could only ever understand an individual in relation to a whole, e.g. a single concrete flower in

---

plant forms. See Goethe (1887–1919, 2.13: 41): 'Attentive to new forms (*neue Gestalten*), I rose from the limited concept of a original plant [*Urpflanze*] to the concept and, if you like, to the idea of a legal, uniform, if not equally shaped formation and transformation of plant life from the root to the seed [*Idee einer gesetzlichen, gleichmäßigen, wenn schon nicht gleich gestalteten Bildung und Umbildung des Pflanzenlebens von der Wurzel bis zum Samen*].' See Breidbach (2006, 17–22).

[19] Goethe (1887–1919, 1.32: 42 f.): 'Im Angesicht so vielerlei neuen und erneuten Gebildes fiel mir die alte Grille wieder ein: ob ich nicht unter dieser Schaar die Urpflanze entdecken könnte? Eine solche muß es denn doch geben! Woran würde ich sonst erkennen, daß dieses oder jenes Gebilde eine Pflanze sei, wenn sie nicht alle nach Einem Muster gebildet wären?'

[20] Goethe (1887–1919, 2.6: 119 f.). See also Breidbach (2006, 132–9).

relation to and as a specimen of the genus flower. According to Goethe, an inductive conclusion is not possible at all, since in observing the case of each individual, one always already bears the whole in mind. For nature, as Goethe remarks in a note on his first acquaintance with Schiller in 1794, 'is to be represented as acting and living, striving from the whole into the parts [*wirkend und lebendig, aus dem Ganzen in die Theile strebend*]' (Goethe 1887–1919, 1.36: 250). Nature is thus not to be understood as a sum of individual things, but the natural individual is always already ordered towards a whole as an element of a development, a process of metamorphoses, which presents itself in the individual.[21] In this respect, the 'original phenomenon' cannot be separated from the individual phenomenon.

It is true that Goethe's philosophical interlocutor Schelling generally shares his method of approaching nature not through constructions on the basis of hypotheses but through empirical observation. At the same time, Schelling is sceptical, as will be shown in the following, about the unbroken phenomenal visibility of the dynamics and processuality of the natural. He considers this tense mediation of the phenomenal and the dynamic to be possible only *ex negativo*, in such a way that what appears is a 'coagulated' document of a non-phenomenal dynamic.

7.2 Schelling's Search for the 'Unconditionally Empirical': Nature as Productivity and the Unconscious

(a) F. W. J. Schelling and J. W. von Goethe can justifiably be described as the 'central figures' of the 'culture of science' that established itself in Jena and Weimar around 1800, with its focus on *Naturphilosophie* (Breidbach and Ziche 2001). On the one hand, Schelling, with his lectures and writings on *Naturphilosophie* first given in Jena in 1798/99, had an influence on nineteenth century *Naturphilosophie* that can hardly be overestimated, especially on A. K. A. von Eschenmayer, G. H. Schubert, H. Steffens, L. Oken and even A. Schopenhauer, among many others (Bach and Breidbach 2005; Höppner 2017; Höfele and Hühn 2021). On the other hand, Goethe was of central importance for Schelling and Romanticism as well as their natural philosophical concepts through his scientific research as well as his literary work (Fröschle 2002). After a first meeting between Schelling and Goethe on May 28, 1798 and the 'optical experiments' they then jointly performed (Goethe 1887–1919, 3.2: 209)—experiments which Goethe only carried out with selected persons[22]—Goethe himself pushed for the appointment of the 23-year-old Schelling to Jena. Goethe thus made a decisive contribution to the establishment of the 'culture of science' in Jena and Weimar around 1800 (Adler 1998; Nassar 2010).

[21] For more detail, see Breidbach (2006, esp. 45–62).

[22] So later also with Schopenhauer. See Rehbock (2014).

As Goethe's diary entries and correspondence show,[23] a scientific dialogue developed between Goethe and Schelling—just as it did later between Goethe and Schopenhauer. This dialogue was initially concerned with the theory of colours and its criticism of Newton (Müller 2014). In contrast to Newton, Goethe sought to understand the colour wheel as the inner law of all phenomena—thus it is not only that the eye strives to bring the physiological and physical colours in general to light, but it is also with regard to the chemical colours adhering to the objects themselves that the harmony and unity of nature can be seen to 'wander' through the colour wheel.[24] In an analogous way, Schelling also uses light to depict a unity and the 'ground of a continuity in nature [*Grund einer Continuität in der Natur*]', of which light is the 'symbol'.[25] In a sense, Schelling continues the holistic tendency that Goethe had already begun by understanding light not as an individual phenomenon, but in terms of its function and significance for a whole. To this end, he gives Goethe's considerations a metaphysical twist, so to speak, by linking his metamorphosis and colour theory with each other. As a letter from January 1801 shows, Schelling extends Goethe's concept of metamorphosis to the whole of nature, relates it to the law of polarity, and finally links it to the theory of light and colour. According to Schelling, light represents the point of unity towards which everything strives, as evidenced by the 'so-called hunger for light of plants [*sogenannte Lichthunger der Pflanzen*]'.[26] In contrast to Newton, both thinkers understand light as a uniform

[23] See the correspondence between Schelling and Goethe in Schelling (1976ff., 3.1–2) as well as the editorial report by T. Kisser in Schelling (1976ff., 3.1: 97–117).

[24] Goethe (1887–1919, 2.1: 219): 'What, however, does not seem to be in accordance with the inorganic, we will find possible when we speak of organic natures, in that in this last realm such a penetration of the circle from yellow through green and blue to purple really occurs [*indem in diesem letzten Reiche eine solche Durchwandrung des Kreises vom Gelben durch's Grüne und Blaue bis zum Purpur wirklich vorkommt*].'—See also the editorial report by T. Kisser in Schelling (1976ff., 3.1: 103 f.).

[25] See Schelling's entry in the hand copy of the *Ersten Entwurfs eines Systems der Naturphilosophie* (*First outline of a system of the philosophy of nature*) from 1799, in Schelling (1976ff., 1.7: 323 [note in Schelling 1976 ff., 1.7: 161,18]): 'Light is not even matter in the process of *becoming*—in the process of development; it is rather *becoming*, productivity itself [*das Licht ist nicht einmal eine werdende—in der Entwicklung begriffene—Materie; es ist vielmehr das Werden, die Produktivität selbst*], which propagates itself in light, the direct symbol, as it were, of general *productivity*. We have deduced productivity as the basis of all continuity in nature. But light is the symbol of all continuity'—Schelling still emphasises the importance of light as the ideal in the late Munich lecture *Grundlegung der positiven Philosophie* from 1832/33: 'The ideal is nothing insofar as it is the opposite of something; but for this very reason it is also something again, itself a finite thing. We will say that the pure being posited as such, which is opposed to matter as nothing, is the light in nature. … Light fills space in an ideal way, just as matter does. Light is therefore the concept of matter, which is posited in matter itself' (Schelling 1972, 192).

[26] Schelling (1976ff., 3.2.1: 306): 'The so-called light hunger of plants, and the undeniable bends and turns towards the light, betray a real *need* of plants to cohere with the sun through light, as it were [*Bedürfniß der Pflanzen, mit der Sonne durch das Licht gleichsam zu cohäriren*].'

structure that cannot be prismatically dissected and that forms the pole of all phenomena that is decisive in terms of the philosophy of nature.[27]

(b) More decisive, however, is another moment in the dialogue between Goethe and Schelling, namely their common front against transcendental philosophy, especially of Fichtean provenance, and its marginalisation of nature. Here, even more clearly than in the discussion about light and colours, the 'meta-physical' twist that Schelling gives to Goethe's natural philosophical reflections becomes apparent. In contrast to Fichte's figure of the '*Thathandlung*' or the self-setting of the ego, Schelling asserts in the *Einleitung zu seinem Entwurf eines Systems der Naturphilosophie* (*Introduction to the outline to a system of the philosophy of nature*) from 1799 'an unconscious productivity that originally related to the conscious, the mere reflex of which we see in nature [*eine bewußtlose, aber der bewußten ursprünglich verwandte Productivität, deren bloßen Reflex wir in der Natur sehen*]', and which should function as the starting point of a science equivalent to transcendental philosophy, namely the philosophy of nature (Schelling 1976 ff., 1.8: 30). Schelling thus supplements and then critically overcomes the transcendental philosophical activity of the ego postulated by Fichte (Schwenzfeuer 2012). As Goethe noted in August 1827, 'Schelling placed the object, the infinite breadth of nature, first [*das Object, die unendliche Breite der Natur, vorangestellt*]', while 'Fichte preferred to take up the subject'.[28] This approach of Schelling's, however, is based precisely on insights that go back to the intensive exchange with Goethe from 1798 onwards, in the course of which Schelling, as already indicated, received Goethe's concept of metamorphosis in particular (Schelling 1976ff., 1.7: 103f., 284–288; Schelling 1976ff., 1.8: 55f.),[29] in order to philosophically 'construct' an autonomous, self-(re-)producing nature that stands opposite the spirit, which is conscious of itself, as its necessary other. The kinship between Schelling and Goethe can be seen precisely in the fact that both, in conscious demarcation from Fichte's

[27] Schelling (1976 ff., 1.7: 95): 'Finally, what is to be said of *light*?—May it be, according to Newton, originally already decomposed into a number of different simple actions, the total impression of which is only white light—or may it be *originally* simple according to *Göthe*, in any case the polarity of the colours in every solar image is proof of a duality prevailing in the phenomena of light, the cause of which is still to be investigated [*Polarität der Farben in jedem Sonnenbild Beweis einer in den Phänomenen des Lichts herrschenden Dualität, deren Ursache noch zu erforschen ist*].' See also the entry in the hand copy in Schelling (1976ff., 1.7: 283): 'What proves more than anything else the relationship of light to electricity are the *prismatic* phenomena, as Goethe has set them out in his contributions. It is clear to me at least from these, and will perhaps soon be clear to others as well, that Newton's theory of white *light* as a composition of seven colored rays separated in a prism is wrong, that the prismatic phenomena are rather something far higher than a mere mechanical or chemical decomposition of light.'

[28] Goethe (1889–96, 6: 182 ['1827, 28. August. Mit Parthey']).

[29] Nassar (2010, 314–319), Breidbach (2006, 211–22).

rather marginal philosophy of nature, see nature *in itself* as generally characterised by 'polarity' and 'heightening [*Steigerung*]'.[30]

In this context, it is striking that Schelling first of all takes up Goethe's early conception, which the latter had repudiated around 1794, of a 'pattern [*Muster*]'[31] underlying the change of form. In the *Introduction to the outline* of 1799, he sees the product of nature 'as conceived in infinite metamorphosis [*als in unendlicher Metamorphose begriffen*]'. In order to avoid the appearance that this metamorphosis is completely irregular, Schelling at the same time presupposes 'a basic type [*einen Grundtypus*]', 'which lies at the basis of all [forms, *Gestalten*]—and which all of them, albeit with manifold variations [*mannichfaltigen Abweichungen*], express'. (Schelling 1976ff., 1.8: 55) In the *Entries in the Hand Copy*[32] of his *First Outline of a System of the Philosophy of Nature,* probably written somewhat later, the 'metamorphosis going to infinity [*ins Unendliche gehende Metamorphose*]' characterised by polarity and heightening appears in a completely different way (Schelling 1976ff., 1.7: 284): Schelling now emphasises, on the one hand, the radical nature of metamorphoses in nature, since in the case of the butterfly, for example, 'just the transition from one state of metamorphosis to another is not a mere partial, but a total change [*überhaupt nicht etwa eine bloße partielle, sondern eine totale Veränderung*]' (Schelling 1976ff., 1.7: 286). On the other hand, Schelling sees a tendency of nature towards the abolition of all individual forms, it 'constantly strives to abolish duality and to return to its original identity [*strebt beständig die Dualität aufzuheben und in ihre ursprüngliche Identität zurückzukehren*]': 'As soon, therefore, as the product has reached the highest summit, it is subject to nature's general striving towards indifference [*unterliegt es dem allgemeinen Streben der Natur nach Indifferenz*]' (Schelling 1976ff., 1.7: 287f.; Höfele 2019, 62f.). In this way, Schelling emphasises the productivity and dynamics of nature, which play with the emergence

[30] See e. g. Goethe (1887–1919, 2.11: 11), where he speaks to Chancellor von Müller of the 'two great driving wheels of all nature: the concept of *polarity* and of *heightening* [*zwei großen Triebräder aller Natur: der Begriff von Polarität und von Steigerung*], the former belonging to matter insofar as we think of it materially, the latter to it insofar as we think of it spiritually; the former is in perpetual attraction and repulsion, the latter in perpetual ascent [*in immerstrebendem Aufsteigen*].' See also Goethe's essay preceding the notebooks *Zur Morphologie* (1817) entitled *Die Absicht eingeleitet* (*The purpose set forth*) in Goethe (1887–1919, 2.6: 8–15, here 10 f.): 'Every living being is not an individual, but a majority; even insofar as it appears to us as an individual, it nevertheless remains an assembly of living independent beings that are equal in idea, in disposition, but can become equal or similar, dissimilar or dissimilar in appearance. These beings are partly already originally connected, partly they find and unite. They divide and seek each other again and thus bring about an infinite production in all ways and in all directions. The more imperfect the creature is, the more these parts are equal or similar to each other, and the more they resemble the whole. The more perfect the creature becomes, the more dissimilar the parts become. In the former case the whole is more or less like the parts, in the latter the whole is unlike the parts. The more similar the parts are to each other, the less subordinate they are to each other. The subordination of the parts points to a more perfect creature.'

[31] See note 18, above.

[32] For the dating see the editorial report by W. G. Jacobs and P. Ziche in Schelling (1976ff., 1.7: 37–40).

and decay of the individual, to such an extent that it can ultimately no longer be described on the basis of Goethean methodology. In this way, Schelling pushes the methodological approach of Goethe's doctrine of metamorphosis beyond itself.

However, as already noted, both pursue a similar methodological concern in that they advocate an 'empiricism' reaching beyond the Kantian-Fichtean application of transcendental philosophical categories to nature and the natural sciences; however, this 'empiricism' is applied in ultimately different ways by each thinker. As far as the adequate consideration of productive nature and its metamorphoses is concerned, both Goethe and Schelling advocate a passivity and self-retraction that in some respects, especially in Schelling's case, anticipates Schopenhauerian thinking of serenity, '*Gelassenheit*' (Hühn 2006). As Goethe put it in 1792/93 in the short essay *Der Versuch als Vermittler zwischen Subject und Object* (*The experiment as mediator between subject and object*), which was probably originally intended as an introduction to the *Beiträge zur Optik* (*Contributions to optics*), the scientist should 'take the data of evaluation not from himself, but from the circle of things that he observes [*nicht aus sich, sondern aus dem Kreise der Dinge*]' (Goethe 1887–1919, 2.11: 22). Here, Goethe develops the idea of a 'renunciatory divestment [*entsagenden Entäußerung*]'[33] in the course of observing natural scientific phenomena and in particular 'primordial phenomena', which, as mentioned, cannot be grasped within the framework of a hypothesis formation that does not 'listen' to nature. As Schelling analogously remarks in his treatise *Ueber das Verhältniß des Realen und Idealen in der Natur* (*On the Relationship between the Real and the Ideal in Nature*), added to the second edition of the *World Soul* in 1806, only 'this divine confusion and incomprehensible abundance of formations [*diese göttliche Verwirrung und unfaßliche Fülle von Bildungen*]' of nature itself would 'introduce the mere observer into the holy Sabbath of nature, after he has given up all hope of comprehending it with the intellect [*nachdem er alle Hoffnung aufgegeben sie mit dem Verstande zu begreifen*]' (Schelling 1856–1861, 2: 378). Only a 'silencing [*Verstummen*]' of reason (Schelling 1856–1861, 2: 378) or, as it is later called in Erlangen in 1821, an 'ecstasy [*Ekstase*]' of the ego (Schelling 1856–1861, 9: 230) as a self-abandonment and '*Gelassenheit*' is capable of allowing nature itself to appear in its absoluteness (Hühn 1994, 206–212; Höfele 2019, 94f. and 237–240). It is precisely with regard to this point of comparison between the two that Goethe and Schelling have unfolded a decisive history of influence, right up to Schopenhauer.

At the same time, however, this also indicates a decisive difference from Goethe: As Goethe notes in *The Experiment as Mediator Between Subject and Object,* he sees it as the *proprium* of research into nature to work through the individual experiments without gaps and 'in the most careful way and thus to form the experiences of the higher kind [*die Erfahrungen der höheren Art auszubilden*]' (Goethe 1887–1919, 2.11: 35), which he finally calls, as already mentioned, a 'generally pronounced main phenomenon, a basic and primordial phenomenon [*im Allgemeinen*

---

[33] Goethe (1987–1919, 2.11: 336). The manuscript's full addition to the previously quoted passage reads: 'The history of science teaches us how difficult this divestment (*Entäußerung*) is for human beings' (336).

ausgesprochene Haupterscheinung ein Grund- und Urphänomen]' (Goethe 1887–1919, 2.1: 72). In contrast, Schelling is already interested in expanding 'empiricism to unconditionality [*Empirismus zur Unbedingtheit*]' in 1799 (Schelling 1976ff., 1.7: 87). That 'the empiricism raised/extended to unconditionality [*Der Empirismus zur Unbedingtheit erhöht/erweitert*] is indeed *Naturphilosophie*. Schelling' (Goethe 1887–1919, 1.42.2: 525) is something that Goethe took note of, but at first probably did not adequately understand in its scope. According to Schelling, the product of nature always refers back to a productivity that he calls 'action [*Aktion*]', which alone shows and reveals itself in the product of nature. This productivity is therefore not directly recognisable, but must be recognised through the product: 'For our knowledge [*Erkenntniß*] does not reach beyond the product, and for the size of the action there can be no other expression than the product itself [*für die Größe der Action kann es keinen andern Ausdruck geben, als das Product selbst*].' (Schelling 1976ff., 1.7: 86f.) According to Schelling, the task of *Naturphilosophie* consists 'in recognising that in these actions which is unconditionally empirical [*das unbedingt-Empirische in diesen Actionen*]. For empiricism extended to unconditionality is, after all, *Naturphilosophie* [*der Empirismus zur Unbedingtheit erweitert ist ja Naturphilosophie*].' (Schelling 1976ff., 1.7: 87) It is necessary to look beyond the finitude and conditionality of empiricism in order to arrive at the ground of appearing which is actually unconditional and no longer appears itself. The method of what Schelling called 'dynamic atomistics [*dynamischen Atomistik*]' (Schelling 1976ff., 1.7: 86) does not, however, merge into transcendental philosophy, as Kant still presupposed in the *Metaphysical foundations of natural science* (Kant 1900ff., 4: 473f.),[34] but rather consists in the 'empirical test [*Probe*]', insofar as from the unconditional, non-appearing 'presupposition all natural phenomena [*alle Naturerscheinungen*] must be derivable' (Schelling 1976ff., 1.8: 35). The explanation from an unconditional proves to be correct only insofar as it fully explains experience.[35]

It is true that Schelling comes close to Goethe's methodology and his theory of the 'primordial phenomenon' around 1800 insofar as he assumes, as he does in the supplement to the second edition of the *World soul* in 1806, an 'imagining of the

[34] Nevertheless, Schelling concedes to Kant that—in contrast to the 'users of his principles'—he always kept to the limits of transcendental philosophy, that he 'nowhere dared to really construct the specific (qualitative) difference of matter from its two fundamental forces' (Schelling 1976ff., 1.7: 279).

[35] This concept of an 'unconditional empiricism' is still represented by Schelling according to its fundamental claim in his late 'positive philosophy', even if Schelling's 'conception of the system' has changed here in many ways. Schelling's 'higher, metaphysical empiricism' (see e. g. Schelling 1856–1861, 13: 114ff.) also shows the claim that behind empiricism, a justification of it must be sought, without this, however, being *a priori* derivable as in transcendental philosophy. In 1842/43, he finally equates—comparable to Schopenhauer—these dynamically understood figures of justification with the phenomena of 'decision and deed [*Entschluß und That*]' that go hand in hand with volition, which alone can 'justify an actual experience' and in this respect represent 'more than can be recognised in mere thinking [*mehr, als sich im bloßen Denken erkennen läßt*]' (Schelling 1856–1861, 13: 114).

whole into an individual [*Einbildung des Ganzen in ein Einzelnes*]' (Schelling 1856–1861, 2: 364), whereby 'in the temporal the eternal [*in dem Zeitlichen das Ewige*]' or 'unconditioned' emerges (Schelling 1856–1861, 2: 373). But he ultimately understands this in such a way that the appearing product refers back to something that can never appear in its productivity and—in the *System of transcendental idealism* of 1800—remains the 'eternally unconscious [*ewig Unbewußte*]' (Schelling 1976ff., 1.9.1: 299; Hühn 1994, 19–41; Höfele 2019, 75–80). In the 1823 Fragments entitled *Problem und Erwiderung* (*Problem and reply*), Goethe admits: 'Nature has no system, it is life and consequence from an unknown centre, to an unknowable boundary [*Leben und Folge aus einem unbekannten Centrum, zu einer nicht erkennbaren Gränze*]'; but for him it only follows from this that the 'contemplation of nature [*Naturbetrachtung*] is therefore endless, one may proceed by dividing it into the minutest details, or by following the track as a whole, according to breadth and height [*man mag in's Einzelnste theilend verfahren, oder im Ganzen, nach Breite und Höhe die Spur verfolgen*]' (Goethe 1887–1919, 2.7: 75). The idea that there could be something 'eternally unconscious', or that never appears, is completely beyond his grasp. Goethe, despite the 'mobility and flexibility' of natural vegetation, still generally assumes an 'original identity' to be described in his doctrine of metamorphosis. Schelling, on the other hand, in view of nature's 'infinite metamorphoses' and its all-encompassing striving for 'indifference', argues for a consistent understanding of nature as 'unconscious productivity', which as such cannot be translated into a 'conscious', conceptual understanding and made to appear adequately.

7.3 Nature as Process: Proximity and Distance Between Goethe, Schelling and Whitehead in the Face of a Nature to Be Grasped Holistically

Although A. N. Whitehead mentions Goethe rather negatively in conversations[36] and only indirectly mentions Schelling (Whitehead [1919] 1964, 47f.), he shares with the two central thinkers of the philosophy of nature around 1800 more than the dissociation from Newton and from the reason-centred transcendental philosophy. With Goethe and Schelling, he further shares in particular the concern for a holistic approach to nature and the search for an understanding of the dynamics and processuality of everything natural.

Whitehead already noticed the closeness of his understanding of nature to that of Schelling's in 1919 when, at the end of the second lecture of *The concept of nature,* he quoted at length from Schelling's *Über den wahren Begriff der Naturphilosophie*

[36] Whitehead (1954, 46, 122 and 181 f). Whitehead's remark that 'it has been occurring to me of late … that Goethe's thinking is too special' (122), however, suggests that Whitehead did appreciate Goethe before this 'late' insight.

(*On the true concept of philosophy of nature*)[37] and remarked that the topics addressed here, 'though they lie outside the range of our discussion, are always being confused with it', since they nevertheless 'lie proximate to our field of thought' (Whitehead 1964 [1919], 48). The fact that Whitehead still thinks he has to distinguish his concept of nature from that of Schelling has to do with the circumstance that he does not want his considerations presented in 1919 to be understood as 'metaphysical', but merely, in a rather Kantian fashion, as a 'necessary prolegomena for philosophy and for natural science', which consists in the 'understanding of types of entities, and types of relations among those entities' (Whitehead 1964 [1919], 48). In 1919, Whitehead wanted to leave out 'any pronouncement as to the psychological relation of subjects to objects or as to the status of either in the realm of reality' (Whitehead 1964 [1919], 47). According to Whitehead, 'natural philosophy' or 'philosophy of nature'[38] is concerned only with nature and not with the mind, for 'everything perceived is in nature' (Whitehead 1964 [1919], 29). In Whitehead's view, however, this does not mean that natural philosophy simply excludes a sub-area of reality; rather, with this thematic focus, he is concerned with avoiding the traditional 'bifurcation of nature into two systems of reality' (30), in which a distinction is made between a nature that causes knowledge and a nature that is perceived in sense cognition by the mind, or, in Kantian terms, between nature as a 'thing in itself' and as an appearance: 'Natural philosophy should never ask what is in the mind and what is in nature'; instead, it is necessary to ask about the 'relations *inter se* of things known' (30). Here, Whitehead is already aiming at his universal-organic cosmology, as he formulated it 10 years later in 1929 in *Process and Reality* (Löw 1990). In this work, he explicitly develops a 'speculative philosophy' in the sense of a metaphysics that claims to explain everything, which assumes an absolute relationality of all entities and in which the bifurcation into nature and mind, or into known things and the fact that they are known, is completely rejected. Here, therefore, Whitehead's motive for demarcation vis-à-vis Schelling is entirely suspended, thus revealing that from the beginning, it was only illusory. As with Schelling, there is no longer the 'necessity of opposing nature as constructed (i.e. experience) to real nature' (Whitehead [1919] 1964, 48; Schelling

[37] Whitehead cites the following passage: 'In the 'Philosophy of Nature' I considered the subject-object called nature in its activity of self-constructing. In order to understand it, we must rise to an intellectual intuition of nature. The empiricist does not rise thereto, and for this reason in all his explanations it is always *he himself* that proves to be constructing nature. It is no wonder, then, that his construction and that which was to be constructed so seldom coincide. A *Natur-philosoph* raises nature to independence, and makes it construct itself, and he never feels, therefore, the necessity of opposing nature as constructed (i.e. as experience) to real nature, or of correcting the one by means of the other.' (Whitehead 1964 [1919], 47f.; Schelling 1976ff., 1.10: 100). He quotes it from a work translated into English by the Russian philosopher Lossky (1919).

[38] Whitehead uses both the expression 'natural philosophy' and 'philosophy of nature', whereby he uses the former more frequently and on his own account, while he uses the latter expression more for the older philosophy of nature. Due to persistent misunderstandings, Whitehead is skeptical of the 'existing technical terminology of philosophy': 'The modern natural philosophy is shot through and through with the fallacy of bifurcation' (Whitehead [1919] 1964, vi).

1976ff., 1.10: 100). In 1929, both are drawn together into a unified phenomenon and understood as merely two aspects of the real: 'Each actuality is essentially bipolar, physical and mental, and the physical inheritance is essentially accompanied by a conceptual reaction'.[39] According to Whitehead, each actuality, and not only organic life, emerges passively from its environment in the form of a physical-causal 'inheritance'. At the same time, as an independent entity, it is able to 'react' more or less 'conceptually' to the environment,—resulting in a 'community of actual things' in the sense of an 'organism', which 'is not a static organism' and consequently 'is combined with that of 'process'' (Whitehead [1929] 1985, 214).

On a *methodological* level, this *ontological* assumption of a dynamic 'community of actual things' corresponds to the establishment of a 'speculative philosophy', which Whitehead describes in 1929 as 'the endeavour to frame a coherent, logical, necessary system of general ideas in terms of which every element of our experience can be interpreted' (Whitehead [1929] 1985, 4). In particular, the explanations of the adjectives 'coherent', 'logical' and 'necessary' characterising the system indicate that Whitehead certainly does not mean to be able to think of a self-transparent closure of a whole. Rather, he understands reality, as seen, as a process that ultimately cannot be fixed. The logical aspect of the system then only refers to the demand for freedom from contradiction, necessity is supposed to guarantee universality, while coherence asserts an ultimately unmanageable organic connection of all elements in the system, none of which can be thought in isolation and abstracted from the whole. With regard to those 'general ideas' or 'metaphysical principles' to be unified in the system, this also implies the following according to Whitehead: 'Philosophers can never hope finally to formulate these metaphysical first principles. Weakness of insight and deficiencies of language stand in the way inexorably' (Whitehead 1985 [1929], 4). In his endeavour to describe the dynamics of nature given in experience, Whitehead's conception of the system can thus be described in some respects as a union of Schelling's and Goethe's approach. On the one hand, in view of the processual reality, he assumes a necessary incompleteness of his system and the withdrawal of the 'metaphysical first principles'. On the other hand, he also assumes that, due to the coherence of each element in the system with the whole, each individual refers to the whole and represents it in a way reminiscent of Leibniz's monads, but without this whole ever being able to exist in the form of a completeness of the system elements.

Despite all their differences, Goethe, Schelling and Whitehead think of nature in its productivity and dynamics as something that cannot be completely defined. Thus they demonstrate their modernity regarding natural sciences and their preoccupation with molecular, atomic and even subatomic structures that are not immediately visible and graspable. In the sense of Whitehead's search for a 'hint of that aspect of self-production, of generation, of φύσις, of *natura naturans*' mentioned at the beginning, Schelling in particular is interested in describing the phenomenon of the

[39] Whitehead ([1929] 1985, 108). See on the parallelism with Schelling in this respect also Höfele (2019, 98 f.).

natural in its freedom and productivity that cannot be attributed to the spiritual alone. Whitehead could therefore have found points of reference for his own process-philosophical thinking not only with regard to Plato's *Timaeus,* but also by referring to the *Naturphilosophie* around 1800.

In this respect, all three thinkers mentioned here are able to provide a *corrective backdrop* for the negative phenomena of the present named in the 'Anthropocene' discourse. With a view to a holism encompassing human beings and nature, which also calls attention to their complex interactions, Goethe, Schelling and Whitehead break through the boundery between nature and that which belongs to the realm of the human in a comparable manner. However, this is not done *from the human perspective,* as in the case of the Anthropocene, but in exactly the opposite way, namely *from nature.* In Goethe's case, this takes the form of a 'contemplation' that goes hand in hand with a withdrawal of the self and '*Entäußerung*'. The natural scientist must withdraw himself in order to allow the natural phenomena to appear from within. In Schelling's, on the other hand, it takes the form of a 'silencing [*Verstummen*]' and passive 'fathoming' that goes back to a ground that is only hinted at in the phenomenal. Schelling thus radicalizes Goethe's approach by seeing nature's own life as so independent of human beings that they are ultimately unable to grasp it intellectually. With this methodological movement, both are concerned with that which Whitehead, in contrast to the Kantian transcendental philosophical approach, understands as the *proprium* of his own philosophy: 'For Kant, the world emerges from the subject; for the philosophy of organism, the subject emerges from the world—a 'superject' rather than a 'subject'' (Whitehead 1985 [1929], 88).

According to all three authors, a productivity and dynamism should become visible that characterises the whole of nature and life, from which the human and the spiritual emerge fluidly and without rupture. Both are integrated into the whole of nature rather than representing the other in relation to it. If the *Naturphilosophie* around 1800 as well as Whitehead's process philosophy are exemplary for the present natural sciences precisely in this respect, as they attempt to think the unity of the natural and the human-spiritual without usurping the former by the latter, but rather reconstituting the orientation towards the natural, this would certainly not be the least of their merits and for this reason alone they would be worthy of an orienting update.

References

Adler, Jeremy. 1998. The aesthetics of magnetism: Science, philosophy and poetry in the dialogue between Goethe and Schelling. In *The third culture. Literature and science*, ed. Elinor S. Shaffer, 66–102. Berlin: De Gruyter.

Bach, Thomas, and Olaf Breidbach, eds. 2005. *Naturphilosophie nach Schelling*. Stuttgart-Bad Cannstatt: Frommann-Holzboog.

Bahr, E. 1998. Ironie. In Goethe Handbuch. Ed. Bernd Witte et al. Vol. 4.1: *Personen, Sachen, Begriffe A–K*, ed. Hans-Dietrich Dahnke and Regine Otto, 443–445. Stuttgart / Weimar: J. B. Metzler Verlag.

Behler, Ernst. 1992. *Frühromantik*. Berlin: De Gruyter.
Breidbach, Olaf. 2006. *Goethes Metamorphosenlehre*. München: Wilhelm Fink Verlag.
Breidbach, Olaf, and Paul Ziche, eds. 2001. *Naturwissenschaften um 1800. Wissenschaftskultur in Jena–Weimar*. Weimar: Verlag Hermann Böhlaus Nachfolger Weimar.
Crutzen, Paul J. 2002. Geology of mankind. *Nature* 415:23.
Crutzen, Paul J. and Eugene F. Stoermer. 2000. The 'Anthropocene'. *Global Change Newsletter* 41:17–18.
Dierks, Jan. 2016. Holismus. In *Handbuch Umweltethik*, ed. Konrad Ott, Jan Dierks, and Lieske Voget-Kleschin, 177–183. Stuttgart: J. B. Metzler-Verlag.
Falkenburg, Brigitte. 2017. Mathematisierung der Natur und ihre Grenzen. In *Naturphilosophie. Ein Lehr- und Studienbuch*, ed. Thomas Kirchhoff et al., 32–41. Tübingen: University of Tübingen Press.
Fröschle, Hartmut. 2002. *Goethes Verhältnis zur Romantik*. Würzburg: Verlag Königshausen & Neumann.
Goethe, Johann Wolfgang von. 1887–1919. *Werke*. 143 vols. in 4 sections, ed. on behalf of the Grand Duchess S. v. Sachsen. Weimar: Hermann Böhlau.
———. 1889–96. *Gespräche*. 10 vols. Leipzig: G. W. von Biedermann.
Guzzoni, Ute. 1995. *Über Natur. Aufzeichnungen unterwegs: Zu einem anderen Naturverhältnis*, 81–115. Freiburg: Verlag Karl Alber.
Ho, S. C. 2012. Kant, Immanuel (1724–1804). In Goethe Handbuch, ed. Bernd Witte et al. Supplementa 2: *Naturwissenschaften*, ed. Manfred Wenzel, 490–492. Stuttgart: J. B. Metzler-Verlag.
Höfele, Philipp. 2019. Wollen und Lassen. In *Zur Ausdifferenzierung, Kritik und Rezeption des Willensparadigmas in der Philosophie Schellings*. Freiburg: Verlag Karl Alber.
———. 2020. The changed role of anthropology in the Anthropocene. In *Le tecnologie 'morali' emergenti e le sfide etico-giuridiche delle nuove soggettività. Emerging 'moral' technologies and the ethical-legal challenges of new subjectivities*, ed. Silvia Salardi and Michele Saporiti, 125–144. Turin: Giappichelli.
———. 2021. Schelling–Goethe–Schopenhauer: Zur holistischenBetrachtung der Natur in der 'Sattelzeit' um 1800. In *Schopenhauer liest Schelling. Freiheits- und Naturphilosophie im Ausgang der klassischen deutschen Philosophie. Mit einer Edition von Schopenhauers handschriftlichen Kommentaren zu Schellings 'Freiheitsschrift'*, ed. Philipp Höfele and Lore Hühn, 163–195. Stuttgart-Bad Cannstatt: Frommann-Holzboog.
Höfele, Philipp, and Lore Hühn, eds. 2021. *Schopenhauer liest Schelling. Freiheits- und Naturphilosophie im Ausgang der klassischen deutschen Philosophie. Mit einer Edition von Schopenhauers handschriftlichen Kommentaren zu Schellings 'Freiheitsschrift'*. Stuttgart-Bad Cannstatt: Frommann-Holzboog.
Höppner, Stefan. 2017. *Natur/Poesie. Romantische Grenzgänger zwischen Literatur und Naturwissenschaft – Johann Wilhelm Ritter, Gotthilf Heinrich Schubert, Henrik Steffens, Lorenz Oken*. Würzburg: Verlag Königshausen & Neumann.
Hühn, Lore. 1994. *Schelling und Fichte. Oder: Über die Grenze menschlichen Wissens*. Stuttgart / Weimar: J. B. Metzler-Verlag.
———. 2006. Der Wille, der Nichts will. Zum Paradox negativer Freiheit bei Schelling und Schopenhauer. In *Die Ethik Arthur Schopenhauers im Ausgang vom Deutschen Idealismus (Fichte / Schelling)*, ed. Lore Hühn, 149–160. Würzburg: Ergon-Verlag.
Kant, Immanuel. 1900 ff. *Gesammelte Schriften*, ed. Berlin-Brandenburgischen Akademie der Wissenschaften. Berlin: de Gruyter.
Karafyllis, Nicole. 2019. Interaktionen in der Technosphäre und Biofrakte. In *Mensch-Maschine-Interaktion. Handbuch zu Geschichte–Kultur–Ethik*, ed. Kevin Liggieri and Oliver Müller, 106–113. Berlin / Heidelberg: J. B. Metzler-Verlag.
Köchy, Kristian. 1997. *Ganzheit und Wissenschaft. Das historische Fallbeispiel der romantischen Naturforschung*. Würzburg: Verlag Königshausen & Neumann.

———. 2017. 'Kampf' um die Naturphilosophie. In Thomas Kirchhoff et al., eds. *Naturphilosophie. Ein Lehr- und Studienbuch*. 57–65. Tübingen: University of Tübingen Press.
Lossky, Nikolai Onufrievich. 1919. *The intuitive basis of knowledge. An epistemological inquiry*. Translated by Nathalie A. Duddington. London: Macmillan and Co.
Löw, Reinhard. 1990. Essay. In Alfred N. Whitehead. *Der Begriff der Natur*. Translated by J. v. Hassell, 159–73. Weinheim: VCH.
Müller, Olaf L. 2014. Goethes Pech mit Schelling. Optimistische Blicke auf ein ideengeschichtliches Fiasko. In *Nature and realism in Schelling's philosophy*, ed. Aandrea Dezi and Emilio C. Corriero, 131–185. Rome: Accademia University Press.
Muraca, Barbara. 2010. *Denken im Grenzgebiet. Prozessphilosophische Grundlagen einer Theorie starker Nachhaltigkeit*. Freiburg: Verlag Karl Alber.
Nassar, Dalia. 2010. From a Philosophy of Self to a Philosophy of Nature: Goethe and the Development of Schelling's Naturphilosophie. *Archiv für Geschichte der Philosophie* 92,3: 304–321.
Newton, Iasaac. 1704. *Opticks: Or, a treatise of the reflexions, refractions, inflexions and colours of light*. London: Samuel Smith and Benjamin Walford.
Newton, Isaac. 1713. *Philosophiae naturalis principia mathematica* (1687). Cambridge: publisher unknown.
Rehbock, Theda. 1995. *Goethe und die 'Rettung der Phänomene'. Philosophische Kritik des naturwissenschaftlichen Weltbilds am Beispiel der Farbenlehre*. Konstanz: Verlag am Hockgraben.
———. 2014. Johann Wolfgang von Goethe. In *Schopenhauer-Handbuch. Leben–Werk–Wirkung*, ed. Matthias Koßler and Daniel Schubbe, 226–230. Stuttgart: J. B. Metzler-Verlag.
Schelling, Friedrich Wilhelm Joseph. 1856–1861. *Sämmtliche Werke*, ed. Karl F. A. Schelling. Stuttgart: Cotta.
———. 1972. In *Grundlegung der positiven Philosophie*, ed. Horst Fuhrmans. Turin: Bottega d'Erasmo.
———. 1976. *Historisch-kritische Ausgabe*, ed. Projekt Schelling–Edition und Archiv der Bayerischen Akademie der Wissenschaften. Stuttgart-Bad Cannstatt: Frommann-Holzboog.
———. 1994. *'Timaeus'* (1794). *Mit einem Beitrag von Hermann Krings: 'Genesis und Materie–Zur Bedeutung der 'Timaeus'-Handschrift für Schellings Naturphilosophie'*, ed. Hartmut Buchner. Stuttgart-Bad Cannstatt: Frommann-Holzboog.
Schneider, Birgit. 2019. Menschen-Maschine-Schnittstellen in Technosphäre und Anthropozän. In *Mensch-Maschine-Interaktion. Handbuch zu Geschichte–Kultur–Ethik*, ed. Kevin Liggieri and Oliver Müller, 95–106. Berlin: J. B. Metzler-Verlag.
Schwenzfeuer, Sebastian. 2012. *Natur und Subjekt. Die Grundlegung der schellingschen Naturphilosophie*. Freiburg: Verlag Karl Alber.
Seibt, Johanna. 2003. *Process theories: Crossdisciplinary studies in dynamic categories*. Dordrecht: Kluwer Academic Publishers.
Steffen, Will, et al. 2011. The Anthropocene: Conceptual and historical perspectives. *Philosophical Transactions of the Royal Society A* 369 (1938): 842–867.
———. 2015. The trajectory of the Anthropocene: The great acceleration. *The Anthropocene Review* 2 (1): 81–98.
Steigerwald, Joan. 2019. *Experimenting at the boundaries of life. Organic vitality in Germany around 1800*. Pittsburgh: University of Pittsburgh Press.
Wenzel, Manfred and C. Schweizer. 2012. Urphänomen. In Bernd Witte et al., eds. *Goethe Handbuch*. Supplementa 2. *Naturwissenschaften*, ed. Manfred Wenzel, 680–682. Stuttgart: J. B. Metzler-Verlag.
Whitehead, Alfred N. 1954. *Dialogues of Alfred North Whitehead, as recorded by Lucien Price*. Boston: David R. Godine.
———. 1964. *The concept of nature. The Tarner lectures*. Delivered in Trinity College, November 1919. Cambridge: Cambridge University Press.

———. 1985. *Process and reality. An essay in cosmology. Gifford lectures delivered in the University of Edinburgh during the session 1927/28* (1929), ed. David R. Griffin and Donald W. Sherburne. New York: Free Press.

Philipp Höfele is a postdoctoral fellow of the Young Academy for Sustainability Research at the Freiburg Institute for Advanced Studies (FRIAS), University of Freiburg, and board member of the International Schelling Society. He is author of *Wollen und Lassen. Zur Ausdifferenzierung, Kritik und Rezeption des Willensparadigmas in der Philosophie Schellings* (Alber, 2019), co-editor of the international journal *Schelling-Studien* (Alber), and of the special issue *Nature in the Anthropocene* (*The Anthropocene Review*, 2022).

Chapter 8
Schelling's Principle of Life and the Unity of Mind and Nature

Lara Ostaric

Abstract In his Critique of Teleological Judgment, Kant addresses an apparent 'contradiction' concerning organic formations: organisms are governed by universal mechanical laws yet are end-directed, that is, they exhibit purposiveness. Mechanical, or 'real', causality pertains to nature, or the world. 'Ideal' causality pertains to the mind. Kant resolves the antinomy with the idea of the 'supersensible, on which we must base nature as phenomenon', and in which both types of causalities originate.

Schelling, I contend, follows closely Kant's formulation of the antinomy. His aim, however, is to show that we can have an empirical *cognition* of these objects, as well to offer what Kant calls a scientific 'explanation' and not a mere 'elucidation' of their constitution and function. Showing this will entail an ontological extension of Kant's own solution to the antinomial conflict, namely, a unity of mechanical and teleological causes by the principle of life, the 'world-soul', which has a constitutive and not a regulative status and which permeates both inorganic and organic nature. This unity is possible only insofar as both inorganic and organic nature exhibit the prior unity of mind and world in the Absolute.

In his Critique of Teleological Judgment, Kant addresses what appears to be a 'contradiction' (Kant 2000 [1790], §70, 259) pertaining to organic formations: organisms are a part of nature governed by universal mechanical laws while at the same time, they are end-directed, that is, they exhibit purposiveness. The former, the mechanical, or the 'real' (§65, 244) type of causality, pertains to nature or the world. The latter, or the 'ideal' (§65, 244) type of causality, pertains to the mind. In his

L. Ostaric (✉)
Philosophy Department, Temple University, Philadelphia, PA, USA
e-mail: lostaric@temple.edu

P. Cheyne (ed.), *Matter and Life in Coleridge, Schelling, and Other Dynamical Idealists*, International Archives of the History of Ideas Archives internationales d'histoire des idées 253, https://doi.org/10.1007/978-3-031-78315-9_8

antinomy of teleological judgment, Kant summarizes the contradiction that ensues with our reflection on organic formations. He resolves the antinomy with the Idea of the 'supersensible, on which we must base nature as a phenomenon' (§78, 281) and in which both types of causalities originate. For Kant, the supersensible substrate of nature is a regulative idea relative to the needs of our discursive understanding and not an object of our empirical knowledge.

Schelling, I contend, follows closely Kant's formulation of the antinomial conflict insofar as for him, just as for Kant, mechanical (real) and teleological (ideal) causes in organisms are distinct and not reducible to one another. He, however, wishes to move beyond Kant and to offer a perspective that is, in his view, truly 'critical'. Put differently, given that organisms are a part of nature, his aim is to show that we can have an empirical *cognition* of these objects, as well as be able to offer what Kant calls a scientific 'explanation' (*Erklärung*) (Kant [1790] 2000, §78, 281) and not a mere 'elucidation' (*Erörterung*, §78, 281) of their constitution and function. Showing this will entail an ontological extension of Kant's own solution to the antinomial conflict, namely, a unity of mechanical and teleological causes by the principle of life, the 'world-soul', which has a constitutive and not a regulative status and which permeates both inorganic and organic nature. This unity is possible only insofar as both inorganic and organic nature exhibit the prior unity of mind and world in the Absolute.

In this essay, I will proceed as follows. In Sect. 8.1, I lay out Kant's antinomy of teleological judgment and in Sect. 8.2 I discuss Kant's solution. In Sect. 8.3, I explain how Schelling follows Kant's understanding of the antinomy insofar as he, like Kant, avoids the dogmatic reduction of real causes to ideal and vice versa. However, Schelling also points out the insufficiency of the Critical approach insofar as for him uniting the ideal and the real is only possible in our cognition of their common origin in the Absolute. In Sect. 8.4, I lay out Schelling's method of reaching the cognition of this common unity in the Absolute, that is, his 'construction' of the principle of life. In the last section, Sect. 8.5, I offer a conclusion and a brief evaluation.

8.1 Kant's Antinomy of Teleological Judgment

Kant contends that determinative judgment cannot generate antinomy because 'it has [for itself] no principles that ground concepts of objects' (Kant 2000 [1790], §69, 257).[1] Put differently, unlike teleological judgment, determinative judgment does not have its own principle (such as the principle of objective purposiveness) according to which it can generate a concept (such as the one of a 'natural end'). Determinative judgment 'merely subsumes under given laws and concepts' (§69,

[1] I modified Paul Guyer and Eric Matthews's translation to show better the emphasis in the original German text that these are reflective judgment's *own* principles, according to which, in its reflection, it generates concepts of objects.

257), namely, those of the understanding. Because it does not have its own principles, which could be contradictory, it cannot generate antinomy.

The principle of reflective judgment, on the other hand, does not determine the object as it is in itself but serves as a rule for reflection on the manifold relative to the needs of our human cognitive capacities and their limitations. Thus, Kant claims that this principle is not 'objective' but a 'merely subjective principle for the purposive use of the cognitive faculties' (§69, 257). This is also the reason why Kant calls both the thesis and the antithesis of the antinomial conflict 'maxims' (§69, 257) of the reflecting power of judgment. For the task of, for example, providing the necessary unity of the particular laws of nature, reflective judgment can employ two maxims: one that is given to it by 'the mere understanding a priori' (§70, 258) and another that is 'suggested to it by particular experiences that bring reason into play' (§70, 258), that is, the experiences of organic formations in nature. Kant therefore formulates the antinomy as follows. Thesis: 'All generation of material things and their forms must be judged as possible in accordance with merely mechanical laws' (§70, 258). Antithesis: 'Some products of material nature cannot be judged as possible according to merely mechanical laws (judging them requires an entirely different law of causality, namely that of final causes)' (§70, 258–9).

Commentators find puzzling the fact that the thesis assigns mechanical principles a merely regulative status and thereby contradicts the conclusions of the Analytic of the *Critique of Pure Reason*.[2] Kant is unfortunately not clear about the meaning of 'mechanical' in the antinomy. However, the meaning of 'mechanical laws' in the thesis of the antinomy is not the one of the second analogy according to which all events in the phenomenal world are causally determined according to universal laws (but not any particular empirical law). His aim is not to regard causal relations as a law that *any object* if it is to be an object of a possible experience must meet. It is also not the meaning of 'mechanism' given in the *Metaphysical Foundations*, that is, the laws that determine the relation between two bodies in space insofar as they are *material bodies in general*.[3] Kant's concern here is to account for the unity or the whole, which presupposes the relation of the parts, or more precisely, to give an account of an organic whole that presupposes the particular function and properties of the parts. And 'mechanism' is a specific way in which *our* understanding relates the parts to the unity of the whole. Our understanding explains the connection of the parts in the whole by the employment of real/efficient causes (e.g., the whole is like a sand dune explicable in terms of the properties and motions of matter) and by the employment of ideal/final causes (e.g., 'the purpose of a watch is to tell time') which further determines the position of the parts within the whole. In the former case, the whole is the unity of its parts. In the latter case, the whole is the unity of its parts but the positioning of the parts depends on the 'representation of a whole' (Kant [1790] 2000, §77, 277), the concept of the object.

[2] See e.g. Butts (1984).

[3] This point is argued in McLaughlin (1990) and Allison (1991). Watkins (2008) outlines their disagreement with Butts (1984).

This particularity is specific to our discursive, 'image-dependent understanding (*intellectus ectypus*)' (§77, 277), and may not be constitutive of nature.

Therefore, it would be a mistake to conceive of the antinomy as the conflict between efficient and final causality. If that were the case, then the antinomy could be easily resolved on its own: 'mechanism', conceived as a series of efficient causes, would be necessary but not sufficient. In special cases, namely, those that present us with organic formations, it would have to be supplemented with the use of final causes (Zanetti 1993, 351). But the whole of an organism is both unlike a mechanical heap of matter and unlike an artifact. The principle of its organization is 'not analogous with any causality that we know' (Kant 2000 [1790], §65, 246). Thus, if the antinomial conflict is to be a genuine conflict that cannot be resolved on its own, then by 'mechanism' in the thesis Kant must understand the connection of the parts in the whole that pertains to the generation of a mechanical heap of matter *and* the one that pertains to the generation of an artifact. The antithesis then implicates the explanatory failure of any causality available to us (both real and ideal) and the necessary need to represent organic formations as if they were the ends of a mind more powerful than ours.

Moreover, if the antinomy is to generate a genuine conflict, both thesis and antithesis must be necessary.[4] Kant's claim that unless the maxim of the thesis is 'not made the basis for research then there can be no proper cognition of nature' (Kant 2000 [1790], §70, 259) indicates that we are constrained to explaining nature in mechanistic terms relative to the limitations of our discursive human understanding. Our understanding, as discussed above, is discursive, meaning, it must 'progress from the parts, as universally conceived grounds, to the different possible forms, as consequences, that can be subsumed under it'. Kant claims further that 'In accordance with the constitution of our understanding … a real whole of nature is to be regarded only as the effect of the concurrent moving forces of the parts' and that we 'represent the possibility of the whole as depending upon the parts, as is appropriate for our discursive understanding' (§77, 277). It is also necessary, given the particularity of our discursive understanding, that when confronted with some special cases, namely organisms, our mechanical way of explaining the connection of particular empirical laws is insufficient for explaining the lawfulness of the whole and that we are compelled to employ the principle of objective purposiveness in order to represent the connection of the parts in the whole as lawful and necessary and not merely contingent.

Thus, given the above, I shall contend that the antinomial conflict is best construed as the conflict between two necessary perspectives on nature that we are asked to take simultaneously when confronted with organic formations: that of the analytic procedure of the understanding (from concepts to given empirical intuition and according to which the whole is the sum of its parts) and the other of the synthetic procedure of reason (the idea of an intuitive understanding that proceeds from

[4] See this point emphasized in McLaughlin (1990, 156–157) and Allison (1991, 28). For Kant's claim that reflective judgment's maxim to research nature according to the mechanistic principle is necessary for the cognition of nature, consider Kant 2000 [1790], §78, 279, 281.

the intuition of the whole as such so that the possibility of the parts is represented as depending on the whole as their ground).[5] The former represents nature as governed by the theoretical-mechanical lawfulness of the understanding presupposed by the method of science and, therefore, as governed by necessity. The latter represents it as governed by the lawfulness of reason, namely, freedom and, therefore, as governed by contingency. This is because if we are required to represent organic formations as ends of nature, then we are also required to represent the whole of nature as *creation*, a system of ends with the human being and the highest good as its final end.[6]

To unite 'Two such dissimilar principles as nature and freedom' (Kant 2000 [1790], 342, 'General Remark on the Teleology')—i.e. the principles of theoretical and practical reason, respectively—in the representation of one single object of nature, we must offer a justification.

8.2 Kant's Solution to the Antinomial Conflict

According to Kant, the solution to the antinomy should unfold as follows: 'Now of course the principle of the mechanism of nature and that of its causality according to ends in one and the same product of nature must *cohere* in a single higher principle and flow from it in common, because otherwise they could not subsist alongside one another in the consideration of nature' (Kant 2000 [1790], §78, 281, my emphasis). Thus, when it comes to the explanation of the possibility (the origin) of *one single thing* (organism) mechanism and teleology exclude each other: a thing in nature can either be a product of mechanical generation (a product of real and ideal causes, a heap of matter or an artifact), or we can think of it as a product of a mind more powerful than ours that we can only represent in '*remote* analogy with our own causality in accordance with ends' (Kant 2000 [1790], §65, 246; my emphasis) as a product of divine creation. Yet, organisms, require for their intelligibility and their explanation both principles. In a 'natural end', parts must be combined into a whole 'by being reciprocally the cause and effect of their form' (§65, 245). The whole or the form of the organisms is the effect of the properties and functions of the parts (i.e. mechanism) but the whole is also at the same time the cause of the function and properties of the very same parts of which it is the effect (i.e. teleology which entails the concept of an end 'without the causality of the concepts of a rational being outside it' (§65, 245). Their simultaneous explanatory employment can be done in a justified manner only if they are regarded as originating (and thus uniting) in one higher common principle. This principle is a 'justifying [*berechtigende*] principle of the commonality of the maxims of natural research' (§78, 281, translation slightly modified). Put differently, regarding both principles as originating in one

[5] I defend this interpretation of Kant's antinomy of teleological judgment in Ch. 8 of Ostaric 2023.

[6] See Kant (2000 [1790], §67, 250 and §§82–4).

higher principle justifies us in taking these principles as having explanatory compatibility which also secures the unity of the organisms' natural lawfulness. The fact that we take these principles as originating in one common principle does not mean that we can derive the possibility of organisms (i.e. explain them) from this common principle. This common principle is the supersensible, which cannot be an object of our theoretical knowledge. Thus, for us, 'from the human point of view' (§76, 273), these two principles must forever remain separate, but transcendental idealism leaves 'at least a possibility that they may be objectively unifiable in one [supersensible] principle' (§76, 273).

Thus, when we consider organisms with respect to their possibility, then, in order to justify the simultaneous employment and explanatory compatibility of both mechanical (theoretical view of nature) and teleological principles (the view of nature from a practical perspective) in our representation of a single organism, we must view them *as if* originating in a single supersensible principle. Thus, reflective judgment's representation of organisms with respect to their possibility creates an image of the world according to which theoretical (mechanical and necessary) and practical (teleological and contingent) perspectives stand in a special relation of 'coherence'. Put differently, reflective judgment creates an image according to which theoretical and practical representations of nature, each governed by two entirely 'dissimilar principles' (Kant 2000 [1790], 342, 'General Remark on the Teleology') (i.e., the law of nature and the law of freedom, respectively) 'must cohere' (*müssen zusammenhängen*) (§78, 281).

8.3 Schelling's Conception of the Antinomial Conflict: The Inadequacy of the Dogmatic and the Critical Approaches to the Antinomy

For Schelling, as for Kant, the antinomy is the opposition between mechanism and organism. In other words, by 'mechanism' he understands the procedure of the understanding which presupposes both the efficient causality and ends that are conceived as 'external' to the object, as in artifacts. Organisms, on the other hand, presuppose a special form of causality that is unique to them. Schelling understands himself as a follower of Kant's Critical philosophy. Therefore, he maintains that in the solution to the antinomial conflict, it is important to acknowledge the distinction of both the ideal and mechanical causes. Hence, the possibility of an organism as a *natural end* must be sought in their unification and not in reducing one to the other and thereby favoring either real causes (nature) or ideal causes (the purposiveness of mind). In his *Ideas for a Philosophy of Nature* (1797), he writes as follows: 'First of all you must concede that here the talk is of a unity' (Schelling 1988 [1797], introduction, 31). This is why Schelling criticizes the two 'dogmatic' approaches to the antinomial conflict: hylozoism, which reduces ideal to real causes, and physico-theology, which reduces real to ideal causes.

In *On the World-Soul*, written in 1798, a year after his *Ideas*, Schelling criticizes hylozoism:

> It is the culmination of all anti-philosophy (*unphilosophie*) to contend that life is a property of matter and to invoke against all the universal laws of inertia that we can find an example of exception, namely, a living matter! Then [...] one already presupposes in the concept of animal matter the cause of life which continuously acts on the animal substance (that is also freely in the animal matter) and then one has with analysis an easy job; but the question here is not an analysis but a synthetic construction of a concept: of the concept of animal life. (Schelling 2000 [1798], AA I.6: 187)[7]

In the opening lines of the passage, Schelling closely follows Kant's criticism of hylozoism, a branch of vitalism, in the third *Critique* according to which the structure and organization of organic beings are explained by appealing to the properties of matter itself (a natural cause), but where each particle of matter was conceived as a 'little life', capable of acting in accordance with ends (Kant 2000 [1790], §65, 246). It is 'anti-philosophy' because it contradicts the proven universal laws of inertia founded on the concept of 'dead matter'. It is also 'anti-philosophy' because it presupposes an 'analytic approach', that is, our cognition of life can be reached merely by analyzing the concept of matter. A proper philosophical approach, one that conceives of cognition in the spirit of Kant's Critical philosophy, must be synthetic. There must be a concept but also an intuition corresponding to it in virtue of which it can be shown that the concept has objective reality.

Later, in his *System of Transcendental Idealism* (1800), Schelling further elaborates on his critique of hylozoism as one of the two 'dogmatic' approaches to the problem of explanatory compatibility of mechanism and teleology in organic formations:

> Now this contradiction, whereby one and the same product is at once a blind product, and yet is purposive, is utterly inexplicable in any system except that of transcendental idealism, inasmuch as every other denies either the purposiveness of the products, or the mechanism involved in bringing them about, and so must do away with this same coexistence. One possibility is to suppose that matter shapes itself automatically into purposive products, whereby it at least becomes intelligible how matter and the concept of purpose interfuse in the products; and then one either ascribes absolute reality to matter, as happens in hylozoism, a nonsensical system, inasmuch as it supposes *matter itself* to be intelligent ... (Schelling 1978 [1800], 216)

Thus, hylozoism would be a 'dogmatic' approach for explaining the contradiction of how a single object can be a product of nature (of 'blind' necessity) and at the same time be directed towards ends or goals, that is, 'purposive'. Hylozoism can be successful in explaining how the two types of causes can co-exist in one single object only insofar as it reduces the ideal causes (purposiveness) to the mechanism. Put differently, for hylozoism, the end-directedness of organic formations is the

---

[7] The cited passages from Schelling's *On the World-Soul* are my own translation. They are cited first by series, followed by the volume and page number of the collected edition of Schelling's works by the Bavarian Academy of Sciences (AA).

working of the matter itself qua intelligence. By claiming that all reality must be reduced to matter, hylozoism is an example of dogmatic realism.

> 'The other possibility', continues Schelling,
>
> is to suppose matter to be absolutely inert, and to have the purposiveness in its products brought about by an intelligence outside it, in such a way that the concept of this purposiveness must have preceded production itself; but then there is no seeing how concept and object can have been everlastingly interfused, or how—in a word—the product can be a work, not of artifice but of nature. For the difference between artifact and natural product resides precisely in this, that in the former the concept is impressed only upon the surface of the object, while in the latter it has gone over into the object itself and is utterly inseparable therefrom. (Schelling 1978 [1800], 216–17)

The other 'dogmatic' approach to explanatory compatibility of mechanism and teleology is the one that reduces mechanical causes to ideal causes, as in physico-theological approaches. While hylozoism fails to explain how an organism is an end-directed formation, physico-theology fails to explain how an organism is a *natural* end, an end-directed formation that is a part of nature, and not a mere artifact of divine intelligence. This is a type of idealist dogmatism.

In both cases, hylozoism and physico-theology, the cognition of organisms is analytic. In hylozoism, we explain 'life' by analyzing the concept of matter. In physico-theology, we explain 'life' from the concept of a divine designer. Both are incapable of explaining how it is that we have an *empirical cognition* of organisms. Following Kant, for such a cognition, we need both concepts and intuition. Already in *Ideas* (1797), Schelling criticizes physico-theology as incapable of explaining our empirical cognition of organisms:

> Let us assume it is through the creative power of a divinity that the whole system of Nature arose and with it all the diversity of purposive products external to us. Have we in fact advanced even a single step farther than before, and do we not find ourselves once again at the same point from which we set up at the beginning? How organic products external to, and independent of, me have actually come to be was not at all what I required to know; for how could I even form a clear idea of that for myself? The question was: how the *representation* of purposive products outside me has got into me, and how, *although it pertains to things only in relation to my understanding*, I am nevertheless compelled to think of this purposiveness as *actually outside me* and necessary. This question you have not answered. (Schelling 1988 [1797], introduction, 34)

Consistent with the limits of Kant's Critical philosophy, Schelling denies the possibility of knowing the *origin* of organisms or of 'how they have actually come to be' (Schelling 1988 [1797], introduction, 34). The latter, because of the special unity of final and mechanical causality, suggests the Idea of the supersensible. As a transcendental philosopher, he takes his objective to be to explain the condition of the possibility of having a 'representation' (*Vorstellung*) of an organism as an object of nature that is purposive but that is at the same time objectively real or 'actually outside me' and the purposiveness of which is 'necessary' (expressing universal lawfulness).[8] This is something that cannot be explained by appealing to the power of a divinity.

While Schelling agrees with Kant that the solution to the antinomial conflict cannot be found by either reducing ideal to real causes or vice versa, he argues against Kant that the possibility of special unification of mechanical and final causes in organisms, or life, is to be sought not 'in our idea, but in the object itself' (Schelling 1988 [1797], introduction, 31). Thus, Kant's solution to the problem—namely, that mechanism and life are unified in the *Idea* of the supersensible and, furthermore, that by means of reflective judgment, we should regard an organism *as if* purposive and *as if* originating in the supersensible—is not a satisfactory solution for Schelling. For Schelling, this unifying principle must be constitutive and not regulative, that is, it must be true of the world.

To build on Kant's Critical philosophy while at the same time moving beyond it, Schelling appeals to an original unity of mind and nature:

> The sheer wonder which surrounds the problem of the origin of organic bodies, therefore, is due to the fact that in these things necessity and contingency are most intimately united. *Necessity,* because their very *existence* is *purposive*, not only their form (as in the work of art), *contingency*, because this purposiveness is nevertheless actual only for an intuiting and reflecting being. For that reason, the human mind was early led to the idea of a *self*-organizing matter, and because organization is conceivable only in relation to a mind, to an original union of mind and matter in these things. ... Here for the first time there emerged from its sacred obscurity that ideal being in which the mind supposes concept and deed, design and execution, to be one. Here first a premonition came over man of his own nature, in which intuition and concept, form and object, ideal and real, are originally one and the same. (Schelling 1988 [1797], introduction, 35)

Thus, an organism exhibits the necessity of mechanism insofar as an organism is an existing object of nature, which both as a whole and in terms of its parts is governed by universal natural laws. While a work of art is also an existing material object subjected to natural laws, the parts-to-whole relation, what the work of art represents, is its formal feature that is not subjected to nature's necessary lawfulness. An organism also exhibits the contingency of purposiveness insofar as an organism's function and properties exhibit end-directedness. The explanatory compatibility of these causes in one single object can be justified only by seeing these objects as individual manifestations of a prior original unity of mind and world (nature) in which our own human nature (as conscious subjects) also originates.

In *On the World-Soul* (1798), Schelling explores further the constitutive status of the unifying principle, the fact that '[o]ne and the same principle connects the inorganic and organic nature' (Schelling [1798] 2000, AA I.6: 70) which determines the way nature is in itself and not merely for us:

[8] I admit that Schelling's *Ideas* is a transcendental project, that is, an inquiry into conditions of the possibility of our cognition of nature. I will show below that the main ideas of his *On the World-Soul* are better understood in light of his later *Identitätsphilosophie*. But this does not exclude the fact that his *On the World-Soul* very much continues the project of his *Ideas* and that to understand his *On the World-Soul* the reader must first read and understand his project in *Ideas*. See on this point Schelling (2000 [1798], AA I.6: 70).

> The consideration of the universal natural changes as well as the progress and persistence of the world of organisms leads a natural scientist towards a *common principle* that fluctuates between the inorganic and organic nature and contains the first cause of all the changes in the former and the ultimate ground of all the activity in the latter. Because it is present *everywhere*, it is *nowhere*, and because it is *everything*, it cannot be anything *particular* or *special*. Because of this reason the language does not have a name for it and the Idea of it was handed down to us by the oldest philosophy [...] only in poetic representations. (Schelling 2000 [1798], AA I.6: 67)

The original antinomy between inorganic nature (mechanical) and organic nature (life) for Schelling, unlike for Kant, is merely an apparent conflict insofar as there is no true opposition between the principles governing mechanism and those governing organisms. Both mechanism and organism are governed by one and the same principle to which the Ancients, and more specifically Plato in his *Timaeus*, referred to 'only in poetic representations' (Schelling 2000 [1798], AA I.6: 67). They called it, namely, the 'world-soul'. In other words, there *is,* in nature, a common unifying ground of the inorganic and organic worlds. However, unlike Kant, who, given the limitation of our cognitive capacities, claims our ignorance of the supersensible principle, Schelling's objective will be to show that we have knowledge of this principle. For Schelling, this unifying ground must be a *natural* principle: 'The unproven claim that the *first* principle of organic nature is *physically* unfathomable is useful for nothing but to crush the courage of a researcher' (Schelling 2000 [1798], AA I.6: 68). If the unifying principle is to serve as an explanatory ground of a natural phenomenon, whether organism or mechanism, it must itself be a part of nature, the existence of which *Naturphilosoph* can confirm empirically. However, just as Kant's supersensible, Schelling's 'world-soul' has an ontological priority in relation to individual objects of both inorganic and organic nature. Unlike determinate individual objects, it is indeterminate (and, therefore, it is 'everywhere' and 'nowhere') and because it is the *ground* of all individual objects it is 'everything' and yet not identifiable with any particular individual object.[9]

Because organisms imply an organization, that is, a rule or a concept that accounts for their inner lawfulness and regularity, only a mind could have access to the principle of the original unity of mind and nature. Put differently, only a mind

[9] Franks shows the influence of not only Plato's *Timaeus* but also of kabbalistic tradition on the development of Schelling's conception of the world-soul. Franks argues, however, that Schelling 'physicalizes what, in kabbalistic texts, are physical symbols of spiritual reality' (Franks 2020, 87). In the Platonic-kabbalistic tradition that Schelling inherits, light, for example, is a symbol of the tendency of the infinite towards revelation, while contraction of the infinite to the center is the symbol of spiritual self-limitation that remains immanent to the created world. Indeed, for Schelling, light is a part of physical reality and the contraction is the physical force of gravity. But Schelling aims to argue that the existence of the world-soul is manifested in these physical phenomena since it is their explanatory ground yet is not identical with them. Although for Schelling, the world-soul is a part of nature, his aim is to maintain the ontological priority of the world-soul in relation to other material individual objects and phenomena.

could have cognition of the Absolute.[10] Moreover, a particular faculty of the mind is credited with this special access to the Absolute:

> Hence the peculiar aura which surrounds this problem, an aura which the philosophy of mere reflection, which sets out only to separate, can never develop, whereas the pure intuition, or rather the creative imagination, long since discovered the symbolic language, which one has only to construe in order to discover that Nature speaks to us the more intelligibly the less we think of her in a merely reflective way. (Schelling 1988 [1797], introduction, 35)

While dogmatic approaches could only prioritize either the material object or the producing subject whereby ideal causes were reduced to real and vice versa, the approach of Critical philosophy acknowledges the distinction between the ideal and the real but these two views of the world for Critical philosophy remain forever separate. In other words, Kant could only argue for their 'coherence', complementary co-existence in one single object which was for Schelling ultimately insufficient. Schelling's *Naturphilosophie* unites the ideal and the real insofar as it presupposes their common origin in the Absolute. And it is the task of Schelling's *Naturphilosoph* and their intellectual intuition expressed in the creative imagination, to trace mind and nature to the Absolute in which they are united. This is the objective of Schelling's method of the 'construction' of the concept of life.

[10] Schelling's *Naturphilosophie* investigates the objective or real aspect of the original unity in the Absolute while his Transcendental Philosophy investigates the subjective or the ideal aspect of this original unity. Each original division of the Absolute further divides again into its ideal and real parts. Thus, we have the division of nature into its ideal part, the principle of life (the 'world-soul'), and its real part, dead matter. The 'world soul' in relation to the Absolute has a positive or expansive, productive (ideal) aspect, but given that it is a product of the Absolute's division (that is, its objective manifestation), it also has a negative, or limiting, receptive (real) aspect. Schelling elaborates on this more explicitly later in his *Identitätsphilosophie*. See on this Frank (2014, 124–36). Frank is especially helpful in pointing out the influence of Oetinger and Hahn, the Schwabian pietists, who argued for the connection between a necessary (natural) and a divine causality through a middle. Their reference was Plato's *Timaeus*, the subject of Schelling's 1794 work under the same title. In his *Timaeus* (31c), Plato speaks of the 'indissoluble … bond' which connects not only what is bound but also itself with what is bound. The creation of the world follows from the original sublation (*Aufhebung*) of the original point of 'indifference' which now actualizes itself through the opposition of the forces of nature. Schelling relates the latter to Kant's antinomial conflict in the third *Critique* and Kant's *Metaphysical Foundations of Natural Science* (1786). On the fact that it is helpful to understand Schelling's *On the World-Soul* in light of his *Identitätsphilosophie* see also Fischer (2021, 549).

8.4 Schelling's Construction of the Concept of Life as a Solution to Kant's Antinomial Conflict

The objective of Schelling's *Naturphilosophie* is not simply to appeal to Plato's poetic imagery of the 'world-soul' but to show how matter is a product of life (and not vice versa).[11] In other words, Schelling's objective is to show how the principle of life is the unifying ground not only of organic nature but also of inorganic nature, portraying thereby nature in its entirety as living: 'This philosophy must accept, therefore, that there is a hierarchy of life in nature. Even in mere organized matter there is *life*, but a life of a more restricted kind' (Schelling 1988 [1797], introduction, 35). The goal of his *Naturphilosophie* is to show that '[n]ature should be Mind made visible' while his transcendental philosophy and his *1800 System* showed that '[m]ind [should be] the invisible nature' (Schelling 1988 [1797], introduction, 35). Thus, the goal of *Naturphilosophie* is to show the unity of mind and nature from an objective perspective while the objective of his transcendental philosophy is to show this unity from a subjective perspective.

In order to show that nature is a mind made visible Schelling proposes a 'construction' of the concept of 'life' (Schelling 2000 [1798], AA I.6: 186). A 'construction' of a concept for Schelling is a demonstration in a geometrical sense, an ostensive and not a discursive proof (Breazeale 2014, 98 ff.). In other words, it is a method of displaying or exhibiting (*darstellen*) in the particular (i.e. a natural phenomenon) the universal (i.e. the principle of life) just as we 'display' or 'lay before our eyes' in a particular geometrical figure (i.e. 'this triangle') necessary and universal truths about all such figures, directly and self-evidently. This is why Schelling contends that to construct the concept of life is for it to be explained as an 'appearance of nature' (Schelling 2000 [1798], AA I.6: 186) or the concept the content of which is given or displayed before us in individual natural objects or phenomena.[12] Furthermore, the method of construction must proceed dialectically, from one step in the construction to a higher one, so that each step displays more fully the content of the concept of life.

---

[11] This claim should dispel any doubt whether Schelling embraced hylozoism. Beiser contends that 'In seeing matter as only the negative side of life, as its lowest degree of organization and development, Schelling had virtually embraced the very hylozoism Kant had condemned' (Beiser 2002, 521). But to show that Schelling remains consistent in his critique of hylozoism, it suffices for now to emphasize that hylozoism attempts to explain life as a product of dead matter while Schelling attempts to explain dead matter as the product of life.

[12] Beiser rightly notes that Schelling's *Naturphilosophie* is grounded in a priori methodology. In Kant's idealist tradition that Schelling follows, to have cognition of an organism is to determine the principles of its possibility, the principle of life, and to determine this is to be able to construct it or reproduce its activity in thought. The philosopher's constructions correspond to nature because the philosopher, through her intellectual intuition, shares in the same productive Absolute as nature. This does not imply that Schelling's *Naturphilosoph* ignores the importance of experience and experimental method. While by means of a theory *Naturphilosoph* sets up an experiment, the results of the latter reject or confirm her a priori construction (Beiser 2002, 525–8).

Schelling conceives of nature as consisting of two fundamental opposing forces.[13] This opposition between the expanding positive force and the limiting, or negative force explains the possibility of the restriction of the absolute productivity of nature (infinity) into some finite product. The activity of nature is conceived as a dialectical progression from identity to opposition and then to a higher form of unity of forces, 'potencies' (*Potenzen*).

In *On the World-Soul*, the first step in the construction of the concept of life consists in the refutation of the two dogmatic positions described above, namely, vitalism and physico-theology.[14] It first shows the falsity of the claim that the ground of life is in the 'negative principle', that is, inside of animal matter itself. This view corresponds to vitalism and above I summarized the argument of Schelling's criticism of this position. The second step consists in showing the falsity of the claim that the ground of life must lie outside of animal matter, that is, in the 'positive principle', a divine artisan of physico-theology. The proponents of this view argue that life is the product of the irritability of muscles and nerves from the outside. But, argues Schelling, '[i]f the ground of life lies completely outside of the animal body, then we must assume that the animal body with respect to the principle of life is absolutely passive ... Absolute passivity', however, 'is a fully meaningless concept' (Schelling 2000 [1798], AA I.6: 192). In other words, there must be something in matter that would explain why matter is responsive or receptive to the positive principle. In other words, the positive principle cannot be entirely external or foreign to matter. This is why Schelling resolves the dialectical conflict with the following proposition: 'The ground of life consists in the opposed principles so that one (positive) is to be found outside of the living individual and the other (negative) in the individual itself' (192). The 'positive principle' is the universal principle. It is 'spread through nature in its entirety and penetrates each being like a common breath of nature' (192). This is the 'spiritual' and 'infinite, absolute' principle (192). This is also the principle that Schelling, in Kantian terms, associates with spontaneity. The 'negative' principle is the material principle of individualization that he associates, in Kantian terms, with receptivity. 'The entire variety of life consists in the *unity* of the positive principle in *all* beings and the *difference* of the *negative*

[13] Schelling's notion of the opposition of forces is modeled after Kant's dynamic conception of matter in his *Metaphysical Foundations of Natural Science* (1786). However, for Kant, the fundamental forces of attraction and repulsion are empirically given, while for Schelling, these forces are metaphysically more fundamental than they are for Kant. While Kant was ambivalent with respect to the possibility of a construction of matter, Schelling argued that the construction of matter from these fundamental forces must be necessary. See more on this in Ostaric 2014. As we will see below, for Schelling, these forces and their interaction are the result of the working of the principle of life, the world-soul, and all natural phenomena, including matter, must be derived from these forces.

[14] I will only be focusing on Schelling's construction of the principle of life in organic forms. He discusses the construction of the same principle in inorganic forms in *Ideas* and in Part I of *On the World-Soul*. There he explains the phenomena of matter (first potency), magnetism, electricity, and chemical processes (second potency), and light (third potency) as the products of the working of the two opposing forces, the positive, expansive force and the negative, limiting force.

principle in *each* being' (193). In other words, matter or the 'negative principle' is the ground of individuation, in virtue of which an organism is one particular thing. The 'positive principle', on the other hand, is the principle in virtue of which this individual entity has a specific form, a rule which unifies its parts: 'The positive principle of life must be unity and the negative must be the variety. As there are many possible unifications of this variety in one whole, there are as many particular organisms, each representing one particular life' (Schelling 2000 [1798], AA I.6: 15).

Next, Schelling addresses each principle, starting with the negative and shows how it cannot be the ground of life without its opposite. With each step of the construction, Schelling criticizes a concept of the natural sciences of his time that is considered to be the explanatory ground of life. The first such concept is the one of 'chemical processes' that in a reductive materialist manner explains growth as the effect of the 'blind working of nature' (Schelling 2000 [1798], AA I.6: 204). But the materialist cannot explain why chemical processes that come to an end, to an equilibrium, are always disturbed anew. The materialist also cannot explain why certain chemical processes are correlated with the same type of changes for a certain part of the body. Finally, the materialist cannot explain why the chemical processes are contained within an individual body (compare Beiser 2002, 540). The negative or the material principle, argues Schelling, always strives towards the balance of forces, which leads to peace and also death.[15] This is why nature continually enlivens forces and '*forces* dead matter to shoot for a *determined form and shape* which to human judgment appears as a purpose of nature'(Schelling 2000 [1798], AA I.6: 204). An organism is expressive of the absolute unity of necessity and contingency. The gradual growth of animal matter that occurs according to the chemical processes that follow the blind forces of nature is expressive of natural necessity. But the fact that this growth takes a specific shape or form is expressive of contingency, that is, something that we think of as the 'end of a personified nature' (204).

Schelling turns next to discuss the 'negative principle' from the perspective of its ideal or positive aspect. Here, Schelling sets out to criticize the concept of a 'life force' which was widespread in natural sciences of his time. For Schelling, the concept of a 'life-force' is a concept of some mysterious force of nature, which leaves out the possibility of explaining an organic formation 'physically', that is, in terms of the properties of matter. While explaining organic formations exclusively by means of 'dead chemical forces' leaves out the possibility of explaining organism by appealing to the 'freedom of nature', that is, its ideal causes, the concept of a 'life force' leaves out entirely the relevance of dead matter, that is, mechanism, or nature's real causes. 'Both', concludes Schelling, 'should be united … nature in its blind lawfulness should be free, and, vice versa, lawful in its freedom' (Schelling 2000 [1798], AA I.6: 215).

[15] For Schelling's construction of matter see his *Ideas*, esp. the first six chapters of Book II. While in *Ideas* Schelling follows transcendental arguments for the construction of matter, in *On the World-Soul* each step in the construction is treated as a hypothesis that requires empirical confirmation. See on this also Fischer (2021, 536).

However, rejecting the independent reality of a 'life force' as an explanation of life was not the reason for Schelling to endorse Blumenbach's concept of the 'formative drive' (*Bildungstrieb*) as a possible explanation of the unity of real and ideal forces in nature. Blumenbach's epigenesis and his notion of 'formative drive' incorporated Newtonian mechanics into his tracing of patterns of generation within different species of organized matter.[16] But he at the same time assumes that matter is organized and predisposed towards life, meaning, that organization is not explicable in terms of properties and function of dead matter. For Schelling, Blumenbach's 'formative drive' is 'an *expression* of every original unity of freedom and lawfulness of all natural formations, but not an *explanatory ground* of this unity itself' (Schelling 2000 [1798], AA I.6: 216). It is

> a barrier for a researching reason, or a cover for a dark quality, which cannot bring reason peace. This concept already presupposes organic matter because every drive must and can be only effective in organic matter. This is why this principle cannot indicate the *cause* of organic formation *but, on the contrary, it presupposes a higher cause of an organic formation.* (216)

For Schelling, the concept of *Bildungstrieb* is a 'barrier for a researching reason' because it cannot serve as an explanation of the organization itself, or how life is generated. This is because, for Blumenbach, *Bildungstrieb* is only specific to organic matter and it presupposes rather than explains the spontaneous change from dead matter into organic matter. In other words, in the concept of *Bildungstrieb* naturalism and mechanical explanation are not sufficiently heeded while for Schelling it must remain integral to the method of *Naturphilosophie*.

The construction then turns to the second side of the dialectical horn, namely, the 'positive principle of life'. Schelling reminds us that the subject of his investigation is the cause of life, but also that

> life consists in a circular process, that is, in a consecutive sequence of processes that constantly turn into themselves so that it is impossible to say which process, in fact, begins life, and which is supposed to be the earlier and which the later. Each organism is one enclosed whole in which everything is simultaneous and where mechanical explanation completely abandons us because in such a whole there is no before and after. (2000 [1798], AA I.6: 237)

Schelling, therefore, contends that we cannot assume that in an organism, one process determines the other but that they 'mutually determine each other' (2000 [1798], AA I.6: 238). Here, Schelling echoes Kant's way of specifying causal relation in organic formations as one of reciprocal causal determination, that is, as the

[16] Blumenbach criticized preformationism as a theory that can adequately explain the generation and development of species. According to the former, all the information for the development of an adult individual in a species is contained in its egg (embryo). To this, Blumenbach objects that preformationism cannot explain the contingent elements in the development of an individual of a species, for example, changes in the development of a rose-bush due to a mosquito bite: 'Should, therefore', comments Blumenbach, 'all these contingent products of a small mosquito preexist in all the parts of all the rose bushes of the world?' (Blumenbach 1781, 31; cited in Jantzen 1994, 638). For the explanation that Blumenbach is referring to wasps and the development of gall apples on a plant that serve to protect their larvae, see Jantzen (1994, 638 n.).

'cause and effect of itself' (Kant 2000 [1790] §65, 245). He, therefore, argues that an organism as the 'ideal whole is the cause of all its parts (it is the cause of itself as a real whole), and as a real whole (insofar as it has parts) it is the cause of itself as an ideal whole' (Schelling 2000 [1798], AA I.6: 208). The whole is the cause of the function and properties of the parts insofar as, for example, the growth and self-maintenance of a deciduous tree are different from the growth and self-maintenance of an evergreen and, therefore, dependent on the kind of plant (the whole) it is. But the whole is at the same time the effect of the properties and functions of the parts insofar as it depends on the functions and activities of these parts—i.e., it depends on the growth, reparation, and self-maintenance of an organism.

Having denied that the positive principle of life can be considered as the first cause in a linear sequence of cause and effect, Schelling identifies the positive principle of life with irritability, a concept that in Schelling's time was associated with the organism's capacity for activity and movement.[17] Unlike his contemporaries who, according to Schelling, treated the phenomenon as 'Qualitas occulta' (Schelling 2000 [1798], AA I.6: 235), an innate power of the organ's tissue to move when stimulated, he attempts to explain the phenomenon of irritability in terms of 'natural causes' (235), that is, in terms of mechanical material causes. This will also help him show that irritability as a positive principle is inseparable from its 'negative' principle.

Schelling further identifies the positive principle as the principle that works through the nerve on the irritable organs. If there are fewer nerves going to the organs, then the organ's capacity for oxidation is decreased. The more the organ's capacity for oxidation is decreased, the more necessary (less subjected to *Willkühr*) deoxidizing process, and, consequently, the more restless its irritability.[18] Schelling illustrates this in the example of the autonomic nervous system that regulates the activity of internal organs. For example, Schelling refers to the heart as the 'unintentional muscle' (*unwillkühricher Muskel*). The opposite, however, also holds: if there are more and bigger nerves leading to an organ, its capacity for oxidation increases, and therefore its capacity for necessity and unintentionality in its expression of irritability decreases. Schelling has in mind our somatic nervous system as an example and he takes the thumb as an example of an organ that is most intentional in its expression of irritability.

If irritability is the positive principle associated with spontaneity (here manifested as the animal's capacity for movement), then *sensibility* is its opposite, that is, its negative principle, associated with receptivity. The animal's capacity for

[17] Schelling's discussion of irritability is influenced by the reception in Germany of the work of the Scottish physiologist John Brown (1735–88) and his theory of excitability but also by the 1752 lectures of the German physiologist Albrecht von Haller whose theory of irritability and sensibility anticipates in some respects Brown's findings. See Jantzen (1994, 375–498).

[18] Schelling seems to be building his theory on the findings of the Swiss chemist and physician Christopher Girtanner (1760–1800) who explained irritability within the context of circulation and intake of oxygen in the body. The higher the intake of oxygen, the stronger its irritability and vice versa. See Jantzen (1994, 500).

movement, its 'drive' (*Trieb*), is directed towards something 'indeterminate' (*ein Unbestimmtes*) (Schelling 2000 [1798], AA I.6: 249) but it becomes a determinate movement only in virtue of its negative, namely, sensibility. In other words, sensibility gives the animal's drive for movement a specific direction insofar as the animal receives stimuli that are specific to its own nature. The higher the irritability, that is, the more capable the animal of voluntary movement, the stronger its sensibility, its capacity for receiving external stimuli, and vice versa. When both principles, irritability and sensibility, are united in one concept, then the concept of *instinct* emerges. The animal instinct is that which unites both the contingency of irritability and the necessity of sensibility. It is a 'drive to a movement determined by sensibility' (249).[19]

Having shown how the concept of 'instinct' unifies both contingency and necessity, Schelling must now show the advantages this concept has when compared to Blumenbach's 'formative drive' (*Bildungstrieb*). He must show how his notion of instinct is not limited to organic nature and entirely severed from inorganic nature. He must show how all nature is one single whole and how organic nature emerges from inorganic nature. He begins to address the problem with the following question: 'How does the universal formative force of matter become the formative drive?' (Schelling 2000 [1798], AA I.6: 253) In other words, the formative forces of matter are only defined by necessity. The formative drive, on the other hand, is defined by both necessity and contingency. Whence this foreign principle that 'interrupts' the formative forces of matter and forces them to produce a determined form [*Gestalt*]? 'The essence of life', contends Schelling in response, 'consists not in a force, but in the free play of forces which is continuously maintained through some external influence' (254). Schelling adopts Kant's logical principle of aesthetic judgment, that is, the free harmony of the faculties, to his *Naturphilosophie*.[20] He identifies the necessary element of the harmony of the faculties with the natural forces and the contingent element with the principle that maintains the play of forces. The principle that maintains the play of forces, argues Schelling, must be a 'material principle' (254).

Just as Kant's solution to the antinomy of teleological judgment was resolved by appealing to the common ground of both mechanism and organism (life) in the supersensible, Schelling's principle of life is a common ground to both mechanism and organism. Various natural phenomena are only explicable, argues, Schelling:

---

[19] While Schelling's focus in *On the World-Soul* remains animal instinct, he identifies human genius as manifesting the instinct of the rational soul: 'Man sees and hears only by means of a higher instinct that directs him towards the great and the beautiful. And this instinct is called genius' (Schelling 2000 [1798], AA I.6: 250). That the concept of genius unites both human spontaneity (i.e., contingency) and human receptivity (i.e. necessity) is already anticipated in Kant's conception of genius (Ostaric 2012).

[20] Schelling's adoption of Kant's free harmony of the faculties is not arbitrary. It represents a model for a unity of contingency and necessity, that is, a unity of a purposive contingent harmony of the imagination with the lawfulness (and hence necessity) of the understanding.

> if we assume a *universal continuity of all natural causes* and a *common medium* through which alone all forces of nature have an effect on sensible beings. Because this principle maintains the continuity between the inorganic and the organic world, it connects the whole of nature into a universal organism so that in it we recognize anew that essence that the oldest philosophy anticipatory greeted as the universal soul of nature and that some physicists of that time identified with the shaping and forming ether (the most noble part of nature). (Schelling 2000 [1795], AA I.6: 257)

For Kant the reflection on organic formations led to the representation of nature as a system of ends with a human being as a final end of nature. This is because if we are required to present organic formations as an end of nature, then we are also required to represent the whole of nature as creation, a system of ends with the human being and the highest good as its final end.[21] For Schelling, resolving the antinomial conflict presented by organic formations does not lead to a representation of nature as a system such that this representation is relative to the limitations of our cognitive capacities. Instead, it leads to our *cognition* of nature as a system within which we can explain the evolution of mechanical, non-teleological forms, and from these the evolution of teleological organic formations, and from these ultimately the evolution of conscious rational agents (Schelling 1984 [1802], 160–85). While Kant's representation of the system of nature leads to the idea of freedom as autonomy and the hope of its realization in the world in the form of the highest good, Schelling's system of nature leads to his notion of *absolute* freedom as a unity of contingency and necessity to be realized in the conscious rational agent as the nature's highest formation, and which approximates the infinite of the absolute in which it originates.

8.5 Conclusion and Evaluation

I have argued in this essay that Schelling follows Kant's formulation of the antinomial conflict insofar as for him, as for Kant, the problem of the antinomy is the problem of how an organism as an object of nature exhibits a principle of organization, the whole, that cannot be exhausted by mechanism. The whole of other objects of nature, such as, a sand-dune or an artifact, can be explained mechanically, that is, as the sum of its parts. But the whole of an organism points to a unique unification of both real (mechanical) and ideal causes in one single object. However, unlike Kant, Schelling does not resort to reflective judgment and the regulative status of the principle of purposiveness as a solution to the antinomial problem. In other words, he does not resolve the antinomy by arguing that we must represent organisms *as if* they originated in a supersensible, a mind more powerful than ours, an intuitive understanding. By arguing the latter, Kant accomplished a mere 'coherence', a complementary co-existence of the two principles, namely, mechanism and teleology.

[21] For Kant's claim that the representation of an organism as a 'natural end' necessarily leads to the idea of the whole of nature as a system of ends, see Kant (2000 [1790], §67, 250).

Motivated by his desire to offer a scientific explanation of life, and not a mere Kantian 'elucidation', Schelling ventures to identify the explanatory ground of the special unity of the ideal and real causes in organic beings.

What follows is, what some have called, Schelling's 'physicalization' (Franks 2020) of the Platonic notion of the world-soul that serves as a 'bond' (*das Band*) between the Absolute and the world of material objects. As such, nature itself has an aspect that is ideal (the principle of life or the world-soul) and one that is real (dead matter). The ideal aspect exemplifies itself in the concrete objects of nature, whether inorganic or organic.

Schelling's solution, however, does not leave us without questions. One question pertains to the nature of the causal relation between the principle of life and individual beings. The ideal part of nature, or the 'soul' of nature's systematic whole, gives the form and shape to the real part: 'The principle of life is only the cause of a determined *form* of being, and not the cause of the being itself (as this is impossible for us to think)' (Schelling 2000 [1798], AA I.6: 254).[22] Schelling further warns that we should not think of this form-giving as a representation 'external' to the organic formation:

> The principle of life did not enter organic matter from outside (as if through an infusion)—(one spiritless but widespread representation)—but on the contrary, built for itself [*sich anbilden*] organic matter. Insofar as it individualizes itself in single beings and, on the other hand, gives these beings their individuality, it became a principle that is inexplicable from the organic formation itself and the effect of which is manifested to a feeling as the ever more intense drive [*Trieb*]. (255)

According to Schelling, we should not think of this causality as an external causality, one that enters organic matter 'from outside'. Instead, Schelling captures the causal relation with the phrase that the principle of life 'builds itself organic matter' (*sich die organische Materie anbilden*). In other words, individual beings are not metaphysically distinct entities from the world-soul. They are the negative aspect of nature, the individualized particular of its positive aspect, the world-soul.[23] In the *Freiheitsschrift* of 1809, for example, Schelling similarly refers to the positive principle of the 'all powerful will' which brings the rational order to nature as 'not through external representation but through in-forming (*Einbildung*)' (Schelling 2006 [1809], 31, Jeff Love and Johannes Schmidt translation modified). If the formation of *Gestalt* or form were an activity that proceeded according to a rule that is

[22] Schelling will attempt to address this in his theodicy in the *Freiheitsschrift*. Theodicy becomes a manner of addressing that which lies beyond our rational inquiry.

[23] Beiser (2002, 543) objects that Schelling does not explain the possibility of 'interaction' between the world-soul and concrete material things, given their heterogeneity. Beiser is, however, mistaken to assume a heterogeneity between the world-soul, the positive principle of life, and its individualization in concrete things. While the positive principle is ontologically prior to its individualization in concrete things, it is itself a part of nature, a 'material principle' that is the outcome of the prior division of the absolute into its subjective and objective parts. Not only was Schelling aware of this problem but also the fact that his *On the World-Soul* approximates his later *Identitätsphilosophie* can be seen as his search for a solution to the classical problem of the 'interaction' between mind and matter.

an 'external representation', the organizing principle of nature would be determinate, one which fully exhausts its phenomenon. But the phrases *sich anbilden* and *Einbildung* indicate that the rule or principle of organization cannot be conceived independently from the object in which it exhibits itself. This also explains why, as with Kant, the rule of organic formations cannot be accounted for by analyzing its parts. The whole of an organism is not the sum of its parts; rather, there is inevitably something that cannot be fully exhausted by our discursive understanding that proceeds analytically from parts to wholes.

Because Schelling's positive principle, the principle of life, is in the end a 'material principle' (Schelling 2000 [1798], AA I.6: 254) and, thus, a part of nature, one may question to what extent Schelling is successful in avoiding vitalism which collapses ideal causes into mechanical. However, given the Absolute's initial division into the ideal and the real part, its real part (nature) is further divisible also into its ideal and its real part. Thus, the 'principle of life', or Plato's 'world-soul', or 'ether', is the 'mind' of nature. The ideal and the real principles do not collapse as in vitalism because although the world-soul is nature, Schelling's metaphysics affords an ontological distinction, a form of ontological priority, between it, as its 'most noble part', and the rest of material nature, the dead matter, or its real part.

References

Allison, Henry. 1991. Kant's antinomy of teleological judgment. *Southern Journal of Philosophy* 30 (Supplement): 25–42.

Beiser, Frederick. 2002. *German idealism: The struggle against subjectivism, 1781–1801*. Cambridge, Mass: Harvard University Press.

Blumenbach, Johann Friedrich. 1781. *Über den Bildungstrieb und das Zeugungsgeschäfte*. Göttingen: Dieterich.

Breazeale, Daniel. 2014. 'Exhibiting the particular in the universal': Philosophical construction and intuition in Schelling's philosophy of identity (1801–1804). In *Interpreting Schelling: Critical essays*, ed. Lara Ostaric, 91–119. Cambridge: Cambridge University Press.

Butts, Robert E. 1984. *Kant and the double government methodology*. Dordrecht: Kluwer Academic Publishers.

Fischer, Naomi. 2021. The unity of nature in Schelling's *World Soul. The Review of Metaphysics* 74:527–552.

Frank, Manfred. 2014. Identity of identity and non-identity': Schelling's path to the 'absolute system of identity. In *Interpreting Schelling: Critical essays*, ed. Lara Ostaric, 120–144. Cambridge: Cambridge University Press.

Franks, Paul. 2020. From world-soul to universal organism: Maimon's hypothesis and Schelling's physicalization of a Platonic-Kabbalistic concept. In *Schelling's philosophy: Freedom, nature, and systematicity*, ed. Anthony Bruno, 71–92. Oxford: Oxford University Press.

Jantzen, Jörg. 1994. Theorien der Irritabilität und Sensibilität. In *Wissenschaftlicher Bericht zu Schellings Naturphilosophischen Schriften. 1797–1800. Schelling: Historisch-kritische Ausgabe Im Auftrag der Bayerischen Akademie der Wissenschaften. Ergänzungsband zu Band 5–9*, ed. Hans Michael Baumgartner, Wilhelm G. Jacobs, and Hermann Krings, 375–498. Stuttgart-Bad Cannstatt: Frommann-Holzboog.

Kant, Immanuel. 2000. *Critique of the power of judgment* (1790). Translated by Paul Guyer and Eric Matthews, ed. Paul Guyer. Cambridge: Cambridge University Press.

McLaughlin, Peter. 1990. *Kant's critique of teleology in biological explanation.* Lewiston: Mellen Press.

Ostaric, Lara. 2012. Kant on the normativity of creative production. *Kantian Review* 17 (1): 75–107.

———. 2014. The concept of life in early Schelling. In *Interpreting Schelling: Critical essays*, ed. Lara Ostaric, 48–70. Cambridge: Cambridge University Press.

———. 2023. *The 'Critique of judgment' and the unity of Kant's critical system.* Cambridge: Cambridge University Press.

Schelling, Friedrich Wilhelm Joseph. 1978. *System of transcendental* idealism (1800). Translated by Peter Heath. Charlottesville: University of Virginia Press.

Schelling, Friedrich Wilhelm Joseph. 1984. *Bruno or on the natural and the divine principle of things* (1802). Translated by and ed. Michael G. Vater, Albany: State University of New York Press.

———. 1988. *Ideas for a philosophy of nature as introduction to the study of this science* (1797). Translated by Errol E. Harris and Peter Heath. Cambridge: Cambridge University Press.

———. 2000. *Von der Weltseele—eine Hypothese der höhern Physik zur Erklärung des allgemeinen Organismus* (1798). *Schelling: Historisch-kritische Ausgabe Im Auftrag der Bayerischen Akademie der Wissenschaften* Band I.6. ed. Jörg Jantzen. Stuttgart-Bad Cannstatt: Frommann-Holzboog.

———. 2006. *Philosophical investigations into the essence of human freedom* (1809). Translated by Jeff Love and Johannes Schmidt. Albany: State University of New York Press.

Watkins, Eric. 2008. Die Antinomie der teleologischen Urteilskraft und Kants Ablehnung alternativer Teleologien. In *Immanuel Kant: Kritik der Urteilskraft*, ed. Otfried Höffe, 241–258. Berlin: Akademie Verlag.

Zanetti, Véronique. 1993. Die Antinomie der teleologischen Urteilskraft. *Kant-Studien* 84 (3): 341–355.

Lara Ostaric is associate professor of philosophy at Temple University. She is the author of *The Critique of Judgment and the Unity of Kant's Critical System* (Cambridge UP, 2023) and the editor of *Interpreting Schelling: Critical Essays* (Cambridge UP, 2014). She is an Alexander von Humboldt Fellow and is Philosophy Editor for the *Society of German Idealism and Romanticism Review.*

Chapter 9
Living and Dead Forms: The Factuality of Meaning in Schelling and Other Naturalists

Daniel Whistler

Abstract The aim of this chapter is to describe a set of post-Kantian accounts of meaning, according to which: (i) meaningful properties are formal ones; (ii) these properties should be described in terms of the natural processes out of which they emerge; (iii) meanings should therefore be considered natural forms and be explained in the same way as any other natural fact. This is a set of naturalist positions that typically construes meaning as a 'living form', but equally—at its most radical—exceeds this organicist framework to think inorganic meanings. I consider this tradition through a detailed interrogation of Lectures Eight and Nine of F. W. J. Schelling's *Historical-Critical Introduction to the Philosophy of Mythology*, arguing that this is one exemplary instance of the confusion of forms and grounds in such a post-Kantian tradition of accounting for meaning. Ultimately, I suggest it is a position that puts into question the traditional category of meaning itself.

We are all well aware of the dangers of organicism and of the particular risks involved in understanding purportedly mental phenomena, like meaning, on the model of the organism.[1] An organicist tradition that makes sense of linguistic, mythic and aesthetic meanings as 'living forms'—as explicitly theorized by Friedrich Schiller, among others—has long been lost as a live option for contemporary theory. Nonetheless, there is more at work here than meets the eye, not least the fact that thinkers of this type were often as interested in inorganic and dead forms as

[1] E.g., in Paul de Man's criticisms of an organicism of meaning as 'the confusion of linguistic with natural reality' (1986, 7), and its 'nostalgia for the natural object' as 'ontological bad faith' (1983, 211).

D. Whistler (✉)
Department of PIRP, Royal Holloway, University of London, Egham, Surrey, UK
e-mail: daniel.whistler@rhul.ac.uk

P. Cheyne (ed.), *Matter and Life in Coleridge, Schelling, and Other Dynamical Idealists*, International Archives of the History of Ideas Archives internationales d'histoire des idées 253, https://doi.org/10.1007/978-3-031-78315-9_9

living ones. This chapter is an attempt to capture some of these minoritarian variants in the thinking-through of meaning as 'living form'.

In what follows I aim to delineate a set of post-Kantian accounts of meaning, according to which:

(i) meaningful properties are formal ones (*formalism thesis*);
(ii) these properties should be described in terms of the natural processes out of which they emerge (*embeddedness thesis*);
(iii) meanings should (therefore) be considered natural forms and be explained in the same way as any other natural fact (*a naturalism of meaning*).

This set of claims thus cohere around a naturalist position that construes meaning as a living form, but equally—at its most radical—exceeds this organicist framework to think meaning as *inorganic* form too. It is a set of positions espoused, at least partially, by a syncretic group of thinkers from very different traditions, including J. W. Goethe, Friedrich Schiller, Novalis, F. W. J. Schelling, Remy de Gourmont, Roger Caillois and even, at times, Immanuel Kant.[2] These thinkers all (to some extent) decorrelate meaning from the human mind, from intentionality and from referentiality—and, instead, employ the notion of a *natural fact* (whether organic or inorganic) to interrogate the value of the concept of meaning and, ultimately, to flirt with the possibility of the extinction of meaning.[3] As is hopefully clear, my implicit intention in drawing attention to these conceptions of meaning as natural fact is to suggest that contemporary versions of linguistic naturalism are not exhaustive; other naturalisms of meaning are possible—some of which occur persistently and distinctively in post-Kantian philosophies.[4]

[2] I have elsewhere (Whistler 2016) traced this naturalist of account of meaning through early German Romanticism and the early Schelling, particularly insofar as their theories of the symbol emerge out of an aesthetic naturalism. The following chapter should be read as a sequel to that earlier work. It is also intended as a partial corrective to an earlier article (Whistler 2015) which emphasized Schellingian literalism.

[3] What I am getting at here is nicely expressed in a modernist context by Willem de Kooning's comment (1951, 6) on Kandinsky: 'Kandinsky understood "Form" as a form, like an object in the real world… and so, of course, he disapproved of it.' The proto-modernist naturalists discussed in this chapter equally understand 'form as a form, like an object in the real world', but, unlike the spiritualist Kandinsky, they affirm it.

[4] Given the significant role the concept of decorrelation will play in the second half of the chapter, the position explored in what follows is best described as a *Naturphilosophie* of meaning—or even a 'speculative' naturalism. Ultimately, it constitutes one answer to the question: what is the philosopher of nature to do with meaning?

9.1 Tautegorical Meaning as Natural Fact

In his 1825 *Aids to Reflection* (1993, 206), S. T. Coleridge introduced the neologism, 'tautegory', to describe a distinctive species of meaning which 'express[es] the same subject but with a difference, in contra-distinction from metaphors and similitudes, that are always allegorical, that is, expressing a different subject but with a resemblance.' Tautegory signifies 'speaking the same' in opposition to allegory or 'speaking otherwise'.[5] Likewise in his 1825 lecture to the Royal Society of Literature, 'On the Prometheus of Aeschylus', Coleridge invokes the category of tautegory again to resist interpretations of the Greek gods a 'mythus, in which symbol fades away into allegory' (1995 [1825], 1280). It was from Coleridge that Schelling explicitly borrowed the concept as part of his philosophising on mythology during the 1840s.

Broadly speaking, this Coleridgean rendering of meaning as tautegorical begins from the idea that tautegories 'express the same meaning' as themselves or, in Schelling's terms, they 'do not mean something else, but rather mean only what they are' (2007 [1842], 136). How to understand what Schelling means by this is one of the tasks of this chapter, but it certainly appears the case for Coleridge, at least, that tautegories are purely self-referential and self-contained or autarchic (i.e., they re-posit the same form).[6] Tautegory for Coleridge involves a faithful reproduction of an initial truth in the form 'A is A', where the positing of the latter 'A' discloses this truth more fully than before. This is what Hazard Adams calls, the 'special sort of transparency in language' engendered by tautegorical structures (1983, 82); at its most mundane, Coleridgean tautegory becomes *tautology*, the repetition of the same proposition.[7]

From the beginning Schelling appropriates the concept of tautegory in order to contest Coleridge's use of it. In the very footnote in which he acknowledges his debt to Coleridge, Schelling is clear that his notion of tautegory operates very differently (2007 [1842], 187). And specifying this difference is another of the tasks I wish to undertake in this chapter. Notwithstanding such differences, from the above introduction to tautegory as self-referential and autarchic the following principles broadly follow:

(1) from tautegory's self-referentiality follows a formalism thesis: the salient properties of tautegories are formal ones, for meanings 'are not first present in

[5] On the question of whether Coleridge himself coined this term, see Tilliette (1984, 36) and Tritten (2012, 261).

[6] What will become clear by the end of this chapter is that, while Schelling—as another proponent of tautegory—will pay lip-service to the self-referential structure of tautegory, this is not what matters for him. Instead, he does something more radical with this species of meaning. This emphasis on the difference between Coleridge's and Schelling's versions of tautegory is already present in the anglophone literature (although for very different reasons)—see Beach (1994, 261).

[7] As on Ewan James Jones's interpretation (2014, 146–86).

another form, but rather only emerge in [one specific] form' (Schelling 2007 [1842], 136);

(2) from tautegory's autarchy follows the implication that tautegories exist independently of and prior to any encounter with the human mind (a *decorrelation* of meaning from the human subject);

(3) and from this follows an embeddedness thesis: tautegories must be understood as part of nature—in terms of processes external to the human mind.

In short, tautegorical meaning is modelled as a natural fact, i.e., it partakes of the factuality of natural phenomena. Appeal to 'tautegory' is one way to establish a naturalism of meaning.

In what follows, I am primarily interested in how F. W. J. Schelling's lecture course on the *Historical-Critical Introduction to the Philosophy of Mythology*, given in Berlin during the 1840s, develops, alters and transforms this concept of tautegory to radicalize precisely such a naturalism of meaning. That is, I am concerned with the specific ways in which this late lecture course manages to embed the concept of tautegory into an overarching project of naturalising mythology, such that meaning becomes one more natural fact among others.[8] This is ultimately an issue of (in the terms of my chapter) the relation between the formalist thesis and the embeddedness thesis in Schelling's lectures—a relation manifest in the seemingly uneasy transition from Lecture Eight of the course on tautegorical forms to Lecture Nine which embeds them in an underlying dynamic process. In these two lectures on philosophy of mythology, Schelling attempts to synthesize a discourse on forms with a discourse on natural grounds, so as to constitute a naturalism of meaning.

While the majority of the chapter is spent interrogating this transition in Schelling's philosophy of mythology, one of its further aims is to contextualize his approach to meaning in terms of the broader group of thinkers mentioned above who equally develop a naturalism of meaning. In short, my ultimate interest in this chapter will be the extent to which the late Schelling's account of mythological meaning is symptomatic of a broader, if relatively subterranean, post-Kantian philosophical trend.

[8] By 'nature' and 'naturalism' in this chapter, I refer to the 'rich naturalism' of the naturephilosophical tradition, in which all necessary processes (including the theogonic process out of which mythology arises) are considered natural (a recapitulation of natural potencies). I thus follow Grant (2006) in seeing Schelling's work as naturephilosophical to the end. And, although this view is at times controversial (see, e.g., McGrath 2016), I take it that the controversy is primarily a matter of whether Schelling's description of post-Christian free acts of revelation that break with the necessary mythological process can still be described under the rubric of *Naturphilosophie*, as opposed to the naturephilosophical resonances of the theogonic process itself.

9.2 The Transition from Lecture Eight to Lecture Nine of Schelling's *Historical-Critical Introduction to the Philosophy of Mythology*

The transition Schelling makes from the conclusions of Lecture Eight of his *Historical-Critical Introduction* to new subject matter in Lecture Nine is announced as follows,

> This investigation had proceeded from the question of what mythology originally means, that is, what it means to them for whom it emerged… This investigation can be considered as closed. But precisely thereby are we summoned to the higher question, of what the process means not in relation to the consciousness subjected to it, but what it in itself means, what it objectively means. (2007 [1842], 142–3)

The shift is defined in terms of a movement from the question of meaning for those involved in the myth-making themselves to a 'higher question' of meaning—the question of what mythology 'objectively means'. Broadly, it rests on a distinction between, on the one hand, 'mythology subjectively meant', i.e., the meaning of myths for historically-situated consciousnesses practising tautegorical interpretation; and, on the other hand, 'mythology objectively meant' which concerns the metaphysical grounds or, more precisely, the natural conditions that give rise to mythological representations.[9] In the terms of this chapter, what seems to be at stake in the transition from Lecture Eight to Lecture Nine is a shift from a formalist thesis to an embeddedness thesis. And in this initial section I want to give a preliminary (if, I will argue, ultimately inadequate[10]) presentation of this shift premised on a seeming problem of coherence between the two theses. That is, at first blush, these two theses seem difficult to reconcile, since, if myths in some way reveal their own natural grounds despite not explicitly talking about them, then they do not seem to be tautegorical, but allegorical; i.e., the embeddedness thesis seems to contravene the literalism demanded of tautegorical interpretation.

In Lecture Eight of the *Historical-Critical Introduction*, Schelling's account of the 'subjective' meaning of mythology culminates with the introduction of tautegory:

> Mythology emerges immediately as such and in no other sense than in which it articulates itself…. Everything in it is thus to be understood as mythology expresses it, not as if something else were thought, something else said. Mythology is not allegorical; it is tautegorical. To mythology the gods are actually existing essences, gods that are not something else, do not mean something else, but rather mean only what they are. (2007 [1842], 136)

[9] Tritten (2012, 284) puts it nicely, 'In beginning by asking for mythology's meaning from the question of how it is to be taken and answering, "Tautegorically!" one has only answered what it means for the human being. Objectively, however, mythology is a theogonic process that has a meaning for God and Being—an onto-theogony.'

[10] To be clear from the beginning: the main reason this initial reading of the lectures will turn out to be inadequate is that it remains too close to a Coleridgean version of tautegory as primarily a matter of self-referentiality and literal interpretation.

For allegorists, it is as if 'something else [is] said' (136), which the original form of presentation distorts or disguises; the allegorist must ultimately subscribe to the view that 'there is truth in mythology, but not in mythology as such' (149). Allegorical form is 'cloaked' (Schelling 2007 [1842], 27), i.e., it conceals what is crucial, the content. Allegory is therefore premised on (i) a discrepancy between form and content in mythology, and (ii) the contingency (and substitutability) of form.

On the contrary, tautegorical meaning is premised on (i) the identity of form and content in mythology and (ii) the necessity of the original form of presentation. In Schelling's own words,

> Form and content, matter and outer appearance cannot be differentiated in [mythology]. The ideas are not first present in another form, but rather only emerge in, and thus also at the same time with, this form. (2007 [1842], 136)

Any translation or alteration of this original form is a falsification, for the formal properties are necessary to the constitution of meaning. In Gabriel's paraphrase, 'Whereas Hegel tried to uncover the necessity of the content of mythology, Schelling insists on the necessity of the form of representation which cannot be sidestepped.' (2009, 61) And what is more, Schelling further mentions that such necessity of form entails an absolute literalism in interpretation—that is, myths 'require nothing other than to be understood literally' (2007 [1842], 137): they must be taken at their word. This is the Schellingian version of the formalist thesis rehearsed above.

However, in the next lecture, Schelling changes perspective to mythology as it is 'objectively meant', and, in so doing, he ends up saying something quite different about how to interpret mythology—something that appears, on first blush, inconsistent with this formalist thesis. For example, Schelling states in Lecture Nine,

> It is not at all the things with which man deals in the mythological process by which consciousness is moved, but rather it is the powers arising in the interior of consciousness itself… The contents of the process are not merely imagined potencies but rather the potencies themselves… the pure creating potencies whose original product is consciousness itself. Thus, it is here where the explanation fully breaks through into the objective realm, becomes fully objective. (2007 [1842], 144)

The 'contents' of the mythological process, Schelling asserts, are 'the potencies themselves' insofar as they act as the material conditions of consciousness. But, evidently, to understand the 'contents' of mythology in terms of these potencies is precisely not to interpret mythology literally; it is an allegorical interpretation which translates mythology into being actually about something other than 'the things with which man deals in the mythological process'. Bluntly put, no system of mythology makes the potencies its explicit subject matter. Rather, unsurprisingly, the 'objective' meaning of mythology turns out to be nothing other than Schellingian metaphysics itself.[11] A few pages after the introduction of tautegory as an interpreta-

---

[11] Hence, Schelling concludes, 'Mythology is known in its truth and thus only truly known, when it is known in the process. But the process that repeats itself in it, only in a particular way, is the universal, the absolute process, and thus the true science of mythology is accordingly the one that

tive ideal, Schelling himself refuses to respect the necessity of mythic form, but renders it in need of transformation into his own philosophical terminology, in order to thereby unlock a hidden meaning. The truths buried within mythology are only revealed once mythology is reconstructed as philosophy—a typically allegorical procedure.

Indeed, according to Lecture Nine, no specific instance of mythological representation can lay claim to truth ['No single moment of mythology is the truth, only the process as a whole' (Schelling 2007 [1842], 147)]. Each individual myth is 'always only a dead remnant—a remnant thus become meaningless.' (148)[12] Myths lack meaning when considered as self-contained stories; according to Lecture Nine, they should not be interpreted literally. On this view, Schelling's formalist thesis is superseded by an embeddedness thesis, by an 'objective' perspective: 'The objective explanation elevates itself… above a subjective explanation… as the ultimately alone victorious one.' (145) The discourse on the natural grounds of meaning trumps any discourse on the forms of meaning. Introduced with such fanfare in Lecture Eight, tautegory is part of a subjective interpretative framework suddenly discarded a few pages later.

9.3 The Confusion of Forms and Grounds

Schelling scholarship contains a number of ways to deal with the apparent discrepancy between the perspectives of Lecture Eight and Lecture Nine, but I want to suggest in this present section that these solutions often promulgate a problematic separation between 'subjective' and 'objective' meanings of mythology that in fact falsifies the structure of Schelling's lecture course.[13]

One example is Markus Gabriel's use of the different interpretative frameworks provided by Lectures Eight and Nine to distinguish between mythological and philosophical consciousness. He writes, 'The condition of possibility of a philosophy of mythology is that mythology is absolutely other to philosophical logos.' (2006, 252) While mythmakers are compelled to find mythological forms necessary and so must interpret them tautegorically, as soon as consciousness becomes free, as soon as there is a deliberate, intentional product of consciousness, then the *philosophy of*

presents the absolute process in it. But to present this process is the task of philosophy. The true science of mythology is for this reason the philosophy of mythology.' (2007 [1842], 151) Mythology finds its correct interpretation in the structures of Schellingian metaphysics.

[12] Schelling's use of the language of life and death to describe meaning is worth bearing in mind considering what follows in this chapter.

[13] On occasion, it should also be noted, this tension is embraced without fear of inconsistency: Beach, for example, emphasizes Schelling's critique of 'translatability' (i.e., that myths can be 'properly "deciphered" and translated into straightforward concepts' [1994, 38]) at the very same time as characterising his project as a 'search for hermeneutical principles' to get at the deeper ontological significance of myths. (1994, 1)

mythology becomes possible—and this demands an allegorical practice of interpretation. The difference between Lecture Eight and Lecture Nine thus corresponds to the difference between mythology and philosophy of mythology. Philosophy is in our control; it is intentional, and so, for the philosopher, the original mythic forms can be disposed of and reconstructed (form is contingent).[14] Gabriel therefore retains a separation between (in my terms) a formalist thesis and an embeddedness thesis on the basis of the 'heterogeneity of mythos and logos' (2006, 255). Accordingly, Schelling appeals to tautegory precisely to indicate the fact that mythology is not philosophy, that the two domains obey very different rules of interpretation.

Another solution to the above apparent incoherence between forms and grounds of meaning ends up picking one 'side' over another. Jean Pépin's 1958 *Mythe et allégorie* exemplifies this approach. Pépin celebrates in Schelling's philosophy of mythology 'the discovery of the immediate truth of mythology':

> Deprived of all signification, [mythology] gains validity by itself alone for itself alone; if the truth of mythology does not reside in some kind of secret message, it itself must be true. Schelling became increasingly aware of the necessity of defining myth by this double character of autonomy and truth. (1958, 58)

The tautegorical structure of myth (its 'autonomy') is, for Pépin, its defining feature, and so the essence of a Schellingian approach to mythology is to be located in its literalism:

> While the nature of signs is obtained only by means of signification, the entire value of myths resides in their own literal meaning; they are the real as significant or signification as real; content and container, form and matter are in them one and the same thing; they allude only to themselves and have broken any link by which the allegorist wished to subject them to signification.... Myth is true in an immediate, intrinsic and unconditional manner. (1958, 60)

Pépin goes on to add to this essential feature of 'autonomy' the further Schellingian principle of the 'historicity [of mythology] in the experience of consciousness' (1958, 60), i.e., mythology as revelatory of the world-forming structures of the potencies. Nevertheless, this second principle is a somewhat belated addition to Pépin's reconstruction: despite a cursory appeal to the conceptual apparatus of Lecture Nine, it is clear that autonomy is the central feature of Pépin's interpretation. Hence, according to Pépin, the structure of Schelling's *Historical-Critical Introduction to the Philosophy of Mythology* can be reduced to a simple three-term

[14] One peculiar result of Gabriel's analysis is that the philosopher (and the discipline of philosophy in general) is necessarily post-Christian, i.e., only possible once consciousness has freed itself from its necessary subjection to the theogonic process. In other words, the logos/mythos dualism comes to be married to a Christ/pagan dualism. Gabriel views his own philosophical interpretation of Schelling as liberated in this way: 'I do not venture to repeat what Schelling and Hegel said… I do not even believe that this is possible… Understanding always entails understanding differently. Or, to borrow Gadamer's famous phrase, 'it is enough to say that we understand in *a different way, if we understand at all*.'' (Gabriel 2009, 19) For further interrogation of Gabriel's own practice in this context, see Whistler (2015).

dialectic (reflected in Pépin's chapter-structure): the question at stake in the lecture course is that of the correct meaning of mythology, and, to answer it, Schelling first rejects the idea that mythology has no truth; he then rejects the idea that mythology contains allegorical truth; before finally finding his answer in the idea that myth possesses autonomous truth, i.e., tautegorical interpretation is the solution to the problem posed by the philosophy of mythology.

Despite their vast differences, both Gabriel and Pépin insist on preserving the separation of the subjective perspective from the objective perspective in a similar fashion to my (inadequate) summary of Schelling's position above. However, this is to misrepresent the structure of Schelling's lecture course, which ultimately does not neatly distinguish between the two perspectives in this way. To take Pépin's reconstruction as my reference point: Lectures One and Two do indeed reject the contentions, first, that 'mythology has no truth' and, secondly, that 'truth is in mythology, but not in it as such' (the allegorical approach) (Schelling 2007 [1842], 3). But from Lecture Three onwards, Schelling's focus changes: he undertakes a 'transition to the examination of the historical presuppositions of mythology' away from the question of the correct meaning of myths (3). Even within Lecture Eight itself, Schelling may well suggest that his introduction of the concept of tautegory provides a solution to the problem of subjective meaning posed in Lectures One and Two; but this is incidental, for what is primarily at stake is a continuation of the discussion of the grounds of myth within Lecture Eight itself. To put it another way: in Lecture Eight, Schelling mixes up the two discourses of forms of meaning and grounds of meaning—and the following passage is exemplary of this 'confusion':

> The succession [of mythic forms] is to be understood as an actual one… It is movement that truly occurs, to which consciousness is in reality subjugated. Even in the specifics of the succession, it is not arbitrariness, but rather necessity, that a particular god and no other precedes or follows some other god … No matter how off they appear to us, in the consciousness, the conditions out of which the representations of the same naturally arise, will constantly be proven. In order to take on a comprehensible and graspable meaning, the emasculation of Uranus, the regicide of Kronos, and other countless acts and events of the history of the gods require nothing other than to be understand literally. (Schelling 2007 [1842], 137)

The demand to understand myths literally (the culmination of the formalist thesis) is a direct consequence of Schelling's description of a transcendental process to which consciousness is subjugated (the embeddedness thesis). Far from being the moment of their greatest separation, Schelling talks about forms of meaning and grounds of meaning at the same time in Lecture Eight. The formalist thesis and the embeddedness thesis here come together—and, I am going to suggest, reveal their basic unity as part of an overarching naturalism of meaning.

Hence, what I want to argue in the continuation of this chapter—against any account of the incoherence or even separation of the formalist thesis and the embeddedness thesis—is that interpretation by tautegory and interpretation in terms of natural grounds are consistent. And this is because the notion of tautegory already includes a grounding moment within itself as well as a formalist moment—in Schelling's terms, it already announces and even fulfils the transition to the perspective of objective meaning. That is, the passage from Lecture Eight to Lecture Nine is internalized within the concept of tautegory itself; it acts as a kind of trojan-horse concept opening out the subjective perspective to objectivity. One way of making this immediately intuitive is in reference to the pervasiveness of grounding structures in Schelling's post–1809 philosophy: any self-referential claim modelled on a principle of identity like tautegorical interpretation (A = A) is immediately a claim that involves a ground-consequent relation (Tritten 2012, 280–2). The formalist thesis has the structure of grounding build into it, as it were. This is one reason why 'mythology subjectively meant' and 'mythology objectively meant' should not be kept apart when reconstructing Schelling's philosophy of mythology. However, I want to go on to suggest that there is an even more specific reason for this unity of the formalist thesis and the embeddedness thesis in the late Schelling's account of meaning: *tautegory renders meaning a natural fact*. This is, I contend, ultimately why, far from marking a moment which is maximally irreconcilable with the argument of Lecture Nine, the introduction of tautegory is the moment at which the 'subjective' tilts over into the 'objective'.

To demonstrate this, I proceed in a circuitous manner by suggesting that Schelling here belongs to a syncretic, post-Kantian group of philosophers who likewise try to think these formalist and grounding moments together to render meaning a natural fact.

9.4 Contexts I: The Two 'Slopes' of Linguistic Meaning

To invoke the names Maurice Blanchot and Remy de Gourmont in this Schellingian context might feel jarring; nevertheless, in the present section, I want to make this leap, so as to suggest that even in thinkers as otherwise distant from each other as the late Schelling and these two theorists, there remains an invariant structure in their accounts of meaning that brings together versions of the formalism thesis and the embeddedness thesis for the purpose of a naturalism of meaning. To turn to them is therefore to enhance our understanding of how this structure plays out in Schelling's philosophy of mythology.

My suggestion is that Blanchot's reflections on language identify, in a manner comparable to Schelling, a dual structure of ready-made forms and natural conditions. Unlike his contemporaries Paul de Man (for whom the naturalization of language constitutes the very structure of ideology)[15] or Hilary Putnam (for whom naturalising meaning is magical thinking; Putnam 1981, *passim*), Blanchot does not avert his gaze from the *factuality* of meaning, but instead confronts it by precisely trying to think together his own variants of the formalism thesis and the embeddedness thesis. In Blanchot's own terms, 'language has two slopes' (1995, 330), and these two slopes correspond relatively precisely to the Schellingian split

[15] See note 1 above.

between 'subjective' tautegorical forms and 'objective' grounds. Thus, on the one hand, to explain meaning is to excavate the grounding (and ungrounding) 'movement which precedes' language (1995, 327). On this 'slope', meaning must be described according to this prior movement as 'matter without contour, content without form, a force that is capricious and impersonal.' (1995, 330) On the other hand, meaning is equally to be explained in terms of 'solidified, petrified and stupefied words' (1995, 331). This second slope entails 'meaning that has become thing', that is on the verge of 'becoming senseless' (1995, 327, 331). This second slope corresponds to a formalist moment—to considering words as ready-made, static forms.[16] Nevertheless, Blanchot's point is that the two slopes ultimately converge—the formalist thesis and the embeddedness thesis cannot be thought apart. He concludes, 'Language is divided between these two slopes. The problem is that even though they are apparently incompatible, they do not lead toward distinctly different works or goals, and that an art which purports to follow one slope is already on the other.' (1995, 332)[17] And the reason for this convergence of the two slopes of linguistic meaning is 'the materiality of language'—or more fully, 'the fact that words… are a kind of nature' (1995, 327). On the one slope, meanings are reducible to natural processes; on the other, they take on an objecthood analogous to other natural phenomena. In both cases, to treat meaning as part of nature is to reconcile a discourse on the forms of meaning with a discourse on the natural grounds of meaning.

One site where the convergence of these two 'slopes' of linguistic meaning is evident is the tradition of what might be called, *Lebensphilologie* (see McLaughlin 2020), the organicist tradition according to which 'good' language is alive and animated, whereas 'bad' language is dead and petrified—along the lines of Nietzsche's treatment of dead metaphors, 'worn out and drained of sensuous force', as 'the petrification and coagulation of a mass of images which originally streamed from the primal faculty of human imagination like a fiery liquid' (Nietzsche 1979 [1873], 86). In this tradition, a special place is reserved for *the cliché* as the form in which linguistic matter dies—clichés are, as Proudhon puts it, 'large corpses lying across our path… dead words',[18] or, on Paulhan's analysis, 'a crust of words' and 'a hypertrophy of matter' in which meaning is 'already dead and reduced to nothing' (2006, 33).

The point is that, on this view, linguistic meaning is treated 'as if it were a natural effect' (2006, 68), and this, I am suggesting, is one symptom of a larger post-Kantian

[16] Notably, Blanchot develops this formalist thesis with reference to Schelling's philosophy of mythology and particularly Pépin's interpretation of it. Thus, after acknowledging Schelling's critique of allegory as 'destroying literal appearance', Blanchot concludes, 'Schelling, refusing all division between figure and meaning, nature and surnature, arrived at his famous idea of a mythology in which the gods signify only what they are.' (1992, 320)

[17] He continues, whichever slope is foregrounded, language 'has already insidiously caused you to pass from one slope to the other and changed you into something you were not before. This is its treachery; this is also its cunning version of the truth.' (Blanchot 1995, 333)

[18] Proudhon, quoted in Paulhan (2006, 29).

effort to naturalize meanings. Remy de Gourmont provides a helpful case study from the nineteenth century: he develops a naturalism of words (he 'studies [meanings] in the same way that an entomologist studies caterpillars' [Paulhan 2006, 18]) that encompasses both a formalist thesis taken to the extreme (in which only the sound of words matters) and an embedding of words in *lebensphilosophische* forces. For example, as above, Gourmont analyses clichés in terms of their 'very materiality' by modelling them as an 'organism inferior' to the 'fragments of life' that constitute valuable language (1928, 280, 304); however, he simultaneously embeds this analysis within an overarching project of a formalist 'aesthetics of the word' (1995, 11), which explains the 'physical beauty of words' in 'complete indifference to the signification of words'—that is, as 'a pure phonetics abstracted from all semantics' (1995, 11, 13). This is a radical formalism: words matter solely on account of the 'purity of their form', their linguistic 'physiognomy' (1995, 12). And, as this term suggests, this is a formalism drenched in organicist terminology: to think linguistic form is to describe 'linguistic forces' (1995, 81); it is to distinguish words that are 'always living', that 'contain a secret force' and that 'testify to the vitality of language' from words 'without living roots'—'foreign bodies that the organism rejects' (1995, 35). Gourmont's concern is ultimately to chart the ways in which meaning as form 'lives, moves, grows and dies', to 'feel… its heat, its movements, its palpitations, its life' (1995, 65, 51).[19]

Whatever else might be of interest in Gourmont's approach to linguistic meaning, it provides some kind of preliminary clue to the possibility of thinking the formalist thesis and the embeddedness thesis together as part of a naturalism of linguistic meaning in just the sort of way that Blanchot is gesturing towards. Gourmont thinks language as part of nature in a way intended to radicalize, rather than dilute, his formalist practice of interpretation.

9.5 Contexts II: *Faits sensés* from Kant to Schiller

Other philosophers who stand historically closer to Schelling also make use of this unification of formalism and embeddedness to explore the idea of meaning as natural fact. And, despite appearances to the contrary, one of them is Immanuel Kant, or, at least, the Kant of the *Critique of Judgment* as interpreted variously by Friedrich Schiller, Eric Weil and Catherine Malabou. All three understand the late Kant's project in the third *Critique* in terms of a naturalization of (some) meanings. Again, Kant's gesture towards naturalism provides a clue to understanding Schelling's own account of meaning in the *Historical-Critical Introduction to the Philosophy of Mythology*.

[19] Gourmont mimics Hölderlin's *Brot und Wein* ('Words, like flowers, must arise'): 'I have seen a word born; it is to see a flower born' (1995, 85).

Hence, Malabou's *Before Tomorrow* argues that Kant's *Critique of Judgment* 'exposed the transcendental to the factuality of life' (2016 [1790], 161)—or, more fully, in the third *Critique*, the whole apparatus of the transcendental philosophy comes up against the meaningful structures of the natural world, the fact that nature's forms should sometimes be encountered as always already containing meaning-like properties—in a way that seems independent from and prior to our understanding of them. In the third *Critique*, the subject encounters nature *ready-made*, with a meaning that appears to already subsist in it. Moreover, Malabou explicitly acknowledges that this transposition of the problematic of the *Critique of Judgment* into the language of meanings in nature resurrects Eric Weil's 1970 *Problèmes kantiennes*, since both read the late Kant as a naturalist of meaning. They both take the 'bridging' project of the third *Critique* (1987 [1790], 5.175) to constitute a reversal of the earlier *Critiques*' insistence on transporting meanings outside of nature (Weil 1970, 59). The third *Critique* undertakes a 'new task' (1970, 64): to think the harmony of meaning and nature *as if* the former were actually encountered empirically. The third *Critique*'s 'new task' is to describe, recognize and interpret meanings *naturalistically*. Weil develops the Kantian naturalization of meaning most fully:

> The *Critique of Judgment*... attempts to comprehend *faits sensés*, not just the meaningless facts that are organized by science and not just (at the level of practical reason) a meaning always postulated and eternally separated from facts... [but also that] meaning is a fact, facts have a meaning—this is the fundamental position of the final Critique. (1970, 64–5)[20]

The task of the third Critique is to seek out and describe the structure of these *faits sensés* in the organic and aesthetic realms.

Such is the context to Malabou's claim that the natural facts of the third *Critique* 'render thought useless through their independence' (2016, 169). Aesthetic and organic objects appear as 'self-meaning'—or, in Kant's original term, 'heautonomous' (1989, 5.185). Meanings are decoupled from subjects.[21] They 'appear closed in on themselves, self-formed, self-normed, dismissing our jurisdictions immediately', and so exhibit 'a rationality that appears to be able to do without reason'—a 'decorrelated' or 'factual rationality': 'rationality in which meaning is given without us' (Malabou 2016, 169). This precisely recalls the structure of tautegorical meaning—a 'meaning... given without us', i.e., to experience it is to experience a *decorrelation of* meaning from the subject. The late Schelling stands in some conti-

[20] I leave the term '*fait sensé*' untranslated in what follows. While the English edition of Malabou's *Before Tomorrow* translates it as 'meaningful fact' (2016, 170), this seems to downplay the idea that such facts have been *meaning-ed*.

[21] When the human subject encounters this meaning, it encounters something alien, even if the human must in some sense still ultimately be the source of this meaning, according to Kant. This is the Kantian circle of reflection or the structure of the 'as if' set out in the third *Critique*. Hence, Malabou uses quasi-Hegelian vocabulary of alienation and appropriation ['Reason encounters itself *as a fact* in nature' (2016, 172)] and Weil turns to a language of 'borrowing' ['Nature, as nature, must be conceived as borrowing... a meaning from thinking' (1970, 66)]. Naturalistic meaning is on loan from the subject; Kant's position is strictly a *quasi*-naturalism of meaning.

nuity with Kant in describing forms of meaning indifferent to human thinking—what Pépin dubs in a passage quoted above, 'the real as significant or signification as real'.[22] Weil will even go on to set out on this basis an interpretative practice of 'modern naivety' or 'post-critical naivety' (1970, 107–9) that roughly echoes Schelling's and Coleridge's avowals of literalism.

Significantly, this Weil-Malabou interpretation of the *Critique of Judgment* as a book about *faits sensés* had already been developed in post-Kantian Germany—in Friedrich Schiller's naturalism of aesthetic meaning. Schiller's reception of the third *Critique* is similarly premised on the property of heautonomy, or the autarchy of meanings—a form is beautiful if it does not 'require anything outside of itself', 'only if it explains itself' (2003 [1793], 167, 155).[23] And this is a consequence of the rigour with which Schiller thinks through Kant's coupling of beauty and the organism: the aesthetic comes to be a form of life located in the natural world. Schiller writes, 'Self-determination resonates back at us from certain appearances of nature, and we call it beauty.' (2003 [1793], 152) Beauty is a product of natural facts already bestowed with meaning, i.e., 'living forms' (1967 [1794], 101, 107). Hence, meanings are to be found already 'realized in the world of facts' (1900 [1793], 131–2) and particularly among those natural forms that appear to have already formed themselves ready-made, independently of the mind's structuring activity—they 'demand no explanation' (2003 [1793], 155).

9.6 Contexts III: Inorganic Meanings

The series Kant-Schiller-Gourmont constitute a set of organic naturalisms of meaning. For them, meaning is living form, or nature come alive. Schelling himself certainly flirts on occasion with this equation of meaning and living form, but he also repeatedly resists the reduction of meaningful nature to the organic.[24] He celebrates inorganic meanings too, and, to this extent, also participates in a counter-tradition of inorganic naturalisms of meaning, in which words arise like stones, rather than like Hölderlinian flowers.[25] For example, Schelling writes that 'each mineral is a real

[22] See, for example, Jean-Luc Nancy's reconstruction of Schellingian tautegory as 'the name for logos structuring itself or, and this comes down to the same thing, the name for the cosmos structuring itself in logos.' (1991, 50)

[23] Hence, Schiller will speak of the Greek gods in ways that stand extremely close to Coleridge's and Schelling's tautegorical interpretation of them: 'The whole figure reposes and dwells in itself, a creature completely self-contained' (1967 [1794], 109).

[24] In line with Schelling's overall project to counter the 'excessive organicism' of the philosophical tradition as Grant reconstructs (2006, 125).

[25] This is Dirk de Schutter's counter to de Man's critique of organicism. He points to the importance of those inorganic aspects of meaning in classical German thinking where the signifier 'turns toward the earth' and takes on 'the materiality of stones' (1989, 107) as personified in the figure of Niobe who, on Schutter's interpretation, 'gives up all human characteristics and accepts the self-effacing existence of a stone' (1989, 106–7).

philological problem', that 'there are… homologous language formations like there are mountain formations' and that one must 'cognize the physical in language, and pursue… language in connection or at least in analogy to the geological' (1856–61: 5.247, 8.453, 8.453). At these moments, Schelling ruptures the identification of meaning with living form to posit an unrestricted naturalism of meaning (an *absolute* naturalism of meaning) that encompasses 'dead' forms.

Moreover, Schelling is not alone in breaking from the organic paradigm for meaning. Novalis, for instance, similarly tries to think the 'similarity of historical geology and mineralogy to philology' (1997 [1797–9]: 125), such that the tautegorical formalism gestured towards in his *Monologen* (according to which words 'play only with themselves, express nothing but their own marvellous nature'; 83) testifies in part to a petrological vision of meaning. A later variant of this inorganicism of meaning, one which emerges out of a very different intellectual background, is to be found in Roger Caillois' investigations into the writing of stones, in which meaning manifests 'a kind of indifference towards what is human' (Yourcenar 1985: xiii). Caillois encounters in the stone a means to decorrelate meaning from the human subject; the stone testifies to a regime of sense impervious to all organic meaning-making. His position sometimes looks a lot like Kant's invocation of *faits sensés* in the *Critique of Judgment*, but extended into the inorganic (it is *as if* rock formations have meaning too): just as for the Kant of the third *Critique*, there is in Caillois' work what Yourcenar describes as an 'inverted anthropomorphism' (1985: xii)—a kind of divesting of the realm of meaning out of the human mind and into nature. And just as for Kant too, these inorganic meanings are understood through a reflective judgment structured by analogy, through 'surprising resemblance[s]', 'accidental figures', 'echoes', 'suggestions' and 'subtle semblance[s]', for 'it is as if [these rocks] were really conscious organisms' (1985: 1, 22, 34, 52). According to Caillois, aesthetic meaning has 'emigrated from the art gallery to the museum of natural history' (1985: 25).

There is even a tautegorical resonance to Caillois' analysis. Like tautegories, his stones 'refer only to their own form' (1985: 70) in a way that both 'precedes and goes beyond' anthropocentric meanings. 'The style of the stone' (1985: 36) in Caillois' analysis exemplifies a version of the formalist thesis (in analogy to the structure of tautegory) and embeds this formalism within inorganic nature. This is a radicalization of the natural fact—*faits sensés* that have become unrestricted.

9.7 'Mythology Objectively Meant'

Much of the above can, I am suggesting, help make sense of Schelling's *Historical-Critical Introduction to the Philosophy of Mythology*: Blanchot and Gourmont suggest ways in which versions of the formalist thesis and the embeddedness thesis can be reconciled as part of a naturalism of meaning; the Schiller-Weil-Malabou reading of Kant's third *Critique* helps determine what such a naturalism of meaning might look like through the concept of *faits sensés*, and Caillois makes clear—in line with

Schelling's project itself—that this naturalism can be extended into inorganic regimes of sense-making. I am contending that these clues bring us closer to an understanding of the compatibility of forms and grounds in the philosophy of mythology.

Everything rests on the concept of tautegory. Tautegorical meaning appears *as an object*, specifically as one natural object among others. A tautegorical meaning is a naturalized meaning, a *fait sensé*. Thus, while Schelling does indeed partially introduce tautegory in Lecture Eight of his course to answer a problem arising from the 'subjective' approach to mythology, i.e., a problem with interpretation, tautegory itself performs a reversal—it marks the moment when a historically-situated consciousness first takes meaning as an object, when 'mythology *objectively* meant' becomes available to consciousness—and leaves the practice of interpretation behind. Tautegory makes a non-subjective perspective on mythology available to the subject, thereby instantiating the moment at which thinking and subjectivity are first decorrelated, when the subject transcends itself, negates itself or, at the very least, abstracts from its own subjectivity.

This is, in fact, a long-standing conceptual structure throughout Schelling's philosophical trajectory—a practice by which the subject divests itself of subjectivity and decorrelates itself, so as to become immersed in and oriented by a natural world independent of the human. For instance, this structure is to be found in the activity of immersive abstraction that opens the system of identity, in the process of depotentiation that starts off the philosophy of nature and the moment of *ecstasis* celebrated in the Erlangen lectures.[26] On this reading, tautegory takes its place as one more methodological redescription of the ways in which subjectivity can be neutralized for the subject.[27] The very idea of a philosophy of nature to which Schelling remains committed and which the *Historical-Critical Introduction* recapitulates in a mythological key demands this methodological structure: the objective genitive of philosophy of nature ('of nature') designates a philosophising immersed in nature, i.e., a thinking as nature. Tautegory, like the other methodological concepts mentioned above, is structured by this objective genitive: thinking is thrown into the natural world, in this case by reconceiving meaning as a natural object, as an object of study suitable for a philosopher of nature. Thus, while the ecstatic structure of Schelling's contemporaneous positive philosophy might throw the subject out of

[26] See further Berger and Whistler (2020, Chapter Five).

[27] The concept also recalls many of the structures found in those philosophical, mystical and scientific sources that inform Schelling's own work—from Eckhart to Baader. One source worth mentioning is Goethe's practice of '*gegenständliches Denken*' (which Trop has recently translated as 'object-oriented thinking' [2023]), for this too is both a radicalisation of the Kantian-Schillerian tradition of *faits sensés* and also a deliberately paradoxical moment of confusion between meaning and object. With this idea, Goethe, in an 1823 essay, identifies a 'thinking that works *objectively*', i.e., that is 'not separate from objects' (1988, 39); hence, his maxim, 'Nature and idea are not to be separated', or more famously still, 'Everything factual is already theory' (1998 [1833], §1070, §575). As Titzmann has emphasized of Goethe, 'Meaning [already] necessarily inheres in all of the world's apparently accidental and meaningless phenomena… Reality must already have its own meaning' (1979, 658).

these necessary natural processes into a history of free, revelatory acts, tautegory is still naturephilosophical: it renders the philosopher ever more identical to the necessary processes to which consciousness submits.

This marks a break from the Coleridgean notion of tautegory, as well as from those Schelling scholars who interpret Schellingian tautegory in this vein as 'saying itself to itself', as Nancy puts it (2016: 154). Tritten is clearest on the radical nature of this break. He stresses how the introduction of tautegory collapses the fact-value distinction to the extent of doing away with any conventional notion of meaning whatsoever. Tautegorical meaning is no longer a matter of utterances, enunciations, propositions or references at all; instead, Tritten writes,

> Tautegory does not recognize a distinction between the fact and its meaning or a fact-value distinction… Its function is also not to express any sort of propositional truth about the world… Unlike propositional truth, there is no question of trying to correlate how the myth is intended with how to take it… One may not separate the meaning of mythology from its emergence because in a natural emergence the question of the truth and meaning of the myth is not asked. Where content and form cannot be separated, one may not ask for the meaning because the meaning is not separate from the very fact. One does not ask for the meaning of facts. (2012: 278, 282)

Tautegory does away with intentional, representational meanings. Schelling might introduce the concept as an answer to the question of the subjective meaning of mythology, but it in fact marks the destitution of its very idea. Tautegory marks the point at which Schelling moves beyond 'meaning' traditionally understood.

Hence, it is here—with a better understanding of tautegory—that the solution to the problem of coherence between subjective and objective perspectives on mythology (or forms and grounds) as established in the first half of this chapter is to be found. Unlike the Coleridgean version of tautegory, Schellingian tautegory does not primarily entail a literalism in interpretation in opposition to the allegorical apparatus of Lecture Nine of the lecture course (for example). Rather, Schellingian tautegory does away with the whole problematic of interpretation in general to, instead, consider meanings as objects—as natural facts to be perceived, not interpreted. When Schelling writes that mythic events 'require nothing other than to be understood literally', he is not so much proposing an interpretation strategy as refusing to propose one. Schelling might more accurately be said to be eradicating the last vestiges of referentiality entirely from his philosophy by way of tautegory. His interest is not in a new way of reading or interpreting myths—but in bringing about the end of reading and of interpretation. To repeat, his philosophy of mythology describes the 'elevation' of the 'victorious' objective perspective over the subjective perspective—and the radical nature of this victory should now be clear: it effectuates the complete transformation of the idea of meaning from that of a signifying proposition to that of an object, i.e., to the immersion of meaning into nature. This is a victory consonant with Schelling's overarching naturephilosophical project.

Just like Schiller, Gourmont, Caillois, Blanchot and even, at moments, Kant, Schelling advocates a naturalism of meaning that encompasses both a formalist moment and a moment of grounding in natural processes. The transition from

Lecture Eight to Lecture Nine does not reveal a latent conflict between literalism and allegory as strategies of interpretation, but is rather a symptom of the increasing immersion of meanings into nature, the increasing disavowal of interpretation and hermeneutics altogether and so the increasing transformation of meaning into a natural fact, an 'elevation' of meaning into philosophy of nature. Such a transformation is clearest in the following passage:

> [The objective viewpoint] completely removes mythology (and with it our investigation) from the sphere in which all prior explanations maintained themselves… The question of how the mythological ideas were meant shows the difficulty or impossibility in which we find ourselves when assuming that they were meant as truth. For this reason then the first attempt is to interpret them figuratively; that is, to assume in them a truth, but a different one that they express immediately; the second attempt is to see in them an original truth, but one that has been distorted. But rather, after the result just obtained one can open up the question of whether the mythological ideas were meant at all, namely, if they were the object of an intent—that is, of a free holding for true. (Schelling 2007 [1842]: 135)

Buried in this passage is a suggestion that Schelling's transition away from the subjective meaning of mythology entails an abandonment of the category of meaning altogether—or, at the very least, the redundancy of all traditional categories of meaning.[28] These traditional versions of meaning presuppose that a conscious choice to intend something has been made; myths, however, Schelling has concluded, do not possess this intentional structure; they are more like stones than they are like propositions. Myths have no meaning on any conventional account.

9.8 Conclusion: Schelling and the End of Meaning

I am now in a position to determine more precisely the terms 'subjective' and 'objective' as they structure Schelling's lectures on the *Historical-Critical Introduction to the Philosophy of Mythology*:

> *Mythology subjectively meant* = mythology as it appears to an interpreting subject as reducible to a proposition composed of a series of references;
>
> *Mythology objectively meant* = mythology from the point of view of nature (i.e., abstracted from the interpreting subject), considered as an object or natural form.

In other words, to understand meaning objectively is to decorrelate meaning from the human subject, to insist on the factuality of a decorrelated meaning—or, following Thacker (2010: xv, 268), on 'meaning without us'. It is to account for meaning

[28] My use of the term 'end' here is provocative considering the fact that Schelling situates mythology at the origin of his history of consciousness and so it is more common to assert, like Nancy, that 'what Schelling demands with his "tautegory"' is 'the inauguration of meaning itself', a moment 'anterior to all representation.' (1991, 53–4) In these beginnings the end is nevertheless visible—mythology 'objectively meant' anticipates the redundancy of any traditional appeal to meaning from a genuinely *naturphilosophische* perspective.

in a way that is resistant to what have often been taken to constitute the necessary features of this concept throughout Western intellectual history (as referential/as intentional/as solely available to a concrete consciousness/as subject to interpretative practices/as non-natural and irreducible to facts, etc.).

In fact, all of the naturalists considered in this chapter ultimately end up with a similarly ambivalent attitude to the very idea of meaning. To consider words, myths or artworks formally, independent of signification and reference, but also as just another part of nature, is to find oneself on the verge of doing away with the category of 'meaning' altogether. In this vein, both Gourmont and Caillois will speak of their objects of study as 'above meaning', as 'removed… from meaning' and 'without meaning' (Gourmont 1995: 11; Caillois 1985: 43, 56, 68). With the victory of the objective perspective and the 'elevation' of meaning into a rigorously decorrelated naturephilosophical framework, Schelling too comes close to announcing *the end of meaning*—the destitution of meaning as a meaningful concept and the advent of philosophy of mythology as a narrative of 'meanings without us', a *Naturphilosophie* of *faits sensés*.

References

Adams, Hazard. 1983. *The philosophy of the literary symbolic*. Tallahassee: University of Florida Press.

Beach, E.A. 1994. *The potencies of God(s): Schelling's philosophy of mythology*. Albany: SUNY Press.

Berger, Benjamin, and Daniel Whistler. 2020. *The Schelling-Eschenmayer controversy, 1801: Nature and identity*. Edinburgh: Edinburgh University Press.

Blanchot, Maurice. 1992. *The infinite conversation*. Translated by Susan Hanson. Minneapolis: University of Minnesota Press.

———. 1995. *The work of fire*. Translated by Charlotte Mandell. Stanford: Stanford University Press.

Caillois, Roger. 1985. *The writing of stones*. Translated by Barbara Bray. Charlotesville, VA: University of Virginia Press.

Coleridge, Samuel Taylor. On the Prometheus of Aeschylus (1825). In *Shorter works and fragments*. vol. 2, ed. H. J. and J. R. de J. Jackson, 1251–1301. Princeton: Princeton University Press.

———. 1993. *Aids to reflection* (1825). Ed. John Beer. Princeton: Princeton University Press.

De Man, Paul. 1983. *Blindness and insight: Essays in the rhetoric of contemporary criticism*. 2nd ed. London: Metheun.

———. 1986. *The resistance to theory*. Minneapolis: University of Minnesota Press.

Gabriel, Markus. 2006. *Der Mensch im Mythos: Untersuchungen über Ontotheologie, Anthropologie und Selbstbewußtseinsgeschichte in Schellings Philosophie der Mythologie*. Berlin: De Gruyter.

———. 2009. The mythological being of reflection: An essay on Hegel, Schelling and the contingency of necessity. In *Mythology, madness and laughter: Subjectivity in German idealism*, ed. Markus Gabriel and Slavoj Žižek, 15–94. London: Continuum.

Goethe, J. W. 1988. *Scientific studies*. Translated by Douglas Miller. New York: Suhrkamp.

———. 1998. *Maxims and reflections* (1833). Translated by Elisabeth Stopp. London: Penguin.

Gourmont, Rémy de. 1928. *Esthétique de la langue française*. Paris: Mercure de France.

———. 1995. *Esthétique de la langue française*. Paris: Ivrea.

Grant, Iain Hamilton. 2006. *Philosophies of nature after Schelling*. London: Continuum.
Jones, Ewan James. 2014. *Coleridge and the philosophy of poetic form*. Cambridge: Cambridge University Press.
Kant, Immanuel. 1987. *Critique of judgment* (1790). Translated by Werner S. Pluhar. Indianapolis: Hackett.
Kooning, Willem de. 1951. What abstract art means to me. The Museum of Modern Art Bulletin 18.3: 4–8.
Malabou, Catherine. 2016. *Before tomorrow: Epigenesis and rationality*. Translated by Carolyn Shread. Cambridge: Polity.
McGrath, Sean. 2016. Is the late Schelling still doing nature philosophy? *Angelaki* 21 (4): 121–141.
McLaughlin, Kevin. 2020. Walter Benjamin's philology of life. *MLN* 135 (3): 746–769.
Nancy, Jean-Luc. 1991. *The inoperative community*. Translated by Peter Connor. Minneapolis: University of Minnesota Press.
———. 2016. *The Disavowed community*. Translated by Philip Armstrong. New York: Fordham University Press.
Nietzsche, Friedrich. 1979. On Truth and Lies in a Nonmoral Sense (1873, pubd 1896). In *Selection from Nietzsche's notebooks of the early 1870s*. Translated by and ed. Daniel Breazeale, 79–100. New Jersey: Humanities Press.
Novalis. 1997. *Philosophical writings* (compiled 1797–9). Translated by and ed. Margaret Mahony Stoljar. Albany: SUNY.
Paulhan, Jean. 2006. *The Flowers of Tarbes: or, Terror in literature*. Translated by Michael Syrontinski. Champaign, IL: University of Illinois Press.
Pépin, Jean. 1958. *Mythe et allégorie. Les origines grecques et les contestations judéo-chrétiennes*. Paris: Aubier.
Putnam, Hilary. 1981. *Reason, truth and history*. Cambridge: Cambridge University Press.
Schelling, F. W. J. 1856–61. *Werke*. Edited by K. F. A. Schelling. Stuttgart: Cotta.
———. 2007. *Historical-critical introduction to the philosophy of mythology* (unpublished lectures, c. 1842). Translated by Mason Richey and Markus Zisselsberger. Albany: SUNY Press.
Schiller, Friedrich. 1900. On the sublime. *Essays aesthetical and philosophical*. 128–41. London: G. Bell and Sons.
———. 1967. *On the aesthetic education of man in a series of letters* (1794). Translated by Elizabeth M. Wilkinson and L. A. Willoughby. Oxford: Clarendon Press.
———. 2003. Kallias or concerning beauty: Letters to Gottfried Körner (1793). Translated by Stefan Bird-Pollan. In *Classic and romantic German aesthetics*, ed. J. M. Bernstein, 145–84. Cambridge: Cambridge University Press.
Schutter, Dirk de. 1989. Words like stones. In *(Dis)continuities: Essays on Paul de Man*, ed. Luc Herman et al., 99–110. Amsterdam: Rodopi.
Thacker, Eugene. 2010. *After life*. Chicago: University of Chicago Press.
Tilliette, Xavier. 1984. *La mythologie comprise*. Naples: Instituto Italiano.
Titzmann, Michael. 1979. Allegorie und Symbol im Denksystem der Goethezeit. In *Formen und Funktionen der Allegorie: Symposium Wolfenbüttel 1978*, ed. Walter Haug, 642–662. Stuttgart: Metzlersche.
Tritten, Tyler. 2012. *Beyond presence: The late F. W. J. Schelling's criticism of metaphysics*. Berlin: de Gruyter.
Trop, Gabriel. 2023. The ethics and politics of force from Hemsterhuis to Günderrode. *Symphilosophie* 4: 227–261.
Weil, Eric. 1970. *Problèmes kantiennes*. Paris: Vrin.
Whistler, Daniel. 2015. The new literalism: Reading after Grant's Schelling. *Symposium* 19 (1): 125–139.
———. 2016. Naturalism and symbolism. *Angelaki* 21 (4): 91–109.
Yourcenar, Marguerite. 1985. Introduction. In Roger Caillois, *The writing of stones*. Translated by Barbara Bray, xi–xix. Charlotesville: University of Virginia Press.

Daniel Whistler is professor of philosophy at Royal Holloway, University of London. He is author and editor of several works on eighteenth- and nineteenth-century European thought, including *The Schelling Reader* (Bloomsbury, 2021), *The Edinburgh Edition of the Philosophical Works of François Hemsterhuis* (Edinburgh UP, 2022–3), and Victor Cousin's *Philosophical Fragments* (Oxford UP, 2025).

Chapter 10
Schelling and Coleridge's Theories of Matter

Dale E. Snow

Abstract Coleridge often claimed that the similarities of his thought to Schelling's was due not to direct influence, much less plagiarism, but their shared intellectual background and inspirations. I explore one sense in which this may have been true and examine how Coleridge employs one of the key insights he shares with Schelling: the idea that matter, as reflected in natural entities, has an educative and even revelatory function. One of Coleridge's longest quotations from Schelling is from a minor polemic of 1806, in which Schelling clearly identifies his quarrel with the science of his time; it is reminiscent of Coleridge's own criticisms of empiricist science. I discuss the neo-Platonic roots of this set of ideas, and how Schelling and Coleridge both argued for, and in their later writings thought through, an understanding of nature which could serve to aid in our recovery from our alienation from it.

Coleridge claimed in *Biographia Literaria* that he was fortunate to have arrived at many of the same insights that Schelling had, but that he had taken a different path to reach conclusions that in some respects seem quite similar. More than a few erudite literary detectives have applied themselves to what might loosely be called the anxiety of influence question. Many of these efforts have been focused on Schelling's better-known works, such as *Vom Ich*, *System des tranzendentalen Idealismus*, and *Über das Wesen der menschlichen Freiheit*. I intend to interrogate a less popular work of Schelling's, and consider a comparatively neglected question, that of the meaning and status of nature. It seems clear that Coleridge ultimately rejected Schelling's *Naturphilosophie*; indeed, he even objected to the very word.[1] However,

[1] 'I *cannot* approve the choice of the Proper Name, Naturphilosophie, because it is 1. A useless Paradox, 2. chosen to make the difference … and his old Master (Fichte's) systems greater than it is' (Coleridge 1983 [1817], 1: 420–11).

D. E. Snow (✉)
Philosophy Department, Loyola University, Baltimore, MD, USA
e-mail: Dsnow2@loyola.edu

P. Cheyne (ed.), *Matter and Life in Coleridge, Schelling, and Other Dynamical Idealists*, International Archives of the History of Ideas Archives internationales d'histoire des idées 253, https://doi.org/10.1007/978-3-031-78315-9_10

I hope to show that Schelling's concept of matter did exert a lasting influence on Coleridge that can be detected as late as the *Opus Maximum*.

10.1 Coleridge

The more compelling question is: What is it that Coleridge found to be of such value that he strongly identified with and/or borrowed from it?[2] I want to begin by examining one of the longest paraphrased texts in the *Biographia Literaria.* This text was not from Kant, who Coleridge greatly admired, nor was it from one of Schelling's major works, but rather a little-known polemical 'pamphlet,' as Coleridge called it, 'The Statement on the True Relationship of the Philosophy of Nature to the Revised Fichtean Doctrine' of 1806 (Schelling 2018 [1807]). The 'Statement' is a fiery final reckoning with Fichte, who after several years of silence had published three works in rapid succession in 1805–06. Schelling took umbrage at several explicit and implicit criticisms of the philosophy of nature in these works, and considered himself duty-bound to respond, which he did in great haste. 'The Statement,' written specifically to set the record straight, and to combat the influence of Fichte's recent publications, had the misfortune to be published just as the Napoleonic Wars were intensifying in Germany. Real war eclipses philosophical conflict, and little contemporary notice of the 'The Statement' was taken in Germany, and less in England. Fichte was enraged by it, but the rambling refutation he composed was not published in his lifetime. Coleridge apparently read the 'pamphlet' in 1815 (Coleridge 1980–2001 [1815], 4: 344).

In *Biographia Literaria,* Coleridge (1983 [1817], 1: 142) explains his disappointment with the philosophers he has studied, from Aristotle to Locke and Hartley, and remarks 'The term, philosophy, defines itself as an affectionate seeking after truth; but Truth is the correlative of Being.'. Here we begin to see what may have drawn Coleridge to Schelling's text, for variations on this idea are a frequent refrain in 'The Statement.' Coleridge briefly traces the evolution of this idea from Plato to Proclus, Ficino, and Böhme. He then turns to Schelling's text, adding that he 'might have transcribed the substance from memoranda of my own, which were written many years before his pamphlet was given to the world' (Coleridge 1983 [1817], 1: 147).

The long passage from 'The Statement', which he part translates and part paraphrases, discusses the misjudgment and mistreatment of those thinkers vilified and persecuted as fanatics [*Schwärmer*] for their 'investigation of the indwelling and

[2] James Engell (2006, 68) sums up the situation well when he observes that 'No exculpation—fast dictation, intermingling and fusing of passages, the bizarre fact that in *Biographia Literaria* Coleridge either mentions by name or describes *every* book from which he takes significant material, or the surprising fact that Schelling, unperturbed at the situation, later had only kind words and praise for Coleridge (as did Ludwig Tieck)—can alter the fact that Coleridge's standard of citation falls far below not only that of modern scholarship but that of his own day as well'.

living ground of all things' (Coleridge 1983 [1817], 1: 148), in other words, for their unorthodox interpretation of the relation of truth to being. Schelling's text seems to be referring to Böhme, although he does not name any thinker in particular. Coleridge names Böhme and adds the names of Johann Tauler and George Fox (149). Coleridge expresses the gratitude that he feels toward these authors:

> For the writings of these mystics acted in no slight degree to prevent my mind from being imprisoned within the outline of any single dogmatic system. They contributed to keep alive the *heart* in the *head*; gave me an indistinct, yet stirring and working presentment, that all the products of the mere *reflective* faculty partook of DEATH, and were as the rattling twigs and sprays in winter, into which a sap was yet to be propelled, from some root to which I had not penetrated, if they were to afford my soul either food or shelter. (1983 [1817], 1: 152).

The stark contrast between the dead letter of the understanding and the revivifying potential of reason is a recurrent theme of Schelling's. In the *Ideas for a Philosophy of Nature* (1797) he writes: '*Mere* reflection is therefore a spiritual sickness in mankind, the more so where it imposes itself in domination over the whole man, and kills at the root what in germ is his highest being, his spiritual life, which issues only from Identity' (Schelling 1988 [1797], 11; see Coleridge 1983 [1817], 1: 158–9, fn. 1).

There are three main theses defended in 'The Statement' which would have been of interest to Coleridge. The first relies upon the previously established idea that the understanding is limited compared to reason; the second is that mechanism and mechanical philosophy is insufficient and indeed deadening to any true grasp of nature; the third is that the concept of matter is reduced to a means by mechanistic philosophy, but it ought to serve as a revelation, and thus in a sense is both a highest and lowest truth. What I am arguing is that Coleridge enthusiastically agreed with the first two claims, but not the third.

An excessive reliance on mechanistic science distances us from nature and from truth, and elements of both of the first two theses can be seen in *Biographia Literaria's* discussion of Fichte's philosophical achievement. Fichte is given credit for going beyond Kant in an effort to set up a more systematic metaphysics. But Coleridge (2000 [1817], 158–9) chides him for overreaching: 'this fundamental idea he overbuilt with a heavy mass of mere *notions*, and psychological acts of arbitrary reflection. Thus, his theory degenerated into a crude egoismus, a boastful and hyperstoic hostility to NATURE, as lifeless, godless, and altogether unholy'. Coleridge's claim is basically a gloss on another passage of 'The Statement', in which Schelling (2018 [1807], 21) explains how pernicious the concept of nature Fichte defends truly is: 'Herr Fichte has taught and maintained [that] … nature is an empty objectivity, a mere world of the senses: it consists of affections of the self, rests upon inconceivable limits, in which it feels itself enclosed, it is in its essence non-rational, unholy, not divine; in every respect dead; the basis of all reality and all knowledge is the personal freedom of man'.

With respect to Schelling's first thesis, by 1818 Coleridge had composed an essay on reason and understanding, published in his work *The Friend*, which he thought made entirely clear his views on the heterogeneity of the two (Coleridge

1969 [1818], 1: 154–61). Writing to the Rev. Joseph Hughes a year later, Coleridge reflects, 'My philosophy (as metaphysics) is built on the distinction between Reason and Understanding' and refers him to the aforementioned essay in *The Friend* (1969, 1: 161; Coleridge's letter quoted in n. 1). The distinction between two very different forms of cognition has been firmly established. Peter Cheyne (2020, 9) argues that he was influenced in this primarily by Kant: 'As Kant limited reason to make room for faith, Coleridge subordinated understanding to reason'. Inspired by Kant's philosophy of science though he may have been, Cheyne is also clear that Coleridge took this distinction in a decidedly un-Kantian direction as he continued to develop it.

Coleridge's proclivity for criticizing the inadequacy of empiricism and mechanistic philosophy for a true knowledge of nature was also an early preoccupation, even before he visited Germany. The scientific picture of the world he encountered in Newton and Locke seemed to him fatally arid, mechanical, and lifeless. Empiricism could not be the final truth of either science or the world, for consistently applied, it actively distanced man from both nature and God. Perhaps Fichte, or at least Fichte's well-known pride and insistence on human superiority to nature, is also a partial target of Coleridge's reproach expressed in the early poem 'The Destiny of Nations' (2001, 1: 281–22; self-quoted in part at 1983 [1817], 1: 122–3):

For what is Freedom but the unfetter'd use
Of all the powers which God for use had given?
But chiefly this, Him first, Him last to view
Through meaner powers and secondary things
Effulgent, as through clouds that veil his blaze
For all that meets the bodily sense I deem,
Symbolical, one mighty alphabet
For infant minds, and we in this low world
Placed with our backs to bright Reality
That we may learn with young, unwounded ken
The substance from the shadow. Infinite Love
Whose Latence if the plenitude of All,
Thou with retracted Beams, and Self-eclipse
Veiling revealest thy eternal Sun.

But some there are who deem themselves most free
When they within this gross and visible sphere
Chain down the winged thought, scoffing ascent
Proud in their meanness: and themselves they cheat
With noisy emptiness of learned phrase
Self-working tools, uncaused effects
Their subtle fluids, impacts, essences
Those blind Omniscients, those Almighty Slaves
Untenanting creation of its God.

There is a properly reverent and awe-stricken approach to nature, reflecting the awareness that nature is the other book of God. Yet what is more characteristic of man than the temptation to misuse his freedom? Those who arrogate to themselves a 'learned' insight into and ability to control nature achieve this at the price of no longer acknowledging, much less celebrating, the truth that it is God who ought to be both first and last for us. 'The Destiny of Nations' was much revised, and it seems to be a view Coleridge upheld all his life.

Over time Coleridge developed his objections to mechanism and materialism, but two central ones are interdependent and mutually reinforcing: a thoroughgoing materialism always leads ultimately to atheism, he argued, and furthermore the passivity attributed to the mind by empirical and materialist epistemology made all real connection between the mind of man and the mind of God impossible. On the mechanist view, man is at best, a passive spectator of the unspooling of an interplay of forces.

In place of this bleak vision, he proposed a dynamic model of nature as constituted by the interaction of ideal powers in relationships of polarity, which certainly has many similarities to systems constructed by Schelling and Heinrik Steffens.[3] It was all-important for Coleridge to keep this vision of nature compatible with Christian doctrine, especially the biblical account of the creation of the world, so he had to reconcile a view of nature as an apparently self-organizing organic unity with a belief in the transcendent Christian God. Cambridge Platonist and transcendental idealist theories of life are incompatible; I would argue that they are destined to come into conflict, as Coleridge's efforts to reconcile them show, and a major battle in that conflict is over the concept of matter.

On the one hand Coleridge could write that nature should be regarded as 'the other great Bible of God' and should be read as 'the expression, an unrolled yet glorious fragment of the wisdom of the Supreme Being' (1972, 6, 70, and fn. 3). Yet he also maintained that all knowledge of nature and matter is through the senses, and thus can never reach the level of true knowledge—it can only be reflected upon, leaving all of creation a mystery. One way around this difficulty is through the symbol; as when Coleridge writes of the mystics' conception of beauty as 'the subjection of matter to spirit so as to be transformed into a symbol, in and though which the spirit reveals itself' (Coleridge 1969 [1808–9], 2: 239). That could perhaps suffice to explain how we encounter natural beings, yet the conundrum of the meaning and status of matter remains, since it cannot be symbolized.

Coleridge faced further difficulties with respect to matter: it cannot be conceived of as being self-subsistent or having an independent reality, since that would be an affront to God's omnipotence. Yet it cannot be wholly dependent either, since that would make it identical with nature and raise the spectre of pantheism. One of the clearest statements of Coleridge's concept of matter is to be found in his marginal notes on Böhme's *Aurora.* He exclaims in apparent exasperation that 'the word matter, *materia*, hyle [Gk] is among the most obscure and unfixed in the whole Nomenclature of Metaphysics—and I am afraid that the knot must be cut—i.e. a fixed meaning must be arbitrarily imposed on the word, as I have done in defining matter as mere videri) (Spirit [Coleridge's own symbol for antithesis] as mere 'visibility' [lit. to be seen, to appear] as opposed to Spirit as 'that which acts and does not appear', the synthesis being Body, 'which is and is seen' (Coleridge 1980–4, 3:

[3] Some of these similarities are sympathetically discussed in Abrams (1984, 159–1).

130). Matter does not seem to be a self-subsistent entity, but to be the accidental guise in which the things of nature appear. When Coleridge speaks of a 'fixed meaning' which is 'arbitrarily' imposed, it is a striking admission that he is simply trying to dismiss the vexing problem of matter, not solve it.

Therefore, it is unsurprising that Coleridge was unhappy with Kant's theory of matter in his *Metaphysical Foundations of Natural Science.* Coleridge's doubts about both the style and substance of Kant's approach are evident. In a comment on Kant's explanation that it is not possible to make the fundamental forces of matter conceivable precisely because they are fundamental, that is, not derived from anything else, Coleridge comments grimly: 'Be it so in the present instance: and in a system which *begins* with Space and physical Powers defined exhaustively by effects and products solely relative to Space and manifestable in and by Space, so it must be. But I have learnt by experience not to suffer myself to be cowed by the term *Grundkraft*, and the impossibility of rendering same comprehensible' (Coleridge 2000, 12, 3: 284).

Yet when Kant goes on at length to characterize the fundamental powers of attraction and repulsion, Coleridge also finds this unsatisfactory, although for a different reason. He asks, of these two putatively equally important forces, how it is that 'if the two opposite forces are equal, how comes it that they do not destroy or suspend at least the other? This surely should have been explained out of the nature of one or both of the forces, or the power of which these are the forces' (1980–2001, 3: 289). This criticism has a strong family resemblance to several made by Schelling. In the *Ideas* (1797) he argues that the conflict between Kant's two powers could have no permanence without the assumption of a third factor. The *Introduction to the Sketch* (1799) returns to this issue using very similar language: 'Kant cannot complete the construction of matter from *two forces alone*. He still requires the third force which fixes the opposition, and which, according to us, is to be sought in the universal striving toward indifference, or in *gravity*' (Schelling 1988 [1797], 189).

However, in the end it was Kant's acceptance of matter as a given that seems to have distressed Coleridge the most. He complains: 'Again and again Matter is assumed as a datum, the *subject* of the powers tho' two of these powers are elsewhere taken as constituting matter' (Coleridge, 1980–2001, 3: 295). No doubt it is this assumption of matter as a datum that makes it possible for Kant to claim that matter can neither be created nor destroyed, another assertion that Coleridge takes exception to. After Kant's proof of the first law of mechanics, Coleridge does not quarrel with the proof, but simply disagrees. 'I turn coward at the thought of my own Courage, while I am about to avow, that this "Proof" appears to me a mere Sand-Rope of Assertions,' which is either atheistic (like Spinoza) and makes matter a synonym for God, or makes substance a synonym for God. He adds, 'I do not, however, deny it as a necessary assumption in mechanics; but will not let it be forced upon me, in Cosmology' (1980–2001, 3: 296).

Kant's theory of matter implies too broad a gulf between nature and spirit and at times hints at the ontological independence of matter; if no substance arises or is annihilated in all changes of nature, but the quantity of matter always remains the same, it seems to Coleridge to at least leave open the possibility of, or potential for, a troubling independence. In other words, he does not directly refute Kant's arguments so much as his conclusions, since they provide support for a concept of matter that claims a substantiality Coleridge regards as unproven and unwarranted.

At the beginning of his marginal comments on Böhme's *Aurora*, Coleridge pleads with the reader not to fall into the same error that Böhme, and at one time Coleridge himself had done, of approaching too closely to pantheism. The editors of the *Marginalia* also point to the last of the 'Essays on Method,' which they report that Coleridge considered one of his most important, on this topic. In it, Coleridge discusses 'the ground-work of all true philosophy' (Coleridge 1969 [1818], 1: 520), which is apprehension of the difference between the contemplation of reason and mere abstract knowledge. This separation must be overcome 'and prepare for the intellectual reunion of the all in one, in that eternal reason, whose fullness hath no opacity, whose transparency hath no vacuum' (522). It is at this point in the essay that Coleridge inserted a passage, sent in several letters to friends in 1818, which contains one of his strongest denunciations of pantheism. He describes it as:

> a paragraph ... unfortunately omitted—it's object being to preclude all suspicion of any leaning toward Pantheism, in any of it's forms. I adore the living and personal God, whose Power is indeed the *Ground* of all *Being*, even as his will is the efficient, his Wisdom the instrumental and his Love the final, Cause of all *Existence*; but who may not without fearful error be identified with the universe, or the universe to be considered an as *attribute* of his Deity. (1969 [1818], 1: 520, n. 1)

This, of course, was Schelling's cardinal sin, as Coleridge makes most colorfully clear when he calls Schelling a 'Hylotheist' (1980–2001, 1: 620).[4] For Schelling it is nature that is absolute and living, rather than the 'living and personal God', and this is what made *Naturphilosophie* so unacceptable to Coleridge. Indeed, Schelling compares unfavorably with Böhme in this respect: 'Schelling and his Followers are in the same case as Böhme—only that Böhme oscillates or rather leaps from Error to Truth or to the neighborhood of the Truth, while Schelling *settles* on the Error' (1980–2001, 1: 619). There are similar claims in the annotations Coleridge made to the *Freiheitsschrift*: 'In truth, from p. 429 I find little but Behmen—which a Reader must have previously understood in order to understand' (1980–2001, 4: 425). It is also noteworthy that this focus on the metaphysical status of the ultimate reality remains a preoccupation for Coleridge in the *Opus Maximum*, showing, as Paul Hamilton says, how Schelling refused to be exorcized (Hamilton 1983, 189). As Christoph Bode (2012, 615) observes, 'Coleridge's Will is not so different from Schelling's *Wollen*, Coleridge's Ground has the same systematic place as Schelling's *Grund* and *Ungrund*'.

Coleridge began to develop his own theory, as can be seen in the *Theory of Life*, probably composed in late 1816, that he hoped would move beyond and avoid the difficulties posed by the *Naturphilosophie.* It is true that well prior to this time

[4] The editors note that this usage of hylotheist—one who identifies God with matter—precedes the first citation given in the *Oxford English Dictionary*.

Coleridge had been aware of and interested in the debate about 'animal vitality'.[5] Here the traces left by his struggles with the concept and nature of matter can be seen in reverse: matter is simply appearance, nothing more. We might say that there is nothing to it, for Coleridge. Indeed, he has so successfully vanquished it, reduced it to brute stuff, that he has his work cut out for him to be able to account for the profusion and variety of actual natural entities, to say nothing of accounting for life, the title subject of his work.

This difficult marvel will be accomplished by the application of the principle of individuation. Levere (1981, 42) points out that Coleridge's definition of life, like Schelling's, drew on scholastic usage: 'I define life as the *principle of individuation*, or the power which unites a given *all* into a whole that is presupposed by all its parts'. This aspect of Coleridge's theory draws extensively from the original geological research performed by Henrik Steffens, a student and follower of Schelling's who saw himself as illustrating and in some respects extending the insights of the *Naturphilosophie*. Although not well known now, in its time Steffens' *Beyträge zur innern Naturgeschichte der Erde* (1801), exerted a wide influence. Coleridge may have felt that Steffens was a kindred spirit when he read these words in the preface: 'have I succeeded in elevating certain tones from nature that chime together with the eternal harmonies?— will that which I devotedly planted, and carefully nurtured, grow upward into a plant of a *nobler nature*?' (Steffens 2018 [1801], unpaginated].[6] Coleridge was plainly fascinated with the wealth of scientific detail from Steffens' field work and observations.[7] Steffens wanted to compose a philosophically rigorous, scientifically based history of the earth, and he composed tables of the proper relationship of the elements to each other which drew on Schellingian ideas about polarity. Coleridge returned again and again to the *Beyträge*, which he seems to have wanted to use as raw material for his own theory, providing a wealth of examples for his contentions about individuation and its development. There are many unattributed statements in the *Theory of Life* which seem to be translations or perhaps paraphrases of Steffens in the *Beyträge*, as well as from his later work, *Grundzüge der philosophischen Naturwissenschaft* (1806). The *Beyträge* drew upon the latest discoveries in chemistry and geology, and engaged in highly specific debates on particular issues; the *Grundzüge* contains more sweeping philosophical conclusions.[8]

---

[5] Nicholas Roe (2001, 10), points to the letter to John Thewall of 31 Dec. 1796, as 'mark[ing] the beginning of Coleridge's preoccupation with the science of life in poems, letters, and in notebooks'.

[6] Steffens (1801, unpaginated): 'ob es mir gelang, einzelne Töne aus der Natur herauszuheben, die mit feinen ewigen Harmonien zusammenstimmen?—ob das, was ich mit stillem Fleiss aussäete, mit andachtsvoller Sorgfalt pflegte, emporwachsen wird zur Pflanze einer edlern Natur?'

[7] Coleridge (1980–2001, 5: 241–372) made extensive commentary, including disparaging remarks on the quality of his German, on six works of Steffens. Coleridge twice remarks that he had made Steffens' discoveries 'years earlier'. See 343, 11 n., and 349.

[8] For example: 'The true speculative sense of human existence is the deepest and highest in nature, it is the innermost, holiest presentation of his true essence; man is not thrown into an alien world of necessity, he is at home in his own world, born into the light, he is not separate from nature, but rather a living organ of an eternal nature, which, like love itself, confirms, affirms, and does not

It is clear from early on in *Theory of Life* that Coleridge will not be receptive to theories which explain life as inherent in nature (too pantheistic) or as some sort of inexplicable quality (1995 [1816?], 500)[9] (too unscientific). He is unequivocal about his conviction that God is the only possible cause of life (503). 'The most comprehensive formula that he could give for life in its ascent was 'the *power* which discloses itself from within as a principle of *unity* in the *many*'' (510). Or in other words, the emergence of individuality. Coleridge's illustration of the ascent to individuality describes a process that mimics what Steffens does. Like him he begins with the metals: 'the form of unity with the least tendency to individuation' (1995 [1816?], 515). The second step is crystals, which are unions of both powers and parts. The third step is to be found in the vast roughly differentiated formations of the residue of animal and vegetable life, such as peat bogs and coral reefs. Steffens concludes in the *Beyträge*: 'In this way, through greater and greater individualization, nature draws closer and closer to the realm of the intelligences, and that which shows itself *there* lies already in the dark fastness of unconscious nature' (Steffens 2018 [1801], 315).

For Coleridge, the transition from the lower to the higher forms is also distinguished not just by the emergence of intelligence and individuality, but also by an ever more pronounced emergence of life:

> By life I everywhere mean the true Idea of Life, or that most general form under which life manifests itself to us, which includes all its other forms. This I have stated to be the *tendency to individuation*, and the degrees or intensities of life to consist in the progressive realization of this tendency. (1995 [1816?], 517)

Another equally fundamental assumption Coleridge exhibits here is that man, as Christian doctrine teaches, is the crown of creation. It is not merely anthropomorphic of him to think that he sees human superiority as consisting most clearly in the human degree of individuation; it is also of great existential import to him given his deep-seated need for redemption. As the editors of the introduction to the *Theory of Life* put it, 'the controversies over which came first, matter or spirit, whether there was a God who created and cared for man, or whether man was simply the highest point of organization reached by matter and then by organic life, were for Coleridge not merely intellectual problems but problems felt in the innermost fibres of his being' (Coleridge 1995 [1816?], 483). Whether because of or despite its Christian

---

inhibit or disturb itself. Freedom and necessity are not in contradiction, they are originally one, and what we call necessity, is the essence, the holiness of nature, it is its own, not imposed from without. [Dem ächt spekulativen Sinne ist das menschliche Daseyn selbst das Tiefste und Höchste der Natur, diese die innerste, heiligste Darstellung seines eigensten Wesens; der Mensch ist nicht hingeworfen in eine fremde, ihn zwingende Welt, er ist heimisch in der eigenen, seine Eigenthümlichkeit, rein gefaßt, klar an das Licht geboren, ist nicht ein von der Natur Getrenntes, sie ist lebendiges Organ einer ewigen Natur, die, wie die Liebe, Alles bestätigt, bejaht, in dem eigenen Wesen nicht hemmt und stört. Freiheit und Nothwendigkeit sind nicht im Widerspruche, sie sind ursprünglich eins, denn was wir Nothwendigkeit nennen, die eigentliche, heilige der Natur, ist die der eigenen, nicht einer fremden.]' Anthologie [aus seinen Werken] (projekt-gutenberg.org/steffens/werke/chap004.html)

[9] Coleridge declares that he will not explain life as an 'occult quality'.

roots, this idea still seems familiar and convincing in the guise of its modern analogues. The indisputable uniqueness of each human being has been so firmly inculcated in us from birth that we may be unprepared to recognize that there are much larger, longer-lived and arguably more complex organisms that could arguably dethrone any mere human from the top of the individuality contest.[10] Therefore all forms of matter, up to and very much including the human body, are rescued from the formlessness and passivity of being material entities by being infused with life, which enables them to develop their unique individuality.

10.2 Schelling

Schelling's earlier *Naturphilosophie* presents life and matter as arising out of a dynamic process, indeed as arising out of their mutual conflict: '*Where* it expresses itself, the principle of life is shown to be an activity that resists every infiltration of matter from outside, every impact of external force; but this activity does not *express* itself unless it is *excited* by external impact; the negative condition of life is *excitation through external influences*' (Schelling 2004 [1799], 62). However, as early as 1806, Schelling seems to want to assign greater importance to the nature and role of matter in his essay, the 'Treatise on the Relationship Between the Ideal and the Real in Nature' which he wrote as a preface to the second reprinting of *Von der Weltseele*:

> The most obscure thing of all, yes, obscurity itself according to some, is matter. However, it is precisely this unknown root out of which arises all the forms and living appearances of nature. Without knowledge of it, physics is without a scientific basis, and science itself loses the bond by means of which the idea is connected to reality. I take matter neither to be something independent of the absolute unity, which might underlie it as a kind of stuff, nor do I regard it as mere nothingness' (Schelling 2015 [1806], 235).

This seems similar to Coleridge's complaint that matter is not self-sufficient, but merely the guise in which nature appears, and needs to be animated by life and distinguished from other living things by the individuality of the forms it takes. Yet even here Schelling is groping toward a more important role for matter than as merely that which is opposed to spirit, in the way that Coleridge often does, describing a complete opposition between matter and spirit, to which God then adds life. Had he gone in a different direction with the idea that nature is 'the other great Bible

[10] Consider the massive life forms of clonal trees: some trees grow as clones of a parent tree through a very old root system. Such is the case with Old Tjikko, a Norway spruce tree (*Picea abies*) that has been growing in Sweden for an estimated 9550 years. But even that old-timer pales in comparison to the clonal cluster of quaking aspens (*Populus tremuloides*) in Utah—genetic testing has revealed that this small forest of clonal trees, named Pando, is about 80,000 years old. Considered to be one organism, its estimated weight is over 6000 tons, making it not only the oldest living thing on Earth but also the heaviest. https://www.thespruce.com/great-trees-of-the-world-1709081

of God,' he may have developed an idea of matter and life closer to the middle Schelling.

Schelling reflects further on matter and life in *The Ages of the World* (1811), which admittedly Coleridge could not have read. This has been claimed to be Schelling's *magnum opus* (Schelling 2000 [1815], vii)[11] and like Coleridge he would never publish this work despite decades of effort on various versions of the project. Joseph Lawrence has argued that *The Ages of the World* represents a final extension of the *Naturphilosophie*, rooted in the criticism of mechanical science but 'culminating in Schelling's wider attempt to universalize the evolutionary point of view' (Schelling 2019 [1811], 13). Schelling argues that even at the earliest stages, we encounter 'the spiritual and the corporeal as two sides of the same existence[;] we can well say that the present moment of their supreme inwardness of feeling is the shared birthplace of what later, as matter and spirit, will be so decisively opposed' (2019 [1811], 91). Thus, their opposition is by no means a simple one, and it is also supremely necessary, for 'if there were not such a point of convergence where the spiritual and physical are fully united, then matter would be incapable of being raised back to mind and spirit' (2019 [1811], 91). This is the problem that Coleridge is attempting to get around with the concept of life; it is the only way he can get inert matter to behave with a semblance of the grace and liveliness the world of nature exhibits.

How does Schelling arrive at the conclusion that matter plays this pivotal role? He declares that it is reserved for his own age to finally discover that path to philosophical objectivity:

> As long as philosophy restricts its concern to what is inward it must lack the natural means of an external form of exposition. But now things have finally changed. After having so long gone astray, philosophical science has recalled the memory of nature and of its own previous unity with her. And this is not all. Hardly were the first steps taken to rejoin philosophy with nature, when the enormous age of the physical became apparent. Far from being what is last, it is much rather what is first. … In due time, the contempt with which ignorant people alone still look down on the physical will pass away.' (Schelling 2019 [1811], 63).

In all three versions of *The Ages of the World* that we have, the stated plan is to treat the past, present and future in that order. However, in all three versions Schelling does not get far beyond the past. Nor does he give much context to his description of the original state of matter, or 'the secret birthplace of the living essence … in its primordial beginnings' (2019 [1811], 74). However, what is clear is that Schelling puts matter at the heart of creation. This new respect accorded the 'enormous age' of the physical is probably explained by the influence of Steffens and the rapidly accumulating evidence that the age of the earth was far greater than had previously been appreciated.

Therefore, the objection Coleridge originally raised against Kant's view, that matter ought not to be conceived of as too other, too independent, has been answered. There is an ineffable or, perhaps better, irreducible remainder in matter, an essential

[11] Jason Wirth's translator's introduction.

quality that cannot be resolved into something else. Yet it is at the vital heart of the created world. Tellingly, a part of Schelling's treatment of matter in *The Ages of the World* makes use of some of the same ideas that Coleridge adapted from Steffens, although with a different emphasis. Schelling remarks that a good observer of nature sees more in things than simply their existence as corporeal entities.

> There is something else in them or about them that first grants them the full luster and sheen of life: something superfluous, as it were, is at play around them that streams through them as a living being [Wesen] that, while ungraspable, is imperceptible. What should this nature that shines through and looks through things be if not the inner spiritual matter that still lies concealed in all the things of this world and only awaits its liberation? In corporeal things we look for it above all in metals (Schelling 2019, [1811], 91).

This is a reference to the starting point of Steffens' *Beyträge,* the metals, and the series which ends with vegetation, the animals, and after the all-important introduction of the principle of individualization, rises to the level of the human, who is said to contain the world in himself. Or as Schelling puts it, 'it is in organic life that we can most clearly see the resurgence of this principle, where it appears as the oil that sustains the green of plants or as the balsam of life that is the source of health. One recognizes it in the undeniable physical radiance that shines through flesh and glimmers in the eye' (Schelling 2019 [1811], 92). It would not be difficult to dismiss these statements of Schelling's as mystical or obscure, but it was Coleridge who thanked the certain mystical writers for reintroducing the heart, and not allowing the head alone to determine reality.

There is a final objection to address: that Schelling's philosophy and his concept of matter are too dependent, too identified with God, in short too pantheistic. By the time of the writing of the *Ages of the World* Schelling's earlier enthusiastic pantheism has transformed itself into a darker emphasis on the irrationality that precedes the rational. As was argued in the 1809 *On Human Freedom*, reality itself has a ground. In the *Ages*: 'The fundamental force that draws things together is the truly original, root force of nature. Darkness and concealment are the character of the primordial age. All life forms itself and comes into being in the night' (Schelling 2019 [1811], 83). This is no longer pantheism but a more decided identification of reality with matter; it could be regarded as the fulfillment or final development of Schelling's hope in the 'Statement' that matter ought to serve as a revelation, and thus in a sense is both a highest and lowest truth. In short, the direction Schelling went in was toward a more central role for matter, not less.

Tyler Tritten has argued that Schelling 'sets the *canonical* reading of Neoplatonism (and Platonism at large) on its head by rejecting the idea that reality descends from superior originals to inferior copies. … The contention is that Schelling's later thought continues to build upon his early reading of Plato's *Timaeus*, but in such a way that it inverts the *direction* of so-called emanationism in Neoplatonism' (Tritten 2018, 110). In other words, reality moves from the lowest level of matter toward higher levels of organization. Matter is not the opposite of spirit but a principle productive in itself, a dynamic process. In the *Ages of the World* it is also a precondition even for the emergence of the Absolute. Schelling has fulfilled his aim of placing matter at both the highest and lowest point of his system.

References

Abrams, M.H. 1984. *The correspondent breeze*. New York: Meyer Howard.

Bode, Christoph. 2012. Coleridge and philosophy. *The Oxford Handbook of Samuel Taylor Coleridge*, ed. Frederick Burwick. Oxford: Oxford University Press.

Cheyne, Peter. 2020. *Coleridge's contemplative philosophy*. Oxford: Oxford University Press.

Coleridge, Samuel Taylor. 1969. *The friend* (vol. 1, 1818; vol. 2, 1808–9), ed. Barbara E. Rooke. Princeton: Princeton University Press.

———. 1972. *The statesman's manual* (1816). In Lay Sermons, ed. R. J. White. Princeton: Princeton University Press.

———. 1980–2001. *Marginalia*. Vols 1 and 2 (1980–4), ed. George Whalley. Vols 3–6 (1992–2001), ed. H. J. Jackson and George Whalley. Princeton: Princeton University Press.

———. 1983. *Biographia literaria, or biographical sketches of my literary life and opinions* (1817), ed. James Engell and W. Jackson Bate. 2 vols. Princeton: Princeton University Press.

———. 1995. Theory of life (c.1816–18). In *Shorter works and fragments*, ed. H. J. and J. R. de J. Jackson, 1: 485–557. Princeton: Princeton University Press.

———. 2001. *Poetical works, part one: Poems (reading text)*. 2 vols. Ed. J. C. C. Mays. Princeton: Princeton University Press.

Engell, James. 2006. Biographia Literaria. In *The Cambridge companion to Coleridge*, ed. Lucy Newlyn, 59–74. Cambridge: Cambridge University Press.

Hamilton, Paul. 1983. *Coleridge's poetics*. Oxford: Oxford University Press.

Levere, Trevor. 1981. *Poetry realized in nature*. Cambridge: Cambridge University Press.

Roe, Nicholas. 2001. In *Samuel Taylor Coleridge and the sciences of life*, ed. Nicholas Roe. Oxford: Oxford University Press.

Schelling, F. W. J. 1988. *Ideas for a philosophy of nature, as introduction to this science* (1797; second edn, 1803). Translated by Errol E. Harris and Peter Heath. Cambridge: Cambridge University Press.

———. 2000. *The ages of the world* (1815). Translated by Jason Wirth. Albany: State University of New York Press.

———. 2004. *First outline of a system of the philosophy of nature* (1799). Translated by Keith R. Peterson. Albany: State University of New York.

———. 2015. *Treatise on the relationship between the ideal and the real in nature* (1806). Translated by Dale E. Snow. International philosophical quarterly 55.

———. 2018. *Statement on the true relationship of the philosophy of nature to the revised Fichtean doctrine* (1799). Translated by Dale E. Snow. Albany: State University of New York Press.

———. 2019. *The ages of the world* (1811). Translated by Joseph Lawrence. Albany: State University of New York Press.

Steffens, Henrik. 1801. *Beyträge zur innern Naturgeschichte der Erde*–Primary Source Edition. CPSIA reprint Breiningsville, PA.

Tritten, Tyler. 2018. On matter: Schelling's anti-platonic reading of the *Timaeus*. *Kabiri* I.

Dale E. Snow is associate professor of philosophy at Loyola University Maryland, USA. Her research is focused on German idealism and the history and philosophy of science. Among other works is her monograph *Schelling and the End of Idealism* (SUNY Press, 1996) and a translation of F. W. J. Schelling's *Statement on the True Relationship of the Philosophy of Nature to the Revised Fichtean Doctrine* (SUNY Press, 2018).

Chapter 11
The Ordure of Things: Coleridge, Schelling, and the Indivisible Remainder

Tim Milnes

Abstract One of the most significant features of Coleridge's thought is the chiastic way in which it figures the relationship between 'life' and 'thought'. While he developed vitalist theories about life in his 'Essay on Scrofula' and *Theory of Life*, he also insisted on the life *of* and *in* ideas. Coleridge's philosophy of/as life thus reflects his twin aims: to idealize organic life *and* to enliven and 'organicize' thought. This philosophical chiasmus, however, harboured a volatile kernel, a void which Coleridge would have encountered in the form of Schelling's 'indivisible remainder', an 'incomprehensible base of reality in things'. This chapter argues that Coleridge's term for this void was the '*caput mortuum*', an ontological excess that threatened to erupt volcanically within his 'living' ideas. The essay concludes that the central problem of Coleridge's 'Life-Philosophy' remains one of what to do with God's waste.

> What is Life? Were such a question proposed, we should be tempted to answer, what is *not* Life that really *is*?
>
> Coleridge, *Theory of Life* (1816?/1819?)

11.1 Introduction: Life, the Third Something, and the Dead Head

One of the most significant, and characteristically Romantic features of Coleridge's philosophical thinking is the chiastic way in which it figures the relationship between 'life' and 'thought'.[1] From one direction, Coleridge attempts to *conceptu-*

[1] On Coleridge and chiasmus, see Christensen (1981, 27). Christensen argues that, in Coleridge, instead of 'the methodological envoy of a dialectical empire [...] the chiasmus figures the promise of dialectic as merely the enabling presupposition of a rhetorical practice [...].'

T. Milnes (✉)
English Department, University of Edinburgh, Edinburgh, Midlothian, UK
e-mail: Tim.Milnes@ed.ac.uk

P. Cheyne (ed.), *Matter and Life in Coleridge, Schelling, and Other Dynamical Idealists*, International Archives of the History of Ideas Archives internationales d'histoire des idées 253, https://doi.org/10.1007/978-3-031-78315-9_11

alise the vital. This culminates, in the twilight of his experimentation with *Naturphilosophie,* with the aborted 'Essay on Scrofula' and *Theory of Life,* which postulates that life is the 'copula' as he puts it, and not the reduction, of irritability, reproduction, and sensibility: 'Life itself is not a *Thing*—a self-subsistent *Hypostasis*—but an *Act* and *Process*' (Coleridge, 1995b, 2: 557). Simultaneously, however, Coleridge's thought moves in the opposite direction, subordinating intellection to life itself. Accordingly, he emphasises the life *of* ideas and *in* ideas, a life that descends from the divine verb substantive, the 'I AM', which encompasses the unquantifiable copula of being and action. Sprung from the self-inaugurating Logos, such dynamic ideas are for Coleridge closer to Plato's notion of 'mysterious powers, living, seminal, formative,' and the antithesis to the mere representations or 'sensuous images' postulated by empirical philosophy since Locke (Coleridge [1817] 1983, 1: 97). In a manner characteristic of Romantic idealist thought, Coleridge's 'vital' philosophy attempts both to idealise the dynamic and to dynamize the idea-as-representation.

The reciprocal action between 'life' and 'idea' reaches its apogee in *Biographia Literaria*'s account of the emergence of a new idea-entity from the dynamic interaction of contrary powers: the 'tertium aliquid,' or 'third something'. As Coleridge explains, 'this tertium aliquid can be no other than an inter penetration of the counteracting powers, partaking of both' (Coleridge 1983 [1817], 1: 300). Significantly, it is at this very moment in the discussion that *Biographia*'s protracted preamble to the transcendental deduction of the principles of imagination comes to a halt, segueing into the self-penned letter from a 'friend,' which advises the deferral of any further treatment of the matter until his projected great work in the 'Constructive Philosophy'. And yet, this troublesome 'third something' retains a central role in Coleridge's thought. Any system that could not account for the creation of new ideas or new modes of being could not be considered as 'vital' or 'dynamic'; conversely, as he argues in *Theory of Life*, only a 'vital philosophy' could explain how 'the two component counter-powers actually interpenetrate each other, and generate a higher third' (1995 [1816?/1819?] 1: 530), thereby revealing natural evolution not as great suspended chain, but a 'ladder' that 'expands as by concentric circles' (Coleridge 1995b, 1: 537).

Conceptualising this growth proved challenging. For Coleridge, the relationship between *thinking* life and the *life of* thought was not one that could be contained within philosophy itself. As Douglas Hedley observes, 'Coleridge belongs to a Christian Platonic tradition which stretches from John Scot Eriugena to Hegel, which … is inclined to identify philosophy with theology' (Hedley 2000, 4). Accordingly, in an 1818 letter to Hugh Rose, Coleridge asks: 'Is there a such, i.e. *any one Science* which has for its specific object to be supplemental of all other Sciences? In its highest sense, Theology … is that science' (Coleridge 1956–71, 4: 863). Coleridge believed that philosophy can truly complete its task only when it becomes something *else*, something more than just philosophy; to achieve this, it must transcend the space of reasons, logic, and dialectic, and embrace Christian faith. As he writes in the 1818 *The Friend*, 'There are spiritual truths which must derive their evidence from within,' but which cannot be intuited without faith; 'and what is faith, but the personal realization of the reason by its union with the will?' (Coleridge 1969, 1: 432) Just as reason is nothing without faith and the will, so philosophy itself cannot function without the supplement of Christianity. Since 'Religion, in its widest sense, signifies the act and habit of reverencing the Invisible,' the same principle, 'which in its application to the whole of our being becomes religion, considered *speculatively* is the basis of *metaphysical* science' (Coleridge 1969, 1: 440). Applied to 'the whole of our being,' religion encompasses not just philosophy, or even theology, but 'life' in its broadest sense. It is for this reason that Coleridge insists in *Aids to Reflection* that 'Christianity is not a Theory, or a Speculation; but a *Life*. Not a *Philosophy* of Life, but a Life and a living Process' (Coleridge 1995a, 202). Viewed this way, Coleridge's twin aims, on one hand, to idealise organic 'life', and on the other, to 'organicise' and enliven ideas, precipitate a kind of disciplinary ambiguity in his thought. Thinking about life without objectifying 'life' necessitates a kind of *Lebensphilosophie*: a philosophy-of-life that is simultaneously life-as-philosophy.

The problem, however, with this chiastic philosophy of/as life is that, in 'life,' it harbours a volatile kernel whose indeterminacy Coleridge was unable to systematise and idealise. The source of the disturbance in Coleridge's philosophy can be traced to a rupture that he encountered in Schelling's middle-period *Philosophical Investigations into the Essence of Human Freedom* (1809). Schelling's vision of 'life' as pure, excessive drive within a conflicted and pulsating cosmos exposed the void at the heart of Coleridge's 'living' ideas, short-circuiting the dynamic plenitude of the synthetic *tertium aliquid* with the abject figure of the non-All: a living 'something' whose organic form bore not fullness, but a wound, a traumatic insufficiency within the real. Greg Ellerman has vividly described the 'paradoxical condition of the unconditioned, this indivisible remainder,' which is both 'contaminant and sustenance' in Schelling's system. The same unquantifiable but ineliminable excessiveness, what Ellerman identifies as 'the foreign body that enables the absolute to develop and that gives reality to all particular things,' lies at the heart of Coleridge's *tertium aliquid* (Ellermann 2015, 40). In this way, as Tilottama Rajan observes, 'life' and life sciences came to 'profoundly threaten the very project of science, if science is logic rather than asystasy.' Increasingly, Coleridge found that there was 'something uncontrollable in life: an excess, a "resistance of Life to Life"' (Rajan 2003, 405). In this essay, I build on the work of Ellerman and Rajan by arguing that another name for this asystatic, indivisible remainder is *waste*, what Coleridge terms the *caput mortuum*, or 'dead head' of existence. As Coleridge's marginalia reveals, what troubles and yet fascinates him most about the Munich period Schelling is the way in which the latter ontologises the Hartleian 'phantasmal chaos' of association, transferring radical indeterminacy from the mind of the subject into Being itself. In doing so, Schelling installs the remaindered 'dead head' *within* the Absolute, locating a void at the heart of the real to which (intolerably for Coleridge) God himself is subjected.

The central problem, then, for Coleridge's philosophy of/as life is not one of what to make of the Absolute as the 'All' or the 'One', but one of how to accommodate the Absolute as the 'non-All.' As I discuss below, one notable consequence of this rupture within the real is the dissolution of the aesthetic boundary between the idealised surplus of the *tertium aliquid* and the ordure of things—in other words, more violent forms of excremental excess and traumatic enjoyment. David Baulch notes that 'in post-Spinozist thought from Robinet to Schelling the vitalization of nature is also the exposure of idealism to the consequences of its own monism. For this monism allows matter to become Spirit, but also implicates Spirit in the processes of life, degeneration, and waste' (Baulch 2004, 576). Seen from this perspective, Coleridge's philosophy of/as life is his response to a peculiarly human problem of what to do with excrement—more specifically, the 'phantasmal chaos' of divine shit.[2] In this darker organicism, just as the wasteful remainder returns to haunt the unified Absolute, so 'life' as the flourishing, progressive *tertium aliquid* cannot be isolated from the abjected *caput mortuum*. The prospect of slipping from the sublime to the abject emerges in Coleridge's writing as a fascination with the obscene figure of pure drive. At times, indeed, Coleridge's own fascination with revolting remainders outgrows his adherence to philosophical and religious principle, in much the same way that 'life'-as-drive threatens to overwhelm the symbolic containment of polar logic and metastasise into the libidinal, violent contingency that elsewhere he strives so urgently to contain.

11.2 The non-All

By 1818 Coleridge had finally identified what had been bothering him about Schelling's philosophy. In a notebook entry from September of that year, he concludes that 'all his [Schelling's] other errors … are referable to the one [fundamental error]—the making *Nature* absolute' (Coleridge 1957–2002, 3: §4449). Two months later, in a letter to C.A. Tulk, he completes a philosophical *volte-face* that echoes his recantation of Hartley and 'all the irreligious metaphysics of modern Infidels' 17 years earlier (Coleridge 1956–71, 2: 706–07). Admitting that Schelling's philosophy is 'extremely plausible and alluring at a first acquaintance,' Coleridge complains that as a system 'it is reduced at last to a mere Pantheism … of which the Deity itself is but an Out-birth' (Coleridge 1956–71, 4: 883). It is important to understand here how for Coleridge, this first error, while fundamental, catapults Schelling into a far more egregious one: the positing of a polarity principle within the Absolute that is *antecedent* to God's will, rendering divinity as a mere 'Out-birth' of substance. Although Coleridge also accused Schelling of materialism, pantheism, and Catholicism, in truth none of these positions held any terrors for him (Plato would always silence the materialists, while the perils of pantheism were ultimately obviated by the principles of polar logic). What *did* unsettle Coleridge, however, was the implication that the condition of possibility for the divine

[2] See Žižek (1996, 36): Žižek observes that there is 'an abundance of "anal", excremental innuendo in Schelling: God secretes the "*abject*" and then pronounces the Word in order to pull Himself out of the shit He has got into'.

personality was a prior indeterminacy within Being itself. Using polar logic, Schelling, it appeared, had placed a void within God.

How did Coleridge arrive at this conclusion? It was in the *Philosophical Investigations into the Essence of Human Freedom* (1809), the last of Schelling's works to appear in the first (and only) volume of his *Philosophische Schriften*, that he had encountered the German philosopher's most sustained engagement with the relationship between absolute Being and divine Will.[3] Throughout the main phases of Schelling's long philosophical career—the *Naturphilosophie* and transcendental idealism of his early period (1794–1800); the phase of the *Identitäts-philosophie* (1801–9); the *Freiheitsschrift* and the *Weltalter* of his middle period (1809–27); and his later, Positive Philosophy (1827–54)—a fundamental question recurs, one which Kant's critical philosophy had left unresolved: what is the relationship between nature and freedom, the conditioned and the unconditioned? In *Concerning the Doctrine of Spinoza in Letters to Herr Moses Mendelssohn* (1785), F.H. Jacobi had challenged transcendental philosophy to answer this question without either exhaustively conditioning the unconditioned (thereby slipping into Spinozism) or stepping outside the conditioning chain of philosophical explanation altogether (this was Jacobi's own strategy, necessitating a '*salto mortale*,' or leap of faith [Jacobi 1994, 189]). Fichte's proposed solution was to make the subject or 'I' the unconditional ground from which nature springs through a reflexive action of willed self-positing, whereby, as he puts it in *Foundations of the Entire Science of Knowledge* (1794), '*To posit oneself* and *to be* are, as applied to the self, perfectly identical' (Fichte 1982, 99). For Schelling, however, Fichte's absolutisation of subjectivity fails to account for the *basis* of the relation between the (unconditioned) 'I' and the (conditioned) 'not-I'. Any truly systematic philosophy could not, he argued, pivot from the subject, but must instead explain the grounding of both the 'I' and nature in the copula of both—that is, in *Being*. At the same time, he dismisses Hegel's argument that relation precedes essence or substance by distinguishing between the reflexivity of human *knowledge* intimated by Fichte's dialectic and the fundamentally non-reflexive nature of the ground that any relation must presuppose as its external term. The question Schelling poses, then, is: what does it mean to think the Absolute as the ultimate (unconditioned) *precondition* of the system, rather than (as it was for Hegel) the system's dialectical *telos*? In 1809, following his attempts to think this ground in terms of nature (which, *qua* organic productivity, elides the division between freedom and necessity), the work of art (which semi-unconsciously embodies a unity of conditioned and unconditioned beyond the grasp of philosophy), and absolute Identity (as the grounding copula of self and nature, ideal and real, rendering nature as an unfolding hierarchy of potencies), Schelling adjusted his frame of reference once again.

Accordingly, in *Philosophical Investigations* Schelling attempts to grasp the nettle of thinking the unconditioned as *absolute freedom*—not in terms of human judgement or relational logic, as in Kant, Fichte, and Hegel, but as the unthinkable

[3] See *CM* 4.344.

precondition of existence. This systematic (rather than critical/transcendental) investigation into freedom entailed speculating not just about human freedom, but also about God's. Consequently, it also involved a return to theodicy, a project that Kant had endeavoured to make obsolete—only now this was revived as an inquiry into evil not (*a la* Leibniz), as mere absence or lack of being, but as the ontologising of the Kantian conception of absolute freedom. Indeed, one of the striking things about Schelling's theodicy is that evil (as absolutely unconditioned freedom) is its fundamental starting point or postulate. Since Being precedes logic, and evil, *qua* freedom is, in Žižek's words, 'the primordial, radically contingent fact, a fact which can in no way be accounted for,' the problem becomes one of how to account for the possibility and persistence not of freedom, but of reason (Žižek 1996, 16).[4] In *Philosophical Investigations*, the challenge Schelling faces is not how to reach into the shadowy domain of Being, but how to *return* from this dark, hidden world to the realm of language, communication, and understanding. From Schelling's perspective, as Žižek observes, 'the terms of the traditional problem of the ontological proof of God had to be inverted: what is truly problematic is not God's existence but his notion' (Žižek 1996, 73).

Schelling structures this relationship (between God's unthinkable freedom and his formal intelligibility) through an ontology of negation. Described in the 1815 draft of *The Ages of the World* as 'the necessary precedent (*prius*) of every movement' and 'the first transition whatsoever from nothing into something,' negation is the basis of the dynamic energy within life (Schelling 2000, 16). Thus, in *Philosophical Investigations*, he claims of God's origins that 'there is no first and last because all things mutually presuppose each other, no thing is another thing and yet no thing is not without another thing' (Schelling 2006, 28). Accordingly, God's notion, which emerges through the self-revelation of the divine Logos as *ratio* (the existing God: the illuminating, absolute whole) is only possible as a counterthrust to the inward gravitational pull of evil (the dark, irreducibly partial ground of God). Nature as ground becomes the condition of possibility for the absolute God whose very existence depends upon his attempt to extricate himself from a chaotic, spiralling death-embrace with it. As Schelling puts it, 'God has in himself an inner ground of his existence that in this respect precedes him in existence; but, precisely in this way, God is again the *prius* [what is before] of the ground in so far as the ground, even as such, could not exist if God did not exist *actu*.' While evil as *ground* contracts into darkness, God, propelled by 'the yearning the eternal One feels to give birth to itself,' emerges into existence, creation, and light (Schelling 2006, 28).

[4] Žižek's 'radically contingent fact' might sound odd here, given that Schelling dismisses the 'complete contingency [*Zufälligkeit*] of individual actions' (48) as a basis for freedom in *Philosophical Investigations* on the grounds that such empirical freedom is 'subordinate' to 'the inner necessity of a free being' (50). Žižek's point, however, is that by introducing the imbalance between existence and ground at its heart Schelling's system incorporates a radical indeterminacy between necessity and contingency. Schelling's 'higher' necessity thus never entirely escapes the spectre of the contingent.

And yet, this emergence comes at the unavoidable cost of God's perpetual entanglement with an ontologically indeterminate mode of being. The possibility of unconditioned freedom cannot be separated from the indeterminate non-All that eternally threatens to drag God back into the dark inwardness of the ground. It is simultaneously God's condition of possibility *and* that which must be rejected for darkness to be overcome and light to emerge. As Žižek notes in his reading of Schelling's *Ages*, 'Since there is nothing outside God, this "crazy God"—the antagonistic rotary motion of contracted matter—has to beget out of Himself a Son, that is, the Word which will resolve the unbearable tension' (Žižek 1996, 32). Subjected to the diastolic and systolic actions of his ground and Being, God as 'rotary motion' is unsustainable: only by releasing part of himself through the Logos or 'I Am' can he self-reveal and thereby become himself.

Schelling's identification in *Philosophical Investigations* of an inherent instability within the dynamic 'life' of the universe marks the culmination of a retreat from rational and demonstrative argument already evident in *System of Transcendental Idealism's* appeal to the semi-unconscious artwork and the attempt in the *Identitäts-philosophie* to redescribe 'philosophy [… as] nothing other than a *natural history of our mind*' (Schelling 1988, 29). This in turn involved a reconception of the function of philosophy, what Žižek describes as the 'move from *logos* to *mythos*' in Schelling's middle period, motivated by his determination to 'break out of the logical structure of reality (which can be presented as a notional system) into the Real of primordial drives (where there is no deduction, one can only tell a story)' becomes clear (Žižek 2020, 2). By installing contradiction as fundamental to any system, *Philosophical Investigations* establishes the non-coincidence of part and whole, predicate and subject as the founding principle of life. Schelling postulates an irresolvable conflict at the base of reality, a 'contradiction of necessity and freedom [without which] not only philosophy but each higher willing of the spirit would sink into … death' (Schelling 2006, 10). Reason, itself merely the negation of freedom, is incapable of encompassing its own *un*-grounding. As Žižek puts it, 'the endless oscillation between contraction and expansion [in Schelling's universe] is propelled by the impossibility of formulating the "stable" relationship between S and P that forms the structure of a propositional judgement' (Žižek 1996, 28). Even in tautological propositions, Schelling observes, there remains an ineliminable difference, a part that is unsubsumed by the whole: thus, 'Whoever says, "The body is body," surely thinks something different with respect to the subject of the sentence than with respect to the predicate' (Schelling 2006, 14).

At this point, we begin to discern what troubled Coleridge about *Philosophical Investigations*: the connection between the asystasy in Schelling's system and the 'life' of unconditioned freedom in an indeterminate ground. For Schelling, any system that is not to revert into subjective idealism on one hand *or* slide into Spinozism on the other must not only make 'life' its central term, it must also confront what is at stake in the implications of 'life' for a truly dynamic philosophy whose condition of possibility lies in the indeterminacy of evil/nature/freedom. By arguing that life itself depends upon the ceaseless, churning conflict of freedom and necessity, part and whole, Schelling exposes an abyss within the Absolute, an 'incomprehensible base of reality in things, the indivisible remainder, that which with the greatest exertion cannot be resolved in understanding but rather remains eternally in the ground' (Schelling 2006, 29). This 'indivisible remainder' in turn becomes, as Žižek characterises it, 'a void within the Real, which, since it precedes difference/indifference, necessitates the introduction of 'a new trans-ontological function … that of "less than nothing"' (Žižek 2020, 61). Neither nothing nor being, nor the synthesis of both, the indivisible remainder cannot be sublated into any symbolic order, regardless of whether this is the *Aufhebung* of Hegelian dialectic or the polar logic of Coleridge's Divine Tetractys.

This, then, is the challenge posed by Schelling's *Philosophical Investigations* to Coleridge's attempt to construct a dynamic philosophy of living ideas. As Lara Ostaric observes, 'Schelling conceives of life as a constitutive principle. Put in other words, it is not something we must assume given the limitations of our cognitive faculties, but that which pertains to the object as it is in itself' (Ostaric 2014, 62). The excessiveness of Schelling's organicism resists Coleridge's symbolic order, not epistemologically, but ontologically, and it harbours potentially devastating implications for the stability of the *tertium aliquid,* Coleridge's prototype for the growth of the productive, living idea. In Coleridge's Noetics, or higher logic of absolute relations, the principles of tetradic logic articulate the descending emanation of the Prothesis into differentiation-in-unity.[5] Instead of a unified Prothesis, however, Schelling postulates a radical disunity at the very core of God's being: thus, there can be no light without darkness, no One without the Less than Nothing, no freedom without rupture. As an appalled Coleridge notes, 'does not Schelling use Freedom often where he means no more than others mean by *Life*—i.e. the power of *originating* motion?' (Coleridge 1980–2001, 4: 443). And yet, for Schelling, 'life' depends not upon difference-in-unity but upon the chaotic vortex of forces caused by the fracture within the real. The growth in the dynamic system that is essential for life stems not from the ineffable plenitude of the One, but from an endless vacillation of the Absolute, the irresolvable conflict between God's ground and being. This schism guarantees that every productive synthesis leaves behind something that cannot be harmonised, unified, or sublated: an indivisible remainder or irreducible excess that simply *grows* as pure drive. For Schelling, every 'living idea' carries this excessiveness within itself.

11.3 Partibility

Having established the importance of this lacuna, we can return to Coleridge's critique of middle-period Schelling. As he delved deeper into Schelling's work following the completion of *Biographia Literaria*, Coleridge was in equal parts perplexed, appalled, and fascinated by what he found. For instance, in the margins of his copy

[5] For a detailed analysis of Coleridge's tetractys, see Chapter 9 of Cheyne 2020.

of Schelling's *Monument to the Scripture of the Divine Things of Mr. Jacobi* (1812), he wrote:

> Spite of all the superior Airs of [th]e Natur-philosophen, I confess [that] in the perusal of Kant I [b]reathe the free air of Good Sense [a]nd logical Understanding, with [th]e Light of Reason shining in it [an]d thro' it—While in the Physics [of] Schelling I am amused with happy [con]jectures and but in his theology [be]wildered by Positions which in their [? *be*]*st* sense are transcendent (*überfliegend*) [but] in their literal sense scandalous … (Coleridge 1980–2001, 4: 372)

This scandalised bewilderment coalesced into three principal areas of concern: materialism, pantheism, and the denial of God's omnipotence. Coleridge registers his suspicions regarding the first of these in the margins of his copy *Philosophical Letters on Dogmatism and Criticism* (1795), noting that 'the more I reflect, the more [am] I convinced of the gross materialism, [which lies under the whole system].' (Coleridge 1980–2001, 4: 412) For Coleridge, Schelling's materialism was the consequence of his dichotomisation of the real and the ideal. By treating them as opposites (by introducing a polarity *within* the Absolute), Schelling creates an indeterminacy between being and form, threatening the ontological hierarchy that elevated ordering spirit over chaotic matter.[6] Accordingly, in his notes to *Philosophical Investigations*, he attributes Schelling's separation of 'this mysterious *Ground* of Existence, which *is* and yet does not exist,' from God's actual existence to 'the material phaænomen of Partibility,' and invokes instead 'the old more reverential Distinction of the Divine Will relatively to the End from the same Will relatively to the Means: the latter of which we term his Wisdom, and to the former appropriate the name of the Divine Will' (Coleridge 1980–2001, 4: 425). What is important to note here, however, is that for Coleridge the most troubling thing about Schelling's account is not its covert materialism, but the way in which it implies a 'phaænomen of Partibility' within the divine Being, creating a gap within the Absolute that is incompatible with conceiving the Prothesis as One.

The same point applies to pantheism, which, though certainly scandalous, did not trouble Coleridge's philosophical dreams to the extent that some have suggested.[7] Certainly, many of Coleridge's difficulties with post-Kantian German idealist philosophy can be traced to his perception that it annihilated God's personality by translating divine personhood into divine substance. In his unfinished *Magnum Opus,* he assembles his most formidable critique of this tendency, defending the divine personality by arguing, first, that Being presupposes an absolute Will, 'which, therefore, is essentially causative of reality,' and secondly, that personality is an 'essential attribute' of Will (Coleridge 2002, 194–95). Moreover, to distinguish divine from mundane personhood, Coleridge differentiates God's *personeity* from human 'personality', claiming that 'the perfection of person is in God, and that personeity, differing from personality only as rejecting all commixture of

[6] For an insightful analysis of this problem, see Hedley (2000, 28).

[7] See, for example, McFarland (1969, 107): McFarland argues that the 'inability either really to accept or wholeheartedly to reject pantheism is the central truth of Coleridge's philosophical activity.'

imperfection associated with the latter, is an essential constituent in the Idea of God' (Coleridge 2002, 177). Having established the perfect personality in the Prothesis, Coleridge deduces the Logos, the 'begetting of the identity in the alterity' by way of interlinking transcendental arguments (God is love; there is no love without communication; there is no communication without otherness or alterity). Thus, language or 'the word itself,' becomes intelligible only as the 'adequate expression of the paternal personeity' (Coleridge 2002, 199).[8]

And yet, against the Munich-period Schelling at least, it is difficult to make the pantheism charge stick. Indeed, a substantial part of Schelling's discussion in *Philosophical Investigations* is dedicated to demonstrating that *only* through a philosophy of freedom-*qua*-indeterminacy can the personality of God be preserved against Spinozism. More specifically, he argues that it is precisely because rationalist philosophies possess 'no concept of personality, that is, of selfhood raised to spirit, but rather only the abstract concepts of finite and infinite' (Schelling 2006, 38) that they can only account for evil as the privation (negation) of the good. By rendering God as abstract substance, rationalism reduces Being to a lifeless concept. Against this, Schelling argues that personality presupposes *life*; thus, while 'In the divine understanding there is a system; yet God himself is not a system, but rather a life' (Schelling 2006, 62). *Life*, however, implies that within all being there is an irreducible spontaneity, an absolute freedom in the non-ground from which God must extricate himself in order to come into being. As Schelling puts it, 'All existence demands a condition so that it may become real, namely personal, existence. Even God's existence could not be personal without such a condition except that he has this condition within and not outside himself' (Schelling 2006, 62). It is worth noting that in this final caveat Schelling introduces a distinction between absolute and contingent personality that closely resembles Coleridge's discrimination between personeity and personality. God *alone* makes the condition (viz. the primal will of nature as non-ground) of his existence *qua* communicative Logos 'into *his own*'; Man's personality and selfhood, by contrast, 'can never rise to full actuality [*zum Aktus*]' (Schelling 2006, 62). Nonetheless, from a Christian perspective, it remains an uncomfortable feature of Schelling's account that without the 'primal will' (Schelling 2006, 37) of nature, within which the positivity of evil resides, the existence of God as personality would be impossible. Accordingly, even once God has emerged from the darkness of the ground, he 'does not now resist the will of the ground or abolish it. This would be precisely as much as to say that God would abolish the condition of his existence, that is, his own personality' (Schelling 2006, 66).

In this detail we can identify what is, for Coleridge, the truly 'scandalous' core of *Philosophical Investigations*. That this is *not* simply the annihilation of God's personality is clear from the fact that Schelling's philosophy of freedom not only

[8] See Cheyne (2020, 284). Cheyne notes that Coleridge's argument 'does not attempt to prove Schelling incoherent,' but instead presents an alternative account of divine self-begetting that eschews 'a pantheism leaving God bereft of personëity,' which in turn was 'untenable to most forms of Christianity, especially Trinitarianism.'

invokes God's personality as a way of countering pantheism, it also distinguishes between absolute and mundane personality in a way comparable to Coleridge's own distinction. The bone that stuck in Coleridge's throat, instead, was the way in which Schelling's account of ontogenesis as pure *life* entails the fracturing of the One, the negation of the All by the *non*-All. For Schelling, God's existence depends upon a dynamic imbalance within the real, a ceaseless roiling of ground and being that implies the irreducibility of part to whole. Accordingly, although the fractured, indivisible remainder 'is nothing else than the eternal ground for the existence of God, it must contain within itself, although locked up, the essence of God as a resplendent glimpse of life in the darkness of the depths' (Schelling 2006, 30). Instead of materialism or pantheism, then, it is Schelling's claim that God is 'a life, not merely a being' and as such subject to profound change, which bewildered and appalled Coleridge, and to which he counters in his marginal comments that 'God, as infinite & self-existing, is the alone One, in whom Freedom & Necessity can be one and the same, from the Beginning' (Coleridge 1980–2001, 4: 441). The capacity of *Philosophical Investigations* to scandalise and baffle Coleridge lay in the way that it exposed the incommensurability of forces that underlies any 'dynamic' philosophy, an asystasy which threatened to lay waste the equilibrium that he sought to establish between the vitality of life and the order of things. While the identification of the divine with matter and/or nature could be refuted, Coleridge found the subjection of God to an indeterminate non-ground, though appalling, more difficult to dismiss precisely because of the way in which it is entailed, in Schelling's demonstration, by the dynamic activity of 'life' in the system. By arguing that a *living* system presupposes the presence of a disruptive void within the Absolute, *Philosophical Investigations* undermines not just the foundations of Coleridge's philosophy of/as life, but also divine unity itself.

11.4 The Caput Mortuum

And yet, as both Hedley and Rajan have noted, even as he attacked Schelling for (as he saw it) simply rehashing Böhme and mixing philosophical argument with 'many even more *over*meaning or unmeaning Quid pro Quos—so many ex. gratiâ, *Thing*-phrases,' Coleridge remained fascinated by the latter's 'uncouth mysticism' (Coleridge 1980–2001, 4: 443).[9] This in turn raises the question: what *was* it about the bewildering and scandalous *Philosophical Investigations* that he found so intriguing? One answer might lie in the connection between the excessive, indivisible remainder and waste: more specifically, the excremental logic of the Christian incarnation. In positing a void within God, Schelling postulates a model of ontogen-

[9] See Hedley (2000, 80). Hedley registers Coleridge's ambivalence to Schelling's 'irrationalist metaphysics,' noting that he was 'most fascinated, but also rather disturbed, by the Munich Schelling.' See also Rajan (2003, 401): Rajan claims that Coleridge was 'still secretly fascinated by Schelling once he had acknowledged reading the Freedom essay and renounced him.'

esis that is more defecatory than emanatory. In its unending struggle with the All, the non-All persists as the figure of the part that cannot be unified with the whole, as the waste in the system that can be neither transcended nor sublated. And yet, as Žižek observes, this vision accords with the Protestant concept of the incarnation, in which Christ as Logos is conceived as a God who freely identifies himself with his own waste, 'with the excremental Real that is man' (Žižek 2006, 187). Here, a striking contrast emerges between Coleridge and Schelling: while in Coleridgean 'Noetic' logic (Coleridge 1995a, b, 180), God as Prothesis descends into the polarities of Logos and Spirit, in Schelling's system (since the Logos is that which God abjects in order to exist as communicative love), expelling defecatory humanity is a process that God *must* undergo if he is to achieve his synthesised existence as Father. In this way, Coleridge's philosophy of/as life is shadowed by Schelling's darker chiastic relationship between the idealised Prothesis and the abjected ground without whose wasteful divisiveness His being would be impossible.

For Schelling, then, just as there can be no God without life as pure drive, so there can be no human subject without divine shit. As Žižek's analysis demonstrates, this installation of waste as the negation of pure spirit has aesthetic as well as theological ramifications. Among the most significant of these is the way in which the aesthetics of the incarnation configures the modern subject in relation to an obscene literalisation of the irresolvable conflict between asystatic part and organised whole. Consequently, the Protestant subject emerges not as the noumenal figure whose unrepresentable essence is displaced into the ineffable experience of the Kantian sublime (or in Coleridge, faith), but as the revolting, monstrous entity whose indeterminate being (thanks to the gash in the real that establishes its condition of possibility) produces what Žižek calls a 'null representation': a literal, rather than a metaphorical failure of reconciliation and embodiment. The disgust evoked by this excremental kernel within the human is, in turn, ultimately a revulsion at the void within God, the abyss through which 'life' as growth without form or purpose contaminates organisation. As Žižek notes, 'the ultimate object of disgust is *bare life itself*' (Žižek 2016, 159).

Nonetheless, it is important to register the emotional ambivalence that lurks within disgust. Taken as abjected substance, pure life both repels *and* attracts. In our fascinated enjoyment of the loathsome and the monstrous we encounter the ground whose excessiveness is the condition of possibility for the real. It is this contradictory complex of emotions that lies beneath Coleridge's ambivalent response (a mixture of philosophical rejection and scandalised fascination) to *Philosophical Investigations*, an ambivalence that manifests itself through his own ontology of waste, or what he terms the *caput mortuum* ('dead head') of existence. Coleridge deploys the term *caput mortuum* and its variants in a generalised way to denote whatever is dead, inert, and redundant in existence and in thought: in short, whatever has been remaindered and abjected. For example, when, in *Biographia Literaria,* he attacks David Hartley's claim 'that the will, and … all acts of thought and attention, are parts and products of … the phantasmal chaos of association,' his charge against Hartley is that this materialistic hypothesis renders the soul as the mere residue of self-organising matter, the 'caput mortuum of the Hartleian process'

(Coleridge [1817] 1983, 1: 116–17). Tellingly, he deploys a similar expression in his marginalia on Schelling's *Philosophische Briefe über Dogmatismus und Kriticismus* when he complains of the 'gross materialism' that underpins Schelling's system, which rests upon '[m]ere [sen]sations, [the] corpus [mo]rtuum [of] vol[a]tilized [me]mory' (Coleridge 1980–2001, 4: 412). It was this same imputed materialism that had led him to suspect, as he put it in the letter to C.A. Tulk, that Schelling had reduced the Deity to an abjected 'Out-birth' of the system (Coleridge 1956–71, 2: 706–7). And yet, the significant common denominator in the two cases is not materialism, but the groundlessness of a 'phantasmal chaos.'

It is in *Theory of Life* (1816?/1819?), however, that the significance of the *caput mortuum* emerges into clear relief. Fittingly, Coleridge's *Theory* is itself doubly abjective: a discarded fragment of quasi-*Naturphilosophie* that he failed to sublimate into the symbolic structure of his metaphysics, it is also the by-product of another aborted project, the 'Essay on Scrofula,' which he had attempted to write with his physician, James Gillman. While writing the 'Essay,' Coleridge came to realise that if one was to understand '*the essence of the Disease itself*' (rather than simply enumerating its symptoms and accompaniments), then this involved 'the necessity of a distinct conception of life itself, or ... of the living principle' (Coleridge 1995a, b, 1: 477–8). In short, one could not answer the question, what is scrofula? without first addressing the question, what is life? In responding to this broader issue, *Theory* seeks to '*account* for Life' by *establishing* God as its prior 'ground or cause' (Coleridge 1995a, b, 1: 503). Without this 'living principle,' the investigation of life risks merely *explaining* it in terms of material organization, thereby prompting the further question: 'what is *not* Life that really *is*?' For Coleridge, there is only one possible answer to this supplementary question: God himself. *Accounting* for life, rather than merely *explaining* it empirically, involves invoking God as the absolute condition of its possibility. Consequently, and in accordance with Coleridge's hierarchical dialectics of Being, life (taken absolutely) conforms to the principle of difference-in-unity or unity-in-multeity, while in the realm of existence (or taken '*eminently*'), it appears as '*the principle of individuation*, or the power which unites a given *all* into a *whole* that is presupposed by all its parts' (Coleridge 1995a, b, 1: 510). As the disclosive, communicative unfolding of God, life's emergence is governed by the logic of polarity: it is 'the copula, or the unity of thesis and antithesis, position and counterposition' (Coleridge 1995a, b, 1: 518–19) in which the 'counter-powers actually interpenetrate each other, and generate a higher third, including both the former, "ita tamen ut sit alia et major [in such a way that it is different and greater]"' (Coleridge 1995a, b, 1: 530). The *ratio* of life, then, is once again that of the *tertium aliquid*, insofar as the latter consists in the dialectical production of difference within a principle of unity: accordingly, 'in the identity of the two counter-powers, Life *sub*sists; in their strife it *con*sists: and in their reconciliation it at once dies and is born again into a new form, either falling back into the life of the whole, or starting anew in the process of individuation' (Coleridge 1995a, b, 1: 520).

In 'falling back into the life of the whole,' however, life inevitably produces *waste*. The lesson of Schelling's *Philosophical Investigations* is that, *qua* indivisible remainder, waste is both the necessary and the impossible precondition for any conception of wholeness or unity. While Coleridge resists this conclusion, there is ample evidence that he was intrigued and unsettled by its implications. In *Theory of Life*, for instance, having traced the 'ladder' (Coleridge 1995a, b, 1: 537) of creation vertically from God to man, and thence from man to animal, vegetable and mineral life, Coleridge finds that life excretes its materiality as dead residue, leaving behind strata of 'schistous formations, which we must here assume as in great measure the residua of vegetable creations, that have sunk back into the universal life, and in the later predominant calcareous masses, which are the *caput mortuum* of animalized existence' (Coleridge 1995a, b, 1: 537–8). Preserved in the sedimented layers within the earth are the aborted products of life, indivisible remainders that could not be sublated or synthesised into a higher level of intensity and perfection. On the surface, Coleridge's account merely notes the position of the *caput mortuum* on the very lowest rung of the ladder of creation. It does not register the possibility that such leftovers are *essential* to the productivity of the system, much less that waste might form the fundamental precondition of the logic of creation behind the *tertium aliquid* itself.

Coleridge's response to this suggestion would doubtless be that life presupposes God, not the reverse. And yet the problem that he faces is that his chiastic philosophy of/as life short-circuits the relationship between explanandum and explanans. Life as explanandum, he insists, must be accounted for according to the explanans of a 'living principle'. The alternative (explaining life in terms of abstract qualities or empirical principles) 'is the philosophy of Death, and only of a dead nature can it hold good' (Coleridge 1995a, b, 1: 530). Invoking a *living* principle, however, reintroduces at the level of the explanans the problem of how to account for the 'life' in the 'principle,' which raises the uncomfortable possibility that life might be at work everywhere, even in the divine *ratio* governing productivity. This, we recall, is precisely Schelling's claim in *Philosophical Investigations* when he argues that God is 'a life, not merely a being' and as such 'subject to suffering and change' (Coleridge 1980–2001, 4: 441). Eternally caught in the gravitational pull of the negating *caput mortuum*, the Absolute as One cannot be isolated from the Absolute as less than nothing: God's plenitude is conditional upon his waste.

There is an instructive contrast to be drawn here between Schelling's philosophy, which preserves the excremental, and Coleridge's, which attempts to expel it. For Schelling, the abjected *caput mortuum is* the indivisible remainder that forms the 'unprethinkable' (Schelling 2000, 12) condition of God's existence. In *Philosophical Investigations* he describes how a chaotic 'non-ground [*Ungrund*]' of 'absolute *indifference* [*Indifferenz*] ... a being against which all opposites ruin themselves, that is nothing else than their very not-Being [*Nichtsein*]' (Schelling 2006, 68–69) finally emerges into existence and difference with the appearance of the Logos or word. The indifferent non-ground then recedes and 'dissolves itself, as the initial yearning does in man when he crosses over to clarity and, as an enduring being, grounds himself, in that everything true and good in this yearning is raised into bright consciousness'. Despite man's self-grounding, the non-ground endures as the indivisible remainder that forms the ground, just as 'the false and unclean, is locked away forever in the darkness as the eternally dark ground of selfhood, as *caput mortuum* of his life process and as potency left behind that can never emerge into actuality [*zum Aktus*]' (Schelling 2006, 70). Rather than being consigned to the wastes of redundant materiality, in Schelling's account the 'dead head' of life, as a 'potency' prior to 'actuality,' preserves the incommensurable otherness within the Absolute from which God never completely manages to extricate himself (since to do so would be to bring an end to life itself). In Schelling's terms, then, the excessiveness of the *caput mortuum* itself constitutes the precondition for the dynamic 'life' of the whole. From a Coleridgean perspective, however, Schelling effectively transplants Hartley's indeterminate, excremental 'phantasmal chaos' from the psychological realm of the associative imagination into the very ground of God's being.

11.5 Conclusion: The Excrescences of Life

Although Coleridge publicly disavowed Schelling's philosophy after 1817, his notes and private comments reveal the extent to which the chaotic ontogenesis it described continued to fascinate him. One instance of this occurs in his gloss to Schelling's explanation of the origins of disgust. Observing the human tendency to be revolted by life forms that pose no physical threat, Schelling attributes this to a malformation in all repellent creatures that stems from a primordial eruption of the ground. Our disgust at such creatures, he reasons, 'can be explained only as the result of an excitation of the irrational or dark principle of the creature that occurred at the original creation' (Coleridge 1980–2001, 4: 430). Indeed, he claims, the *caput mortuum* of extinct species in the earth is evidence that the 'inchoate ground' of nature 'may have attempted a creation on its own which, however, in the end always sank back into chaos because the bond of love was lacking' (Coleridge 1980–2001, 4: 431). Though God's loving creation brought lifeforms capable of flourishing into existence, the malign agency of the ground remains visible in repugnant creatures both extinct and existing. To this account Coleridge adds the following annotation:

> [I h]ave. often [dr]eamt of [th]is, as [a] Theory of [Hy]datids, [Ta]pe-worms and other [ex]crescences [of] Life. Supp<ose> [a] brief [me]tathesis of [? the] Bildungstrieb = nisus formativus, from its [ap]propriate Organs in Male or Female, to the [St]omach, or Liver. (Coleridge 1980–2001, 4: 431–2)

Having attended Blumenbach's lectures on physiology in Göttingen in 1799 (Coleridge 1983, 1: 207), Coleridge was familiar with the notion of the *Bildungstrieb* or *nisus formativus*, the vital power which, it was hypothesised, endowed matter with order and form, and his own medical research was influenced by the idea that all disease could ultimately be traced to a malformation of this force. In this comment, however, Coleridge toys with a more radical possibility. By associating his 'Theory of Hydatids' (larvae of tapeworm) with a dream or fantasy, Coleridge allows himself to imagine a scenario where Schelling's account is *true*, so that in certain creatures (such as tapeworms) one might witness the abjected indeterminacy of the ground creeping insidiously into existence. What Schelling's middle-period philosophy suggests, as Rajan aptly phrases it, is that 'degeneration is at the heart of life' and that, fundamentally, there can be no creation without excretion, no flourishing without entropy (Rajan 2003, 409).[10] This in turn implies that Life as unified growth presupposes the distorted metathesis of the *Bildungstrieb*, or 'life' as an excrescence—that within every organic *tertium aliquid* there is a void, an inert *caput mortuum* of less than nothing that remains simultaneously the former's condition of possibility and its impossibility. For Coleridge, then, the true scandal of Schelling's philosophy of freedom is not pantheism or materialism, but the bewildering claim that order is somehow grounded in ordure, and that, in the indivisible remainder, Being itself is riven by an abjected, 'phantasmal' ground that grasps the subject with the fascinating and uncanny knowledge of its own impossibility.

References

Baulch, David M. 2004. The 'perpetual exercise of an interminable quest': *The biographia Literaria* and the Kantian revolution. *Studies in Romanticism* 43 (4): 557–581.

Cheyne, Peter. 2020. *Coleridge's contemplative philosophy*. Oxford: Oxford University Press.

Christensen, Jerome. 1981. *Coleridge's blessed machine of language*. Ithaca: Cornell University Press.

Coleridge, Samuel Taylor. 1956–71. *Collected letters of Samuel Taylor Coleridge*, ed. Earl Leslie Griggs, 5 vols. Oxford: Clarendon Press.

———. 1957–2002. *The notebooks of Samuel Taylor Coleridge*, ed. Kathleen Coburn, et al. 5 vols. London: Routledge.

———. 1969. The Friend, ed. Barbara E. Rooke, 2 vols. Princeton: Princeton University Press.

———. 1980–2001. *Marginalia*, ed. H.J. Jackson, George Whalley, et al. 6 vols. Princeton: Princeton University Press.

———. 1983. *Biographia literaria*, ed. James Engell and Walter Jackson Bate, 2 vols. Princeton: Princeton University Press.

———. 1995a. In *Aids to reflection*, ed. John Beer. Princeton: Princeton University Press.

———. 1995b. *Shorter works and fragments*, ed. H. J. Jackson and J. R. de J. Jackson, 2 vols. Princeton: Princeton University Press.

———. 2002. *Opus maximum*, ed. Thomas McFarland, with the assistance of Nicholas Halmi. Princeton: Princeton University Press.

Ellermann, Greg. 2015. Late Coleridge and the life of idealism. *Studies in Romanticism* 54 (1): 33–55.

Fichte, Johann Gottlieb. 1982. *The science of knowledge*. Translated by Peter Heath and John Lachs Cambridge: Cambridge University Press.

Hedley, Douglas. 2000. *Coleridge, philosophy and religion: Aids to reflection and the mirror of the spirit*. Cambridge: Cambridge University Press.

Jacobi, Friedrich Heinrich. 1994. *The main philosophical writings and the novel allwill*. Translated by George di Giovanni. Montreal: McGill-Queen's University Press.

McFarland, Thomas. 1969. *Coleridge and the pantheist tradition*. Oxford: Oxford University Press.

[10] See also Ellermann (2015, 41): '[t]he problem [for Coleridge] is that an absolute *life* cannot be isolated from its deformed or diseased manifestations.'

Ostaric, Lara. 2014. The concept of life in early Schelling. In *Interpreting Schelling: Critical essays*, ed. Lara Ostaric, 48–70. Cambridge: Cambridge University Press.
Rajan, Tilottama. 2003. The unavowable community of idealism: Coleridge and the life sciences'. *European Romantic Review* 14: 395–416.
Schelling, Friedrich Wilhelm Joseph von. 1988. *Ideas for a philosophy of nature as introduction to the study of this science*. Translated by Errol E. Harris and Peter Heath. Cambridge: Cambridge University Press.
———. 2000. *The ages of the world*. Translated by Jason M. Worth. Albany: State University of New York Press.
———. 2006. *Philosophical investigations into the essence of human freedom*. Translated by Jeff Love and Johannes Schmidt. Albany: State University of New York Press.
Žižek, Slavoj. 1996. *The indivisible remainder: On Schelling and related matters*. London: Verso.
———. 2006. *The parallax view*. Cambridge, MA: Massachusetts Institute of Technology Press.
———. 2016. *Disparities*. London: Bloomsbury.
———. 2020. *Sex and the failed absolute*. London: Bloomsbury Academic.

Tim Milnes is professor of romantic literature and philosophy at the University of Edinburgh. He is the author of *The Testimony of Sense: Empiricism and the Essay from Hume to Hazlitt* (Oxford UP, 2019), *The Truth about Romanticism: Pragmatism and Idealism in Keats, Shelley, Coleridge* (Cambridge UP, 2010), and *Knowledge and Indifference in English Romantic Prose* (Cambridge UP, 2003). His next book, *Romantic Waste: Excess and Emptiness in English Literature 1789–1832* is under contract with Stanford UP.

Part III

Chapter 12
The Limits of Analogy in Coleridge's Philosophy of Nature

Lydia Azadpour

Abstract Coleridge displays a distinct ambivalence about the role of analogy in his philosophy of nature, which has given rise to opposing scholarly accounts of his views on analogy. The goal of this chapter is to untangle the various uses, contexts, and meanings Coleridge gives to analogy in his philosophy of nature. This essay argues that Coleridge's concerns regarding the notion of analogy reflect his general move to distinguish himself from what he and other contemporaries viewed as problems with *Naturphilosophie*, which he managed by limiting the role and scope of analogical reasoning in his account. In turn, Coleridge's interest in analogy took a form particular to him. He aimed to retain the homogeneous root nature of powers, while maintaining qualitative distinctions in kind between aspects of different powers. This, the essay argues, is most clearly expressed in his dual representation of the plant–animal relation as simultaneously antithetical and hierarchical.

> I no more confound magnetism with electricity, or the chemical process, than the mathematician confounds length with breadth, or either with depth…
>
> Coleridge, *Hints Towards the Formation of a More Comprehensive Theory of Life*

Coleridge makes plain his engagement with the philosophy of nature in his c.1816 *Theory of Life* and other philosophical works of the following decade. Like many of the thinkers he engages with, Coleridge presents a philosophy of nature that involves the representation of a scale of life, which seems to invite the use of analogical thinking to understand nature and living beings. But unlike many of those thinkers, Coleridge displays a distinct ambivalence about the role of analogy in his work. He often employs analogical reasoning throughout *Theory of Life* and other works on philosophy of nature. Equally, he repeatedly berates the 'German

L. Azadpour (✉)
Department of Philosophy, University of Nottingham, Nottingham, Nottinghamshire, UK
e-mail: lydia.azadpour@nottingham.ac.uk

P. Cheyne (ed.), *Matter and Life in Coleridge, Schelling, and Other Dynamical Idealists*, International Archives of the History of Ideas Archives internationales d'histoire des idées 253, https://doi.org/10.1007/978-3-031-78315-9_12

natural scientists' for 'abuse of analogy' in his marginalia,[1] with the intention correcting what he believes to be misuse of the term or the practice. But he also claims that more progress is to be made in natural science from looking for 'antithesis' than analogy.

Coleridge's apparent ambivalence has given rise to differing pictures of what he thought about the use of analogy. To give three examples: Byron Hawk gives analogy a central role in Coleridge's *Theory of Life,* by writing that for Coleridge, 'any explanation of life force has to operate through analogy', and that Coleridge 'was using argumentative analogy in relation to function' (Hawk 2007, 143). By contrast, in his paper addressing Coleridge's later work from the 1820s, 'Coleridge, Natural history, and the "Analogy of Being"', A. J. Harding claims that Coleridge's philosophy mobilises the *analogia entis* to reconcile scientific developments with his theological commitment to an intelligent divinity (Harding 2000, 144). Equally, George R. Potter (1925, 384–5) claims that Geoffroy Saint-Hilaire's ideas 'exerted… a very strong influence on Coleridge's growing appreciation of organic nature's unity of structure', citing Geoffroy's claim that different classes have analogous (what we would nowadays refer to as homologous) organs. Of course, analogies have been used and theorised in many ways—both in metaphysical and in epistemological modes. The waters are further muddied by the lack of agreement about the correct understanding of analogy; and the use of the same word can obscure instances where people are in fact operating with different concepts. In short, there are a range of approaches by scholars to the role of something called 'analogy' in Coleridge's thought, and a variety of statements from Coleridge himself about the value of analogy in the context of the philosophy and science of life.[2]

Given this complexity, this paper focuses on the role and limits of analogy in thinking about nature for Coleridge in 1816.[3] I will claim, first, that Coleridge presents an attempt to limit the scope and the epistemological role of analogy 'proper' in natural science. The understanding of analogy he proposes has very specific conditions: its scope is limited to apply only to goal-directed phenomena in different genera. Further, his views coincide with the popular eighteenth-century concern that analogy should be used only as an 'aid' to understanding, and further, that analogical arguments do not, of themselves, furnish proof. That said, Coleridge also presents analogy as an important complement to other methods to think about the scale of life: comparison of differences, discussion of resemblances, deductive reasoning ('unfolding' powers), and the use counterposing and attending to differences.

Nevertheless, I would also note Coleridge's frequent tendency to refer to 'analogy'—perhaps more loosely than his own account of analogy would ostensibly

[1] Such comments appear throughout Coleridge's marginal notes. E.g. on Adolph Eschenmayer see Coleridge (1992), 545; on Georg August Goldfuss, 1992, 860.

[2] For Coleridge's engagement with the sciences, see e.g. Modiano (1985). For a discussion of the relationship between Coleridge and *Naturphilosophie,* especially regarding the reception of Schelling, see Whiteley (2018), 207–36.

[3] The investigation has been restricted to its use with respect to the scale of life and scientific enquiry.

permit. These instances and his more general remarks about the scale of nature will lead to the further question regarding his ontology and its relation to the use of analogy: does Coleridge think that powers on the different levels of being can, or should, be considered as analogous? If so, what is the basis of this consideration, and how does it relate to his explicit definition of analogy? I will examine Coleridge's account in *Theory of Life* with the aim of unentangling various notions of analogy, and in so doing hope to go some way to explaining the diversity of scholarly opinions summarised above regarding Coleridge's use of analogy. Ultimately, I argue that the concerns Coleridge voices regarding the notion of analogy reflect his general move to distinguish himself from (what he and other contemporaries viewed as) key problems with *Naturphilosophie*. By putting Coleridge's view of the relative value of analogy as a method for understanding nature within the more general context of *Naturphilosophie*, the paper will show that Coleridge's use of analogy takes a form particular to him.

The ambiguity surrounding Coleridge's discussions of analogy is particularly important given his keen interest in the philosophy and science of nature of his time, which gave varied yet important roles to analogies and analogical arguments. This investigation will therefore build on the recent growth in scholarship on analogy in late eighteenth- and early nineteenth-century philosophy and science, in which scholars have emphasised its prominent role in in late eighteenth- and early-nineteenth century debates about scientific research and philosophy of nature, in figures such as Johann Wolfgang Goethe, Johann Gottfried Herder, Friedrich Schelling, Immanuel Kant, Buffon (Georges-Louis Leclerc), Friedrich Schlegel, Novalis (Georg von Hardenberg), and Johann Friedrich Blumenbach.[4] Recent scholarship demonstrates not only the import of analogy, but the presence of plurality in the forms of analogy and analogical reasoning as well as the varied epistemic uses proposed for it.[5]

The paper will have the following structure. First, (1) I will address Coleridge's explicit remarks concerning the 'correct' understanding of analogous phenomena in *Theory of Life*. Here analogy is cast as an operation that can only be employed in a teleological context, between things 'different in kind' that have the 'same' end. I

[4] See van den Berg (2018), 67–76, Breitenbach (2014), Callanan (2008), 747–72; Hahn (2014), 196–208, Nassar (2015), 240–57, Whistler (2013), 235–58.

[5] Hein van den Berg, for example, writes that 'analysis of the works of Buffon and Blumenbach shows that eighteenth-century biological theories were fundamentally based upon analogy and analogical arguments' (van den Berg 2018, 71). In philosophy, as Nassar describes, Herder gave analogy a foundational role in cognition, where 'it is only through *seeing one thing through or as another* that something an object of knowledge for us at all' (Nassar 2015, 246). Novalis also made analogy a key part of his method. See Nassar (2015), 240–257. For Friedrich Schlegel too, analogy was of great importance to philosophy. Schlegel discussed analogical reasoning in his philosophical lectures, claiming in his lecture on '*Syllogystik*' that, because what can be proven with certainty is only the negative part of philosophy (uncovering errors and so on), analogy and other forms of reasoning that give probable conclusions are of greater import. Analogy, he says, may even be 'more indispensable for philosophy than the syllogism', so long as we don't believe it to furnish truth. Instead, its benefit 'resides in its practical validity'. Schlegel [1805–6] 1836, 152–6.

will go on (1.2) to address his criticisms of the prevalence of analogy and his remarks on its misuse among his contemporaries. He seems to air the more general complaint—especially in his later marginal notes—that it is used too frequently without proper care. I will then (1.3) focus on the restrictions given to analogy in Coleridge's account, in particular on his remarks on the use of analogy as an aid, as well as on the conditions he suggests for the best use of analogy, and on his endorsing of different methods for better scientific inquiry and philosophy of nature. I will also briefly address here the backdrop of the reception of 'Naturphilosophie' and Coleridge's relation to this multifarious category and its reception. The second part of the paper (2) will situate Coleridge's 'correct understanding' of analogy within his broader ontological commitments and try square his claim that analogy can only correctly be used between different kinds with his metaphysical claim that comparisons are only possible at all in light of the fundamental homogeneity of powers in nature. He claims that powers at different intensities or levels of being are neither completely identical nor wholly different to each other. In 2.2 I will show that Coleridge's difficulty—that of retaining the homogeneous root nature of powers, whilst maintaining qualitative distinctions of kind between aspects of different powers, is most clearly expressed in his dual representation of the plant-animal relation as simultaneously antithetical and hierarchical.

12.1 Coleridge's (Ostensible) Understanding of 'Analogy'

Coleridge indicates his view of the correct understanding of analogy in passing in *Theory of Life*. There, Coleridge responds to the following criticism:

> there is no resemblance, no analogy, between Electricity and Life; the two orders of phenomena are completely distinct; they are incommensurable. Electricity illustrates life no more than life illustrates electricity ([1816] 1995, 528).

Coleridge replies:

> I feel some curiosity to discover what meaning the author attaches to the term analogy. Analogy implies a difference in sort, and not merely in degree; and it is the sameness of the end, with the difference of the means, which constitutes analogy ([1816] 1995, 530–1).

Coleridge seems to want to pull apart the equivocation between analogy and similarity made by his interlocutor here. He goes on to suggest that the criticism levelled against him—that he wrongly equates life and electricity via analogy—has missed the mark. Because of his understanding of what an analogy is, he hadn't ever claimed that electricity and life *were* analogous:

> no one would say the lungs of a man were analogous to the lungs of a monkey, but any one might say that the gills of fish and the spiracula of insects are analogous to lungs. Now if there be any philosophers who have asserted that electricity as electricity is the *same* as Life, for that reason they cannot be *analogies* to each other; and as no man in his senses, philosopher or not, is capable of imagining that the lightning which destroys a sheep, was a

means to the same end with the principle of its organization, for this reason, too, the two powers cannot be represented as analogies ([1816] 1995, 531).

From the discussion in the above quotations three points can be distinguished that clarify the conditions for correct use of analogy: 1) analogy involves two sets of means–end relationships of different sorts; 2) it is the *means* of the sets of means-end relations that must differ in kind. Finally, although the means must differ between the analogous items, 3) the ends must have some form of identity between them.

With respect to the first point, according to Coleridge's understanding of analogy, analogies hold between different 'sorts' or kinds. Between these different kinds of things, what is important are the sets of *means–end relationships* involved, rather than questions of—for example—resemblance the lack of resemblance. When thinking about analogies in the natural world, then, we have to consider the goals or functions played by what is under consideration. A lung, for example, is something that fulfils the role (or end) of respiration in the context of the animal of which it is a part.

The second point is that it is the 'means' under consideration in particular that must differ in some way for an analogy to be an analogy. The difference in kind Coleridge first mentions must be reflected in a difference in means. In the third passage given above, it seems that for Coleridge we can't correctly think of the lungs in different species of mammals as analogous, and this is the case because they are both already considered lungs. In this scheme, analogy only works if the things being compared are *different in kind*, and because human and monkey lungs are not 'different means' (they are both lungs)—they don't qualify as analogous. The example may be initially misleading in that monkey and human lungs are in one sense obviously not identical; they are in different species, but for Coleridge it seems they are both nevertheless considered to be *lungs* and are in that way not of different sorts. The crux seems to be what Coleridge would count as a difference in kind. We have seen that the lungs of different species, monkey and human, are not analogous. So to return to the question of life, if someone claimed that electricity *as* electricity were life, that too wouldn't be an analogy because the means in both cases would just be electricity; their identity would preclude analogy as a method for thinking about the electricity in the two cases.

The third point is that his definition of analogy finds it applicable only to sets of means–end relations where *there is an identity between ends*. To return to the lungs example, put simply:

	Means	End
Human	Lungs	Respiration
Fish	Gills	Respiration

An analogy might be asserted between gills and lungs as each of these types of organ, within their own systems, have the same end—they perform the same function, respiration. On this picture, analogies are not just drawn between some two objects taken in isolation (a lung, a set of gills). The analogy is only intelligible given the relation of the relation of the (in this case, organ) to its parent system.

In sum, the conditions for two items to be analogous, on this view, are 1) that analogy holds between two different sets of *means–end relationships,* 2) that analogy pertains where the 'means' are what must differ, and 3) the 'ends' in the two sides of the analogy are identical in some way. This seems to be part of a more general theory of analogy for Coleridge, because he not only expresses concerns about analogy throughout the *Theory of Life,* he also uses the same characterisation to discuss analogy in a literary context elsewhere.[6] In addition, when compared to other prevalent accounts of analogy of his day, this definition of analogy seems quite peculiar to Coleridge. It's a much more restrictive definition of analogy than – to cite one example from an 'influential eighteenth logic textbook' given by Hein van den Berg – Reimarus' account of analogy as 'a similarity between different objects' (Van den Berg 2018, 68).

Coleridge's account here echoes Aristotle's discussion in the *Historia animalium,* where Aristotle makes the following remark:

> Some animals, however, have parts which are not specifically identical, nor differing merely by excess and defect: these parts correspond only "by analogy": of which an example is the correspondence between bone and fishspine, nail and hoof, hand and claw, feather and scale: in a fish, the scale is the corresponding thing to a feather in a bird (Aristotle I: 486b).

Here Aristotle contrasts the type of difference that holds between e.g. nail and hoof, to differences *within* species or genera, which would be manifest in either identity of 'form' or of 'excess or defect': it is a difference which is not merely a question of quantity, but of kind. Coleridge's views on analogy do not coincide with everything Aristotle has to say about analogy, however.

With respect to Aristotle's analysis of analogies in science, Mary Hesse, drawing on Nagel, distinguishes 'formal analogy' and 'substantive analogy' (Hesse 1965, 328).[7] Formal analogies draw on a similarity of relation, or proportionality (scales and feathers stand in the same relation to the organisms they belong to, for example), whereas substantive analogies indicate a feature or property in common (fish spines and mammal bones share a certain 'osseus nature'). According to Hesse, the most interesting cases in Aristotle's work are those that combine both forms of analogy (Hesse 1965, 332). In his official account of analogy, however, Coleridge seems not only to be more interested in formal analogies, but to suggest that a substantive analogy is not an analogy at all. The idea that despite their location in different systems, the hoof and the nail are thought to perform the same function, Coleridge reformulates as sameness of ends, difference of means. In this way, whilst he seems to incorporate functionalist modes of thinking (which can be placed in a broadly

[6] Coleridge gives the same definition of analogy in a literary context in 1811. In the ninth of his lectures on Shakespeare and Milton, he distinguishes 'likeness' from 'analogy' regarding the relations between Shakespeare and ancient Greek tragedies. He writes there that 'they are analogies because by very different means they produce the same end'. Coleridge [1811] 1987, 348.

[7] Besides these strands of analogy, Aristotle also has a further consideration in his metaphysics, which I will briefly return to in Sect. 12.2.

Aristotelian tradition) into his own description of analogy, with its focus on means–ends relations, his explicit understanding of analogy is unlike that of Aristotle.[8]

Having addressed the conditions outlined by Coleridge for the correct use of analogy in *Theory of Life,* I now turn to further instances in the text where he criticises the way others employ it.

12.1.1 Coleridge's Criticisms of the Use of Analogy

In addition to the criteria laid out for analogy just discussed, Coleridge can be found offering criticisms of the use of analogy—especially in the German naturalists. The frequency and wildness with which analogies were made was a common criticism of *Naturphilosophie* in Coleridge's time.[9] Coleridge complains of the '100 instances of the abuse of analogy among the later German naturalists', for example, when discussing Goldfuss.[10] He also complains about misuse of analogy in science in his marginal notes to Eschenmayer:

> where are the proofs that the infusory animals are any more the product of putridity, than the fish are of the rivers, or caterpillars of leaves? All the attempts at proof, that I have yet seen, are drawn from ignorance against knowledge and analogy ([1818] 1992, 545).[11]

The most straightforward reading of this comment is that some of the supposed proofs that infusoria result from putrefaction are based on analogies, and hence not proofs in the proper sense—analogies do not furnish proof. But a complementary complaint seems come from the claim that infusoria result from putrefaction derives from a lack of sufficient evidence or knowledge on Eschenmayer's part. Occasions where there is insufficient proof are the kinds of occasions where people most be

[8] The other obvious influential point of reference of the era is Kant. I set aside here Kant's employment of analogies in the *Critique of Pure Reason* and the *Critique of Judgment* due to the different epistemological frameworks at issue. However, there are some points of similarity between Coleridge and Kant. Coleridge seems to be in agreement with Kant's comment in the *Prolegomena to Any Future Metaphysics* that analogy has to hold between relations, and further, that it should denote a *precise* similarity of *relation* between 'dissimilar things'. See Kant, *Prolegomena to Any Future Metaphysics*, ed. Hatfield 2004, 108 [4:358]. For a full discussion of Kant's uses of analogy, see Callanan (2008), 747–72.

[9] For example, Nicholas Jardine cites Cuvier's 'report of 1810 to Napoleon's Council of state' that claims that *Naturphilosophie* confuses 'the moral with the physical and the metaphorical with the logical.' Jardine (1996), 230. The misuse of analogy will also be one of the criticisms Hegel levels at Schelling, and *Naturphilosophie* more generally, for connecting things that are unconnected. Hegel claims that without providing a logic that explicitly links them, these remain 'surface analogies'.

[10] 'What fair analogy is there between the Ganglia (there are many) and the Brain?' Coleridge (1992), 860.

[11] From an 1818 comment on Eschenmayer's *Psychologie in drei Theilen als empirische: reine und angewandte* (1817). Coleridge's comment does not directly address Eschenmeyer's main point here (which has to do with whether form or composition is important in the organism).

most tempted to introduce analogy—but in an inappropriate way.[12] 'Ignorance against knowledge and analogy' should be interpreted in line with his comments noted above, as a concern for how analogy is to be used in science, i.e., for experimental thinking and comprehension, not as of itself a proof of a theory or an explanation. It can form *part* of a method of comprehension and can be used as an aid in science, but a real explanation of a power requires following its dialectical movement: 'to explain a power', Coleridge writes, 'is to unfold or spread it out'[13]—deduction remains the paradigm for explanatory science (at least regarding powers) but analogies and attending to similarities and differences can help in the comprehension of what is unfolded. It was common at the time to discuss limiting the use of inductive arguments in science, where analogy is taken to be of a piece with induction. In this respect Coleridge appears to align himself with Kant (see Van den Berg 2018, 72 for discussion of the preference for deduction over induction as scientific paradigm).

12.1.2 Comments on the Use of Analogy in *Theory of Life*

For the most part, when Coleridge apparently endorses the use of analogies, he assigns them a heuristic value. For example, he asserts that we shouldn't cast the use of analogy with chemistry aside when considering physiology, writing that it can be an 'aid' when thinking about that subject ([1816] 1995, 498–500).[14] An aid helps us understand something more complicated via the analogy with something simpler, and, according to his earlier framing of analogy, we can think about the relation of means to end in the more complex by thinking about the simpler pair. Most often Coleridge seems to value analogies primarily to help facilitate the *understanding* of more complex things, rather than to provide proof of something. Crucially, he goes on to claim, of the question of the analogy, that:

[12] Coleridge's claim would be, for example, that the analogy between infusoria and putrefying meat and fish and rivers fails because it does not follow the correct usage of analogy.

[13] In a note in *Theory of Life,* Coleridge also uses a thought experiment about two machines to illustrate, I think, a point about the limitations of attending to differences between otherwise similar objects as unable to give a complete explanation of the beings under consideration. If we try to understand things merely by distinguishing them, we don't form an understanding of them, but rather a taxonomy. He writes that noting the difference between two machines could help distinguish them, and facilitate communication, but would not be an understanding that exceeds 'the nomenclature of botany'. This echoes a commonly repeated point among the *Naturphilosophen*, who distinguished between what they considered a true understanding of the kind of being under consideration from the lesser knowledge afforded by a merely formal taxonomy. Coleridge [1816] 1995, 505–6.

[14] But this analogical use, as an aid, would crucially not reduce the higher science to the lower. Because Coleridge thinks that the objects of physiology express powers at a higher intensity (he notes that 'we recognise that we can define the bounds of chemistry through its failures').

> whether the powers which manifest themselves to us under certain conditions in the forms of electricity, or chemical attraction, have any analogy to the power which manifests itself in growth and organization, is altogether a different chain of reasoning. If it be indeed a tree of knowledge, it will be known by its fruits, and these will depend not on the mere assertion, but on the inductions by which the position is supported, and by the additions which it makes to the insight into the nature of facts it is meant to illustrate ([1816] 1995, 503).

In addition to the claim that the inductions supporting the analogy are important, this remark suggests that Coleridge takes a somewhat pragmatic approach to the selection of relevant analogies in the respect that the value of the 'insights' offered is mentioned as a factor here. It also shows that he considers whether or not an analogy is useful is whether or not it be 'illustrative' of chains of facts. Here, understanding some causal story is complimented by analogies—the analogy is not of itself a definition or an explanation. This complements his claim elsewhere in the *Theory of Life*, that a real explanation or definition of a power unfolds it deductively.[15]

12.2 Analogy and Ontology

Thus far, Coleridge's ideas for the use of analogy seems coherent: he has offered conditions for what can count as analogy (teleological, different means, same ends), and emphasised its role as primarily illustrative. Like many of his contemporaries, he expressed concerns about the frequent, overly hasty uses of analogy.

However, another strand of discussion of 'analogy' can be found in *Theory of Life*—a discussion that is not presented in the same terms as his means–ends model. This strand of analogy concerns the *scala naturae,* and the relations between different levels of its composition. According to this conception of analogy, the use of analogies for thinking about powers at different levels in Coleridge's scale of nature is made possible by an aspect of continuity in nature's structure itself.

The idea of an underlying continuity as a condition for analogy is introduced in his discussion of the notion of life and its scope. Coleridge proposes that expanding the notion of life will have an advantage: namely, that it 'fills' an 'arbitrary chasm' between the sciences of physics and physiology, which 'justifies us in using the former as means of insight into the latter'. This, he claims, is a use that 'would be contrary to all sound rules of ratiocination if the powers working in the objects of the two sciences were absolutely and essentially diverse' ([1816] 1995, 508–9). What this suggests is that the use of models from physics for understanding physiology only makes sense on the basis of an underlying continuity of some kind between

[15] Elsewhere, Coleridge takes a different tack. He remarks that the 'geometric regularity' of architecture made by insects 'bears an analogy' to the mineral process of crystallization. Because of the cursory nature of this remark, it is hard to see whether it conforms to his previously stated understanding of analogy or not, and furthermore, what exactly the supposed sameness of ends in this case would consist in (a structure, or perhaps, an activity itself?). In any case, this is an offhand conjecture, and is not proposed as a certain truth or as furnishing proof of anything. [*c.*1817] 1995, 545.

the objects of these two sciences. The continuity in that they are both constituted by the actions of, at root, the same powers, but at different levels of intensity:

> there is an ascending series of intermediate classes, and of analogous gradations in each class. To a reflecting mind, indeed, the very fact that the powers peculiar to life in living animals *include* cohesion, elasticity, &c. (or, in the words of a recent publication, "that living matter exhibits these physical properties," would demonstrate that, in the truth of things, they are homogeneous, and that both the classes [sc. 'living matter' and 'physical properties'] are but degrees and different dignities of one and the same tendency ([1816] 1995, 511).

This claim—that a homogeneity at root is expressed in analogous gradations between levels of phenomena—more explicitly connects Coleridge's perception of similarities or patterns at the phenomenal level with his underlying conception of the structure of nature or, what is the same here, life. Coleridge combines two models for representing the structure of nature in *Theory of Life*: an ascending ladder, and concentric circles of development.

Coleridge claims that 'the progress of Nature is more truly represented by the ladder, than by the suspended chain' ([1816] 1995, 537).[16] Nature as an ascending ladder represents a hierarchical series of kinds of an increasingly individuated kinds of phenomena. This scale is understood to be generated by the activity of powers, in particular, the polar opposition of two primary powers. The series of kinds of beings is generated by the actions of powers at increasing levels of intensity.[17]

Following Schelling, he understands these ascending levels as higher potentiations of the same powers.[18] Coleridge claims that life expands as 'by concentric circles' ([1816] 1995, 509)—the result is a picture of 'ascent modified by collateral expansion' (1999, 500). Coleridge uses the language of 'assimilation' to explain this:

> the more intense the life is, the less does *elasticity*, for instance, appear *as* elasticity.... The lower powers are *assimilated*, not merely *employed*, and assimilation presupposes the homogeneous nature of the thing assimilated; else it is a miracle, only not the same as that of a *creation*, because it would imply that additional and equal miracle of annihilation ([1816] 1995, 511).

The lower power is therefore thought to be assimilated in the higher power, but not *as* the lower power, rather, as something higher. This is influenced by Schelling's understanding of the levels of potentiation in nature. For our purposes, the

---

[16] Coleridge claims that nature: 'does not ascend as links in a suspended chain, but as the steps in a ladder; or rather she at one and the same time *ascends* as by a climax, and expands as the concentric circles on the lake from the point to which the stone in its fall had given the first impulse' Coleridge [1816] 1995, 509.

[17] Barfield discusses Coleridge's understanding of nature as composing levels of powers, see Chaps. 2 and 3 in Barfield (1972).

[18] This was a frequent argumentative feature of late eighteenth- and early nineteenth-century *Naturphilosophie* and can be found in various elaborations in Schelling and Eschenmayer, among others. For example, Schelling writes that 'it is certain that the galvanic phenomena are identical with the electrical in their ultimate principle, although galvanism and electricity are themselves diverse phenomena; for through galvanism electricity is raised, as it were to a higher function'. Schelling [1799] 2004, 151–2.

importance of this ontology is that, on the one hand, it enables Coleridge to renounce reductivism and claim that he does not, for example, 'confound magnetism with electricity'. It is important to understand these powers as *different*, because they are on different levels of potency. This is illustrated by mathematical images: we can understand the same power in magnetism as in reproduction as 'what the root is to the cube of that root' ([*c*.1817] 1995, 556), and in terms of dimensions: 'I no more confound magnetism with electricity, or the chemical process, than the mathematician confounds length with breadth, or either with depth' ([*c*.1817] 1995, 532).

On the other hand, the assimilation of the lower in the higher 'circles' allows for some continuity between the levels of life and the recapitulation of patterns at higher intensities.[19] It is on the basis of this ontological structure—that nature is a ladder of concentric circles—that Coleridge suggests the epistemological strategy of studying 'the complex in the simple'.[20] He proposes that we can better understand a more complex instantiation of something through looking at its simpler manifestations. Again, this is possible because life is made up of a 'perpetual reconciliation, and as perpetual resurgency of the primary contradiction' ([1816] 1995, 537). We can use a simple manifestation to think through a more complex one because this polarity of powers is repeated at different levels of potentiation ([1816] 1995, 548–9).

But what is the 'homogeneous nature' that constitutes the continuity between the levels of life on this account, aside from the different instantiations at different intensities? Or, what is it in virtue of that we could say that they are analogous? If this question is answered in terms of the means used by the powers rather than their ends, that would suggest Coleridge's official account of analogy does not apply here. In any case, for Coleridge there is clearly a sense in which things are 'analogous' that are neither identical, nor wholly different from each other. It is in this sense, then, that Coleridge articulates a strand of analogy in his ontology, one that allows him to consider the relation of powers at different intensities as connected yet distinct. I would argue that this type of analogy is not best understood in terms of the debates surrounding *analogia entis* in the Thomist tradition, not least because Coleridge does not make any explicit reference to such discussions—it could be a misleading term for what Harding has described as an interpretation of nature that is scientifically up to date with his contemporary developments but framed within

[19] This is not extended beyond natural phenomena. Whilst Coleridge can consider his scale of life as a scale of degrees insofar as its manifestations exhibit the power of life, this is also accompanied by the frequent assertion that the 'chasm' between animals and humans is 'unbridgeable'. Due to his theological commitments, he is unwilling to assert continuity between humans and other animals apart from a bodily continuity. He writes that 'In the lowest forms of vegetable and animal world we perceive totality dawning into *individuation*, while in man, as the highest of the class, the individuality is not only perfected in its corporeal sense, but begins a new series beyond the appropriate limits of physiology' ([1816] 1995, 516). See also e.g. his later comments on Schubert's natural history: 'to man, as man, to that which constitutes his kind, I find no approximation. The elephant and ape seem as distant as the ant and the bee'. Coleridge [1827] 1999, 530–531.

[20] Coleridge [1816] 1995, 509. He continued to discuss the idea of anticipation of the higher in the lower, in butterflies, polyps, etc., on through the 1820s.

governance by a divinity.[21] There are clear similarities to neo-Platonic traditions, though, as Harding has also pointed out (Harding 2000, 144).

12.2.1 Antitheses and Counterpoints

Coleridge's ontology of expansion and ascent, composed of increasing potentiations of polar opposition, also indicates limits for the use of analogy. In some cases, Coleridge uses this understanding of the structure of nature to justify the search for 'counterpoints' as a method distinct from and preferable to analogy when conducting scientific enquiries. These limits to the role of analogy appear in the context of his construction of the plant and animal kingdoms. Here, emphasis on the difference between plant and animal is accompanied by an avowed rejection of the utility of analogy:

> From this point flow, as in opposite directions, the two streams of vegetation and animalization, the former characterised by the predominance of magnetism in its highest power, as reproduction, the other by electricity intensified—as irritability, in like manner. The vegetable and animal world are the thesis and antithesis, or the opposite poles of organic life. We are not, therefore, to seek in either for analogies to the other, but for counterpoints ([*c*.1816] 1995, 538).

This suggests that, for Coleridge, despite the continuity of plant and animal kingdoms as far as life is concerned, they are in fact not just different, but are related in a special way, which means that analogy is not the best method to use to understand them. Rather, because of their polar opposition, we should attend to corresponding 'counterpoints' between them. This is more constrained than any analogical relation, as they are held as in some way complimentary or inverse. According to this prescription, we might note corresponding opposite features between plants and animals, as for example in his 1818 commentary on Schelling's *Von der Weltseele*—in which he notes the 'contrast again between plant and animal' and that plants release oxygen and animals perform the inverse operation by absorbing it ([1818] 1995, 764–5).

The importance of emphasising the opposition between plant and animal persists in Coleridge's mental landscape into his later work—his later comments on

---

[21] I should briefly respond to Harding's claim that Coleridge's aim was to present the natural world using a version of *analogia entis*. Harding writes that 'what Coleridge was attempting in the 1820s was nothing less than a reconciliation of the findings of contemporary science with belief in a universe governed by an intelligent divinity. Philosophically, it was also a late flowering version … of a centuries old concept, the scholastic concept of the analogia entis'. First, Harding focuses on Coleridge's later notebooks from the 1820s, rather than *Theory of Life*, so his not intended to cover the same material as this paper. I do not disagree with the content of Harding's reading, but rather his description of it as a version of 'analogia entis', insofar as this could suggest a Thomist understanding of the term (which I take to be the most common); whilst Harding doesn't give an explicit redefinition along new lines, he seems to use it to connect design to purposiveness in nature, which in my view is a distinct philosophical issue. See Harding (2000), 144.

Schubert's natural history make clear some of the difficulties of 1816 in understanding the plant–animal relation both horizontally and vertically. He suggests that Schubert should preserve clear distinction between the plant, insect, and animal kingdoms:

> Dear Schubart! [sic] but what, *what,* WHAT, I say, do these visualities, these pictures of both by altitudes, *mean?* the *verschlungen* und *die umfangenden* &c &c The true problem is: to state the inferiority of the vegetable antecedence & creation as to preserve & imply its counterposition to the animals—i.e. so to place V= Veg., I. Insect; and A = anim[al] that they may be at once…

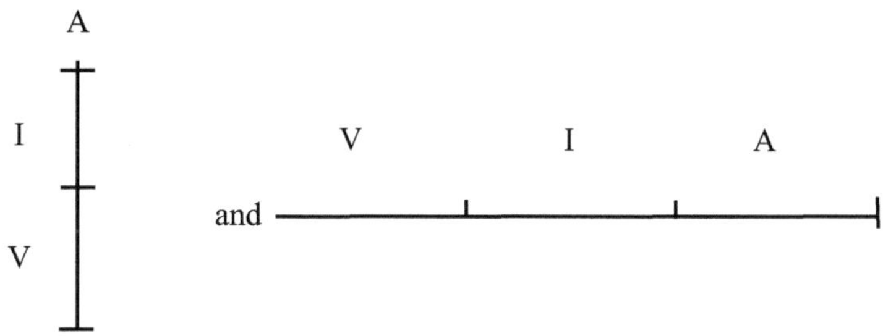

> (Coleridge 1999: 493)

The diagram shows that for Coleridge, both schemas are valuable: we can represent lifeforms in a horizonal scale from vegetable to animal, for which considering 'antitheses' are illuminating—but we can also represent plant and animal in a vertical continuum of increasing individuation where reproduction, irritability, and sensibility predominate respectively, but with the lower powers assimilated in the higher.

Coleridge's diagram above demonstrates his later grappling with this issue. Here he wants to preserve the vertical hierarchy, in which vegetable, insect, animal are an ascending series, alongside the horizontal scheme which shows them as antithetical. The issue with plant–animal analogies in 1817 seems to be not that we can't conceive them, but that they aren't the most helpful way to comprehend what he thinks is going on ontologically.[22]

[22] Where plants 'express reproduction', animals express sensibility, a different dominant power. Coleridge interprets the eyes of insects as 'eye-facets' that 'form the sense of light, rather than organs of seeing'. He is so concerned that these phenomena represent opposing powers that he is unwilling to grant insects 'the independent existence of sensibility *as* sensibility' and wants to cast them as beings that express irritability. He writes that 'their almost paradoxical number at least, and the singularity of their forms, render it probable that they impel the animal by some modification of its irritability, herein likewise containing a striking analogy to the known influence of light on plants, rather than as excitements of sensibility'. Coleridge [1816] 1995, 544.

12.3 Conclusions

Coleridge seems to be committed outwardly to an attempt to limit the scope and role of analogy (as an 'aid' in understanding) in natural science. With respect to the scope, he explicitly advances an understanding of analogy that is drawn between two sets of causes and effects ('means' and 'end'), where the entities that the analogously related cause–effect pairs relate to must be different 'in kind'.

The role of analogy both complements and can be contrasted with deductive explanation, or 'definition', which 'unfolds' powers. Coleridge's proposal seems to be that the appropriate use of analogy—based on an underlying ontology that uses the potencies of being—can help guide our understanding, but only in conjunction with his larger overall framework of methods, and only to help comprehend rather than to prove arguments. This is not a merely outward concern with analogy to head off criticisms that have been levelled at other philosophies of nature. He is also critical of the plethora of extravagant analogies in unpublished work, not because they fail to meet his own description of what analogy is, but because for Coleridge they are misleading and unscientific.

Nevertheless, despite the rigid conditions he places on the correct use of analogy, Coleridge sometimes himself points out in passing 'analogous' things, without doing much more than noting vague resemblances. These passing analogies are made between various natural phenomena, and do not extend beyond the natural to the supernatural, that is, they are not analogies between different strata of being. Finally, when discussing the ontological basis of analogical gradations that allow for vertical analogies to be drawn and lay the groundwork for recapitulations in the scale of life, Coleridge appears to be wrestling with dual commitments. Because the emphasis on difference in kind is so important to Coleridge's initial conception of analogy in (the posthumously published) *Theory of Life*, it may seem strange that he goes on to stress at the same time an underlying continuity between the powers of nature: these levels of being are neither completely identical, nor wholly different to each other. The difficulty of retaining the homogeneous root nature of powers, whilst *maintaining qualitative distinctions in kind* between aspects of different powers, is most clearly expressed in his dual representation of the plant–animal relation as simultaneously antithetical and hierarchical.

References

Aristotle. 1965. Historia animalium. Trans. A. L. Peck. Cambridge, MA.: Harvard University Press.

Barfield, Owen. 1972. *What Coleridge thought*. London: Oxford University Press.

Berg, Hein van den. 2018. Kant and the scope of analogy in the life sciences. *Studies in History and Philosophy of Science, Part A* 71:67–76.

Breitenbach, Angela. 2014. Biological purposiveness and analogical reflection. In *Kant's theory of biology*, ed. Ina Goy and Eric Watkins, 131–149. Berlin: De Gruyter.

Callanan, John J. 2008. Kant on analogy. *British Journal for the History of Philosophy* 16 (4): 747–772.

Coleridge, Samuel Taylor. 1987. In *The collected works of Samuel Taylor Coleridge, Bollingen Series LXXV, volume 5: Lectures 1808–1819 on literature I*, ed. R.A. Foukes. Routledge and Keegan Paul and Princeton University Press.

———. 1995. Theory of life. *c*.1816–18. In *The collected works of Samuel Taylor Coleridge, Bollingen series LXXV, volume 11, Part I: Shorter works and fragments*, ed. H.J. Jackson and J.R. De J. Jackson. Routledge and Keegan Paul and Princeton University Press.

———. 1992. In *The collected works of Samuel Taylor Coleridge, Bollingen Series LXXV, volume 12: Marginalia II Camden to Hutton*, ed. George Whalley. Routledge and Keegan Paul and Princeton University Press.

———. 1999. In *The collected works of Samuel Taylor Coleridge, Bollingen Series LXXV, volume 12: Marginalia IV: Pamphlets to Shakespeare*, ed. H.J. Jackson and George Whalley. Routledge and Keegan Paul and Princeton University Press.

Hahn, Hans. 2014. Die Analogie bei Goethe und Novalis: Zauberstab oder Irrlicht? *Publications of the English Goethe Society* 83 (3): 196–208.

Harding, Anthony John. 2000. Coleridge, natural history, and the 'analogy of being'. *History of European Ideas* 26:143–158.

Hawk, Byron. 2007. *A counter-history of composition: Toward methodologies of complexity*. Pittsburgh: University of Pittsburgh Press.

Hesse, Mary. 1965. Aristotle's logic of analogy. *The Philosophical Quarterly* 15 (61): 328–340.

Jardine, Nicholas. 1996. Naturphilosophie and the kingdoms of nature. In *Cultures of natural history*, 230–245. Cambridge: Cambridge University Press.

Modiano, Raimonda. 1985. *Coleridge and the concept of nature*. London: Macmillan Press.

Nassar, Dalia. 2015. Analogy, natural history and the philosophy of nature. *Journal of the Philosophy of History* 9 (2): 240–257.

Potter, George R. 1925. Coleridge and the idea of evolution. *Publications of the Modern Language Association of America* 40 (2): 379–397.

Schelling, Friedrich. 2004. In *First outline of a system of the philosophy of nature. 1799*, ed. Keith R. Peterson. Albany: SUNY Press.

Schlegel, Friedrich. 1836. Lehre von der Analogie. In *Friedrich Schlegel's Philosophische Vorlesungen aus den Jahren 1804 bis 1806*. Bonn: E. Weber.

Whistler, Daniel. 2013. Post-established harmony: Kant and analogy reconsidered. *Sophia* 52 (2): 235–258.

Whiteley, Giles. 2018. *Schelling's reception in nineteenth century British literature*. Cham: Palgrave Macmillan.

Lydia Azadpour is a teaching associate at the University of Nottingham. Her research is on philosophy and the life sciences in the eighteenth and nineteenth centuries. Her publications include 'Kielmeyer's Economy of Extinction' (2020) and 'Lichens and Philosophy' (forthcoming), and she co-edited the volume Kielmeyer and the Organic World: Texts and Interpretations (2020).

Chapter 13
An 'Alphabet of the Philosophy of Physical Dynamics': Coleridge's Theory of the Metaphysical Foundations of Matter

Dillon Struwig

Abstract This chapter presents an account of the philosophical framework Coleridge worked out to explain the relationship between the diverse physical configurations of matter and the eternal, immaterial, hierarchically ordered powers he held to determine the fundamental features of the material world. It traces the development of Coleridge's theory of matter through an examination of his extensive critical commentary on Kant's seminal *Metaphysical Foundations of Natural Science*, the starting point for the 'dynamical' account of matter favoured by the *Naturphilosophen*. I focus on Coleridge's search for a set of terms that would make clear and consistently maintain the ontological distinctions between (i) matter, material substance, and body, and (ii) the 'fundamental forces' or 'spiritual powers' whose divinely grounded causal activity is necessarily manifested through all such physical entities. I argue that this aspect of Coleridge's view results from his engagement with the concepts of 'substance' and 'appearance' initially put forward by Kant and subsequently refined by Coleridge in *Opus Maximum* and related marginalia and manuscript fragments.

> No inquiry has been surrounded, for the philosophers of every age, by so much darkness as that concerning the nature of matter. (Schelling 1988 [1797], 179)

> The word, matter, materia, ὕλη [*hū́lē*], is among the most obscure, and unfixed in the whole Nomenclature of Metaphysics. (Coleridge 1980–2001, 3: 130, *c*.1816–1818)

D. Struwig (✉)
Department of English Studies, Sophia University, Tokyo, Tokyo Prefecture, Japan
e-mail: d-struwig@sophia.ac.jp

P. Cheyne (ed.), *Matter and Life in Coleridge, Schelling, and Other Dynamical Idealists*, International Archives of the History of Ideas Archives internationales d'histoire des idées 253, https://doi.org/10.1007/978-3-031-78315-9_13

13.1 Introduction

This chapter presents a preliminary account of Coleridge's theory of matter. I examine the set of terms he worked out in attempting to clarify the relationship between matter in its diverse forms and the immaterial, hierarchically ordered ideal or spiritual powers taken to determine the fundamental features of all physical phenomena, and from which the material elements of sensible objects must be kept distinguished. In particular, I focus on how the development of Coleridge's position can be traced through his thoroughgoing critical commentary on Kant's seminal *Metaphysical Foundations of Natural Science* (2004 [1786]), especially its second chapter on the principles of dynamics, perhaps *the* key reference point for the 'dynamical' theories of matter favoured by the *Naturphilosophen*.[1] I take as my starting point two observations by Raimonda Modiano, whose analysis of Coleridge's critical response to Kant's matter theory in *Metaphysical Foundations* in the context of his extensive engagement with the polar, powers-based ontologies of Schelling and Steffens remains the most comprehensive account to date: (1) 'Coleridge felt that Kant's doctrine of matter was not metaphysical *enough*' (Modiano 1985, 156) and (2) 'Coleridge found the main drawback of this theory to be its disregard of the relationship between matter and spirit' (Modiano 1985, 158).[2] As I show, in addition to the Trinitarian metaphysics, pentadic logical method, and speculative cosmogony detailed by Modiano (1985, Pt IV) and Trevor Levere (1981, Chap. 5), Coleridge's reflections on Kantian and post-Kantian theories of matter, forces, and powers also prompted him to devise a conceptual framework of matter-spirit relations based on a fundamentally different set of ontological assumptions about the nature and classification of physical phenomena. Given that Coleridge's worries regarding the potentially problematic philosophical and theological implications of (i) Kant's account of the permanence of substance and (ii) Schelling and Steffens' 'error' of 'placing polarity in the absolute' have been thoroughly analysed,[3] I focus on aspects of Coleridge's matter theory and terminology that have received comparatively little attention: His reworking of Kantian matter and appearance terminology, especially its designations of predicates and properties, and his use of grammatical models or analogues to elucidate both the conceptual relationships and the ontological distinctions between different classes of material phenomena.

I begin by examining Coleridge's criticisms of Kant's concepts of matter and forces in relation to his own definitions of matter and spirit in the *Metaphysical Foundations* marginalia (Sects. 13.2 and 13.3). I then discuss how Coleridge

[1] See e.g. Beiser (2002), 506–50. See also Stern's introduction to Schelling's *Ideas for a philosophy of nature* (1988 [1797]).

[2] On Coleridge's responses to the *Naturphilosophie* of Kant, Schelling, and Steffens, see Modiano (1985), 151–60, 160–173, and 173–86 respectively; for the general historical and intellectual background of Coleridge's nature-philosophy, see 138–51.

[3] See Modiano (1985), 158–9 (i), 169–70, 192–3 (ii). See also Levere (1981), 6–7, 66–9, 104–8, 113–14 (ii).

characterizes properties and predicates in his matter terminology (Sect. 13.4) and reconceives Kantian *Grundkräfte* (fundamental forces) as spiritual powers grounded in 'the workings of the divine consciousness' (Modiano's phrase, 1985, 159) (Sect. 13.5). After defining Coleridge's Kant-inspired term of 'Apparency', I explain how his concepts of appearance, existence, and subsistence are related to his division of material entities into the classes of matter (M_1), material substance (M_2), and body (M_3) (Sect. 13.6), giving close attention to the grammatico-ontological framework he uses to articulate the relations between each category (Sect. 13.7; these terms and classes are defined at Coleridge 1980–2001, 3: 277–9). Finally, I consider how the polar account of the relationship between reality and appearance in his *Logic* (Coleridge 1981 [*c*. 1819–22], 129–130), along with a marginal note on the 'three terms in objective nature' (Coleridge 1980–2001, 6: 132), may help further clarify Coleridge's conception of the ontological framework underlying his matter theory and its key terminology (Sect. 13.8). First (Sect. 13.1), I provide some additional background information on Coleridge's readings and annotations of Kant's *Metaphysical Foundations* and an overview of their differing ideas of what is required for a metaphysics of nature and an accompanying analysis of the a priori elements of our concept of matter.

Coleridge probably first closely read Kant's *Metaphysical Foundations* in 1809–10, a period during which, as his notebooks show, he began to take a more serious interest in inquiries into the nature of matter and how its most basic elements were constituted, whether by 'fundamental forces' or 'absolutely impenetrable' atoms (or monads).[4] In particular, he explored problems related to 'Attraction' and 'Repulsion', the interactions between them, and the possibility of reducing both to a single fundamental force or power (see e.g. Coleridge 1980–2001, 3: 276–80, 281–96; compare Coleridge 1957–2002, 3: §3962). There is also evidence that Coleridge returned to *Metaphysical Foundations* frequently during 1816–18, while working out his system of nature-philosophy and carefully distinguishing it from the positions of Schelling and Steffens; and again in 1818–19, when the preparations for his course of lectures on the history of philosophy led Coleridge to reconsider certain questions about the infinite divisibility of matter and the possible existence of metaphysical simples or 'minima', as well as to reassess his earlier views on how Kant's treatment of such concepts as matter and motion, as '*geometrically represented*', should be interpreted in the context of *Metaphysical Foundations* and Kant's 'construction of Nature' (Coleridge 1980–2001, 3: 298, 301).[5] Indeed,

[4] For Coleridge's *Metaphysical Foundations* marginalia, see Coleridge (1980–2001), 3: 269–303. Coleridge owned and annotated a 1787 2nd edition of Kant's *Metaphysische Anfangsgründe der Naturwissenschaft* (published in Riga). See also e.g. Coleridge (1957–2002), 3: §3587 on Newtonian gravity and Leibnizian monads (September 1809); 3: §3962 on Boscovich's matter theory, ideal powers, and the unity of forces in nature (July 1810); 3: §§4058, 4118 on the 'infinite divisibility of matter' (*c*.1811).

[5] On Kant's metaphysics of nature and matter theory, as developed in *Metaphysical Foundations*, see Friedman (2013). Compare Levere (1981), 123–5; Modiano (1985), 151–60, which both also discuss Coleridge's responses to *Metaphysical Foundations*. For Coleridge on 'minima', see

internal textual evidence suggests that Coleridge's most detailed presentation of his powers-based theory of the metaphysical substructure of the material world, in 'Appendix A' of the *Opus Maximum* (Coleridge 2002 [*c.*1819–22], 347–87; see also 'Fragment 4' on matter and spirit, 291–345), may have been composed in conjunction with a careful rereading of *Metaphysical Foundations*, dated 'August, 1819', that prompted the retrospective note where he modifies his previous position on the role of geometrical construction in Kant's philosophy of nature (see Coleridge 1980–2001, 3: 301, 297–8). My analysis focuses primarily on these 1816–18 and 1818–19 marginalia,[6] considering how the core elements of Coleridge's matter theory and its terminology can be identified and elucidated by examining the arguments outlined in these annotations. Given this temporal and conceptual scope, I leave aside considerations regarding Coleridge's earlier (pre-*Naturphilosophie*) views on matter and spirit, as shaped by thinkers ranging from Cudworth and Berkeley to Hartley and Priestley.[7]

Before considering Coleridge's *Metaphysical Foundations* marginalia in detail, it will be useful to discuss how his approach differs from Kant's project for a metaphysical account of matter. In *Metaphysical Foundations*, Kant seeks to articulate the a priori principles that make possible not only the contents of our empirical cognition of physical phenomena, such as material substances and bodies, but also the conceptual framework of the Newtonian mathematical physics that describes and explains such phenomena. Accordingly, the 'metaphysical foundations' Kant is after are understood in terms of his strict and narrow sense of 'metaphysics' as the universal and necessary principles 'grounded in' the forms of human sensibility and intellect: the transcendental principles of Kant's 'revolutionary metaphysics of experience' (Friedman 2013, 32–3, discussing Kant 2004 [1786], 8–9 [Ak. 4: 472–3]), insofar as these a priori principles and their potential applications are required to explain the possibility of our fundamental concepts of matter and motion (and by extension, the utilization and expression of such concepts within the mathematical framework of the physical sciences). As Friedman argues, Kant's objective is to 'explain how the modern mathematization of nature – the application of mathematics to the empirical or phenomenal world of our sensible experience – first becomes possible' (2013, 33). The elements of this approach can be better understood by briefly examining some key statements from Kant's *Metaphysical Foundations* 'Preface'. First, Kant indicates how his method relates to the 'doctrine

Coleridge (1980–2001), 5: 734 (*c.*1818–19); Coleridge (1957–2002), 3: §4266 (September 1815), §4566 (June 1819).

[6] These date ranges are based on thematic parallels between Coleridge's *Metaphysical Foundations* marginalia, his annotations of *Jarbücher der Medicin* (1805–8), the *Naturphilosophische* journal edited by Schelling and A. F. Marcus (Coleridge 1980–2001, 3: 114–31, *c.*1816–18), and his matter theory-related marginalia in Tennemann's *Geschichte der Philosophie* (see e.g. Coleridge 1980–2001, 5: 719, 733–4, which I conjecturally date *c.*1818–19 based on similarities to Coleridge 1957–2002, 4: §4522, April 1819, a discussion of Zeno's paradox, the nature of motion, and the infinite divisibility of matter that is evidently related to Coleridge's preparations for his 1818–19 lectures on the history of philosophy, and §4566, June 1819).

[7] On this earlier period, see e.g. Wylie (1989). See also Levere (1981, Chap. 1).

of body', his theory of 'corporeal nature', defined as 'the objects of the *outer* senses' ('*extended* nature', those spatially arrayed objects or 'appearances' given in outer intuition) (see Kant 2004 [1786], 3 [Ak. 4: 467]):

> in order to make possible the application of mathematics to the doctrine of body, which only through this can become natural science, principles for the *construction* of the concepts that belong to the possibility of matter in general must be introduced first. Therefore, a complete analysis of the concept of a matter in general will have to be taken as the basis, and this is a task for pure philosophy [i.e., in this context, Kant's transcendental '*metaphysics of corporeal nature*']. (Kant 2004 [1786], 8 [Ak. 4: 472])

Given these requirements, Kant contends that 'all natural philosophers who have wished to proceed mathematically in their occupation have always, and must have always, made use of metaphysical principles (albeit unconsciously)', even in those cases where such thinkers have sought to deliberately exclude all metaphysical assumptions or principles from their proposed explanations of physical phenomena (Kant 2004 [1786], 8 [Ak. 4: 472]).[8] As Kant clarifies, this is not because of the need to use speculative metaphysical constructs (e.g. Leibnizian monads or Cartesian vortices) to explain physical phenomena. Rather, it is because, on Kant's account, the very possibility of the idea of nature as orderly and governed by natural laws that are amenable to mathematical methods of investigation cannot be explained without reference to certain basic a priori principles that make possible not only sensible experience but also the mathematical and empirical concepts required for the construction of theories of matter, motion, and other physical phenomena:

> All true metaphysics is drawn from the essence of the faculty of thinking itself, and is in no way fictitiously invented on account of not being borrowed from experience. Rather, it contains the pure actions of thought, and thus a priori concepts and principles, which first bring the manifold of *empirical representations* into the law-governed connection through which it can become *empirical* **cognition**, that is, experience. Thus these mathematical physicists could in no way avoid metaphysical principles, and, among them, also not those that make the concept of their proper object, namely, matter, a priori suitable for application to outer experience, such as the concept of motion, the filling of space, inertia, and so on. (Kant 2004 [1786], 8–9 [Ak. 4: 472–3])

Based on such passages, one can see why Modiano claims 'Coleridge felt that Kant's doctrine of matter was not metaphysical *enough*' (1985, 156). Kant is interested in the highly specialized problem of how the kind of scientific knowledge in which Newtonian mathematical physics consists is possible, along with the related question of what kinds of a priori principles would be required to secure the scientific status and systematicity of such knowledge (for further discussion, see Friedman 2013, 1–33). Indeed, the opening page of *Metaphysical Foundations* makes clear that Kant's 'metaphysics of corporeal nature' necessarily 'exclud[es] all

---

[8] The contrast Kant has in mind here is between explicitly metaphysical approaches (which make use of mathematical principles), such as Cartesian and Leibnizian physics, and more empirically-minded approaches, such as Newtonian mechanics or Boylean chemistry (which also make use of mathematical principles but tend to disavow explicit metaphysical assumptions). In Kant's view, neither approach can be explained without referring to certain a priori principles, especially those needed to explain how the mathematization of nature is possible.

nonsensible objects' (e.g. putative immaterial forces or spiritual powers in nature that are held to account for the behaviour of physical phenomena or the features of material bodies) (Kant 2004 [1786], 3 [Ak. 4: 467]). Kant is interested only in the 'a priori sources' of our basic concepts of matter and motion (Kant 2004 [1786], 9 [Ak. 4: 472–3]).

Interestingly, as with his view that Kant's transcendental philosophy is complete when considered solely as a theory of the discursive cognition of finite minds (but falls short due to its lack of an account of divine transcendent reason), Coleridge seems willing to grant that the Kantian account of nature is accurate when limited to the fields of phoronomy and mechanics, restricted to 'the objects of *outer* intuition': the motions and interactions of material entities as explained by 'a system which *begins* with Space and physical Powers defined exhaustively by effects and products solely relative to Space and manifestable in and by Space' (Coleridge 1980–2001, 3: 284).[9] Thus, while Coleridge would accept the objectives described above as legitimate aims for such a Kantian metaphysics of nature, he would deny that it provides a full or adequate metaphysical account of matter (or our concept of it), given his much broader and more traditional conception of the content and methodology of metaphysics. Put another way, Coleridge would likely grant that Kant's *Metaphysical Foundations* provides us with a legitimate account of the conceptual and logical framework and foundation of Newtonian mathematical physics (especially as a doctrine of the motions of material bodies in space and time), but he would deny that this approach can ever be sufficient as an account of the underlying nature of matter—or even of the basic a priori elements of our concept of matter. As Levere (1981, 62, 95–6, 123–5) and Modiano (1985, 155–60) have explained, Coleridge's denial is based on his rejection of the limits Kant imposes on our possible knowledge of fundamental forces (e.g. as 'spiritual powers' that casually interact with and determine the fundamental nature of material phenomena). In what follows, I attempt to add conceptual detail to the picture of Coleridge's metaphysics of nature and theory of matter first sketched by these pioneering scholars.

13.2 Coleridge's Response to Kant's Theory of Matter

A key reference point for mapping out the development of Coleridge's matter theory, as it began to take shape from around 1816 onwards, is his critical response to Kant's 'dynamical explication of the concept of matter' as 'the *movable* insofar as it *fills* a *space*' in the second chapter of *Metaphysical Foundations* on the principles of dynamics, where 'To *fill* a space' is defined as 'to resist every movable that strives through its motion to penetrate into a certain space' (Kant 2004 [1786], 33 [Ak. 4:

[9] This claim is discussed further in Sect. 13.5. See also Coleridge (1980–2001), 3: 274–5, 295–6.

496]). Here Kant advances a step further from the phoronomic[10] explication of the preceding chapter, where matter is considered only as 'the *subject* of motion', from a perspective that 'abstracts ... from all inner constitution', ascribing 'no other property' (or predicates) to matter 'aside from *movability*' (Kant 2004 [1786], 15 [Ak. 4: 480]). As I will show, Coleridge's initial objections are informed by what he takes to be the problematic implications of these Kantian explications, particularly in relation to the problem of determining the essential properties 'contained' within our a priori 'concept of a matter in general', the conditions on which its 'inner possibility' rests.[11]

What the dynamical explication 'adds' to this basic concept of matter is 'a property relating as cause to an effect', what Kant calls 'the power to resist a motion within a certain space' (Kant 2004 [1786], 33 [Ak. 4: 496]).[12] When matter is considered in those cases where 'merely the space of its own extension is to be *diminished*', this 'filling of space' or 'power to resist' must be thought as the 'moving force' that 'keeps a certain space free from the penetration of any other movable, when its motion is directed towards any place in this space' (Kant 2004 [1786], 33 [Ak. 4: 496]). In such cases, Kant adds, while 'One uses the expression *to occupy a space*, that is, to be immediately present in all points of this space, in order to designate the *extension* of a thing in space', 'it is not determined in this concept what effect arises from this presence, or even whether there is any effect at all' (Kant 2004 [1786], 34 [Ak. 4: 497]). Thus, if we are to distinguish matter, as presented in the spatial relations of outer intuition, from other extended entities such as geometrical figures, we must add to 'the concept of occupying a space' – which leaves 'undetermined' whether that which is present in a given space exerts some effect, such as attractive or repulsive force, on the spaces or 'movables' surrounding it – the 'more specific determination' of '*filling a space*' (Kant 2004 [1786], 34 [Ak. 4: 497]). This line of thought leads to Kant's first proposition of dynamics: 'Matter fills a space, not through its mere *existence*, but through a *particular moving force*' (Kant 2004 [1786], 34 [Ak. 4: 497]).

Coleridge's initial response to the opening explication and proposition of Kant's dynamics was not favourable. He appears to charge Kant with incorrectly identifying geometrical points with actual physical points or parts of material bodies in

[10] Phoronomy, the study of pure motion, i.e. without reference to force or mass. See Kant (2004) [1786], 15–32 (Ak. 4: 480–95) for further details on Kant's approach. For an interpretation, see Friedman (2013, Chap. 1).

[11] Kant (2004) [1786], 49 (Ak. 4: 511) provides the following definition: 'A property on which the inner possibility of a thing rests, as a condition, is an essential element thereof. Hence repulsive force belongs to the essence of matter just as much as attractive force, and neither can be separated from the other in the concept of matter.' At Kant (2004) [1786], 56–7 (Ak. 4: 518) Kant describes 'the sole universal characteristics of matter, which are comprehensible a priori', noting that 'the possibility of matter itself rests on these two properties'.

[12] Kant here points out that 'the basis for the resistance of matter exerted in all directions, and what this resistance is, must still be investigated', as he will do in the subsequent propositions and explications of this chapter (see Kant 2004 [1786], 35–74 [Ak. 4: 498–535]). For further discussion, see Friedman (2013, Chap. 2).

space, and takes issue with Kant's reference to treating a particular space, or an object presented as extended within this space, as 'a complex of several spaces' (Coleridge 1980–2001, 3: 279). Coleridge also contends that matter in its most general sense, as defined by Kant, may be conceived as occupying or existing within a given space without necessarily being taken to fill it 'exclusively', the implication being that this basic concept should not in fact be thought as including any sort of 'causal power' as an 'essential element':

> A space occupying, or *taking up*, or (in the German phrase) *taking in*, a space ['einen *Raum einnehmen*'] is so very near nonsense, that I hesitate in admitting its logical legitimacy even as an abstraction. Geometrical figures have an *objective* reality, I grant; but then they are only *subjectively* objective, and cannot even be thought of but as theorems or products of mind *in* the mind.[13] Now Matter must be that which appears and is contemplated as *outward*. It is manifest as opposed to Spirit or Power, which is manifest*ed*, cannot *appear* itself but is either self-known (= mind) or is inferred from its representative.—I do not therefore see the necessity any more than I can admit the propriety, of giving the name, matter, to a portion of empty Space, or of treating *a* space as an aggregate of spaces, or of talking of "a point of space", the absolute negation of Punctuality being the essential Character of Space. For I can readily conceive a matter present throughout a given space and, so far *taking it in* or existing in it, without at the same time exclusively *filling* it—which latter added to the former constitutes Body: i.e. Matter + Spirit or rather perhaps M = Sp. + Sp. = Matter. (Coleridge 1980–2001, 3: 276)[14]

Coleridge's view that Kant 'giv[es] the name, matter' to portions or points of space, and that this misidentification underlies the arguments of *Metaphysical Foundations*, is evident in his comment concerning the proofs presented in the chapter on mechanics:

> But I do not understand Kant's drift—in one passage he represents his Laws, as mere geometrical equations, and then I comprehend him fully—in another … he seems to assert its bonâ fide objective Truth, and then I am all in the dark again. (Coleridge 1980–2001, 3: 302)[15]

As I explain later, Coleridge's confusion derives partly from an initial misconception about Kant's views on the ontological relationship between the properties of material substances in themselves and the spatial relations of classical geometry, in accordance with which they are presented (to us) in outer intuition. For now, I focus

[13] 'Subjectively objective' is Coleridge's term for intersubjective validity (in this context, the principles and content of mathematics are described as intersubjectively valid for human cognizers, along Kantian lines, a position that Coleridge discusses at further length in both *Statesman's Manual* and *Logic*: see Coleridge 1972 [*c*.1816], 32–3; Coleridge 1981 [*c*.1819–22], 203n).

[14] On the relationship between matter and spirit, compare 'And both these powers, though spirit as opposed to Matter, yet revealing themselves *materially* and, in so doing, constituting body, are conceivable only in relation to Matter as Phaenomenon, i.e. they not merely need matter in order to manifest their existence, but they cannot be conceived of—I do not say merely not imagined, for that is not involved in the definition of matter, but they cannot even be thought of except as the actualization of matter, except as integrated thereby into Body—in short, except as body and in relation to Body' (Coleridge 2002 [*c*.1819–22], 304).

[15] See Coleridge (1980–2001), 3: 301 for the note dated 'August, 1819' where Coleridge revises this assessment. For a survey of these criticisms, see Levere (1981), 123–5; Modiano (1985), 151–60.

on how this note anticipates Coleridge's key statements on the nature of matter and spirit in his extensive *Metaphysical Foundations* marginalia.

13.3 A Critique of Kantian Forces: Coleridge's Definitions of Matter and Spirit

Another important reference point here is Coleridge's claim that 'Matter must be that which appears and is contemplated as *outward*. It is manifest as opposed to Spirit or Power, which is manifest*ed*, cannot *appear* itself but is either self-known (= mind) or is inferred from its representative' (Coleridge 1980–2001, 3: 276). This statement presents Coleridge's powers-based matter theory in outline, articulating the distinction between matter as 'manifest' and 'Spirit or Power' as 'manifest*ed*' that he develops further in various manuscript fragments and marginalia composed throughout 1816–18 (Coleridge 1980–2001, 3: 276–9; for similar definitions, see Coleridge 1980–2001, 3: 129–30, *c.*1816–18; Coleridge 1957–2002, 3: §4412, May 1818; for entries indicative of terminological continuity, compare Coleridge 1957–2002, 4: §§4674, 4740, *c.*1820, and 4923, *c.*1822). It also draws attention to the connection between the concept of appearance and the content of outer intuition, the spatially ordered manifold in which 'that which appears and is contemplated as *outward*' is presented. Moreover, Coleridge's contention that one may 'readily conceive a matter present throughout a given space and, so far *taking it in* or existing in it, without at the same time exclusively *filling* it' (Coleridge 1980–2001, 3: 276), by implicitly removing the characteristic of 'filling space' from the 'essential elements' or properties on which the 'inner possibility' of the a priori 'concept of a matter in general' is taken to rest, hints at a departure from Kant's 'dynamical explication' and its assumptions concerning the 'moving forces' supposedly inherent in all matter, considered as conditions through which this concept first becomes thinkable at all (see Kant 2004 [1786], 33–6 [Ak. 4: 496–9]). This break, I will argue, results from Coleridge's attempts to rework the subject-predicate terminology of Kant's definitions of matter and substance in the process of constructing a new theory of matter-spirit interactions based on a fundamentally different set of assumptions regarding the ontological foundation of the diverse material entities presented in the spatial manifold of appearances.

Ostensibly, Coleridge's movement towards the detailed statement of his theory of matter and its characteristic terminology was prompted by Note 1 to Proposition 2 of the chapter on Dynamics. P2 is 'Matter fills its space through the repulsive forces of all of its parts, that is, through an expansive force of its own, having a determinate degree, such that smaller or larger degrees can be thought to infinity' (Kant 2004 [1786], 36 [Ak. 4: 499]). The subsequent proof forms the basis of Kant's 'balancing argument' concerning the nature of the relationship of 'action and reaction' between the 'fundamental' attractive and repulsive forces held to be constitutive of, or at least 'essential' to, matter (Kant 2004 [1786], 36–7 [Ak. 4: 499–500];

see Friedman 2013, 180–94 for further analysis). Here he contends that beyond any given degree of 'moving force' by which a matter fills a space (likewise the force by which it 'endeavours' to draw other matters towards this space), 'both greater and smaller degrees can be found to infinity' (Kant 2004 [1786], 36 [Ak. 4: 499]; note that although Kant here lays the ground for his 'balancing argument', he does not begin to deal with attractive force until II.P5 at Kant 2004 [1786], 46ff [Ak. 4: 508ff]). The first note to this proof explains how '*the force of something extended in virtue of the repulsion of all of its parts*', which Kant defines as 'an *expansive force*' through which matter 'fills a space', 'a force resisting the penetration (that is, the approach) of others', is bound up with our a priori 'concept of a matter in general' (Kant 2004 [1786], 36–7 [Ak. 4: 499–500]). According to Kant,

> [this] expansive force of a matter is also called *elasticity* [*Elasticität*]. Now, since it is the basis on which the filling of space rests, as an essential property of all matter, this elasticity must therefore be called *original* [*ursprünglich*], because it can be derived from no other property of matter. All matter is therefore originally elastic. (Kant 2004 [1786], 36–7 [Ak. 4: 500])

The second note indicates how this proof can be applied to '*compressing* force', the 'greater moving force' that 'can be found' beyond any given 'degree of expanding force', namely, the repulsive force exerted on one matter by another, 'which can drive it from the space it fills [also by repulsive force] into a decreased space' (Kant 2004 [1786], 37 [Ak. 4: 500]).

Coleridge's response begins by stating some slight differences with Kant regarding how the nature of '*elasticity*' may be conceived. However, his core concern, which informs all subsequent criticisms of Kant's matter theory and its terminology, is with Kant's employment of the terms 'essential property' [*wesentliche Eigenschaft*] and 'original' [*ursprünglich*]:

> Kant's προτον ψευδος [*proton pseudos*, fundamental error] seems to me to lie in his definitions of *matter*, both the phoronomic and the dynamic: in the former assuming moveable *points* of Space = contrad. in terminis [a contradiction in terms], and then calling them *matter* = a misnomer: in the latter making repulsive expansion an essential property of *all matter* and thus confounding matter with Body, or at least with *material substance*. (Coleridge 1980–2001, 3: 277, in response to Kant 2004 [1786], 36–7 [Ak. 4: 499–500])

Coleridge's criticism of Kant's 'phoronomic definition' of matter[16] derives from an initial misapprehension concerning how Kant intends the principles of the relativity of space and motion articulated in his phoronomy, and employed in working out the concept of what is now called a 'reference frame', to be applied in analysing the relationships between opposing forces or causally interacting bodies in space in his dynamics and mechanics. As Coleridge later recognized, Kant's aim is not to provide a metaphysical account of absolute and relative space and the laws of motion, since he seeks to explain 'not the Truth of Nature, but the forms under which it may

[16] See Kant (2004) [1786], 15, I.E1 (Ak. 4: 480): '*Matter* is the *movable* in space. That space which is itself movable is called material, or also *relative space*; that in which all *motion* must finally be thought (and which is therefore itself absolutely immovable) is called pure, or also *absolute space*.'

be *geometrically represented*' (a process Coleridge terms the conversion of 'all relations into relations of Space as measured by Time, or figures generated by Motion in Space').[17] But what matters here, especially for the subsequent development of Coleridge's matter theory, is his claim that Kant goes wrong in his assumptions concerning the 'essential properties' that may be predicated of '*all matter*' and taken as constitutive of this concept in its most general sense. Since this key point of Coleridge's critical marginalia is not covered in detail by Modiano or Levere,[18] I consider it at greater length in the next section.

13.4 Properties and Predicates: Coleridge's Basic Matter Terminology

The note discussed above, written in the margins of II.P2.N1, includes Coleridge's instruction to turn to the back flyleaves of his copy of *Metaphysical Foundations*, where the arguments introduced here are developed in more detail (Coleridge 1980–2001, 3: 276–9). There we find Coleridge contending that 'We must, in order to preclude all confusion and equivocation, distinguish the terms Spirit (or power), Mind, Soul, Matter, Material Substance, and Body' (Coleridge 1980–2001, 3: 277). 'By Equivocation', he explains, 'I mean the confusion of two or more different Meanings in the same term' (Coleridge 1980–2001, 3: 277). As soon becomes apparent, Coleridge felt this error to be evident in Kant's employment of such terms as 'matter', 'substance', and 'fundamental force' throughout *Metaphysical Foundations*, particularly in those cases where it seems 'Matter [is] assumed as a datum, the *subject* of the powers/tho' two of these powers are elsewhere taken as constituting matter!' (Coleridge 1980–2001, 3: 295). Coleridge's point appears to be that Kant shifts between treating *Grundkräfte* as (i) the 'essential properties' of matter, its irreducible or most basic predicates, and (ii) as powers somehow constitutive of matter (Coleridge 1980–2001, 3: 288 complains that Kant appears to have 'confounded ... the *constituent Powers* [of matter] with the proper forces (or properties) of the Constitutum [thing constituted]'). Indeed, Coleridge contends that in *Metaphysical Foundations* 'Kant is less happy in his discriminative definitions than in any other of his Works', observing that matters could 'scarcely be otherwise in the systematic portion of the Volume from the προτον ψευδος [*proton pseudos*] in the conception of matter' (Coleridge 1980–2001, 3: 294; Modiano 1985, 156–9

[17] Kant (2004) [1786], 15–18 (Ak. 4: 480–83); Coleridge (1980–2001), 3: 301 (this note is dated 'August, 1819', by which point Coleridge appears to have partly revised his earlier (possibly *c*.1811–18) negative assessment of Kant's phoronomical treatment of matter. For further discussion of Kant's employment of 'reference frames' in *Metaphysical Foundations*, see Friedman (2013), esp. Chap. 1.

[18] Modiano discusses Coleridge's criticisms of Kant's account of the properties of matter (see Modiano 1985, 156–7) but does not examine Coleridge's analysis of predication criteria in the *Metaphysical Foundations* marginalia.

discusses Coleridge's response to such terminological issues). In Coleridge's view, it is because Kant fails to keep his basic definition of 'matter' suitably distinct from other closely related terms that such conceptual confusion is allowed to proliferate, even from the first 'phoronomical definition'. This problem prompted Coleridge to develop his own independent set of matter terminology. Nonetheless, he grants that, despite his criticisms of *Metaphysical Foundations*, 'Honor and Thanks are due to Kant for this first attempt', since 'Even tho' the success had been less than it is, the Attempt, the Idea, would have demanded the admiration and gratitude of every Philosopher' (Coleridge 1980–2001, 3: 291; further discussed at Modiano 1985, 157–8).

Coleridge begins his attempt at developing more rigorous distinctions between what he considers to be the grounding terminology of any matter theory with the following set of definitions: 'Spirit' is defined as '= that which cannot *appear* but thro' the medium of a representative [and] is manifested or inferred'; 'Mind' as '= that which is self-known and may be inferred—but cannot appear. = Sp[irit] self-known'; and 'Soul' as '= Spirit as the principle of Individuality', with the 'Corollary' that 'Rational Soul = Mind + Individuality' (Coleridge 1980–2001, 3: 277). 'Matter' is then introduced, in implicit opposition to these immaterial entities that by definition 'cannot *appear*' themselves, except as 'manifested' in or 'thro' the medium of a representative', and which are 'inferred' as the ground or cause of particular material or corporeal objects, ranging from elastic fluids and etheric fields to solid material bodies (which, on Coleridge's account, may act as 'the medium of a representative' for such immaterial powers). Matter in the most general sense is defined as '= that which simply appears, a subject having no other predicate than "to appear"' (Coleridge 1980–2001, 3: 277). Accordingly, for Coleridge, 'Matter is therefore = φαινομενον ψιλον [*phainomenon psilon*, mere appearance]', taken here as equivalent to the more familiar Latin term 'PHÆNOMENON' (Coleridge 1980–2001, 3: 277). 'Material substance' is then explicated as

> = a subject, of which "to appear" is a predicate but not the only Predicable, so however that fixibility [what Kant terms '*matter between determinate boundaries*'[19]] be denied—i.e. when the term is meant to be distinguished from Body:—but when from matter only [… then] we may define Material Substance = Phænomenon Actuale [Actual Appearance], M[atter] + yx—An appearance, not adequately defined by the predicate Apparency. (Coleridge 1980–2001, 3: 277–8; compare Coleridge 1980–2001, 3: 129–30, *c.*1816–18; Coleridge 1957–2002, 3: §4412, May 1818)

Finally, 'Body' is defined as '= Phænomenon actuale *ponderabile* [Actual *weighable* appearance] or perhaps Phænomenon reale [real Appearance]. M[atter] + yxz' (Coleridge 1980–2001, 3: 278). This set of distinctions, Coleridge asserts, 'is the true order in an ascending Series' and entails that 'all MS is M, but not all M. be MS; all Body be MS, but not all MS be B' (Coleridge 1980–2001, 3: 277; on the relationship between the 'actual' and the 'real', see Coleridge 1980–2001, 3: 345–6, *c.*1820–24; Fig. 13.1 visualizes of the relationships between 'M', 'MS', and 'B').

[19] See Kant (2004) [1786], 64 (Ak. 4: 525), Kant's definition of 'A **body**, in the physical sense' which 'therefore has a figure'.

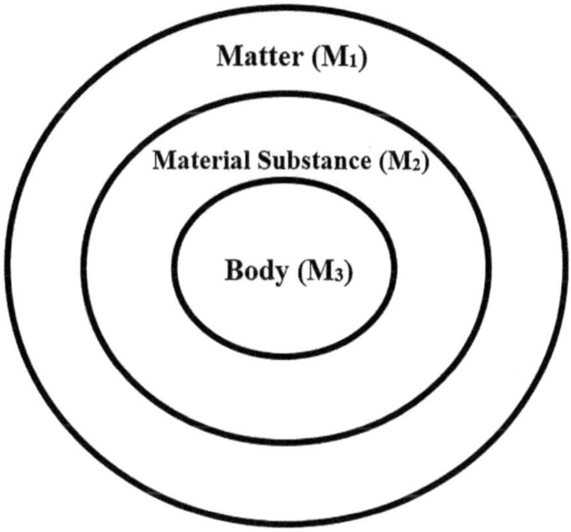

Fig. 13.1 Visualization of Levels of Generality and Class Relations in Coleridge's Theory of Matter. As Coleridge explains, 'all MS [Material Substance (M_2)] is M [Matter (M_1)], but not all M. [M_1] be MS [M_2]; all Body [M_3] be MS [M_2], but not all MS [M_2] be B [Body (M_3)]' (Coleridge 1980–2001, 3: 277)

Commenting on how these terms should be taken to relate to one another, Coleridge observes: 'In Physics we have no proper concern with Mind and Soul, which may therefore be removed: and Spirit we may take inclusively in M[aterial] S[ubstance] and Body, as the yx and yxz which is + (*more than*) M[atter]' (Coleridge 1980–2001, 3: 278). In the second part of this paper, I follow these instructions, focusing on only those terms that Coleridge saw as directly relevant to the field of matter theory. In doing so, I attempt to unpack Coleridge's unexplained term 'Apparency', considering the work he intends it to do and how it relates to Kant's concept of 'Appearance', itself central to Coleridge's matter theory and terminology, particularly the subtle ontological distinctions he seeks to draw between different classes of material phenomena. Before taking up this problem, however, it is necessary to return to Coleridge's statements concerning what it is possible to predicate of matter in its most general sense. For in these remarks Coleridge articulates the assumptions in which his reworking of Kantian matter theory and appearance terminology is grounded. I outline the relevant Coleridgean principles below to help lay the conceptual groundwork for the next section.

13.5 *Grundkräfte* as Spiritual Powers: Conceptualizing Coleridge's Metaphysics of Matter

It should now be evident how Coleridge's definitions and distinctions above relate to the earlier note on Kant's dynamics, where he states that 'Matter must be that which appears and is contemplated as *outward*. It is manifest as opposed to Spirit or Power, which is manifest*ed*, cannot *appear* itself but is either self-known (= mind) or is inferred from its representative' (Coleridge 1980–2001, 3: 276). As noted above, Kant's 'dynamical explication' presupposes the phoronomic account of matters considered merely as points moving through space, with regard to no properties other than their 'movability', adding to it a 'power' of repulsion or 'expansive force', understood as 'a property relating as cause to an effect' (Kant 2004 [1786], 33 [Ak. 4: 496]). However, while Coleridge retains this conception of matters as spatially arrayed objects given in outer intuition ('appearances', taken as 'the real of sensible intuition', as at Kant 2004 [1786], 16 [Ak. 4: 481]), he strictly delimits the attribution of any further properties to matter, and rejects definitions of the term based on the assumption of a fundamental power or force as an 'essential element' inherent in our a priori 'concept of a matter in general'. Although his 'spiritual powers' may seem to mirror Kant's '*Grundkräfte*', being also conceived of as basic properties 'relating as cause to an effect' in matter, Coleridge differs markedly in emphasising that such properties may be predicated only of certain more specifically determined classes of material phenomena (designated as 'material substance' and 'body'), in relation to which they stand (in the manifold of appearance) as causal or constitutive powers (see Coleridge 1980–2001, 3: 277–9).

Any talk of 'original' properties taken to be 'essential' to matter is thus ruled out, at least in our initial definition of this term. In short, for Coleridge, if we analyse the basic elements of our most general concept of matter, we find that it 'contains' or refers to only 'that which simply appears', the objects of cognition considered merely as presented or arrayed in space, with no property or predicate ascribed to them aside from their mere 'Apparency'. Although all material entities presented in the spatially ordered manifold of appearance are classed under 'Apparency', this basic concept can nevertheless tell us nothing about their possible properties or underlying causes. At most, it offers a starting point for inferences to some immaterial 'Spirit (or power)', understood as 'a property relating [to matter] as cause to an effect', with the material or corporeal object in question considered as the physical 'medium' or 'representative' whereby such immaterial powers, which cannot otherwise become apparent/sensible, are necessarily 'manifested' (Coleridge 1980–2001, 3: 276; compare Coleridge 2002 [*c*.1819–22], 304; Coleridge 1980–2001, 3: 130).

This account entails that Coleridge, whilst accepting that matter may be said to fill a space 'not through its mere *existence*, but through *a particular* moving force', rejects Kant's claim at II.P2.N1 that this concept taken on its own, as referring to 'that which appears' and is 'contemplated as outward' or as 'present in space', allows us to ascribe such 'moving force' to the 'inner constitution' of 'a matter in general' (Kant 2004 [1786], 15 [Ak. 4: 480]). If we wish to know 'the basis on which the filling of space rests', we must not seek it in any 'essential property of all matter' (Kant 2004 [1786], 36 [Ak. 4: 500]; see Kant 2004 [1786], 15–17 [Ak. 4: 480–82] for a statement of the Kantian position Coleridge rejects). Rather, we refer to its '*constituent* Powers', those properties which, while they cannot be predicated of 'all matter' in general, may still be ascribed to specific classes of material phenomena 'not adequately defined by the predicate, Apparency' (on 'Apparency', see Coleridge 1980–2001, 3: 277–8; on '*constituent* Powers', see Coleridge 1980–2001, 3: 280, 282–4, 288, 292). To do otherwise risks 'confounding matter with Body, or at least with *material substance*' (Coleridge 1980–2001, 3: 277). As Coleridge puts it,

> I can readily conceive a matter present throughout a given space and, so far *taking it in* or existing in it, without at the same time exclusively *filling* it—which latter added to the former constitutes Body: i.e. Matter + Spirit or rather perhaps M = Sp. + Sp[irit] = Matter. (Coleridge 1980–2001, 3: 276; compare Coleridge 1980–2001, 3: 130 definition of 'Matter as mere *videri* ["visibility", lit. to be seen, to appear]')

Contrary to Kant's position, then, to conceive of matter in the most general sense requires only that we think of an extended entity as present in a given space, merely existing in and occupying this space without filling it.

From Coleridge's perspective, the problem here is *not* that 'the concept of *occupying* a space', which 'leaves undetermined what effect arises from this presence [of an extended object in space], or even whether there is any effect at all', requires some 'more specific determination' such as '*filling a space*' before we can arrive at an adequate concept of the material objects present in space (Kant 2004 [1786], 36 [Ak. 4: 500]; a determination that Kant asserts to be contained in our a priori 'concept of a matter general' as the 'essential element' or 'property on which the inner possibility of [it] rests, as a condition' at Kant 2004 [1786], 49 [Ak. 4: 511]). Rather, as we have seen, it is that our basic concept of 'a matter in general' itself cannot supply this 'more specific determination': to designate a matter as that which fills a space, we require concepts containing those power predicates that Coleridge classes under the general term 'Spirit'. And these 'spiritual powers', he contends, 'we may take inclusively in M[aterial] S[ubstance] and Body, as the yx and yxz which is + (*more than*) M[atter]' (Coleridge 1980–2001, 3: 278). In short, *matter* is the subject term of which only 'Apparency' may be predicated; *material substance* and *body* are those terms which, being 'not adequately defined by the predicate, Apparency', may also be taken as 'the *subject* of [manifested] powers', namely, 'Spirit' as the power predicate or 'yx and yxz which is + (*more than*) M[atter]' (Coleridge 1980–2001, 3: 278; see also Coleridge 1980–2001, 3: 295). While Kant's theory aims 'to prove a priori the uniting of [two fundamental forces] in the concept of a matter in general', as the 'determination' or 'condition' on which our conception of extended material objects as filling a space necessarily rests (Kant 2004 [1786], 49 [Ak. 4: 511]), Coleridge's theory, no less desirous of such 'a priori proof', seeks to present this conception as grounded in the uniting of matter and 'Spirit (or power)' in the general concept of material or corporeal substance.

Kant contends that 'repulsive force belongs to the essence of matter just as much as attractive force, and neither can be separated from the other in the concept of matter' (as those 'essential elements' on which the 'inner possibility' of this concept rests) (Kant 2004 [1786], 49 [Ak. 4: 511]). Reasoning in a similar way, though arriving at a rather different conclusion, Coleridge argues that 'Spirit' or the immaterial '*constituent* Powers' of matter

> not merely need matter in order to manifest their existence, but they cannot be conceived of [..., indeed,] they cannot even be thought of except as the actualization of matter, except as integrated thereby into Body [i.e. actualized or manifested through a material body acting

as a 'medium' or 'representative']—in short, except as body and in relation to Body. (Coleridge 2002 [*c*.1819–22], 304; compare Coleridge 1980–2001, 3: 276–9, 282–4, 288)

In short, just as Kant claims that attractive and repulsive force cannot 'be separated from the other in the concept of matter' (Kant 2004 [1786], 49 [Ak. 4: 511]), Coleridge contends that the concepts of immaterial or spiritual powers and the material entities (ranging from fields and fluids to solid bodies) through which they are necessarily actualized or manifested cannot be separated from each other (i.e. are two mutually necessary components) in our conceptualizations of both matter and spirit, which are most clearly united in the Coleridgean concepts of material substance (M_2) and body (M_3).[20]

In introducing 'Spirit (or power)' as an additional term without reference to which no properties or predicates of 'matter' (viewed as distinct from the less general terms 'material substance' and 'body') beyond mere 'Apparency' can be defined or applied, Coleridge makes evident his significant differences with Kant concerning the statements it is possible to make about the nature and constitution of matter, both as a general concept and as a possible object of experience. For while Coleridge certainly accepts the claim that there are fundamental forces or powers active in nature that are somehow constitutive of matter, related to material phenomena 'as cause to an effect', as well as agreeing with Kant that these can never be directly observed but 'must be inferred … from data of experience', he disregards Kant's views concerning the limited scope of 'metaphysical investigation behind that which lies at the basis of the empirical concept of matter' (Kant 2004 [1786], 73–4 [Ak. 4: 534]).[21] Although Coleridge denies that *Grundkräfte* can be attributed to matter as properties inherent in it, or as 'essential elements' on which the 'inner possibility' of our a priori conception of matter rests, he is not motivated by anything like Kant's warning that 'it lies altogether beyond the horizon of our reason to comprehend original forces a priori with respect to their possibility' (Kant 2004 [1786], 73–4 [Ak. 4: 534]). Rather, it is precisely because he rejects the view that fundamental forces constitute a limit 'beyond which our reason cannot go' that Coleridge can confidently assert that the two powers of attraction and repulsion, or whichever forces our investigations of matter ultimately reduce these to, must be conceived of not as 'original' properties of matter, but as its '*constituent* Powers'.

[20] The case of matter in general (M_1) is possibly more complex in this context, but Coleridge seems to hold that any material phenomenon that 'has no other predicate than "to appear"' (i.e. anything classified as M_1) must still be assumed to have some immaterial cause or correlate, given that all 'Matter as Phaenomenon' is in some sense a manifestation or actualization of spiritual power (as argued at Coleridge 2002 [*c*.1819–22], 304 and Coleridge 1980–2001, 3: 276–7). See Table 13.1 for an overview of Coleridge's basic $M_{1–3}$ terminology (discussed further in Sect. 13.6).

[21] See also 'metaphysical investigation behind that which lies at the basis of the empirical concept of matter is useful only for the purpose of guiding natural philosophy, so far as this is ever possible, to explore dynamical grounds of explanation. For these alone permit the hope of determinate laws, and thus a true rational coherence of explanations' (Kant 2004 [1786], 74 [Ak. 4: 534]). For Kant, this is 'all that metaphysics can ever achieve towards the construction of the concept of matter'. See Coleridge (1980–2001), 3: 284 for Coleridge's response (also discussed below).

Coleridge is willing to grant Kant's claim that such forces ultimately 'cannot be explained', but he denies that this contention entails that the nature and origin of fundamental force necessarily lie beyond the limits of human conceivability. Rather, in Coleridge's view, when we recast 'the reduction of given, apparently different forces to a smaller number of forces and powers that explain the actions of the former' in the schematically arranged terms of a genetic series, we are able, by speculative metaphysical means, to circumvent what Kant asserts is the impossibility of rendering comprehensible to human reason those fundamental forces in which this process of reduction finally terminates (for this view, see Kant 2004 [1786], 63, 73–4 [Ak. 4: 524, 534–5]). Coleridge grants that Kant's position is the only philosophical possibility for 'a system which *begins* with Space and physical Powers defined exhaustively by effects and products solely relative to Space and manifestable in and by Space' (i.e. a theory of the powers and properties of matter limited solely to appearances in terms of its explanations and the metaphysical knowledge it can generate) (Coleridge 1980–2001, 3: 284). But, as Coleridge subsequently explains, he sees no reason to accept such limitations in his own system:

> But I have learnt by experience not to suffer myself to be cowed by the *term*, Grundkraft [Kant's 'fundamental force'], and the impossibility of rendering the same comprehensible. The major syllogism I fully admit— Grundkr[äfte] cannot be explained. But I must look again and again before I admit the minor—A + B [i.e. the powers of attraction and repulsion in Kant's theory of matter] are Grundkräfte—and (of course) the Conclusion. (Coleridge 1980–2001, 3: 284; see also Coleridge 1980–2001, 3: 288, 292)

Much of Coleridge's work in 'Fragment 4' and 'Appendix A' of *Opus Maximum* focuses on achieving this goal (i.e. identifying the spiritual or physical powers that are more fundamental than attraction and repulsion) in the context of his speculative cosmogony and philosophy of nature (for 'Fragment 4', see Coleridge 2002 [*c*. 1819–22], 291–345; for 'Appendix A', see Coleridge 2002 [*c*. 1819–22], 347–87). However, Coleridge's employment of speculative methods to arrive at a 'philosophic genealogy' purporting to account for the cosmic origins and hierarchical, polar order of 'all powers that now are'[22] lies outside the scope of the current paper. We must turn instead to a consideration of how his understanding of the relationship between matter and its '*constituent* Powers' informs the basic distinctions that Coleridge seeks to draw between the most general class of matter, of which no power-like properties may be predicated, and its 'more specific determination' as material substance or body, the two subclasses of physical phenomena and corporeal objects 'not adequately defined by the predicate, Apparency' (Coleridge 1980–2001, 3: 278). As we shall see, this set of distinctions and the ontological framework underlying it depends in turn upon parallel sets of relations that Coleridge presents as subsisting (and maintaining a separation) between different classes or 'modes' of reality and appearance (see e.g. Coleridge 1980–2001, 3: 345–6 for

[22] See e.g. Coleridge (2002) [*c*.1819–22], 337–8; Coleridge (1980–2001), 3: 138–9. For further discussion of Coleridge's 'philosophic genealogy' of 'spiritual powers' (Coleridge 2002 [*c*. 1819–22], 302–4), see Levere (1981), Chap. 5, esp. 129–34. Compare Modiano (1985), 185–90, 201–3 on 'ideal powers' in 'Coleridge's System of *Naturphilosophie*'.

Coleridge on existence, modality, and modes of reality). To better understand the context of such terminological relationships, we must return to the connection described earlier between Coleridge's term 'Apparency' and Kant's concept of 'Appearance'. This approach will help to further contextualize and clarify Coleridge's utilization of property and existence predicates to distinguish between the three basic categories of material entities in his system.

13.6 Apparency, Existence, and Subsistence: Coleridge's Ontological Ordering of Matter

Coleridge expresses some doubts concerning Kant's 'phoronomical definition' of matter on account of its supposed assumption of 'moveable *points* of Space' (see Coleridge 1980–2001, 3: 277, responding to Kant 2004 [1786], 36–7 [Ak. 4: 499–500]). This aside, however, he employs the term 'appearance', with reference to objects presented in a spatially ordered manifold, in much the same way as Kant does in his sections on phoronomy. Here Kant claims that if we 'explicate the concept of matter, not through a predicate that belongs to it itself as object, but only by relation to that cognitive faculty in which the representation can first of all be given to me, then every *object of the outer senses* is matter', a purely a priori or 'merely metaphysical explication thereof' (Kant 2004 [1786], 16 [Ak. 4: 481]). In this case, 'Space' is taken as 'merely the form of all outer sensible intuition', while '*Matter*, as opposed to *form*, [is taken as] that in the outer intuition which is an object of sensation, and thus the properly empirical element of sensible and outer intuition, because it can in no way be given a priori' (Kant 2004 [1786], 16 [Ak. 4: 481]). That matter, in the most general sense of this concept as an '*object of the outer senses*' (presented in space), is closely related to Kant's term 'appearance' is evident from his account of the relationship between human sensibility and the objects that 'affect' it in the first *Critique* (also assumed throughout *Metaphysical Foundations*). There Kant states that

> The effect of an object on the capacity for representation, insofar as we are affected by it, is *sensation*. That intuition which is related to the object through sensation is called *empirical*. The undetermined object of an empirical intuition is called *appearance*. (Kant 1998 [1781, 1787], 172–3 [B34])[23]

The link with *Metaphysical Foundations* is made clear by the subsequent remark: 'I call that in appearance which corresponds to sensation its *matter*, but that which allows the manifold of appearance to be intuited as ordered in certain relations I call the form of *appearance*' (Kant 1998 [1781, 1787], 172–3 [B34]).[24] Although Kant

[23] I follow the standard practice of referring to the original pagination of 1st and 2nd German editions, cited as A and B respectively. Coleridge owned and (lightly) annotated a 1799 5th edition B-version copy of Kant's *Kritik der reinen Vernunft* (published in Leipzig). See Coleridge (1980–2001), 3: 241–9 for Coleridge's *Critique of Pure Reason* marginalia.

[24] That Coleridge accepted this general account is evident at Coleridge (1957–2002), 3: §3605, August–September 1809 and Coleridge (1981) [*c*. 1819–22], 12n, 111–12, 151–4, 263–6.

later expands upon this bare account of 'the concept of matter, not through a predicate that belongs to it itself as object, but only by relation to that cognitive faculty in which the representation can first of all be given to me' (Kant 2004 [1786], 16 [Ak. 4: 481]), going on to consider '*original*' forces or powers that must be thought as an 'essential property of all matter' (Kant 2004 [1786], 36–7 [Ak. 4: 500]), the conceptual interconnection between his terms 'matter', 'object of outer intuition', and 'appearance', which refer to the most general class(es) of extended objects in space, remains evident throughout *Metaphysical Foundations*.[25]

That Coleridge adopts a similar sense of the term 'appearance' is evident from his subsequent fuller description of matter in its most general sense, or 'M_1', the first stage of his 'ascending series' of material phenomena (though, as we have already seen, Coleridge differs from Kant regarding our possible knowledge of the fundamental forces or constituent powers understood to be the cause or ground of such appearances). M_1 is defined as

> Phænomenon, sensu universali physico—Conditio universaliter assumenda Mundi sensibilis [Appearance in a universal physical sense—the Condition universally assumed of the sensible World] even as Spirit is the universal subpositum or subponendum [supposition or what is universally to be supposed]. Mundi sensibilis hic subsumptio, ea co-assumptio sine quibus Mund. Sens. omino non concipi potest [This subsumption of the sensible world is that co-assumption ... without which the Sensible World is altogether beyond conceiving]. (Coleridge 1980–2001, 3: 278; all translations of Coleridge's Latin in square brackets provided by editors)

That is, without an immaterial substratum of 'spiritual powers', upon which the existence of all objects in the material world is assumed to depend as its ground or cause, and from which their properties are taken to be derived, with such powers being related to objects as cause to effect, the very possibility of the 'sensible World' of physical phenomena is inconceivable. As Coleridge frequently notes, this claim is the grounding assumption of his own 'Construction of Nature', deriving from his conception of the relationship subsisting between the ideal and real 'poles' of the cosmos, with the sensible world of 'outward phenomena' being understood as the 'midpoint' between immaterial powers and material objects.[26] This assumption draws Coleridge into conflict with Kant over how the term 'substance', when associated with this 'substratum' or 'Condition universally assumed of the sensible World', should be defined.[27] In what follows, however, I focus on how Coleridge's view of the conceptual connections between the terms 'matter' and 'appearance' relates to the ontological distinctions he seeks to draw between different classes of

[25] On the connections and structural parallels between *Critique of Pure Reason* and *Metaphysical Foundations*, see Michael Friedman's 'Introduction' to Kant (2004) [1786], esp. x–xxx. See also Kant's own remarks at Kant (2004) [1786], 10–12n, 17 (Ak. 4: 474–7, 482).

[26] For a detailed exposition of Coleridge's polarities-based 'construction of nature' and 'science of relations', see Coleridge (1956–1971), 4: 868–76, a September 1817 letter to the Swedenborgian C. A. Tulk. See also Coleridge (1956–1971), 4: 789–90, December 1817. See Levere (1981), Chap. 5; Modiano (1985), Pt IV for further discussion.

[27] For Coleridge's critical response to Kant's account of the concept of substance, see e.g. Coleridge (1980–2001), 3: 296. For further discussion of Coleridge's view, see Modiano (1985), 158–9.

material phenomena. As we shall see, these distinctions are grounded in Coleridge's 'co-assumption' of a substratum of 'Spirit', partly constituted by a hierarchical order of ideal powers, under which the sensible world itself is subsumed (see Coleridge 1980–2001, 3: 276–9; compare Coleridge 2002 [*c.*1819–22], 302–4).

The two subsequent stages of Coleridge's ascending series facilitate the 'more specific determination' of this general concept of an appearance or extended object by the addition of the property and existence predicates that allow us to 'distinguish $M_{[1]}$ [Matter]. M_2 [Material Substance]. M_3 [Body]' (Coleridge 1980–2001, 3: 278).[28] These terms are defined as follows: 'M_3 or Body = phænomenon reale [real appearance], containing its own properties and capable of being contemplated as subsisting by itself', with 'M_2 or Material Substance = phænomenon Actuale [Actual appearance], having properties in itself, but only contemplable as existing in *another* yet not as a property of the latter, but itself remaining another' (Coleridge 1980–2001, 3: 278; for an overview of these terminological relationships, see Tables 13.1 and 13.2). If he offers no specific instances of the classes of objects designated by the terms Matter and Body, Coleridge does provide a hint concerning how these matter terms relate to one another and how they may be applied with his 'Example' of Material Substance: 'Electricity, and (in the theory of the French Chemists)

Table 13.1 Coleridge's matter terminology: basic definitions and predicate specifications (Coleridge 1980–2001, 3: 277–9)

Material phenomena class	Appearance class	Predicate specifications
Matter (**M_1**)	'= φαινομενον ψιλον [*phainomenon psilon*, mere appearance]'	'= that which simply appears, a subject having no other predicate than "to appear"' (i.e. all sensible objects or phenomena in general)
Material substance (**M_2**)	'= Phænomenon Actuale [actual appearance]'	**(1)** 'a subject, of which "to appear" is a predicate but not the only predicable' **(2)** 'M[atter] + yx—An appearance, not adequately defined by the predicate Apparency' (M_2 lacks 'fixibility' or some 'determinate boundary', e.g. a magnetic field, but is inferred from its interactions with M_3)
Body (**M_3**)	'= Phænomenon actuale *ponderabile* [actual *weighable* appearance]' or 'Phænomenon reale [real appearance]'	'M[atter] + yxz' (M_3 has 'fixibility' or a 'determinate boundary', as in any distinct corporeal object, e.g. a stone or an iron rod, that can be directly observed)

[28] For an analysis of existence predicates (in English) that argues 'existence in natural language is semantically conveyed by existence predicates, not existential quantifiers', also allowing for reference to 'different modes of being', see Moltmann (2020). I believe Coleridge's position is consistent with Moltmann's analysis, at least in his matter theory terminology.

Table 13.2 Coleridge's grammar of matter: detailed definitions and existence predicates (Coleridge 1980–2001, 3: 277–9)

Material phenomena class	Appearance class (full definition)	Grammatical equivalents and existence predicates
Matter (**M$_1$**)	'Phænomenon, sensu universali physico—Conditio universaliter assumenda Mundi sensibilis [Appearance in a universal physical sense—the Condition universally assumed of the sensible World]'	**(1)** '= Verb impersonal of the sensible World = videtur) (est [that which *is seen* as opposed to that which *is*]' **(2)** 'Matter = Verb impersonal = videtur [is seen/appears] and construed *relatively* or interdependently as forming a sentence' **(3)** '= Verb neuter' (i.e. intransitive)
Material substance (**M$_2$**)	'= phænomenon Actuale [Actual appearance], having properties in itself, but only contemplable as existing in *another* yet not as a property of the latter, but itself remaining another' (e.g. a magnetic field whose existence is inferred from its interaction with iron filings)	**(1)** '= Participle = existens [that which *is existing*]' **(2)** '= Verb mediæ vocis [in the middle voice] = existet, se manifestat [exists, manifests itself]' **(3)** '= Verb transitive'
Body (**M$_3$**)	'= phænomenon reale [real appearance], containing its own properties and capable of being contemplated as subsisting by itself' (e.g. an iron rod)	**(1)** '= Noun Substantive = Res, sive Phænomenon subsistens [a Thing, or a subsisting appearance]' **(2)** '= Verb substantive = subsistit [subsists]' **(3)** '= Noun Substantive, το ειναι [*to einai*, the being], hoc Esse [this being]'

Calorique' (Coleridge 1980–2001, 3: 278).[29] This terminological set appears to suggest an ontological and conceptual order proceeding from that which is merely present or extended in space (M$_1$), to that which fills space as a type of rarefied fluid or etheric field observable only through its interactions (as a 'medium thro' which immaterial powers act or become 'manifested') with or its effects upon the objects 'embedded' in it (M$_2$), to that which fills space by virtue of its constituent powers and inherent properties, a solid corporeal object bounded by a determinate figure, 'containing its own properties and capable of being contemplated as subsisting by itself' (M$_3$) (Coleridge 1980–2001, 3: 277–8; compare Kant 2004 [1786], 64 [Ak.

[29] As the Coleridge (1980–2001) editors explain (note **8**[9]), 'Calorique' was a 'term introduced by the French reformers of the chemical nomenclature in 1787 to denote a material substance (commonly thought of as an elastic fluid) to which the phenomena of heat and repulsion were attributed' (Coleridge 1980–2001, 3: 278n). Kant discusses 'caloric [*Wärmestoff*]' at Kant (2004) [1786], 71 (Ak. 4: 532) in the context of an overview of the chemical interactions between different types of matter.

4: 525], which defines 'A **body**, in the physical sense' as '*a matter between determinate boundaries* (which therefore has a figure)'). The next section explains how Coleridge articulates the relationships between these three categories by using grammatical analogues to ontologically differentiate classes of material phenomena.

13.7 Grammar and Ontology in Coleridge's Matter Terminology

The specific ontological order underlying Coleridge's 'ascending series' becomes apparent from the relationships between these terms when rendered in the form of his metaphysical grammar.[30] 'Matter' is recast as '= Verb impersonal of the sensible World = videtur)(est [that which *is seen* as opposed to that which *is*]', 'Materia [Material]' as '= Participle = existens [that which *is existing*]', and 'Body' as '= Noun Substantive = Res, sive Phænomenon subsistens [a Thing, or a subsisting appearance]' (Coleridge 1980–2001, 3: 279). Thus, just as an impersonal verb refers to no specific person or thing, 'Matter' designates simply 'that which appears' without reference to particular extended objects or the characteristic properties of more specifically determined material phenomena. It is a general term meaning only 'Appearance in a universal physical sense—the Condition universally assumed of the sensible World', as distinguished from that which *is* or *is existing*. It appears Coleridge is attempting here to indicate that matter, taken generally as 'mere appearance' (M_1), has a 'lesser' (or, at least, less specifically determined) ontological status than material substance, since it is seen but cannot be said to exist—this is evidently a consequence of Coleridge's restriction of what may be predicated of M_1 to 'Apparency' alone, which presumably entails that existence predicates cannot be applied to any appearances not included within those classes of material phenomena designated by M_2 and M_3 (see Coleridge 1980–2001, 3: 277–9). What is taken, then, to make M_2 and M_3 distinct from M_1, in this more exclusive sense of the term, is not only the attribution of additional properties to these material phenomena, but also the application of existence predicates indicative of a distinct ontological status, specifying that such material entities or corporeal objects *exist* or *subsist* rather than merely *appear*. Therefore, just as the participle or noun substantive allow for more specific determinations of the relations between grammatical terms (and thus for reference to more complex states of affairs or more specific objects) than can be designated by the impersonal verb alone, the terms Material Substance (M_2) and Body (M_3) allow for the more specific determination of the general class of sensible objects designated by M_1 (Coleridge 1980–2001, 3: 277–9).[31]

---

[30] For Coleridge's conception of 'philosophical grammar', see Coleridge (1981) [*c.*1819–22], 16–21, where he discusses the ontological implications and connotations of different parts of speech.

[31] In this sense, Coleridge's approach is similar to Kant's *Metaphysical Foundations*, which 'characterizes matter as the movable in space' in the first chapter, with each subsequent chapter

Coleridge further elucidates this grammatical arrangement of his matter terminology in a subsequent remark: 'Or we might (and perhaps still more aptly) treat Nature as a language the parts of speech of which are all Verbs, the infinite moods however becoming Substantives, or least capable of being construed as such' (Coleridge 1980–2001, 3: 279). In this grammatico-ontological scheme, we have, in descending order: 'Body = Verb substantive = subsistit [subsists]', 'Material [Substance]' as '= Verb mediæ vocis [in the middle voice] = existet, se manifestat [exists, manifests itself]', and 'Matter = Verb impersonal = videtur [is seen/appears] and construed *relatively* or interdependently as forming a sentence' (Coleridge 1980–2001, 3: 279). As above, the relation between particular terms, or between a term and its object (as distinguished from those terms lacking an object), enables the more specific determination of diverse classes of appearance. Again, Matter is the most general term and can designate nothing more than a something that merely appears,[32] unless it is 'construed *relatively*', considered in relation to some other less general term, together with which it can be thought of as 'forming a sentence' that tells us something further about particular matter(s) or appearance(s) 'not adequately defined by the term, Apparency' (Coleridge 1980–2001, 3: 278–9). When taken alone, however, 'M[atter] = Verb neuter', in the sense of an intransitive term that lacks any reference to a direct object (Coleridge 1980–2001, 3: 279).

In a similar manner, 'Material' (i.e. 'M_2 or Material Substance'), taken as a reflexive verb, must be understood in terms of its relation to some other subject or object term, without reference to which it cannot signify its meaning or function. This relationship is seemingly the 'grammatical' rationale behind Coleridge's definition of Material Substance (M_2) as that which is 'only contemplable as existing in *another* yet not as a property of the latter, but itself remaining another' (Coleridge 1980–2001, 3: 278). Put another way, although acting as the medium through which some 'spiritual power' is understood to be manifested, the existence of Material Substance must itself be apprehended through its activity in relation to some other thing, through/in which it 'manifests itself': while 'remaining another', an entity with a distinct ontological status, M_2 must nevertheless be contemplated as 'existing in *another* [material entity]', as an electric or magnetic field that exists independently of particular corporeal objects, yet is only indirectly observable through its interaction with or effects upon these objects. Accordingly, if 'MS = Verb transitive', then it is the operation of such reflexive terms that provides a (grammatical) model or analogue for the way in which Material Substance is thought to 'manifest' itself: it is a term which requires, or must be placed in some relation to, a direct object (analogous here to a material body) with reference to which its function is understood, or in relation to which the precise nature of its 'activity' becomes apparent (Coleridge 1980–2001, 3: 276–9). Thus, Body (M_3), as '= Noun Substantive', is

'begin[ning] from the idea that matter is the movable and then proceed[ing] to add a further specification' (Friedman 2013, 34).

[32] Compare Kant's definition of 'appearance' as 'The undetermined object of an empirical intuition' (Kant 1998 [1781, 1787], 172–3 [B34]). Coleridgean matter (M_1) is 'undetermined' in the sense that it has no predicates other than 'Apparency'.

taken to designate some particular object, 'Res, sive Phænomenon subsistens [a Thing, or a subsisting Appearance]', a distinct or determinate material body conceived as 'containing its own properties and capable of being contemplated as subsisting by itself' (Coleridge 1980–2001, 3: 279). Presumably, Body (as a substantive noun) is that direct object in relation to which the operations of the reflexive or transitive term Material Substance (M_2) are to be apprehended or understood (Coleridge 1980–2001, 3: 278–9).[33]

13.8 Coleridge's *Logic* and the Ontological Framework of His Theory of Matter

To understand why Coleridge draws such ontological distinctions between the three different classes of appearance or material phenomena in his theory, we must turn to his account of the relationship between reality and appearance in the *Logic*. This comes at the end of a lengthy discussion of the respective idealist and materialist responses (as Coleridge represents these positions) to the question: 'What are the things [i.e. objects in the physical world] in and of themselves independent of our sensations, representations, and conceptions?' (Coleridge 1981 [*c.* 1819–22], 112–31). Coleridge begins his account with some remarks on 'the definition, and possibly the etymology, of the word "thing"', speculatively linked to the word in various Northern European dialects for 'hearth, that is, a smaller inclosure within another space likewise inclosed' (i.e. 'ing'), which he suggests can be taken as 'the universal exponent of whatever is inclosed, bounded' (Coleridge 1981 [*c.* 1819–22], 114–15). This definition of 'thing' recalls Coleridge's definition of 'body' in his *Metaphysical Foundations* marginalia as 'a Thing, or a subsisting Appearance' taken as distinct from 'material substance', an unbounded or figureless material phenomenon 'only contemplable as existing in *another*' (Coleridge 1980–2001, 3: 279).[34] Much as a body is 'contemplated as subsisting in itself' by virtue of 'containing its own properties', a thing (or object in general) conceived as 'what is inclosed', when 'considered abstractly, must be known by that which incloses it'; and these 'bounds', Coleridge states, 'are the qualities and properties [of the material object] or the correspondent acts or effects in the percipient' (Coleridge 1981 [*c.*1819–22], 115; compare Coleridge 1957–2002, 4: §4740, November 1820). This definition is significant in relation to Coleridge's subsequent distinction between the different degrees of reality attributed to diverse classes of objects. Moreover, it establishes a correspondence between the 'qualities and properties' of things

[33] See also Coleridge (1981) [*c.*1819–22], 16–21, 88–90 on the grammatical and ontological aspects of substantives. For an overview of Coleridge's grammatical terminology and existence predicates, see Table 13.2 below.

[34] Consider, for instance, a magnetic field whose existence is inferred through its observable actions on iron filings (here, the filings are the 'other' in which the existence of the field is 'contemplable').

considered as they are 'in themselves' and certain 'acts or effects in the percipient', with both treated as possible objects of (human) knowledge (see Coleridge 1981 [*c*.1819–22], 129–31 for Coleridge's further comments on '*phaenomena*' and 'things in themselves'). This connection hints at those significant metaphysical differences with Kant that become evident in Coleridge's reworking of Kant's *Metaphysical Foundations* matter terminology and his rejection of Kant's basic definitions of matter and substance (see e.g. his assessments of Kant's position at Coleridge 1980–2001, 3: 276, 282–4, 291–6).

Coleridge concludes his discussion of the problem of our possible knowledge of things in themselves by contending that 'the opposition of real to not real, or mere nothing, is verbal and belongs to the grammar and the dictionary, not to logic; or to philosophy'. Since 'reality can have no contrary, ... in accurate language we should say that "unreal" is [merely] the contrary word to the word "real"'. This statement is elucidated as follows:

> All opposites, like the extreme points of all lines, must have a mid-term common to both. Now in the present case reality itself is that mid-term; for what can we think of antecedent to reality out of which reality might proceed or be evolved? But this granted, we can at once perceive the nature of the two opposites, viz. substantial reality and apparent reality. ... Equally evident is it that all which we receive through the senses must belong to the latter kind, and indeed we comprise them all under the common name of *phaenomena*, that is, 'appearances' ... that belong to mankind generally and constitute the common world of the senses. (Coleridge 1981 [*c*.1819–22], 129–130)

This passage, composed in the early 1820s, sets out an ontological scale which parallels, and could arguably be read as grounding, Coleridge's account of the differences in ontological status distinguishing the three basic classes of material phenomena (M_{1-3}) from one another. He not only reprises the Kantian definition of appearance presented in the *Metaphysical Foundations* marginalia, but also sets out the opposing poles or 'extreme points' of the 'line' of reality in a way that appears to correspond to, and conceptually dovetail with, the arrangement of the earlier matter terminology (worked out *c*.1816–19),[35] as I explain below.

Accordingly, matter in the most general sense (M_1), considered as 'mere appearance' or that which *is seen* as opposed to that which *is* (i.e. either *existing* [M_2] or *subsisting* [M_3]) corresponds to 'apparent reality'. The more specifically determined material substance (M_2), taken as 'actual appearance' or that which *is existing* or *exists* and *manifests itself*, likewise appears to correspond with 'reality' as the 'mid-term' of this conceptual scheme. Similarly, body (M_3), considered as an 'actual weighable appearance' or that which *subsists* (as a distinct 'thing, or subsisting appearance' that is 'capable of being contemplated as subsisting by itself'), can be thought of as corresponding to 'substantial reality' (Coleridge 1981 [*c*.1819–22], 129–130; compare Coleridge 1980–2001, 3: 277–9).

---

[35] On dating Coleridge's *Metaphysical Foundations* marginalia, see Coleridge (1980–2001), 3: 269. On dating Coleridge's *Logic* manuscript (dictated *c*.1819–22), see (1981) [*c*.1819–22], 'Editor's Introduction', xxxix–lv. See also my explanation in note 6 above.

Admittedly, Coleridge's initial definitions of material substance as 'actual appearance' and body as 'real appearance' may pose problems for this reading, raising the possibility that he could have classed both M_2 and M_3 under 'reality' (or 'apparent reality'), with only the immaterial '*constituent* Powers' assumed to underlie all such material phenomena as ground or cause being taken to belong to the realm or general class of 'substantial reality'. However, while this alternative appears to be supported by Coleridge's sharp division between matter and constitutive 'spiritual powers', it does not help to account for his definition of body as a 'subsisting appearance' or material phenomenon 'capable of being contemplated as subsisting by itself' (Coleridge 1980–2001, 3: 278–9). Indeed, Coleridge introduces this further specification precisely because his matter terminology seeks to distinguish just as sharply between matter in general, as 'mere appearance', and body or material substance, as that which contains certain properties or manifests particular powers that cannot be predicated of M_1, as it does between matter and spirit. Moreover, while Coleridge takes 'all which we receive through the senses' as those appearances that 'constitute' the 'common' sensible world, and consequently considers all material phenomena as falling under the general class of 'apparent reality', it is also evident that he regards objects classed as material substances or bodies, and conceived as the 'medium' or 'representative' through which unobservable immaterial powers necessarily manifest themselves (or from which their causal action is inferred), as belonging not only to the realm of the apparent, but also as containing properties that place them at least partly in the domain of the real (or actual) and the subsistent, understood as a separate class of material or corporeal entities ontologically distinct from 'mere appearances' (as defined at Coleridge 1980–2001, 3: 277).

The parallel ontological status of each of these different classes of reality and material phenomena should be apparent from the above terminological arrangement. However, it is worth considering another passage in which Coleridge's statements offer support for this interpretation of the conceptual and ontological order underlying his matter and appearance terminology. In this annotation prompted by William Charles Wells' essay 'upon single vision with two eyes' (1792, repr. 1818), Coleridge presents the basic terms of his matter theory in a polar arrangement that mirrors his description of 'reality' as 'a mid-term common to both' 'apparent reality' and 'substantial reality', considered as 'extreme points' placed at opposing ends of a single line (itself perhaps taken to represent an ontological continuum). Coleridge, possibly thinking of Kant's attempt to arrive at the 'pure forms of sensibility' by abstracting from, or 'removing in thought', all 'matter' or empirical elements of sensation, reflects on how, before 'enter[ing] on an investigation' of optics, it is first 'necessary to settle' such questions as 'what an Organ of sense is' and 'What are the conditions of a physical Abstraction' (Coleridge 1980–2001, 6: 132, *c.*1818–19; compare analysis of subjective and objective elements of cognition at Coleridge 1981 [*c.*1819–22], 139–43; for Kant's approach, see Kant 1998 [1781, 1787], 172–3 [B33–5], 193–7 [B74–82]). He then contends that to facilitate such an inquiry,

> we must assume three terms in objective nature instead of one. viz. Matter; Material Substance; and Body. And the middle term, containing Light, Magnetic, Electric, and Galvanic Ethers, &c [which] are the conditions of all physical abstraction & therefore the objective Correlatives of all organic Sentiency. (Coleridge 1980–2001, 6: 132)

Sadly, Coleridge does not elaborate on this 'mere Hint' concerning how to investigate the nature of the human visual system and its role in the processes of perception and cognition. However, this remark does offer insight into some aspects of Coleridge's matter theory not covered in his definitions in the *Metaphysical Foundations* marginalia. In particular, it provides clues concerning how Coleridge understands the relations between his matter terminology and the ontological distinctions underlying it.

First, that Coleridge presents 'Material Substance' here as the 'middle term' between 'Matter' and 'Body' lends additional support to the terminological parallels proposed above. Given his tendency to schematize conceptual relationships of all kinds in such polar terms, the treatment of both Material Substance and Reality as a 'mid-point' or 'middle term' between two opposing terms with which they form a metaphysical triad may not seem very significant. But placed in the context of a terminological scheme in which both middle terms designate a class of objects with an ontological status conceived of as intermediate between two others, the philosophical import of this parallel becomes evident. Accordingly, if we grant that there is a meaningful correspondence between the ontological status of (i) material substance, thought of as 'actual appearance' or that which exists and manifests itself, and (ii) reality, thought of as a 'mid-term common to both' the classes of 'apparent reality' and 'substantial reality', it becomes possible to interpret such parallels as being grounded in an ontological scheme or continuum arranged as follows: At one end, one has *what appears*, or matter as 'mere appearance' (M_1), at the 'mid-point', *what is actual*, or material substance as 'actual appearance' (M_2), and at the 'opposite' end, *what is subsistent*, namely, body as 'subsisting appearance' (M_3), with the distinct ontological status of each class tied to the properties it is conceived as 'containing' or the powers it is observed or inferred to manifest. Alternatively, on a more cautious reading, we can think of these arrangements as two parallel conceptual schemata, the first explaining how we categorize different classes (or degrees) of reality, the second how we categorize different classes of appearance (with reference to distinct degrees of being and increasing levels of specification or property-predication).[36]

Even if one denies the possibility of deliberate parallels or conceptual interconnections here, it is evident that when we consider Coleridge's polar arrangement of what Kant calls 'the real of sensible intuition' alongside the distinctions he seeks to draw between different classes of material phenomena, we are presented with an order of various classes or modes of reality. In my view, this framework offers a plausible analogue for the ontological relations underlying Coleridge's matter

[36] Coleridge's accounts of matter terminology at Coleridge (1980–2001), 3: 130 and Coleridge (1957–2002), 4: §4655 could be interpreted in this way.

terminology, as well as further insights into how he conceived of these 'three terms in objective nature' as related to (and distinguished from) one another, at both the conceptual and the ontological levels (Coleridge 1980–2001, 6: 132). More importantly, as I hope I have shown in this paper, the careful examination of Coleridge's key *Metaphysical Foundations* marginalia in the wider context provided by other closely related annotations and manuscript materials can help illuminate the crucial phases in the philosophical development of his theory of the metaphysical foundations of matter (or matter-spirit relations) and its conceptual and terminological underpinnings.

References

Beiser, Frederick C. 2002. *German idealism: The struggle against subjectivism, 1781–1801*. Cambridge, MA: Harvard University Press.

Coleridge, Samuel Taylor. 1956–1971. In *Collected letters of Samuel Taylor Coleridge*, ed. Earl Leslie Griggs, vol. 6. Oxford: Clarendon Press.

———. 1972. In *Lay sermons (1816)*, ed. R.J. White. Princeton: Princeton University Press.

———. 1981. In *Logic (c.1819–22)*, ed. J.R. de J. Jackson. Princeton: Princeton University Press.

———. 1980–2001. In *Marginalia*, ed. George Whalley and H.J. Jackson, vol. 6. Princeton: Princeton University Press.

———. 1957–2002. In *The notebooks of Samuel Taylor Coleridge*, ed. Kathleen Coburn, vol. 5. Princeton: Princeton University Press.

———. 2002. *Opus maximum* (*c.*1819–22). Ed. Thomas McFarland, with the assistance of Nicholas Halmi. Princeton: Princeton University Press.

Friedman, Michael. 2013. *Kant's construction of nature*. Cambridge: Cambridge University Press.

Kant, Immanuel. 1998. Critique of pure reason (1781, 1787). Tr. and ed. Allen Wood, Paul Guyer. Cambridge: Cambridge University Press.

———. 2004. Metaphysical foundations of natural science (1786). Tr. and ed. Michael Friedman. Cambridge: Cambridge University Press.

Levere, Trevor. 1981. *Poetry realized in nature*. Cambridge: Cambridge University Press.

Modiano, Raimonda. 1985. *Coleridge and the concept of nature*. London: Macmillan.

Moltmann, Friederieke. 2020. Existence predicates. *Synthese* 197:311–335.

Schelling, F. W. J. 1988. Ideas for a philosophy of nature (1797). Tr. Errol E. Harris and Peter Heath, with an introduction by Robert Stern. Cambridge: Cambridge University Press.

Wylie, Ian. 1989. *Young Coleridge and the philosophers of nature*. Oxford: Clarendon Press.

Dillon Struwig teaches at Sophia University in Tokyo. His research focuses on Coleridge's philosophy of cognition, theory of metaphysical methodology, and philosophy of nature, especially as articulated in Coleridge's *Logic* and *Opus Maximum* manuscripts, philosophical marginalia, and later notebooks. Much of this work seeks to facilitate a deeper examination of Coleridge's response to and interpretation of Kant's critical philosophy.

Chapter 14
Coleridge and the Science of Life

Andrew Cooper

Abstract This chapter examines Coleridge's *Theory of Life* in the context of a dispute in British medical institutions concerning the scientific status of comparative anatomy. It argues that *Theory of Life* is centered on the provocation that comparative anatomy is properly organized under the idea of life. It then demonstrates how Coleridge's method for a comparative science impacted several key figures in British bioscience, including Joseph Henry Green, William Whewell, and Richard Owen. It concludes that Coleridge's ability to adopt, transform and apply the insights of German physiology to the scientific debates in London's intellectual circles played a greater role the development of comparative anatomy in Britain than has previously been recognized.

> [I]t will not surprise us, that Plato so often calls Ideas LIVING LAWS, in which the mind has its whole true being and permanence; or that Bacon, *vicè versâ*, names the laws of nature, *ideas*. (Coleridge 1969, 1: 492)

14.1 Introduction

It has been well noted that Coleridge's encounter with dynamic theories of physics and chemistry at the close of the eighteenth century led him to reject his youthful commitment to David Hartley's empiricism and to develop a new method for natural science inspired by Plato, Bacon and Kant (McFarland 1981; Sloan 2003; Knight 2017). Through his encounter with Thomas Beddoes' circle of chemists and physicians in Bristol, Coleridge became convinced that corpuscular philosophy, in the guise of epistemic modesty, fatally strips reason from nature and thereby reduces the material world to 'an immense heap of little things' (Coleridge 1956–71, 1:

A. Cooper (✉)
Philosophy Department, Warwick University, Warwick, Warwickshire, UK
e-mail: Andrew.J.Cooper@warwick.ac.uk

P. Cheyne (ed.), *Matter and Life in Coleridge, Schelling, and Other Dynamical Idealists*, International Archives of the History of Ideas Archives internationales d'histoire des idées 253, https://doi.org/10.1007/978-3-031-78315-9_14

349). Following his return from Germany in 1799, he set out to demonstrate that the corpuscular school 'has received a mortal blow from the increasingly dynamic spirit of the physical Sciences' (Coleridge 1993 [1825], 395). For Coleridge, this dynamic spirit, manifest in Galvini's discovery of animal electricity and Lavoisier's metric system of elements, vindicated a new conception of matter imbued with inner forces. The challenge was to persuade Britain's gentlemen of science that these developments presuppose a rationally ordered cosmos.

While scholars have identified the significance of physics and chemistry to Coleridge's writings on scientific method, his impact on comparative anatomy and zoology has received far less attention.[1] This is partly due to the interpretive difficulties surrounding *Theory of Life*, his most extended treatise on the subject. Yet it is also due to a widespread view that the romantic conception of nature Coleridge sought to animate in Britain was an eclectic fusion of ideas borrowed from German sources, out of step with British anatomy and physiology.[2]

In this paper I reexamine Coleridge's *Theory of Life* in the context of a dispute in British medical institutions concerning the scientific status of comparative anatomy. While there is no conclusive evidence that *Theory of Life* was widely read, I argue that Coleridge's involvement in Britain's scientific intuitions ensured that the argument it presents nevertheless had a decisive impact on British comparative anatomy. I begin in Sect. 14.2 by situating Coleridge's early interest in German physiology in the contested landscape of British natural science. In Sect. 14.3, I examine the interpretive issues surrounding *Theory of Life* and propose a contextual method for assessing its significance. In Sect. 14.4, I summarize its main arguments, giving particular focus to Coleridge's central claim that comparative anatomy is properly organized under the idea of life. In Sect. 14.5, I demonstrate how the method outlined in *Theory of Life* impacted the work of several key figures in British bioscience, including Joseph Henry Green, William Whewell and Richard Owen. I conclude that Coleridge's ability to adopt, transform, and apply the insights of German physiology to the scientific issues debated in London's intellectual circles played a greater role in the development of comparative anatomy in Britain than has previously been recognized.

14.2 British Anatomy and German Philosophy

To understand Coleridge's argument in *Theory of Life*, it is vital to begin with the broader intellectual milieu in which British anatomists and surgeons played a central role in a Europe-wide discussion concerning a unified science of the animal

[1] There are several exceptions to this neglect, including Desmond (1989, 262) and Rajan (2019, 42). However, these studies do not question the accepted line that Coleridge's romanticism was surpassed by Darwin's materialist conception of comparative anatomy.

[2] For a typical account of this view see Beach (1942, 39). My argument builds on recent work that challenges this view, including Sloan (2007) and Richards (2013).

body. In 1786–7, the Scottish surgeon John Hunter delivered a series of lectures at London's Royal College of Surgeons in which he called for a new movement of surgical practice.[3] Hunter's lectures sketch an experimental science of practical anatomy based on a 'simple principle of life' (Hunter 1839, 20). This principle, he explains, is independent of organic structure and resident in entities that can instigate their own beginning. In such entities, matter is subject to higher order laws that cannot be reduced to their mechanical properties, manifesting a vital power that resides in the blood. Hunter's audacious goal was to replace the Galenic theory of humours with a materialist theory of the animal body, an affront to the established medical profession that saw his radical views remain on the margins of British natural philosophy. The dynamic conception of life presented in his lectures resonated more closely with work of German-speaking physiologists such as Albrecht von Haller.[4] Like Haller's two life forces, irritability and sensibility, Hunter's vital power operated as a Newtonian experimental postulate, irreducible to manifest physical properties. This placed an interpretive burden on his readers, for, like all Newtonian forces, the metaphysical status of Hunter's principle is highly ambiguous. While Hunter's principle might open a new classificatory scheme of animal bodies, it remained unclear how it could be used in the treatment of diseases.

In the mid-1790s, the Edinburgh physician John Brown rose to prominence in German medical departments for proposing a 'science of life' in which a basic force is examined in the same manner as the forces studied in physics and chemistry (Brown 1803, xxvii).[5] In *Elementa Medicinæ* (1780), Brown (1803, 34) defined life as a 'forced state' in which the power of 'excitability' is dynamically balanced with external 'exciting' effects. He insisted that while 'we know not what excitability is, … a certain portion is assigned to every being upon the commencement of its living state' (Brown 1803, 7). The living body can thus be understood as a local economy in which excitability forms the quantifiable basis of medical practice. Disease is a state of imbalance that occurs through a deficiency or excess of excitability, which can be measured and placed on a scale of severity. Medical interventions serve to stimulate or depress the organism's intrinsic activity, assisting it to return to its desired state of balance.

The possibility of a Brunonian science of medicine had an enormous impact on the progressive medical departments of Bamberg and Göttingen during the 1790s,

[3] Hunter's lectures were not published until 1835. His call for a new movement of surgical practice had a greater impact though his oversight of medical training as surgeon to St Georges Hospital and later as general surgeon. His work on gunshot wounds was taught at Edinburgh's Royal College of Surgeons, where John Brown was likely to have encountered Hunter's work. His extensive anatomical collection housed in the Hunterian Museum at London's Royal College of Surgeons, and the establishment of the Hunterian Oration in 1813, ensured that his dynamic conception of living bodies continued to be debated in the London medical community in the early- to mid-nineteenth century.

[4] The connection between Hunter's principle of life with Haller's life forces was made by Abernethy in his Hunterian Oration of 1819, with a view to motivating Hunter's importance to comparative anatomy. Abernethy (1819, 27).

[5] For a summary of Brown's reception in Germany, see Tsouyopoulos (1988).

where physicians including Christoph Girtanner, Christoph Heinrich Pfaff and Andreas Röschlaub attempted to refine the systematic character of medical practice. Girtanner (1797, v) claimed that Brunonianism could, with certain modifications, become 'nothing less than a complete system of the science of medicine'. Röschlaub (1802, 246) argued that Brown 'elaborated the only true principles for the whole organic doctrine of nature', for he was the first to understand that 'life is neither absolutely passive nor absolutely active'. This Germanic conception of Brunonianism inspired Thomas Beddoes to publish a revised English translation of Brown's *Elementa Medicinæ* in 1795, which received far more attention in Britain than the original publication (Knight 2017, 98). In his biographical introduction to the translation, Beddoes presented Brown's work through the systematic interpretation developed by Girtanner and Röschlaub.[6] In 1798, Beddoes established the Pneumatic Institution in Bristol with the intention of 'applying chemistry to the elucidation of animal nature, principally by pursuing the connection between the properties of elastic fluids, and the conditions of life' (Beddoes 1799, 3). With the young chemist-physician Humphry Davy, he attempted to apply Joseph Priestley's experiments on air in surgical practice to test how various excitants might restore diseased patients to health. In the first proceedings of the Institution, Beddoes (1799, 40) records 'S. T. Coleridge' among the early participants in experiments with nitrous oxide.

Coleridge had arrived in Bristol to work alongside Beddoes and Davy in 1795. In 1798, he travelled to Göttingen—presumably on Beddoes' recommendation—to discover more about Brunonian medicine and its philosophical underpinnings (Vickers 2004, 54–5). Under the supervision of Christian Gottlieb Heyne and Johann Blumenbach, Coleridge developed a working knowledge of German physiology and metaphysics (1956–71, 1: 518). His stay in Göttingen coincided with several dramatic events in the German academy, including Fichte's forced resignation from Jena and Schelling's proposal for a new philosophy of nature on Brunonian principles. Both events alerted Coleridge to the significance of Kant's critical philosophy for the construction of a new experiential science, and yet it was not until he returned to England in June 1799 that he was able to undertake 'the most intense Study' of Kant's works (1956–71, 2: 706; see also 1957–2002, 1: §887). Coleridge was particularly taken by Kant's separation of *Verstand* (the understanding, or representational thinking) and *Vernunft* (reason, or conceptual thinking), which undercut the associationist conception of thought in which ideas are formal impressions left on the intellect. On Kant's theory of mind, the understanding is an active and spontaneous power that, following the structure of Brown's excitability, functions as a qualitative response equal to sensory stimulation. Under the stimulation of sense data, the understanding applies determinate concepts in judgment to construct the manifold of perception as intelligible objects. Reason dynamically produces ideas that transcend possible objects, projecting the inferential relations in which the understanding's cognitions can find their location and ground.

[6] For a discussion of Beddoes' role in the transposition of Germanic ideas to the British Isles, see Cooper (2021, 474–478).

While Coleridge retained Kant's distinction between the understanding and reason, he agreed with Fichte and Schelling that transcendental philosophy remained one-sided. For Kant, the understanding prescribes the laws of nature studied in physics. The ideas of reason govern the organization of the understanding's cognitions, yet they have no direct purchase on nature. Fichte and Schelling claimed that while transcendental philosophy vindicated the *a priori* knowledge of natural laws, Kant failed to stitch the causal order of the understanding back together with reason's dynamic power. Fichte's solution was to ground the theoretical side of Kant's philosophy in the practical activity of the transcendental I. Schelling agreed that the activity of thought must provide the basis of the philosophical system, yet he rejected Fichte's claim that nature is a product of mind. He argued that the mind must grasp *itself* as a product of nature, such that reason's ideas are part of the natural order.[7] For Coleridge, Schelling's *Naturphilosophie* provided the foundation for a dynamic conception of the intellect that could recover nature from the abstraction of corpuscular philosophy. It demonstrated that natural science requires the synthetic work of philosophy, which searches for the ideal in the real (Coleridge 1957–2002, 4: 769).

Coleridge's encounter with Germanic ideas was framed within a broader discussion taking place within British medical communities, especially the Pneumatic Institution and the Royal College of Surgeons.[8] In contrast to Hartley's empiricism, in which reason simply associates ideas in the mind, the idealist current in German and British physiology offered to Coleridge a vision of philosophy that influences scientific practice. Building on Schelling's *Naturphilosophie*, Henrik Steffens argued in *Contributions to the Inner Natural History of the Earth* (*Beyträge zur innern Naturgeschichte der Erde*, 1801) that nature, grasped philosophically, is moved by a fundamental principle of polarity. Matter consists in the opposition of length and breadth, magnetism in opposite poles, electricity in positive and negative channels, and physiological movement in receptivity and activity. For Steffens, the 'forced state' of Brunonian medicine is not an isolated condition of living bodies but a natural achievement in a system of ascending complexity of which the mind, as the highest rung, has become conscious. A note from May 1819 indicates that Coleridge was heavily influenced by Steffens' programme of natural history:

> Receptivity, or Excitability, at one pole, and Agency or Excitancy … at the other, are the opposite states in which the **one** Activity, of Nature, which is the Substance of both, and their identity, reveals itself. There can be no *Product* without an Antecedent Power, that produced it, and is known to *be*, and to be that particular Power, by that ⟨particular⟩ Product. In relation to Existence, the Power determines the Product; in relation to our *knowledge* of its Existence, the Product determines the Power. (1957–2002, 4: §4538)

[7] In the *First Outline*, Schelling (1858 [1799], 3: 91) presents his programme of *Naturphilosophie* as a direct outworking of his research on Brunonian medicine with Röschlaub: 'Most partisans did not understand the scientific meaning of Brown's principles, … with one exception, Röschlaub, whose works everyone who has a sense for scientific medicine must study'.

[8] For a survey of Coleridge's encounter with Brunonianism, see Vickers (2004, Chap. 2).

Following Steffens, Coleridge identifies the Brunonian forces of excitability and excitancy with the cognitive forces of receptivity and spontaneity, signalling his intention to ground the mind in nature understood as a dynamic process. The continuity of mind and nature serves as the metaphysical foundation for scientific investigation. That is, it vindicates the quasi-transcendental assumption that the laws of nature are intelligible, for laws are the objective side of the mind's ideas (Coleridge 1976 [1829], 12). The Platonist pairing of law and idea collapses the Kantian distinction between nature and freedom, such that an idea simply *is* a law considered subjectively. The natural philosopher is justified in anticipating a governing power for every natural product, for:

> an idea … is in order of thought always and of necessity contemplated as antecedent. In the idea or principle, Life, for instance—the vital *functions* are the result of the organization; but this organization supposes and pre-supposes the vital *principle*. The bearings of the plants on the sun are determined by the ponderable matter of which they consist; but the *principle* of gravity, the *law* in the material creation, the *idea* of the Creator, is pre-supposed in … the very concept of the existence, of matter itself. (Coleridge 1976 [1829], 20)

Coleridge insists that ideas anticipate the results of empirical research by directing inquiry in search of the lawfulness of empirical objects. Empirical objects can then be examined in terms of these idealizations; the functions of living beings, for instance, can be examined *as* functions only by presupposing a principle of life. This entails that comparative anatomy is not only concerned with the structural adaptations of living beings in a given environment, but also with the principle of actuality that gives rise to those adaptations. To grasp the driving force of anatomical change requires a principle of individuation, which realizes the potential of lower levels of organization. Yet what is such a principle, and by what method might it be discovered?

14.3 Theory of Life

In *Theory of Life*, Coleridge sets out to determine the principle of life and to show how it can unify comparative anatomy as a science. Before turning to his argument, it is important to address several interpretive difficulties that have hindered serious engagement with the text. The first concerns authorship. *Theory of Life* was first published as *Hints Towards the Formation of a more Comprehensive Theory of Life* in 1848 by Seth Watson, who was virtually unknown to the community of scholars associated with Coleridge's estate. Watson included a brief forward to the original publication in which he notes that the manuscript had rested first in the possession of James Gillman in Hampstead and then with him in Cambridge, yet he said nothing about when the text was written or whether he or Gillman had worked on it in collaboration with Coleridge. Someone alerted the publishers to this at the last minute, for a leaf was inserted into some copies of the first edition stating that it had been brought to their attention that, after the initial print run, the work 'might with more propriety be considered as the joint production of Mr. Coleridge and the late

Mr. James Gillman of Highgate'. Nevertheless, *Theory of Life* was included in the 1885 volume of W. G. T. Shedd's *Complete Works of Samuel Taylor Coleridge*, much to the irritation of an early reviewer, who claimed that 'we have yet to learn what is the sufficient authority for the assertion that this treatise ever came either from the writing pen or speaking tongue of S. T. Coleridge' (Anon 1853, 264). The matter was not resolved until the twentieth century, when scholars discovered a letter from Coleridge's daughter Sara in which she explains that it is not a work of collaboration but directly employs the ideas and phrases form her father's published works and goes beyond the works of Gillman and Watson.[9]

Neither Watson's introduction nor Sara's letter shed light on the date of composition. Several proposals have been made, ranging from 1816 to 1831.[10] Those who claim that the text was composed in the final months of 1816 call on Coleridge's frequent reference to Steffens in his notes during that time (1957–2002, 4: §§4652, 4662, 4775–6).[11] Yet given that *Theory of Life* favours Steffens' graded system of ascent over Schelling's idealism, and given that Coleridge had a standing order for Steffens' works in June 1817 (1957–2002, 4: 738), it seems that 1819—the year following Coleridge's break with Schelling—is a more likely candidate.[12] In a note from 1820, Coleridge calls on Steffens to clarify the differences between his own system and that of Schelling, which 'are not momentary but essential' (1957–2002, 4: §4778). 'Let me not fail to declare that in the different several works of H. Steffens', he writes, 'especially in his *Beyträge zur innern Naturgeschichte der Erde*, the Spirit within me bears witness to the same Spirit in him' (§4778).

This 'same spirit' Coleridge encountered in Steffens' work raises an additional question of originality. In the 1930s, Henri Nidecker published a series of papers in *Revue de Littérature* in which he identifies 18 passages from *Theory of Life* taken almost directly from Steffens' *Beyträge* (Nidecker 1932, 870–871). While Coleridge embellishes Steffens' text at certain points, often to dramatize the uniqueness of the idea in its historical setting, his phrasing clearly mirrors Steffens's original formulation. In the following sections I aim to steer between two interpretive strategies found in the literature. The first is dismiss Coleridge's borrowings as having no bearing on the text. C. U. M. Smith (1999, 40), for instance, argues that they are of 'no great importance to those of us interested in the nineteenth-century alternative

[9] Namely, she criticises Watson's decision to publish the text without consultation on the grounds that the lack of clarity left the authorship of the text in doubt. Sara's letter is reproduced in Haeger (1976, 23).

[10] The reviewer in *Review of English Studies* argues for a date shortly after Coleridge's letter Gillman in 1816 that touches on several issues pertinent to *Theory of Life*. See anon. (1928, 443–4). Ashe (in Coleridge 1885, 364) claims that the text must be dated before 1831. Jackson and Jackson, the editors of *Shorter Works and Fragments*, state that *Theory of Life* 'seems to have been written rapidly in Nov–Dec 1816' (1995, 1: I 481).

[11] See for instance Levere (1981, 216).

[12] For instance, in a marginal note of his copy of Steffens' *Grundzüge*, Coleridge (1980–2001, 5: 360) rejects Schelling's claim that the gradient from 'the Insect, the Fish, the Bird, the Beast is *blosser Schein*'. I thus agree with Haeger (1976, 38) that 1819 is the most probable date of composition.

to Darwinism and the influence of Coleridge's ideas on the development of comparative anatomy'. After all, does not Coleridge speak of truth as a 'divine ventriloquist' in *Biographia Literaria* (1983 [1817], 81)? Yet Coleridge's borrowings *are* significant for understanding the nineteenth-century alternative to Darwinism, precisely because he refined his position through Steffens' critique of corpuscular philosophy. The second strategy is to read the text as testament to Coleridge's awareness of his ailing creativity and drug-ridden malaise, which, to use Joseph Beach's (1942, 50) words, drove him to 'literary dishonesty and a vanity that cannot bear to credit another with anything that may possibly be passed off for one's own'. Beach's disparaging critique rests on a problematic assumption that originality is the sole criterion of intellectual merit. Not only does this assumption accept Coleridge's dubious and self-styled conception of genius, it also overlooks the fact that the transmission of ideas can be a major intellectual achievement.[13] The historical significance of Coleridge's account of life can be better understood, I suggest, if we consider how he creatively redeploys Germanic ideas to defend British comparative anatomy against threat of materialism.

14.4 Life as the Principle of Individuation

Coleridge envisaged *Theory of Life* as a Hunterian Oration, an annual lecture held at the Royal College of Surgeons bound by statute to promote Hunter's pioneering ideas and to apply them to the collection held at the College's anatomical museum. The prestigious Hunterian Lectureship gave the holder the opportunity to develop a programme of comparative anatomy based on Hunter's principles. Hunter's disciple John Abernethy was appointed to the lectureship in 1814, which he used to demonstrate how Hunter's principle of life is 'adequate to explain the phenomena' of medical science (Abernethy 1817, 2). When invited to give the Hunterian Oration in 1819, Abernethy used the opportunity to settle an ongoing debate with the surgeon-anatomist William Lawrence, who ridiculed all vital principles and life forces as poetic personifications of nature.[14] To defend Hunter's legacy from Lawrence's materialism, Abernethy (1819, 27) presented his teacher as Britain's Haller, a faithful Newtonian who did not consider the principle of life as a metaphysical postulate but as a placeholder for an unknown force that one can 'uphold … as a good theory till a more satisfactory one is found'.

Abernethy was sympathetic to Coleridge's critique of materialism. He had attended Coleridge's lectures at the Crown and Anchor in 1818, which he cited in the oration against Lawrence (Levere 1981, 46). Several notes from April 1819 suggest that Coleridge attended Abernethy's oration, or at least that he read the

[13] Here I agree with Class (2012, 4–5), who claims that it 'is time for the studies of Coleridge's reception of German philosophy to relinquish "individuality" and "originality" not as a subject of investigation but as a criterion for our assessment of the writers' respective significance and merit'.

[14] For an account of the debate, see Desmond (1989, 117–21).

transcript published later that year (Coleridge 1957–2002, 4: §§4518, 4521). Yet Coleridge was unconvinced that Abernethy's defense of Hunter provided a genuine alternative to materialism. *Theory of Life* presents a revised account of Hunter's life principle, perhaps one that Coleridge hoped would be delivered at the College's invitation.

Coleridge opens the text with the conventions of a Hunterian Oration. He praises Hunter's pioneering work, anticipates future developments of Hunterian scholarship and demonstrates how these developments might apply to the anatomical collections at the College. Yet Coleridge had long been critical of Hunter for restricting the life principle to the blood, which, in keeping with Haller's Newtonianism, assumes an ontological break between living and non-living matter.[15] He laments the fact that Abernethy failed to develop the consequences of his conclusions systematically and carry them 'yet further back, to their ultimate principle' (Coleridge 1848, 19/1995, 1: 487). To complete this task, and to demonstrate how comparative anatomy could be organized under a single principle, Coleridge examines three strategies that physiologists have used to define life over the past half-century, each of which turns out to be hopelessly circular.

The first is to translate the word *life* into 'other more learned words; and this paraphrase of the term is substituted for the definition of the thing' (Coleridge 1848, 22/1995, 1: 489–490). Coleridge's target here is not only Lawrence but also the French physiologist Xavier Bichat, who defined life as 'the sum of all the functions by which death is resisted' (Bichat 1827, 1–2). Bichat had come into vogue among in 1815 when, following the Napoleonic wars, British medical students were able to travel to Paris to attend his famed lectures on materialist medical science (Knight 2017, 97). His definition leads physiology to a dead end, Coleridge contends, for it is nothing other than the circular claim that 'life consists in being able to live' (Coleridge 1848, 22/1995, 1: 489).

The second strategy is to take 'some one particular function of Life common to all living objects—nutrition, for instance; or, to adopt the phrase most in vogue at present, assimilation, for the purposes of reproduction and growth' (Coleridge 1848, 23/1995,1: 490). The functionalist approach isolates a relevant capacity found in every instance of life, and then reasons back to some basic organ or system in which its telic relation is intelligible. Yet this strategy is also circular, Coleridge argues, for it presupposes the organized state to which the isolated function contributes. It thus accepts an ontological break between living and dead matter, and thereby blocks the physiologist from grasping the generative process in which functions arise. Here we begin to catch a glimpse of Coleridge's alternative determination of life. In contrast to a lexical definition, a physiological determination 'would be a history, not a definition' (Coleridge 1848, 25/1995, 1: 492). A history would give expression to 'the *law* of the thing' in which 'all the properties and functions

[15] Coleridge was critical of Hunter's principle from as early as 1796. In a letter to Thelwall, he notes that Hunter held 'that the *Blood* is the Life—which is saying nothing at all—for if the blood were *Life*, it could never the otherwise than Life' (Coleridge 1957–2002, 1: 295).

are admitted by implication', giving 'insight into the necessity and *generation* of the phenomena of which it is the law' (Coleridge 1848, 25/1995, 1: 493).

Before elaborating on what kind of law he has in mind, Coleridge interrogates a third strategy of defining life: to search out 'some property characteristic of all living bodies', such as 'anti-putrescence' or 'the power of resisting putrefaction' (1848, 27/1995, 1: 494). At first glance, this strategy looks more promising than the first two, for a property denotes something internal and irreducible to mechanical analysis. Yet like the first two strategies, Coleridge argues that it confines the idea of life to organized beings discernible by us. While it is preferable to both Bichat's materialism and Haller's mechanism, for it opens the language of powers, at best it 'describes the *vis vita* by one of its many influences' (Coleridge 1848, 27/1995, 1: 495). That is, it defines an emergent property and fails to work back to 'the law or principle of action'.

The thrust of Coleridge's argument is methodological. The three strategies of defining life commence with the representational thinking of the understanding, which attempts to define a concept by analysis of its content. Brunonian medicine takes us beyond the representational stasis of the understanding, Coleridge insists, for it considers life as an idea manifest in the dynamism between stimulus and response. Siding with the German physiologists, who attempted to elevate Brown's principle into a positive philosophy, Coleridge identifies the principle of life with the mind itself, that is, with the locus of activity that is undertaking the inquiry.

Coleridge discovered Steffens' *Beyträge* to be the perfect vehicle to advance British comparative anatomy, for it identifies the principle of life as the culmination of a history of errors made by natural philosophers (Steffens 1801, 37–8). Steffens argued that these errors stem from a well-established method that examines life as objectivity, an occult quality or unknown power that is knowable only through its effects. This method assumes the separation of ideas, understood as subjective reconstructions of objectivity, from nature, such that the naturalist must somehow leap from mind to the objective world. Instead, Steffens begins with nature's subjective side—what is immediately given to the mind as a dynamic, generative power—which he aspires to reconcile with objectivity. Coleridge (1848, 40/1995, 1: 509) saw that by avoiding the 'arbitrary distinction' between mind and nature, Steffens 'fills up the arbitrary chasm between physics and physiology, and justifies us in using the former as means of insight into the latter, which would be contrary to all sound rules of ratiocination if the powers working in objects of the two sciences were absolutely and essentially diverse'. By replacing the hard distinction of kind with a continuous gradient of degrees, he returns to the 'essential vitality of nature, that she does not ascend as links in a suspended chain, but as the steps in a ladder' (Coleridge 1848, 41/1995, 1: 509). The assent of nature is universal such that even 'the life of metals, as the power which effects and determines their comparative cohesion, ductility, etc.' exists on the same scale as 'the Life which produces the first attempts of organization, in the almost shapeless tremella, or in such fungi as grow in the dark recesses of the mine' (Coleridge 1848 40/1995, 1: 508).

In contrast to Hunter's conception of life as an experimental postulate, Coleridge (1848, 42/1995, 1: 510) aims to complete the Brunonian vision of a medical science by showing how the idea of life can be grasped as 'the internal copula of bodies', that is, 'the power which discloses itself from within as the principle of the unity in the many'. Once more he follows Steffens by transgressing Kant's attempt to limit reflecting judgment to the mere estimation of a vital power manifest in the dynamic relation of part and whole. Yet Coleridge's method grants a more explicit role to reason than can be found in Steffens' *Beyträge*. While the understanding cuts, divides and defines, Coleridgean reason grasps the law of generation as a *causal* principle, the internal copula that explains the *generation* of living beings. This opens a distinct definition of life as 'the *principle of individuation*, or the power which unites a given *all* into a *whole* that is presupposed by all its parts' (Coleridge 1848, 42/1995, 1: 510). Coleridge explains that the link that combines the whole and the parts, and acts throughout both, is 'the *tendency* to *individuation*'.[16] Where corpuscular philosophy reduces the diversity of properties to a homogenous substance, Coleridge's principle of life opens a philosophy of ascent. As we reflect on the steps between physics, chemistry and physiology, the lower powers are not destroyed by higher forms of life but are assimilated within them, such that in the transition from chemistry to physiology, elasticity appears as the responsiveness of muscle fibres. Electricity and irritability are 'but degrees and different dignities of one and the same tendency' (Coleridge 1848, 43/1995, 1: 511). Life can thus be presented as a series of steps:

1. Life as the mere unity of powers,
2. Evolution of the simplest forms of totality,
3. The vast formations studied in geology and history,
4. The 'tendencies of the Life of Nature to vegetation or animalization'. (Coleridge 1848 48/1995, 1: 516)

Note that it is only on the fourth step that we encounter organic unity. This step requires its own cascade of divisions, ranging from the lowest vegetable to the highest animal. In the human being, the tendency to individuation is not only perfected but begins 'a new series beyond the appropriate limits of physiology'. Life is the tendency to individuation, such that 'the degrees or intensities of Life … consist in the progressive realization of this tendency' (1848 49/1995, 1: 517).

Having defined the principle of life in terms of individuation, Coleridge sets out to show how nature and its laws can be determined beneath it. A law, he explains, is the cause of union, from which we can derive the necessary unification of the phenomena for which it is a law. The most general law is polarity, or 'the essential dualism of Nature, arising out of its productive unity, and still tending to reaffirm it, either as equilibrium, indifference, or identity' (Coleridge 1848, 50/1995, 1: 518). The productivity of the polar relation is thus incompatible with mathematical

[16]This locution is Coleridgean. Steffens speaks of the '*Production des Individuums*', which is a tendency in nature toward the '*Production der Gattung*', and the '*centripetale Tendenz* der ganzen Natur', which is revealed most fully in the human being. But he does not refer to a tendency to individuation (Steffens 1801, 310).

calculus. It is manifest in the copula that unifies the three dimensions of matter expounded by Schelling and Steffens:

Power of length (magnetism/reproduction)
Power of surface (electricity/irritability) — Life as the copula of constituent forces
Power of depth (chemical affinity/sensibility)

Length, breadth and depth are the three dimensions of matter that correspond the linear axis of magnetism, the electrical charge given to bodies, and the inner chemical effects of a galvanic cell.[17] Together they give rise to three-dimensional matter, and in turn to the cosmos, such that life is properly understood as their unity. In contrast to materialism, this dynamic foundation grounds a programme of comparative anatomy as a series of transitions united under the principle of life. The transition of vegetables to insects is marked by the intensification of sensibility, and the transition from insects to fishes by the intensification of the reproductive force. Each of the following transitions—to birds, quadrupeds and finally to humans—is marked by increasing individuation of the living forces. While nature unfolds according to a purpose, this purpose is not preordained but rather a process of becoming that moves toward ever-increasing complexity.

14.5 The Philosophy of Biology

In *Theory of Life*, Coleridge takes over Abernethy's confrontation with Lawrence's materialism to advance Hunter's one-sided principle with 30 years of research conducted in Germany since Hunter's original lectures. Drawing from Steffens, he attempts to convince his British contemporaries that Hunter's principle requires a philosophical grasp of life if it is to explain nature's inner dynamism and creative power. In this final section I suggest that while there is no conclusive evidence that *Theory of Life* made it beyond the hands of Gillman and Watson, the text nevertheless offers a window into his conversations with members of the Royal College and the British Association for the Advancement of Science (BAAS) in the years following 1819. In agreement with Philip Sloan, I reject John Stuart Mill's (1869, 403) notion of a 'Germano-Coleridgean Doctrine' as a way of framing this conversation, which conveys a unified ideology in opposition to mainstream British natural philosophy. Yet I reject Sloan's (2003, 40) notion of a 'Coleridgean methodology', which glosses over the distinct research programmes advanced by each of Coleridge's interlocutors.[18] Instead, I examine the work of three prominent members of the BAAS to identify a structural similarity with the argument of *Theory of*

[17] For an account of Coleridge's use of length, breadth and depth in relation to Schelling and Steffen's readings of Kant's *Metaphysical Foundations*, see Levere (1981, 116–121).

[18] For a politized version of Sloan's position, see Desmond (1989, 260–275).

Life, indicating that it had at least some of its intended impact, even if it was not widely read.

14.5.1 Joseph Henry Green

During his time as a young surgeon at St Thomas's Hospital in London, Joseph Henry Green was a member of Coleridge's Thursday seminar on literary criticism and German philosophy. In 1817, he travelled to Berlin to study philosophy under Karl Solger, and on return he met weekly with Coleridge to discuss the writings of Kant and Schelling (1957–2002, 2: 688). In 1824, Green was appointed to the Hunterian Lectureship at the Royal College of Surgeons, which gave him the opportunity to give five series of lectures between 1824 and 1828 in which he assimilated recent work in British medicine with German philosophy and physiology (Sloan 2007, 156). He undertook the task in collaboration with Coleridge (1957–2002, 5: 495), and used the opportunity to sketch a philosophically-based method for comparative anatomy that bares several parallels with the second half of *Theory of Life*.

In 1840, several years after his lectures at the Royal College of Surgeons, Green was invited to give the Hunterian Oration, published in the same year as *Vital Dynamics*. His oration presents a revised methodology for comparative anatomy, which includes a detailed recommendation of Coleridge's philosophical work to the members of the Royal College. Green (1840, xiv–xv) praises Coleridge for affirming the 'all important distinction of the Reason and Understanding', and thus for saving philosophy from Locke and Hartley's associationism. He then proceeds to outline a new method that bears the hallmarks of Coleridge's philosophy of natural science. Natural science, Green (1840, xxvii–xxviii) states, is 'a scheme of Causes and Laws in the unity and with the connections of Reason', and, citing Coleridge's *On the Constitution of the Church and State*, he defines ideas as principles of speculative reason that form the subjective side of laws operating objectively in nature.

In *Vital Dynamics*, Green builds on Coleridge's method to develop an alternative to John Herschel's account of induction as generalization. Science must pursue an 'anticipatory' method, he declares, in which ideas direct our search for laws that determine the occurrence of effects according to their grounds (Green 1840, 11). These laws are not available to the understanding, which cuts and divides. They can be discovered only by the synthetic power of reason, which is able to grasp opposites in their unity. Coleridge had stated that 'The anatomist himself really seeks for an *Idea*—not to learn what this or that limb—Hand for instance—is—but to learn what *a* Hand is' (1995, 2: 1311). This is to say that the idea is antecedent to the inquiry and drives the understanding to search for a law that transcends every particular instance. In Green's (1840, 21) terms, 'the method of the sciences must be dynamic; that is, by contemplating nature as a scheme of causes and laws with the connections, and in the unity, of reason'. The idea of life is a 'law' that grounds comparative anatomy in physiology, Green writes, drawing the diverse range of living beings into a natural system:

> by comparing, I say, the various groups of the animal kingdom, we shall find that they may be ranged in an ascending scale, of which the degrees are marked by a relative balance and proportion of the vital forces, and in which the ascent is determined by the evolution of life into Sensibility, and by the subordination of sensibility as the highest force and most essential form of living existence. (Green 1840, 33)

Building on his collaboration with Coleridge, Green develops a method in keeping with German anatomists such as Kielmeyer and Goethe, who identified classificatory categories according to the balance and proportion of the life forces.[19] After presenting his comparative method, Green lays out a series of transitions in anatomy from protozoa to insects, and from molluscs to vertebrates, in which each level is marked by increasing internal organization. The ascent is unified by 'the opposition and harmony of two great tendencies;—on the one hand that of nature to integrate all into one comprehensive whole … and on the other hand, the tendency to integration in the parts' (Green 1840, 39). Here Green's analysis is undoubtedly Coleridgean. He examines 'the advancing perfection of nature towards Individuality' and claims that it reaches its highest degree of individuation in the human (Green 1840, 42). The insight that all living form participates in a dynamic system, Green (Green 1840, 57) explains, enabled the German anatomists to discover 'the pattern and pre-existing model, according to which, and the genetic process itself by which, organic forms are constructed'. This pattern is not a fixed mould, applied like a stamp to passive matter. It is rather an 'inward type' manifest in every outward expression of organic life, grasped by reason's inward eye (Green 1840, 38 n.). The upshot of Green's programme of research is radical in the context of British bioscience. No longer is the purposiveness of organic form understood as the product of a divine idea as efficient cause. The idea of life is grasped through careful empirical examination of the form of activity under which the vast diversity of organic life can be classified. Once discovered, Green declares, this idea can show that the 'varied organic forms are but modifications of one simple primary form'.

14.5.2 *William Whewell*

Green's lectures offered a platform for Coleridge's theory of life some 20 years after Coleridge had first developed a systematic account of anatomy. In *Vital Dynamics*, he also praises the work of William Whewell, professor of minerology at Cambridge, who met regularly with Coleridge and Green at the BAAS during the early 1830s. Green (1840, 21) saw that Whewell's historical and philosophical examination of the inductive sciences offered a much-needed alternative to Herschel's inductivist methodology, for it demonstrates that 'the method of science must be dynamic; that is, by contemplating nature as a scheme of causes and laws with the connections, and in the unity, of reason'.

[19] E.g. Kielmeyer (1793, 43–46) and Goethe (1947, IX 121–125). For an account of Kielmeyer's and Goethe's methodologies in their context, see Cooper (2020, 2022).

Whewell too had a longstanding interest in German philosophy. In 1825, he visited Freiburg to study minerology under Friedrich Mohs, where he records an encounter with 'the metaphysical genius in Germany' that demonstrated to him that the 'relation of Causation is a condition under which we think of events' (Whewell 1847, 172–3). In *The Philosophy of the Inductive Sciences* (1840), Whewell (1847, 25) explains that ideas connect 'impressions of sense according to relations of space, time, number, likeness, cause etc'. These ideas are not Lockean imprints on a passive mind but cognitive activities that prescribe the 'laws' that objects necessarily follow. At the heart of the philosophy of science, for Whewell, lies the 'antithesis of philosophy', a discordant energy between ideas and things, theories and facts, subject and object. The task of inductive science is to harmonize the two sides of the antithesis:

> As in man's contemplation of nature, there is always some act of thought which depends upon himself, and some matter of thought which is independent of him … the combination of these two elements, the subjective or ideal, and the objective or observed, is necessary, in order to give us any insight into the laws of nature. (Whewell 1847, 30)

The antithetical nature of philosophy is particularly evident in what Whewell (1847, 543) terms 'The Philosophy of Biology', which encompasses the sciences concerned with 'living things'. Scientific theories for Whewell work toward clearer and clearer delineations of their subject matter, and the task of the philosopher is to define these delineations and show how they form the history of a science. The philosopher of biology recognizes that the sciences concerned with living things have a distinct methodology and mode of explanation, based on 'a particular *Idea*'.

The structural similarities between Whewell's philosophy of biology and Coleridge's *Theory of Life* are striking. Like Coleridge, Whewell (1847, 574–575) presents a survey of failed attempts to define the idea of life, including Bichat's definition that life is the system of functions by which death is resisted and the attempt to list various living functions. Each of these attempts, he explains, presuppose the idea it aims to determine, for they identify the defining marks of living nature through an exclusively empirical method. In the major discoveries in biology—the circulation of the blood, digestion, assimilation and generation—we can discern the basic principle of the *philosophy* of biology: 'Observation taught much; and among other things, she taught that there was something that could not be observed, but which must, if possible, be conceived' (Whewell 1847, 581).

Having identified the problems encountered by each of the main positions, Whewell (1847, 573) proposes an alternative that draws from Kant's transcendental examination of judgments of organic structure in *Critique of the Power of Judgment*: *X* qualifies as a living being if *X* is judged under the idea of an 'organised product of nature … in which all the parts are mutually ends and means'. The 'great metaphysician' was the first to see that the idea of life cannot be discovered in a particular science; it is rather the enabling condition of domain of inquiry concerned with living things. Whewell (1847, 587) then attempts to show that Kant's definition 'is capable of being made the basis of sound knowledge' if the philosopher of biology provides sufficient determination to the 'Idea of an Organic or Living Being'. A living being is one whose powers are such that it appropriates to itself new substances and makes them part of itself: '*Organic Life is a constant Form of a circulating Matter, in which the Matter and the Form determine each other by particular laws (that is, by Vital Forces)*' (Whewell 1847, 588).

Whewell's philosophy of biology stands in radical opposition to the British physicotheologians, who studied the teleological structure of living beings as the product of an external idea. The idea of life cannot be posited as an external causality but must 'be supplied by the student of organization out of his own mind' (Whewell 1847, 621). This is not to imply that the organising principle of biology is subjective. Echoing Green's comparative anatomy, Whewell (1847, 627) rejects the notion of design in favour of 'type', understood as an inward pattern or idea grasped intuitively by speculative reason. Type introduces an immanent kind of normativity to the examination of organic form. While the natural forces studied in physics and chemistry (gravity, elasticity and affinity) never act in a diseased manner, the forces studied in biology do not always act as they should. Using Brunonian terms, Whewell explains that organic forces can diminish or become overexcited, giving rise to diseased states. They 'have as their object to conform the living being to a certain type' and can fail for multiple reasons. Whewell concludes that while the biological sciences are still in their infancy, the groundwork has been accomplished. Once the study of living beings is based in 'natural relations' rather than mechanical abstractions, our knowledge of organic functions 'have tended more and more to the character of exact and rigorous science' (Whewell 1847, 579).

14.5.3 Richard Owen

Green's Hunterian Oration was attended by Richard Owen, who soon went on to become one of the most prominent anatomists and zoologists of the time. Owen also attended Green's lectures on comparative anatomy in 1826, where he noted that 'For the first time in England the comparative anatomy of the whole Animal Kingdom was described. … The vast array of facts was linked by reference to the underlying Unity, as it had been advocated and illustrated by Oken and Carus' (Owen to John Simon, quoted in Green 1865, 1: xiv-iv). Owen was mentored by Green during the 1820s, and the two surgeons remained on close terms over the following decades.[20] Owen edited Green's *Vital Dynamics*, and urged Green to include a methodological appendix on the use of type in comparison. He received the Hunterian lectureship in 1836, and was elected as president of the BAAS in 1856.

Owen was also mentored by Whewell, who entrusted his mentee with the weighty task of reviewing the proofs of *Philosophy of the Inductive Sciences* in 1840 (Sloan 2003, 54). Like Whewell, Owen adopted a broadly Coleridgean conception of ideas

[20] Green thanks Owen in several places in *Vital Dynamics*, and notes Owen's input in his decision to include an Appendix on transcendental anatomy.

in the study of comparative anatomy.[21] This is particularly evident in the lectures he gave at the Royal College during the early 1840s, in which he presents a theory of the 'vertebrate type' and its relation to homology. Owen explains that the vertebrate type is discovered inductively by generalising from anatomical research, comparing a wide range of vertebrate skeletons to derive an idea, whose source lies in the divine mind. In 1849, Owen developed his type theory in a lecture at the Royal Institution entitled 'Discourse on Limbs'. The task of the anatomist, he explains, is to define the 'Bedeutung' of the limbs in vertebrate animals, 'that essentially which it retains under every modification of size and form, and for whatever office such modifications may adapt it' (Owen 1849, 2). Owen's (1849, 2–3) use of the German term is illuminating; *Bedeutung* ('denotation' or 'sense') refers to the 'archetype or primal pattern' that constitutes 'the basis supporting all the modifications of such part for specific powers and actions in all animals possessing it'. Structural changes can be observed synchronically between existing varieties and diachronically between fossilized remains. The archetype does not, however, provide a causal explanation. 'To what natural laws or secondary causes the orderly succession and progression of such organic phenomena may have been committed', he states, 'we are as yet ignorant' (Owen 1849, 86). Of what we can be sure is that nature 'has advanced with slow and stately steps, guided by the archetypal light amidst the wreck of worlds, from the first embodiment of the Vertebrate idea under its old Ichthyic vestment until it became arrayed in the glorious garb of the human form'.

In his presidential address to the BAAS delivered in Leeds in 1858, Owen laid out the historical dimension of his project in a way that builds on the idealist methodology found in Green and Whewell and yet points to a causal explanation of morphological change. The appearance of higher forms of life, including reptiles, birds and mammals, must be explained by the 'axiom of the continuous operation of the Creative power, or of the ordained becoming of living things' (Owen 1858, li). Drawing on his extensive work cataloguing the geographical distribution of animals in the past, Owen (1846, 179) proposed that the distribution of organic form ceases to correlate with what we find today the further back in time we go. The reason for this difference, he claims, lies in major changes occurring in geological history, which altered the relative positions of land and sea. Changes in environmental conditions even account for extinction, as larger animals are unable to adapt to changing geological conditions. The result is a research programme in which the comparative anatomist, guided by the *Bedeutung* of organic form, can search for the mechanical causes responsible for the continuity, alteration and extinction of various species through the multi-generational processes of propagation, adaptation and the struggle for life.

[21] Desmond (1989, 254–60) traces the influence of Coleridge's methodology on Owen's comparative anatomy.

14.6 Conclusion

Coleridge's *Theory of Life* forged a link between Hunter's lectures on the principles of surgery, Beddoes' application of Brunonian medicine to various diseases, and the scientific approach to comparative anatomy developed in German medical departments. Similar connections can be discerned in the work of Green, Whewell, and Owen, indicating the parallel development of a philosophy of mind capable of uniting a new science of comparative anatomy under life considered as an idea. While it is impossible to define Coleridge's influence in this development with precision, there is sufficient evidence to conclude that his ability to adopt, transform and apply the insights of German philosophers and medical professors to the leading scientific issues debated in London's intellectual circles impacted the consolidation of comparative anatomy in Britain.

By the end of the 1840s, several assumptions were in place that would have a deep impact on the development of comparative anatomy in the decades leading up to Darwin's *On the Origin of the Species* (1859). The transcendental approach to typology had become firmly established in the Royal Institution, the College of Surgeons, the BAAS and the Universities of London and Cambridge. Darwin read Whewell's *History of the Inductive Sciences* in 1838 and reviewed Owen's *On the Nature of the Limbs* and 'Presidential Address' shortly after their respective publications.[22] When his work on comparative anatomy is read against the historical backdrop of comparative anatomy in the mid-nineteenth century, it is clear that Darwin developed his account of morphological change within a British context that was far more conversant with German sources than is often recognized, with Coleridge providing a vital link in this cultural translation. Yet as Owen noted in *On the Nature of the Limbs*, no one had yet been able to establish the laws governing the macro-level structural change. While Owen entertained the hypothesis that a reptile had changed into a mammal, and an ape into a human, he acknowledged that such a hypothesis remained unsubstantiated. Darwin's principle of evolution by natural selection would provide a ground-breaking solution, explaining how environmental pressures act on variations within a species such that viable traits are selected over those that prove to be less conducive to survival (Richards 2013, 125; Rupke 2009, Chap. 5). Yet the broader methodological framework was already in place by which the comparative anatomist reflects on the staggering variety of living form as the product of a dynamic, self-constituting system in which the human mind is very much at home.

[22] Darwin refers to Owen 14 times in *On the Origin of the Species*. For references to the texts mentioned here, see Darwin (2008, 242 and 320).

References

Abernethy, John. 1817. *Physiological lectures: Exhibiting a general view of Mr. Hunter's physiology and of his researches in comparative anatomy, delivered before the Royal College of Surgeons*. London: Longman, Hurst, Rees, Orme, and Brown, Paternoster-Row.

———. 1819. *The Hunterian oration for the year 1819*. London: Straham and Spottiswoode.

Anon. 1853. Coleridgeana I: The new edition of Coleridge. *Literary world* 322:263–265.

———. 1928. Coleridgeana. *Review of English Studies* 4:433–434.

Beach, Joseph Warren. 1942. Coleridge's borrowings from the German. *ELH* 9 (1): 36–58.

Beddoes, Thomas. 1799. *Notice of some observations made at the medical pneumatic institution*. London: Briggs and Cottle.

Bichat, Xavier. 1827. Physiological researches on life and death. Tr. F. Gold. Boston: Richardson and Lord.

Brown, John. 1803. In *The elements of medicine of John Brown*, ed. Thomas Beddoes. Portsmouth: The Oracle Press.

Class, Monika. 2012. *Coleridge and Kantian ideas in England, 1796–1817: Coleridge's responses to German philosophy*. London: Bloomsbury.

Coleridge, Samuel Taylor. 1848. *Hints towards the formation of a more comprehensive theory of life*. London: Churchill.

———. 1885. *Miscellanies, aesthetic and literary*. London: George Bell and Sons.

———. 1957–2002. In *The notebooks of Samuel Taylor Coleridge. Vols 1–3, ed. Kathleen Coburn, 1957–73; vol. 4. Ed. Kathleen Coburn, Merton Christensen, 1990*, ed. Kathleen Coburn and Anthony Harding, vol. 5. Princeton: Princeton University Press.

———. 1956–71. In *The collected letters of Samuel Taylor Coleridge*, ed. E.L. Griggs, vol. 6. Oxford: The Clarendon Press.

———. 1969. In *The friend (vol 1, 1818; vol 2, 1808–9)*, ed. Barbara E. Rooke. Princeton: Princeton University Press.

———. 1976. In *On the constitution of the church and state according to the idea of each (1829)*, ed. Kathleen Coburn. Princeton: Princeton University Press.

———. 1993. In *Aids to reflection (1825)*, ed. John Beer. Princeton: Princeton University Press.

———. 1983. In *Biographia literaria, or biographical sketches of my literary life and opinions (1817)*, ed. James Engell and W. Jackson Bate, vol. 2. Princeton: Princeton University Press.

———. 1980–2001. In *Marginalia. Vols 1 and 2 (1980–4). Ed. George Whalley, vols 3–6 (1992–2001)*, ed. H.J. Jackson and George Whalley. Princeton: Princeton University Press.

———. 1995. In *Shorter works and fragments*, ed. H.J. Jackson and J.R. de J. Jackson, vol. 2. Princeton: Princeton University Press.

Cooper, Andrew. 2020. Force and law in Kielmeyer's 1793 speech. In *Kielmeyer and the organic world*, ed. Daniel Whistler and Lydia Azadpour, 81–98. London: Bloomsbury.

———. 2021. Reading Kant's *Kritik der Urteilskraft* in England, 1796–1840. *British Journal for the History of Philosophy* 29 (3): 472–493.

———. 2022. The economy of the *Bildungstrieb* in Goethe's comparative anatomy. In *The concept of the drive in classical German philosophy*, ed. Manja Kisner and Jörg Noller, 83–105. London: Palgrave Macmillan.

Darwin, Charles. 2008. *On the origin of species (1859)*. Oxford: Oxford University Press.

Desmond, Adrian. 1989. *The politics of evolution: Morphology, medicine, and reform in radical London*. Chicago: University of Chicago Press.

Girtanner, Christoph. 1797. *Ausführliche Darstellung des Brownischen Systemes der praktischen Heilkunde, nebst vollständigen Literatur und einer Kritik desselben*. Göttingen: Rosenbusch.

Goethe, Johann Wolfgang. 1947. In *Goethe: Die Schriften zur Naturwissenschaft (Leopoldina Ausgabe)*, ed. Dorothea Kuhn, vol. 21. Weimar: Hermann Böhlau Nachfolger.

Green, Joseph Henry. 1840. *Vital dynamics: The Hunterian oration delivered in the theatre of the Royal College of Surgeons on the 14th February, 1840*. London: William Pickering.

———. 1865. *Spiritual philosophy: Founded on the teaching of the late Samuel Taylor Coleridge*. London: Macmillan and Co.
Haeger, J. H. 1976. Coleridge's 'bye blow': The composition and date of 'Theory of life'. *Modern Philology* 74 (1): 20–41.
Hunter, John. 1839. *Lectures on the principles of surgery*. Philadelphia: Haswell, Barrington, and Haswell.
Kielmeyer, Carl Friedrich. 1793. *Über die Verhältniße der organischen Kräfte*. Stuttgart: Marburg an der Lahn.
Knight, David. 2017. Coleridge and chemical philosophy. In *Coleridge and contemplation*, ed. Peter Cheyne, 91–104. Oxford: Oxford University Press.
Levere, Trevor. 1981. *Poetry realized in nature: Samuel Taylor Coleridge and early nineteenth-century science*. Cambridge: Cambridge University Press.
McFarland, Thomas. 1981. A complex dialogue: Coleridge's doctrine of polarity and its European contexts. In *Romanticism and the forms of ruin*. Princeton: Princeton University Press.
Mill, John Stuart. 1869. Obituary on Samuel Taylor Coleridge. Westminster review (1840). In Dissertations and discussions. 2 1: 403. London: John W. Parker and Son.
Nidecker, Henri. 1932. Notes marginales de S. T. Coleridge, IV. En marge de Steffens. *Revue de littérature* 12:856–871.
Owen, Richard. 1846. On the geographical distribution of extinct mammalia. *Athenaeum* 14:178–179.
———. 1849. *On the nature of limbs*. London: John van Voorst.
———. 1858. Address. *British Association for the Advancement of Science, report 1858*, xlix–cx.
Rajan, Tilottama. 2019. Immunitary foreclosures: Schelling and British idealism. *International Journal of Philosophy and Theology* 80 (1–2): 39–56.
Richards, Robert. 2013. The impact of German romanticism on biology in the nineteenth century. In *The impact of idealism: The legacy of post-Kantian thought*, ed. Nicholas Boyle, Liz Disley, and Karl Ameriks, 105–133. Cambridge: Cambridge University Press.
Röschlaub, Andreas. 1802. *Untersuchungen über Pathogenie, oder Einleitung in die Heilkunde*. Frankfurt am Main.
Rupke, Nicolaas A. 2009. *Richard Owen: Biology without Darwin*. Chicago: University of Chicago Press.
Schelling, F. W. J. 1858. Erster Entwurf eines Systems der Naturphilosophie [1799]. In *Sämmtliche Werke*, vol. 3. Stuttgart and Augsburg.
Sloan, Phillip. 2003. Whewell's philosophy of discovery and the archetype of the vertebrate skeleton: The role of German philosophy of science in Richard Owen's biology. *Annals of Science* 60:39–61.
———. 2007. Kant and British bioscience. In *Understanding purpose: Kant and the philosophy of biology*, ed. Philippe Huneman, 149–171. Rochester: University of Rochester Press.
Smith, C. U. M. 1999. Coleridge's 'Theory of life'. *Journal of the History of Biology* 32:31–50.
Steffens, Henrik. 1801. *Beyträge zur innern Naturgeschichte der Erde*. Vol. 1. Freyberg.
Tsouyopoulos, Nelly. 1988. The influence of John Brown's ideas in Germany. *Medical History* 32 (Suppl 8): 63–74.
Vickers, Neil. 2004. *Coleridge and the doctors, 1795–1806*. Oxford: Oxford University Press.
Whewell, William. 1847. The philosophy of the inductive science founded upon their history. New edition, 3. London: Parker.

Andrew Cooper is associate professor of philosophy at the University of Warwick. He is author of *The Tragedy of Philosophy: Kant's Critique of Judgment and the Project of Aesthetics* (SUNY Press, 2016), *Kant and the Transformation of Natural History* (Oxford UP, 2023) and *Amalia Holst* (Cambridge UP, 2024), and has published numerous articles on Kant, post-Kantian philosophy, and philosophy of science.

Chapter 15
Bloody Speck: How S. T. Coleridge Turned the Embryological Punctum Saliens into a Metaphysical Principle

Peter Cheyne

Abstract Following Aristotle's and William Harvey's exciting observations of the first form of the heart beating into life in the newborn, Samuel Taylor Coleridge used this imagery of the punctum saliens pulsing into life to convey an emerging existence or form 'leaping' from one level of being into another. This chapter explores how the embryological theory of the punctum saliens became related, in Coleridge's thought, to dynamical idealist theories of matter arising from opposed physical forces to become a metaphysical principle. These idealist theories of matter developed through Cambridge Platonism, Leibnizian dynamics, German transcendental and absolute idealism, and British romanticism. They viewed energies in life and mind in circulation, like the blood in Harvey's theory of the beating heart. Coleridge co-opted the embryological punctum saliens for his metaphysics of the emergence of forms from one level of organization, or even from one hypostasis of being, to another. This essay charts the concept of the punctum saliens from Aristotle's observations to Harvey's theory of circulation, then to Coleridge's metaphysics of matter and emergence, drawing from Gottfried Wilhelm Leibniz's and Immanuel Kant's dynamical theories of matter. The chapter illuminates not only an important strand in the history of scientific and philosophical ideas but also some notable poetry from the late eighteenth century, Coleridge's 'The Rime of the Ancient Mariner' (1798) and George Canning and John Hookham Frere's 'The Loves of Triangles' (April–May 1798), a parody in the *Anti-Jacobin* periodical that curiously admired what it ostensibly lampooned, mimicking Erasmus Darwin's *The Loves of Plants* and Coleridge's 'The Eolian Harp'

> For matter itself is but Spiritus in coagulo; and organized matter the coagulum in the act of *being* restored …
>
> Coleridge

The original version of the chapter has been revised. A correction to this chapter can be found at https://doi.org/10.1007/978-3-031-78315-9_20

P. Cheyne (✉)
School of Letters, Hiroshima University, Hiroshima, Japan

P. Cheyne (ed.), *Matter and Life in Coleridge, Schelling, and Other Dynamical Idealists*, International Archives of the History of Ideas Archives internationales d'histoire des idées 253, https://doi.org/10.1007/978-3-031-78315-9_15

This chapter takes one small point, in a literal sense, from Samuel Taylor Coleridge's philosophy, as an example of how he drew from the history of philosophy and science in developing his own metaphysics. That point is the *punctum saliens,* or the leaping point in a line of blood, first observed and described by Aristotle as a source both of life and of metaphysical wonder. The chapter examines how Coleridge adapted the embryological concept of the *punctum saliens*, originating from Aristotle and following William Harvey's observations of the incipient heartbeat, to describe metaphysical emergence.[1] It explores how Coleridge linked the *punctum saliens* with dynamical idealist theories of matter, which saw physical forces as fundamental to life and mind. Drawing also from Cambridge Platonism and German idealism, the chapter traces the evolution of these ideas through Coleridge's metaphysical framework of form and emergence. Showing how Coleridge's thought was shaped by its scientific context, the chapter demonstrates how biological ideas about life and blood offered a model for his thinking about the most fundamental laws of the mind and the universe.

As Aristotle first observed, the formation and pulsing of the emerging embryonic heart is a constant source of awe and wonder. While many living species from protozoa to starfish, let alone plants and funghi, thrive without a heart or even blood, the first stirrings of the heart remain a potent symbol of the emergence of animal life at the top of the multi-layered strata of material nature. Furthering the observations of medieval writers such as Albert the Great and early and mid-Renaissance physiologists including his own tutors, William Harvey wrote reverently of this 'leaping point', or *punctum saliens*, in almost metaphysical terms, as the commencement of life shimmering between being and non-being.[2]

Harvey's fame, as he developed his observations and theories of the punctum saliens into his theory of the circulation of the blood, brought into the general imagination this fascinatingly synthetic concept of the heart as both mechanical pump and metaphysical marvel. Soon after, James Harrington (1992 [1656]), for instance, took the pulsing heart to represent a universal organizing principle, which he used to describe the bicameral Houses of Parliament within an ideal system of governance.

---

[1] This work was supported by JSPS Kakenhi grant number 19K00107 and by a Shimane University SDG grant for my Living Ideas Cross-faculty Research project. The work also benefitted from a sabbatical visiting fellowship at Clare Hall, Cambridge, where I presented and discussed a shorter version to the history and philosophy of science special interest group..

[2] *Punctum saliens* is the Latin translation of Aristotle's term for the incipient embryonic heart, the primitive heart tube that he observed as a 'leaping point'. Although the English idiom 'the salient point' and the German *der springender Punkt* have come to refer to a rather prosaic 'key relevant feature of something', both derive from Aristotle's exciting observation of the first form of the heart beating into life in the embryo. Samuel Taylor Coleridge used this imagery of the punctum saliens pulsing into life to convey the arising of an emerging existence or form leaping from one level of being into another, such as passivity transforming into action or matter returning, in the dynamical view, to the flow of 'unpaused' energy or spirit.

Section 15.1 of this chapter interrelates the relevant details of the history of the punctum saliens concept in embryology to help understand its later metaphysical use in Coleridge's thought. Section 15.2 will give an account of how and why he transformed this embryological concept into a metaphysical principle at the heart of his theory of emergent properties in a stratified development of increasingly complex physical forms. Viewing energies of life and mind in circulation, like the blood in William Harvey's theory of the beating heart, Coleridge co-opted the notion of the incipient heart as a punctum saliens for his metaphysics of the emergence of forms from one level of organization to another. This reading will shed light not only on Coleridge's metaphysical thinking but also on his poetry, as it opens up a helpful and suggestive dimension to the mysterious appearance of the ghost-ship, carrying the allegorical figures Death and Life-in-Death, in 'The Rime of the Ancient Mariner'. The section then further explores Coleridge's metaphysical punctum saliens in the context of his *Theory of Life* (1816?/1819?) essay, with reference to related thinkers including Friedrich Schelling and J. F. Blumenbach.

Section 15.3 will then relate the principle of the embryological punctum saliens, via Coleridge's ingenious use of it as a metaphysical principle, to the dynamical theory of matter proposed by G. W. Leibniz and developed by Immanuel Kant, which saw matter not as fundamentally corpuscular but as emerging from non-material monadic points holding together a surrounding, spherical field of energy, producing the quality of (near) impenetrability, and thereby producing the stratum of material existence from an underlying energic reality. This section illuminates not only an important strand in the history of scientific and philosophical ideas but also some notable poetry from the late eighteenth century: George Canning and John Hookham Frere's 'The Loves of Triangles' (April–May 1798), a parody in the *Anti-Jacobin* periodical that curiously admired what it ostensibly lampooned, mimicked Erasmus Darwin's *The Loves of Plants* (1789) and Coleridge's 'The Eolian Harp' (1795–6). 'The Loves of Triangles' elaborates a constructivist cosmology from Pythagorean geometrical principles, such as the tetractys, which saw a monadic point ('the One') creating duality, the dyad then producing multiplicity, which in turn produces the stable cosmos. This was an antiquated view that is nonetheless analogous to the modern Big Bang theory of cosmogenesis, which itself was first proposed as a kind of parody versus the Steady State theory. The section then concludes by showing how Coleridge's version of what he called the universal theory of polarity allows for emergent levels and qualities to arise by crossing through the midpoint or 'point of indifference' of natural or cultural polarities—e.g. light and gravity, positive and negatives poles of electromagnetism, conservatism and progressivism, and church and state—via a punctum saliens, the 'living' point that leaps through the polarity to commence a new emergence.

The chapter concludes with Sect. 15.4, which argues how Coleridge's commitment to dynamical idealism, within his philosophy of ideal realism, holds that matter emerges from the interplay of opposing spiritual or energetic forces, which he viewed as expressions of underlying unity. Drawing from Leibniz and Kant, Coleridge argued that matter is a product of the synthesis of forces like attraction and repulsion, rather than a fundamentally mechanical substance. He believed that these forces, which coexist in tension, give rise to material forms and phenomena, reflecting transcendent powers. Influenced by Leibniz's dynamics and his theory of monads and Kant's related theory of matter arising from forces, Coleridge saw matter as emerging from universal powers and evolving through dynamic oppositions. These dynamcal idealist views align with modern physics, which holds that matter and energy are essentially equivalent, with energy, according to current cosmology, being the antecedent, with matter being ultimately an expression of underlying forces.

15.1 The Embryological Punctum Saliens: Aristotle to Harvey

A telling fact within the history of ideas is that the earliest observation and identification of the embryonic heart and its first stirrings came from the father of systematic metaphysics, Aristotle. Indeed, he was also the father of systematic biology, and historians of philosophy often remark on the connection between biology and metaphysics, which Aristotle made vivid in what has been glossed as his recognition that 'animals exist, develop, and change in ways more complicated than any other class of thing' (Connell 2001, 304). In an essay on translating the *Poetics*, Coleridge scholar George Whalley said that Aristotle's:

> biological investigations provide commanding instances of a process that he recognizes in everything he sees and … thinks about, embodied in the notions of potency and act [… and that] active … *dynamis* is *energeia*, actuality, activity from within that drives towards… attaining its own end (telos): that is, *dynamis* is self-realising. (Coleridge… said that 'Every thing that lives, has its moment of *self-exposition*'.) (Whalley 1997, 22, quoting Coleridge, 1957–2002, vol. 3: §4397 [10 March 1818])

'[W]ondering: this is where philosophy begins and nowhere else', Plato has Socrates say, and Aristotle firmly agreed, discovering a fount of wonder in the emerging heart and the development of the embryo (Plato 1997a, b [*c*.369 BC], 155d). Not only did Aristotle find the punctum saliens wondrous in itself, as a believer in the heart as the seat of the soul (unlike Plato, who placed the higher mind in the brain), Aristotle would have marvelled as he observed what he held to be the first pulses of the vessel of all thought and feeling, including wonder itself, throb into life. '[O]wing to wonder' he says in the *Metaphysics*, 'men both now begin, and at first began, to philosophize' (Aristotle 1999 [*c*.330 BC], 982b). Two millennia later, Harvey found his own, similar observations to be equally charged, although in his blend of vitalism and mechanism, the new animal life commenced a stage earlier, with the blood that the incipient heart begins to pump. In their shared problems of form and matter, beginnings and ends, and often porous boundaries, biology and metaphysics profoundly stir the imagination. As the biochemist and historian Joseph Needham (1931, 7) noted:

> Of all the strange things in biology, the most striking of all is the transmutation inside the developing egg, when in three weeks the white and yolk give place to the animal with its

> tissues and organs, its batteries of enzymes and its delicately regulated endocrine system. This coming-to-be can hardly have failed, in the minds of those most intimately acquainted with it, to lead to thoughts of a metaphysical character.

Carefully removing shell from eggs at different periods of incubation, Aristotle discovered that after the yolk has travelled up to the narrow end of the egg, just after 3 days for hens' eggs, a red spot appears in the albumen and starts pulsing (Aristotle 2002 [*c*.350 BC] 665a33 ff.; 1910 [*c*.350 BC] 6.3).[3] This speck, he concluded, is the first self-moving part of sanguineous creatures. He correctly identified this as the incipient heart:

> ὅσον στιγμὴ αἱματίνη ἐν τῷ λευκῷ ἡ καρδία. Τοῦτο δὲ **τὸ σημεῖον πηδᾷ** καὶ κινεῖται ὥσπερ ἔμψυχον. [The heart is like a bloody spot (*sēmeíon*: mark, speck, point) in the egg-white. That **point leaps** and moves as if alive (or ensouled).] (Aristotle 1910 [*c*.350 BC] 6.3.1–2, Bekker 561a13, bold added)

In these studies, Aristotle found it notable that it is the heart which appears first, before all other organs. This is the observation that led him to find in the heart the seat of life and the soul, diverging from Plato's judgement, in the *Timaeus*, that the brain is the central organ of sensation (1997 [*c*.360 BC], 76a–e).

Due to its easy access, the development of the bird's egg, especially that of the hen, became a classic model in embryology that is still widely used today for research and teaching. Aristotle's descriptions in the fourth century BC and the further advances made in the Hippocratic corpus (Chap. 10, Hippocratic corpus 2014 [fourth-century BC]), which correctly identified the heart as a muscle, were advanced by medieval authors, such as Albert the Great, with his *Book on the Animals* (*c*.1265). However, it was not until the end of the Renaissance, in the 1560s, with the Italian naturalist Ulisse Aldrovandi, that embryology took off on a sustained research trajectory. Aldrovandi's student in Bologna, Volcher Coiter, continued this work, followed by Hieronymus Fabricius, at the University of Padua, succeded by Fabricius's famous pupil Harvey, making increasingly detailed headway and founding a research tradition that, as a historian of science describes, 'can be traced up to the great German-Baltic embryologists of the early nineteenth century and, beyond, to the experimental embryologists of the nineteenth and twentieth centuries' (Schmitt 2020).

Already current in the works of Coiter and Aldrovandi, Harvey's was the first use in an English text of the term *punctum saliens*, in *On the generation of animals*, translating Aristotle's description.[4] The 'leaping point' first observed by Aristotle became prominent in Harvey's first published work, the groundbreaking *De motu cordis* (1628). There, he called it the 'punctum saliens' and a 'palpitating' and 'bloody point', following Aristotle's 'bloody spot in the egg-white', which 'point

[3] *Parts of animals*, 665a33, again refers to the heart as a 'point'—'φαίνεται γὰρ ἐν μὲν οἷς ᾠοῖς ἐνίοτε τριταίοις οὖσι **στιγμῆς** ἔχοντα μέγεθος [just about the size of a **point**, sometimes as early as the third day]'—but does not mention 'leaping' or any movement (bold added).

[4] *On the generation of animals* (1653) was first published 2 years earlier in Latin, as *Exercitationes de generatione animalium* (1651).

leaps … as if alive', contending that almost all animals except the most vegetal have at least this rudimentary heart:

> even so in certain animals not destined to attain to the highest perfection in their organization, such as bees, wasps, snails, shrimps, crayfish, etc., we only find a certain pulsating vesicle, like a sort of red or white palpitating point, as the beginning or principle of their life. (Harvey 2016 [1628], 29)

Harvey captivatingly described the incipient heart in:

> the first rudiments of the chick as a little cloud in … the fourth or fifth day of the incubation, with the shell removed and the egg immersed in clear tepid water. In the middle of the cloud there was a bloody point so small that it disappeared during the contraction and escaped from view, but in the relaxation it reappeared again as a red pinpoint. So between the visible and invisible, between being and not being, it gave the beginning of life by its pulses. (29)

With such enraptured prose, Harvey transports into metaphysical contemplation, perceiving the 'bloody point' amid a nebula as 'the commencement of life', where an emergent creature vacillates 'betwixt the visible and invisible', contracting and relaxing 'betwixt being and not being'.

Harvey's marvel at the genesis of the heart was just as keen in his later work as when he first saw it:

> At that time, the foetus in the egg passes from the life of a plant to that of an animal. Then already the limbus or hem of the colliquament begins to turn purple and is outlined with a tiny line of blood, and almost in its centre there leaps a capering bloody point which is yet so exceedingly small that in its diastole it flashes like the smallest spark of fire, and immediately upon its systole it quite escapes the eye and disappears. (Harvey 1989 [1651], 96)

Between his two major works, *De motu cordis* (1628) and *On generation* (1651), Harvey shifted from prioritizing the heart as a little god and the 'foundation of all life' to ranking the blood higher, contra Aristotle, and considering that fluid, not the heart, the bearer of life.[5] Appearing first was definitive for Harvey to make the 'tiny line of blood' the 'prime part' in the transition from vegetal to animal life, the source of 'innate heat' in warm-blooded animals, the birds and mammals, and 'the common bond between soul and body'. He identified the soul in the blood as the transmitter of no less than:

> the first cause of all things, by whatever name this has been designated—the Divine Mind of Aristotle; the Soul of the Universe by Plato; the natura naturans by others; Saturn and Jove by the ancient Greeks and Romans; by ourselves … the Creator … of all that is in heaven and earth, on whom animals depend for their being, and at whose will and pleasure all things are … engendered. (1989 [1651], 402)

Harvey's sense of religious awe at the import of these observations is remarkable. The Jovian power, for example, continues an observation made earlier in the book,

[5] Coleridge opposed this view, arguing that to say '*Blood* is the Life … is saying nothing at all—for if the blood were Life, it could never be otherwise than Life' (1956–71, 1: 294–5, to John Thelwall, 31 December 1796), criticizing 20 years later, in *Theory of Life*, theories that identified life with 'one particular function of Life' or 'any single property' of it (490, 492).

where he conveyed the excitement of seeing the punctum saliens appear 'like a spark darting from a cloud' (1989 [1651], 237).

Harvey was clearly as excited as Aristotle and others at what he saw as the metaphysical implications of these biological observations. Although Aristotle and Harvey both saw great significance in the punctum saliens, the English physiologist came to view the blood as having greater symbolic as well as chronological priority. His tracking back to the tiny line of blood that hems the colliquament (etymologically, 'a thing melted', Harvey's New Latin term, for the earliest embryo) was born of empirical observation through a magnifying lens, unavailable to Aristotle. Yet Harvey's findings and the terms and theory in which he couched them also suggest a constructive, geometrical metaphysics of the transition to animal life whereby the 'soul in the blood' primes the embryonic body through a leaping point in a line of blood that commences the circulation; schematically put, a point propels a line into a circle. This tiny line was for Harvey an ensouled, bodily intermediary that advances beyond vegetal growth to the animal pulse, transmitting the universal and intelligent first cause to replicate in microcosm the creation of the universe.

Opposing the atheist connotations thought to arise from Cartesian dualism and Hobbesian mechanistic materialism, Harvey belonged to a group of early-modern natural philosophers, including Robert Boyle (Greene 1962), who resurrected the neo-Platonic cum Aristotelian teleological theory of the universal 'spirit of nature'. This 'spirit of nature' was adopted by the Cambridge Platonists, commencing with Henry More, who posited an 'hylarchic principle' that intervenes between God and the world. Ralph Cudworth extensively developed this cosmic principle with his concept of 'the Plastick Life of Nature', the artisanal spirit that enacts divine reason throughout the universe. Coleridge would later consider this universal dynamic to be a 'Plastic and vast, … intellectual breeze' sweeping over 'animated nature' as the 'Soul of each', a divine force which nature passes into the human mind to become the 'shaping spirit of Imagination' ('The Eolian Harp', 1795–6, 2001, 1: 234, ll. 44–8; 'Dejection: An Ode', 2001, 1: 700, l. 86). Expounding his universal, plastic principle, Cudworth (1678, 683–5), cited the 'Book of Generation' by 'the Learned Harvey', who reasoned that as the cosmos-governing, 'divine … plastic … force' is:

> the same intelligence or spirit which incessantly actuates the mighty mass of the universe, and compels the same sun from the rising to the setting … so also is there a vis enthea, a divine principle, inherent in our common poultry, showing itself now as the plastic, now as the nutritive, and now as the augmentative force, though it is always and at all times present as the conservative and augmentative force, and now assumes the form of the fowl, now that of the egg; but the same virtue continues to inhere in either to eternity. (Harvey 1989 [1651], 285–6)

It is through More and Cudworth that Coleridge turned to that 'spirit of nature' that courses through all things, which the poet-philosopher later exclaimed to be:

> O the one Life within us and abroad,
>
> Which meets all motion and becomes its soul…
>
> (Coleridge 2001 [1795–6], 1: 233, ll. 26–7)

Following Coleridge, William Wordsworth (2013 [1798], ll. 45–9) imagined:

> a motion and a spirit, that impels
>
> All thinking things, all objects of all thought,
>
> And rolls through all things.

These lines from *The Prelude* follow Wordsworth's allusion to the Harveyan 'motion of our human blood/Almost suspended', as we 'become a living soul … with an eye made quiet by the … deep power of joy'.

Influentially, then, Harvey described what he saw as the plastic and augmentative principle of the universe at work in animal embryos. From the line of blood, the first appearance of the newly generated life, arises a point, which then leaps, dilating and contracting as if between being and not being. This leaping point becomes the foundation, the 'keel' from which the vessel of the body is formed (Harvey 1989 [1651], 338).[6] The 'punctum pulsans' develops around itself 'the rest of the body' and its organs through 'a spirit diffused through the whole', within which, Harvey took as 'beyond all doubt', was an 'operative soul, which is all in the whole, and all in each individual part' (327). This idea of the genetic pattern of the developing creature as a soul pursuing a telos was not uncommon in pre-Mendelian biology, but Harvey gave it an interesting dialectic twist. In a complicating train of thought, he reasoned that:

> It is clear, therefore, that the chick is formed from the egg, as it were by a contrary, namely the aliment, and as if by an abstraction, and from a non-entity. For the first particle of the chick, viz.: the blood or punctum saliens, is constituted out of something which is not blood, and altogether its contrary, the same subject-matter always remaining. (326)

This formation 'by a contrary' foreshadows, as we shall see in the next section, Coleridge's metaphysical use of the punctum saliens, as Harvey considered this pulsing, leaping point to arise through a polar dynamic between the alimentary substance of the egg (the yolk) and a 'nonentity' that is nonetheless a power that operates by 'abstraction'.

15.2 Punctum Saliens in Coleridge's Dynamical Idealism

Harvey's theory of the circulation of the blood had far-reaching cultural as well as medical impact. Finding a rich symbolic connection in Harvey's discovery, James Harrington put it to the service of the political imagination. Impressed by Harvey's *vis enthea* and the experiments that detailed its workings out, Harrington saw in the dynamics of life a universal, vivifying or energizing power that organizes beyond the individual creature. His theory of the state, expressed in his major work, *The*

[6] '[A]s the man proceeds from the boy; the edifice of the body … is raised on the punctum saliens as a foundation; as a ship is made from a keel'.

Commonwealth of Oceana (1656), drew from Harvey's new physiology (Cohen 1994). The principles of this natural philosophy were so ingrained in reality, thought Harrington, that they apply alike to the circulation of the blood via the pumping heart and to the bicameral Houses of Parliament that enact an ideal of political deliberation. For him, going beyond a convenient analogy, 'the parliament is the heart', with its 'two ventricles, the one greater and replenished with a grosser matter, the other less, and full of a purer', together pumping 'the vital blood of Oceana in a continuous process of circulation'. *Oceana* was a favourite in Coleridge's library, and Coleridge refers to Harrington both for his 'worldly wisdom' and as a luminary of the Anglican Platonism that he identified as 'the spiritual platonic old England', whose champions, from William Shakespeare, Edmund Spenser, John Milton, Francis Bacon, with Harrington, to Jonathan Swift and William Wordsworth, conveyed a profounder though less easily articulated wisdom that had become occluded. For Coleridge, this English Platonism counterbalanced the socially and philosophically atomizing forces of the newer, empiricist, 'commercial G. Britain' represented by John Locke, David Hume, Joseph Priestley, William Paley, and others (1957–2002, 2: §2598 f80v [May–August 1805]). Harrington's taking Harvey's punctum saliens and the circulation of blood as an archetypal organizing schema paved the way for Coleridge using the punctum saliens as a metaphysical principle operating between levels of essence, being, and actualization.

Following Harvey and Harrington, the punctum saliens for Coleridge was not simply a metaphor but more a Goethean *Urphänomenon,* like the:

> leaf, there lies hidden the true Proteus that can conceal and manifest itself in every shape. Any way you look at it, the plant is always only leaf, so inseparably joined with the future germ that one cannot think the one without the other. (Goethe 1962, 366 [entry: 17 May 1787])

Yet in the direction it took with Coleridge, the punctum saliens represented beyond phenomena to become the image of a metaphysical and logical form. The embryological punctum saliens enacted for him a jetting from *physis* to *bios*: a point and propellant of evolution from one stage to the next, where properties that had no existence at the lower level emerge to actualize a higher level of organization in an emergentist metaphysics of matter returning to spirit, or the energy it always was, but in a state of relative 'pause'.

In an essay on Coleridge's *Theory of Life*, covering 'his discussion of polarity as the deepest law of organic life' and its 'universal … tendency to individuation', Needham gleaned the valuable germs in Coleridge's 'anticipation of the idea of Emergent evolution … and his biological Platonism' from what the eminent biochemist considered the electrochemical chaff which Coleridge collected from his Royal Institution attendances (Needham 1926, 700–1). Surprisingly, for a specialist in embryology, Needham did not comment on the concept of the punctum saliens in *Theory of Life*, despite Coleridge's co-opting that embryological term for a metaphysical usage of profound import. In Coleridge's new, metaphysical sense, 'punctum saliens' described the leap from the midpoint or equipoise of opposed poles in the immanent plane of nature towards the transcendent, unified power out of which these forces irrupt.

Coleridge's was assuredly an emergentist view, and in the *Logic* (Coleridge 1981 [*c*.1818–22], 37) he argued that even if human mental activity does have a neurophysiological basis (implicitly using David Hartley's neuro-associationist position as an example), human thought, especially in its higher levels, could not be reduced to neurophysiology, which reveals nothing of higher cognitive functions and the activity of thinking more generally.[7]

The term 'evolution', in this sense, is not restricted to the development of species; rather, it signifies the progression of latent potential into outward, physical forms and further complexity, later also denoting the development of energies. From 1816–24, Coleridge often wrote of the evolution of biological life forms, for instance, of the 'ordinance of Nature in the Sea—which sets a bound to the evolutions of life & the types of Life' (1957–2002, 4: §5178 [1824]). This interest became increasingly overtaken by his noetic thinking (i.e. concerning universal, constitutive ideas) on 'the progressive evolution of intellectual Insight' and how 'Ideas exert an informant Power, and evolve or procreate the formae formantes [forming forms], as Energies', as his metaphysics of physics and biology gave way to pure metaphysics and systematic theology (1957–2002, 4: §5411 [July 1826]; 5: 5813 [March 1828]). Coleridge did, however, write a short note in June 1830 on 'the evolution of particular Life' and the relation of hydatids to cancer (5: §6340).

Coleridge converged several strands into a metaphysical constellation that interwove Harvey's punctum saliens, as new life emerging from a cloud between being and not being, and Harrington's ideal of political organization as the pulsing heart of state, with the 'fulgurations' of Leibniz's constructive monadology and Kant's theory of matter as necessarily generated from opposed forces. Medicine was another strand. Writing in a notebook in his sixtieth year, Coleridge recalled cramming a medical dictionary in his 'Ladhood' and how this entwined with a passion for metaphysics:

> about this time [aged *c*.14], my Brother Luke, or the Doctor, … dedicated to the medical profession, came to town, to walk the London Hospital under Blizard—Dr Saumarez … Every Saturday, that I could … to the London Hospital trudged I—O the bliss if 1 were permitted to hold the plaister, or to attend the dressings …. I became wild to be apprenticed to a Surgeon—English, Latin, yea, Greek Books of Medicine read I incessantly—Blanchard's Latin Medical Dictionary I had almost by heart—Briefly, it was a wild dream—which gradually blending with gradually gave way to a rage for Metaphysics, occasioned by the Essays on Liberty & Necessity in Cato's Letters——and then by Theology … (1957–2002, 5: §6675 [1832])

The medical dictionary that Coleridge said he all but memorized was that of eighteenth-century Dutch physician Steven Blankaart. Coleridge owned the Latin

[7] Elsewhere, I showed how Coleridge, especially in *The Friend* (1969 [1818 edn]), influenced the emergentism of George Henry Lewes, the British literary critic and holist philosopher who coined the term (Cheyne 2020, 156–7).

edition of 1754 (Coffmann 1987, 26), where an entry, in the English translation, reads:

> *Punctum Saliens*, In the growth of an Egg you see a little Speck (or Cloud as it were) in the innermost Tunick of it, call'd *Ammos*, which growing gradually thicker, acquires a kind of slimy matter, in the middle whereof you see first this *Punctum Saliens*, (a little Speck that seems to leap) afterward the rude Body of an Embryo, just like a shapeless kind of Maggot, which tends every day more and more to perfection. (Blankaart 1684 [1679], 245; the Latin Coleridge read is at Blankaart, 1754 [1679], 711)

In this passage, the 'little Speck', which for Harvey pulses 'betwixt being and not being', becomes nebulous, 'or Cloud as it were', then begins to move 'in the middle' of 'a slimy kind of matter' to take a more perfect shape. That 'little speck' amid 'slimy ... matter' re-emerges in Coleridge's 'Rime of the Ancyent Marinere' (1798; 2001, 384) as the sign of the vessel carrying 'Life-in-Death':

> Upon the slimy Sea.
>
> ....
>
> I saw something ...
>
> At first it seem'd a little speck
>
> And then it seem'd a mist:
>
> It mov'd and mov'd, and took at last
>
> A certain shape, I wist.
>
> A speck, a mist, a shape, I wist!

In a wonderful discovery, Gerry Briggs and Kiran Toor (2017, 21) connect this imagery in the 'Rime' to Blankaart's embryological definition of the punctum saliens. This scene of a portentous, mysterious emergence seems to combine Harvey's descriptions and Blankaart's gloss of the punctum saliens into a poetic apparition of a supernatural presence emerging within the natural, as the little speck veers into life stirring within the non-alive, and Death and Life-in Death play a game of chance with a human soul.

In December 1799, a year after the 'Rime' was published, Humphry Davy introduced Coleridge to the polymathic London surgeon Anthony Carlisle. In the following year, progressing Alessandro Volta's experiments, Carlisle made his name in discovering electrolysis, with William Nicholson, by decomposing water into separate bubbles of hydrogen and oxygen using Volta cells. Carlisle had earlier studied under John Hunter, whose 'genius' Coleridge would later praise in the introduction to his *Theory of Life* (1816?/1819?; 1995a, 1: 486), especially for being guided by what the poet-philosopher judged to be the 'true idea of Life'. In 1804, Carlisle was elected Fellow of the Royal Society, and in his Croonian Lecture 'On Muscular Motion' (given 8 November, that year), he argued that the 'the rudiments of the punctum saliens' exist before incubation. He also confirmed there that the punctum saliens, 'during its first actions', exists before any muscular fibres or the vascular system, with the blood flowing back and forth in the same vessels (Carlisle 1805, 5). From this, Carlisle inferred that the organism at the commencement of life in complex animals is similar to the ultimate structure of the simpler organisms, the punctum saliens being like a microscopic animalculum, a tiny creature with an apparent life of its own, as Harvey among others had noted.

In 1810, Carlisle, whose continued connection with Coleridge is evidenced by his 1809 subscription to *The Friend*, treated him for opium addiction on the suggestion of Basil Montagu (Coleridge 1956–71, 3: 296 n.; Vickers, 2004, 100). There is a reasonable possibility that the two would have discussed the punctum saliens or that Coleridge would have read Carlisle's lecture pertaining to the subject. 'On Muscular Motion' would certainly have interested Coleridge, and it was read by at least one person in his circle, Robert Southey, who was a correspondent of Carlisle's since *c.*1795, and who was glad to read the lecture soon after its publication (Southey 2023, letter 1064 [1 May 1805, to John Rickman]). Unfortunately, Coleridge and Motagu broke off permanently in Autumn 1810, after the physician related to the poet-philosopher that Wordsworth had given up on him and now considered him an 'absolute nuisance', causing Coleridge to fall into anxiety and depression (1956–71, 3: 376, to John Morgan [23 February 1812]). Coleridge had shared some of anxious concerns with Montagu but later learned that this patient–doctor confidentiality had been betrayed, hearing of 'Mr Carlisle's most unprofessional Detail to a Woman, who made it the subject of common Table Talk, of every thing, I had confided to him as a Surgeon' (1956–71, 3: 408, to William Wordsworth [11 May 1812]).

Composed and printed in 1815, the first volume of *Biographia Literaria* shows Coleridge still heavily dependent on Friedrich Schelling. One of the passages in that volume that he lifted from the German idealist is: 'Matter has no *Inward*. We remove one surface but to meet another. We can but divide a particle into particles' (Coleridge 1983 [1817], 1: 133). Yet this statement on 'dead matter', useful though it was as Coleridge compiled an argument against reductive materialism, demonstrates a view that Coleridge increasingly found one-sided as he pursued the connection of life and matter through his developing dynamical idealism.

In later writings, from as early as *The Statesman's Manual* (1816), Coleridge grew dramatically in the originality of his philosophical thought. *Theory of Life*, however, still contained some passages lifted from Schelling's early writings on *Naturphilosophie*, and relied heavily on the *Naturphilosophen*, especially Henrik Steffens. It was not until summer 1818, that Coleridge began to develop his philosophical thinking in greater distinction from Schelling, in a turning point that leaves behind much of his struggle in the first volume of *Biographia Literaria* (and in the *Theory of Life*): the plagiarisms, the abandoned deduction of the imagination, the difficulty in pulling together a comprehensive transcendental idealist metaphysics that incorporates rather than denies empiricism. In a letter to J. H. Green, 30 September 1818 (1956–71, 4: 874), Coleridge describes his growing distance from Schelling and *Naturphilosophie* as he realized they saw the physical and the ideal as *equal* opposites, whereas Coleridge insisted on the *primacy* of ideas and powers. In

August and September 1818, Coleridge became convinced that Schelling's Identity Philosophy period amounted to pantheism and what he had grown to see as an untenable equivalence of mind and matter as equal opposites, rather than what Coleridge believed to be the truer view of matter as subordinate to forces, forces to powers, and only then to reach any equivalence, namely, between powers and ideas, which he held to be 'correlatives'.[8] Coleridge was not claiming a superiority over Schelling, however, as he also pointed out that Schelling himself had left his own *Identitätsphilosophie* 'in the Lurch', in favour of the *Freiheitsschrift*, or *Freedom Essay*, that Coleridge deeply admired.

This greater independence was prompted by Coleridge recognizing the need to reconstruct his '*ideal Realism*' (Coleridge 1983 [1817], 1: 303) without the susceptibility to pantheism and mind–matter equivalence that he found both in Schelling and the German idealist's sources in Jacob Böhme. Note that while Coleridge also maintained that matter derives from and returns to ideality (i.e. spirit, mind, energy, power)—and in that sense he held a view of matter–energy equivalence—he did not accord them an equal valence: the material for him was, or ought to be, subordinate to the energic or ideal. Böhme's mystical writings were influential on Schelling, Hegel, and Coleridge, who all followed the Lusatian mystic in positing the arising and continuous evolution of nature from the interpenetration of opposed divine energies, where 'divine spirits', primal powers and qualities, manifest themselves in the material world that they create as a constantly transforming cosmos of extensions and quantities, where 'before' (Böhme wrote of this Big-Bang-like 'unfolding' with temporal metaphors, giving conceptual access to the eternal and transcendent) there was only the anterior cosmos of intense reality and qualities. Here, quality precedes quantity, the former creating the latter, upending the Galilean orthodoxy. In Coleridge's gloss of Böhme's cosmology:

> presupposing … the intense Reality, he begins with Qualities, & maketh Quantity one of the Results—Most wisely! For Quantity is image & symbol/. (1980–2001, 1: 563, annotating Böhme)

In Behmenist cosmology, the eternal, involved, intense, subjective, qualities and energies transform into temporal, evolved, extended, objective quantities (Cheyne 2020, 125–61).

The following passage from 1828, annotating a volume of Johann Friedrich Blumenbach, whose lectures on physiology Coleridge attended while enrolled at Göttingen University (1798–9), provides a glimpse into Coleridge's developed view:

> To elicit the qualities of matter [and] … to arrive at the ends of philosophic Arrangement, … [one must consider] nature … elevating matter into organization and *manifest* vitality … to compel the elementary stuffs to reveal their interior Being, and their finer affinities, is an evident End in the organising process, without reference to which it would be impossible to reduce a multitude of phenomena in the lower orders of Organic Nature to any intelligible connection. (1980–2001, 1: 537 [January 1828], annotating Blumenbach's *De generis*)

[8] For ideas and powers as 'correlatives', see Coleridge, 1987, 2: 147–8 (17 February 1818) and *The Friend (1969 [1818]*, 1: 497 n. 2 (autograph note, Copy D, 23 June 1829).

Nonetheless, this later view retains what Coleridge most valued in Schelling, namely, the notion that matter is an outer manifestation of inwardness and a store of 'interior Being', both thinkers holding that matter is ontologically dependent on spirit or energy. The shared ground retained between Schelling and the later Coleridge is a view of the evolution via polar opposition in natural history, where the material world progressively actualizes the powers through which, according to dynamical idealism, matter arises. The Behmenist theory is one of involution and evolution, that is, of manifestation of the involved, intense reality through dialectic evolution and the emergence of new properties approaching the ideal through complexes of interpenetration, self-referential systems, and mind.

15.3 Dynamical Geometry, Constructive Physics

Harvey's accounts of his embryological and other anatomical observations were often filled with wonder and the sense of a transcending metaphysical import befitting his adherence to a physio-mechanism mixed with vitalism. René Descartes's *Treatise on man* (1985a [1662; written 1632], gave early (and partial) support to Harvey's initially controversial theory and generalized the English physician's theory of the circulation of the blood into a reductive theory of biology as mechanics. Despite his physiological mechanism, Descartes's mind–body dualism led him to believe that all human muscular motion is voluntary, which entailed that he could not accept Harvey's theory of the heart as an automatic pump. His alternative theory, in *La description du corps humain*, was that the heart is a furnace, the swelling blood forcing open the heart valves and ensuring circulation (Descartes 1985b [1664; written 1648]), prompting further experimental research by Harvey to prove the details of his own theory. Applied to animals, Descartes's reductive mechanics was total, conceiving all non-human creatures as brute automata with no sensation or inner life. His own introspection and meditations upholding the reality of inner life in humans, he devised a mind–body dualism that admitted *res cogitans* (thinking stuff, mind) but bracketed it from *res extensa* (extended stuff, body), thus continuing his otherwise universal mathematical mechanism.

Although Descartes's mathematical physics correctly conceived momentum as the product of extended matter and velocity, it did not amount to a comprehensive theory of force. Leibniz opposed Descartes's substance-dualism with his monistic energism, a far-reaching view of a universe explained by conatus and perception in mind *and* matter in terms of energy, forces, and fields around atom-like non-spatial points. In contrast to the Cartesian view, Leibnizian substance is essentially action, not extension. A well-advanced forerunner, as Albert Einstein acknowledged, of what became the special theory of the relativity of spacetime, Leibniz was the first to theorize the relative quantities of time, space, motion, and extension, and he maintained this while Isaac Newton's rival insistence on the absolute values of space and time dominated European science. Further, Leibniz was at the vanguard of the energy–mass equivalence theory: viewing space, time, and motion as relative, he saw that what remains as real and absolute is force, as potential or kinetic energy. Thus, Leibniz conceived and elaborated a fundamentally dynamical physics, the mathematical receiving energic content: size became mass, speed became velocity, and physics became a science of force. Geometry, as the science of points, lines, shapes, and solids, is constructive. Physics, as the science of forces, is dynamical. Their synthesis would produce a dynamical geometry and a constructive physics. Modifying Leibniz's monadology, Kant used the mathematical concept of a point to explain physical dynamics in a way that synthesized Cartesian mathematization with Leibnizian and Newtonian notions of force. Furthering the dynamical idealist theory of matter, Kant's theory also inspired constructivist metaphysics. In this space of fusion, where Pythagorean geometrical transcendentals synthesize with a dynamical theory where the material emerges out of the ideal or energic, Coleridge was in his element.

An unexpected impulse for Coleridge's Pythagorean, constructive geometry and physics appears to come from a parodic poem, in the *Anti-Jacobin*, by George Canning (future prime minister and under-secretary of state in William Pitt the Younger's Tory government) and John Hookham Frere, who, from 1816, would become a good friend of Coleridge. Although Coleridge, with Southey, Wordsworth, and Charles Lamb, were by 1798 disillusioned with what had become of the French Revolution, Canning and his vigilant band of anti-Jacobins, including Frere, parodied what they took to be their style of intellectual dissidence, adding Erasmus Darwin and William Godwin into the portmanteau of 'Mr Higgins of St. Mary Axe'. A prominent group of satirical poems in the *Anti-Jacobin* were printed under this name, which likely derives from Nememiah Higginbottom, the pseudonym for the satirical sonnets of Coleridge, born at Ottery St Mary, whose River Otter runs not 10 miles east of the River Axe (Magnuson 1998, 48). St Mary Axe is also a street in the City of London; the humorous confluence of names might have cemented the choice of this pseudonym. Chiefly parodying Darwin's *The Loves of the Plants* (1789), the third verse paragraph appears to parody Coleridge's 'The Eolian Harp' (1795–6). Coleridge's poem has the air-harp enchant a mood imitating 'twilight Elfins … on gentle gales from Fairy-Land', only to be pierced by the 'more serious eye' of his fiancée, which 'Darts' 'a mild reproof' to the 'Bubbles' of the poet's 'unregenerate' musings that 'glitter as they rise and break' (2001, 1: 233, ll. 21–1, 49, 56). The corresponding lampooning lines in 'The Loves of the Triangles' has the 'fair sylphish forms' 'Dart the keen glance … to viewless harpings' 'and titter as they prance' (Canning and Frere 1798, 116, ll. 23–6).

'The Loves of the Triangles' parodies the cosmically creating kind of constructive geometry that enthused Coleridge. Coincidentally, Coleridge's excitement with constructive geometry—from the *Theory of Life* (1816?/1819?) and continuing to *Aids to Reflection* (1825), and beyond—coincided with the time that he and Frere began their friendship, despite their inauspicious relationship in the *Anti-Jacobin*.[9] The poem describes:

How slow progressive *Points* protract the *Line*,

As pendant spiders spin the filmy twine;

How lengthen'd *Lines*, impetuous sweeping round,

Spread the wide *Plane*, and mark its circling bound:

How *Planes*, their substance with their motion grown,

From the huge *Cube*, the *Cylinder*, the *Cone*.

(Canning and Frere 1798, 117–9, ll. 39–44)

A long satiric footnote adds that these lines explain how 'Lines are supposed to be generated by the motion of Points;—Planes by the lateral motion of Lines;—and Solids from Planes'(117 n.). The footnote then considers that:

> We may conceive the whole of our present Universe to have been originally concentrated in a single point: we may conceive this primeval point, or *Punctum Saliens* of the Universe, evolving itself by its own energies, to move forwards in a right Line, *ad infinitum*, till it grew tired—After which, the right Line, which it had generated would begin to put itself in motion in a lateral direction, describing an Area of infinite extent. This Area, as soon as it became conscious of its own existence, would begin to ascend or descend, according as its specific gravity might determine it, forming an immense solid space filled with Vacuum, and capable of containing the present existing Universe. (118 n.)

It is striking how this theory, satirized yet elegantly described in the *Anti-Jacobin*, resembles the now established Big Bang theory of cosmogenesis. This resemblance holds even in the detail that the coiner of the term was indulging in parody. Cosmologist Fred Hoyle, a co-founder of the Steady State theory of the universe, invented the phrase 'Big Bang' in a British Broadcasting Corporation radio talk in 1949 as a succinct and memorable descriptor for Georges Lemaître's rival 'primeval atom' theory (1931), which held that all the matter in the universe exploded into existence at one instant from one non-material point: the most archetypally primal and salient point—the archetypal punctum saliens—, which initiated all time, space, and matter in the expanding universe (Kragh 2013).

Intriguingly, Canning and Frere's description is itself a gracefully composed account of Pythagorean metaphysics, and it required asides such as 'till it grew tired' and 'an immense solid space filled with vacuum' to indicate the parody. As another commentator notes, 'the *Anti-Jacobin* parodies, which are torn between sympathy and satiric bite', are 'almost too accomplished' and seem to 'indicate that the authors feel a genuine admiration for … and find themselves attracted to what they must

[9] Coleridge's first direct contact with Frere may have been when he sent him a copy of *Sibylline Leaves*, with a letter, 2 July 1816.

disapprove' (Mortensen 2004, 29). The *Anti-Jacobin*'s Pythagorean footnote goes on to lampoon the notion of the evolution of species, and the poem itself mocks other theories and technologies hypothesized by Darwin that would later become established, such as domestic uses for electricity and the future importance of aircraft in warfare. That latter parody perhaps borrows from Jonathan Swift's satire, in *Gulliver's Travels* (1726), of highly specialized scientists bombarding enemy regions from the flying island-citadel of Laputa. 'The Loves of the Triangles' is reminiscent of Swift's parody of Laputa in another respect, that of creating captivating accounts of the very things intended to be mocked. For instance, Swift's description of the scientists at the academy of Lagado who spent 'eight Years upon a Project for extracting Sunbeams out of Cucumbers, which were to be put in Vials hermetically Sealed, and let out to warm the Air' (Swift 2003 [1726], 167–8), was intended to appear ridiculous to common-sense readers (for whom the value of a research scheme depends on economic advantage), yet the idea and its realization would be a beautiful endeavour, reaching the height of sublimity and wonder.

The concept of the point as a 'speck' or node, as something actual and initial, indicates an intersection of the mathematical and the empirical (observables in physics and biology). Since the Pythagoreans and Euclid, the geometry of a point producing a line, a circle, and other two-dimensional shapes, and from there to construction in three dimensions, is often present in constructivist metaphysics. Examples developing this Pythagorean geometry, whose transcendent, constructing points, lines, and planes Coleridge commended, include the Platonic solids in the *Timaeus* through which the Demiurge created the world; Plotinus's explanations of emanationism and the productions of the universe through nature contemplating the One; Nicholas of Cusa's neo-Pythagorean accounts of the unfolding of Creation; and Kant's geometrical constructivism of pure intuition and spatial perception.

Coleridge distinguished between the Aristotelian '*point* of view (the reflective, namely)' and 'Plato's the *constructive*'. Where 'Plato sought in the Supersensual the solution of ... itself and of the Sensual—Aristotle began with the Sensual, deducing the Supersensual by hypothesis', as we saw in his observations of the punctum saliens (Coleridge 2000, 245 [18 January 1819, lecture at Crown and Anchor Tavern, Strand, London]). The reflective point, like reflection in general, is conceived in terms of immanence. Reflection is an operation of things on the same immanent level. Construction, by contrast, is in this sense a form of transcendence, projecting existence onto a different level. Such thoughts relate to transcendence in the constructivism of the 'sacred geometry' of Pythagoreans and Hermetics. Coleridge's dynamical sense of geometric construction analogized the embryological observation and concept. In Pythagorean geometry, a transcendent point on a higher plane generates a line below, while the indifference-point, or midpoint, which is immanent to the generated line (projected from the transcendent point), represents (but is not) the transcendent, constructive point.

Each line is assumed to have arisen as a pair of poles (*BC*) where there is always a midpoint (*a*), and the series of points along the line were constructed, or projected, from a transcendent point (*A*). One's understanding of this midpoint is what allows extension to have a rational character, i.e. for the line to end and not go on forever.

The (always immanent) indifference-point and the transcendent point (which Coleridge also called the 'Identity-point') are not coincident: they are on different planes, and the lower represents the higher by a symbolism of participatory representation. Coleridge (1993 [1825], 179) recognized in this divide a bulwark against pantheism. The immanent point of indifference between the poles of the line is like the scar of the navel, the omphalos that is the sign of a connection to its antecedent transcendental. In this analogy, the transcendental point is the mother via the umbilicus, with the immanent navel scar being the null point representing a prior living link. When one point in a series is viewed as especially salient and vital to the development of the series, there is, by analogy, a punctum saliens. Coleridge used the term punctum saliens in his *Theory of Life* to imply both the embryological and the dynamical philosophical meanings. Mathematically, as well as biologically, where there is a power, there is a root.

The indifference-point for Coleridge is thus the product and subsequent representative of the transcendent 'Identity', and not the producer of actuality, as it is for identity-philosophy-period Schelling. This provides an insight into contemplation of transcendent powers, for where literalist, imagistic fancy sees the indifference-point as a nullity, symbolic, reason-directed imagination sees the representative of transcendent power in the immanent. Exactly between the two poles, the indifference-point represents the unity of the higher power, which is otherwise polarized across two opposing forces. This dynamic has important implications for Coleridge's *Theory of Life*, where he contrasts the polar 'equatorial point' conceived as 'rest, or mere neutralization' with the higher view of it from the vantage of reason, where:

> the oneness of space and time, is the predicate of all real being. But as little can we conceive the oneness, except as the mid-point producing itself on each side; that is, manifesting itself on two opposite poles …
>
> To the fancy alone it is the null-point, or zero, but to the reason it is the *punctum saliens*, and the power itself in its eminence. (1995a [1816?/1819?], 1: 521)

Here, Coleridge adopts the embryological term *punctum saliens* to illustrate the easily overlooked, higher meaning represented by the *punctum indifferens*. What sensation-directed fancy assumes to be the dead-point of indifference, reason-oriented imagination sees as an immanent point that can leap into the transcendent.[10] A related analogy comes to mind: the punctum caecum, or retinal blind spot, relays no information to the brain from its position precisely because it is the spot from where all information from the rest of the retina is relayed. Its immanent meaning of nullity belies its significance as the vital link between eye and brain. Another example of a null spot from an immanent perspective revealed as a punctum saliens, when viewed in terms of transcendence, is the 'aching hollowness in the bosom, a dark cold speck at the heart', where, for Coleridge, 'a sort of *sensation*

[10] One might also consider the indifference-point–transcendent point relation in the Christian sense where 'The stone the builders rejected has become the cornerstone' (Luke 20:17).

of unbelief' is a 'negation of the truth' (1993 [1825], 24 and 1957–2002, 5: §6273, fol. 17 [April 1830], discussed at Cheyne, 2020, 308–9.

In Coleridge's logic, punctum saliens is the positive interpretation of the punctum indifferens, the neutral theoretical point between two opposed forces or qualities. Seen as the indifference-point, it is the dead centre, yet as the punctum saliens, it is the living, leaping point—at the emergence, *in ovo*, of life in organic matter. From this central point, the poles protract: 'the process itself, in which THE ONE reveals its Being in two opposite yet correlative Modes of Existence, I designate by the term, Polarity, or Polarizing' (1957–2002, 4: §4538 fol. 167 [May 1819]).

While contraries have an excluded middle (as Aristotle derived from his law of non-contradiction), opposites necessarily have a middle position, a principle Coleridge learned from Kant: 'Note here the difference between opposites and contraries.

O.[pposites] always have an Equator—C.[ontraries] never. ——————————·)
('1980–2001, 3: 327, *c*.1816, annotating Kant, *Vermischte Schriften*).

That is, between truths and falsities, there is no middle position; yet with manifest powers and beings, there are continua, with extremes and middles. Coleridge's bisected line represents opposites necessarily straddling a 'Mesothesis', or 'Equator'. The ')(' sign represents contraries, with an excluded middle. Opposites involve and require one another, developing along polarities, with his examples including electrical and magnetic positive and negative poles, expansive light and contractive gravity, subjectivity and objectivity, and socio-cultural permanence and scientific and political progression. Here, Coleridge develops Kant's distinction between real and logical opposition, whereby, as a dynamical opposition of forces or powers: 'real opposition, is that where two predicates of a thing are opposed to each other, but not through the law of contradiction' (Kant 1992 [1763], 211; Ak. 2: 171). While the two contraries of a logical contradiction cannot exist simultaneously, in real opposition, the two opposed powers must co-exist. Because these opposed powers are essential to Coleridge's metaphysics, he terms his philosophy dynamic (δύναμίς, power).

As opposite forces remain essentially opposed and so cannot reunite into non-opposition and reconvert into originary power, their stasis comes instead through

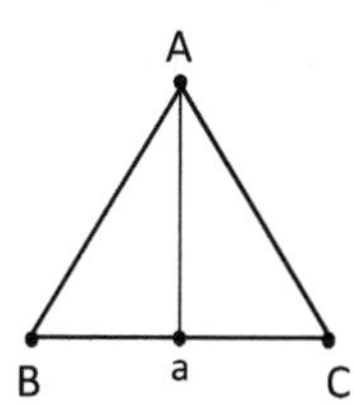

Fig. 15.1 The point of identity (A) and the point of indifference (a)

synthesis, producing the 'pause' or 'coagulation' of matter and concrete natural and cultural forms. Coleridge's ontology of productive powers or energies commences with the ideal prothesis that becomes manifest through two opposite, interdependent forces. When these forces reconcile, it is in the production of a concrete synthesis, the surrounding things of experience. Between the opposed forces is the midpoint, the indifference-point, that represents in a lower order the prothesis (or 'Identity'; 1957–2002, 4: §4835, fol. 64^{v} [1821]), but does so negatively, as a null-point, the 0 between + and -. This indifference-point is 'ideally the whole, really or rather materially nothing' ([1828] 1995b, 2: 1383). While the prothetic power is the transcendent X beyond experience, manifest not in itself but only in its mode as interconnected opposites, the indifference-point is the 0 between the +1, 2, 3, etc. and the -1, 2, 3, etc., and is thus a kind of shadow representative of the transcendent power, the shadow of the absolute in the immanent realm. While the Schelling of the identity-philosophy period saw indifference as the attribute of the absolute, Coleridge finds indifference only at the level of manifestation, at the centre of polarity, and categorically not as characterizing any absolute itself, indifference being the shadow, or signalling nullity, of the transcendent.

The Coleridgean indifference-point is at once nothing, yet real, and entirely immanent. The paradox of this characterization dissolves with the explication that the indifference-point is relationally real but actually nothing. Like an Eleatic paradox, any actual point, however infinitesimally small, that is picked out as the indifference-point between the opposed poles, will actually be positive or negative, and not the true zero. Thus there is no existent, actual indifference-point, yet the point is theoretically and relationally real. The nullity of the indifference-point, for Coleridge, perfectly contrasts with the transcendent 'Identity', which is the plenitude of originary reality from which the line is projected: the negative is the imprint of the positive. The indifference-point for Coleridge is thus the product and subsequent representative of the transcendent 'Identity', and not the producer of actuality, as it is in Schelling's *Identitätsphilosophie* (Fig. 15.1).

As Coleridge noted, the indifference-point symbolizes the 'Identity' as a lower-level counterpart:

> The Punctum Indifferentiæ is not ~~the~~ = A, or the primary Root or Involute of the Productive Power, in the sense of = as "*the same as*" but = A in that sense only in which a Son standing in the place of his Father at the head of a second Table in his Father's House is said to be equal to his Father. (1957–2002, 4: §4835, fol. 64^{v} [1821])

Around the same time, near the beginning of his *Logic* ([*c*.1821–3] 1981, 17), he diagrammed this relation of the transcendent productive power to its immanent representative (see Fig. 15.1), showing:

> the mid-point of the line, the *punctum indifferentiae* representing the *punctum identitatis*, even as the whole line represents the same point as produced or polarised.
>
> In this simple diagram *A* (= the point of identity) is supposed to generate the perpetual eradiation of the line *BC* … while the point *a* expressing the indifference [of *B* and *C*] …

A little later in this period, in a long footnote in *Aids to Reflection*, Coleridge argued for transcendent powers represented in immanent points in an illustration that commences with the observation that ancient geometers understood the line to be 'a point *produced*, the two extremes being its poles', with the initial point represented in the indifference-point at dead centre (1993 [1825], 179 fn.) Yet the producing point, while represented, is not itself immanent to the line, in the Pythagorean scheme, for:

> In order to render the constructions of pure Mathematics applicable to Philosophy, the Pythagoreans, I imagine, represented the Line as … radiated, by a Point not contained in the Line but independent, and … transcendent to all production, which it caused but did not partake in. *Facit*, non *patitur* [It *makes*, not *suffers*]. This was the Punctum invisibile, et presuppositum [invisible and presupposed point]: and in this way the Pythagoreans guarded against the error of Pantheism, into which the latter schools fell. The assumption of this point I call the logical PROTHESIS. (1993 [1825], 181 fn.)

Nature cannot produce the polarities, in Coleridge's view, for productive nature *is* the polarities. Manifesting the transcendent, unitary ideas and powers of the divine *Logos*, nature is the actuality of opposed forces, out of which the phenomenal realm of experience, matter, and concrete syntheses in turn arises. Coleridge held to the etymology of the word nature, the Latin *natura* being the future participle of *nasci*, 'to be born':

> Natura, that which is *about to be* born, that which is always *becoming*. (1993 [1825], 251)

Nature produces syntheses *from* the polarities. For Coleridge, syntheses are relatively stable, yet essentially evolving products that appear at the next ontological level down from polar manifestation. The syntheses produced from the poles comprise the level of concretes and phenomena. The producer of the polarities themselves, he argued, must therefore be a transcendent principle, namely, the identity of *B* and *C* in the transcendent point *A*, which is therefore able to produce *BC* without itself partaking in and being subject to it, in distinction from *a*, at its exact midpoint, which does partake in the line.

The mathematical concept of a point constructing lines, planes, and spaces was used to explain dynamics in physics and was taken in a metaphysical direction by constructivist idealists. However, as Kant's synthesis showed, building on the conflicting theories of Newton and Leibniz, dynamical physics was essentially intertwined with metaphysics, in its reliance on non-phenomenal energies and processes as the basis of matter and phenomena. Similarly, the embryological concept of the punctum saliens lent itself, for Coleridge, analogically but also as a particularly representative *Ur-phenomenon*, to metaphysical use. Thus this bloody speck, leaping at the midpoint along a line of blood, appearing to pulse in and out of existence, between being and non-being, can represent at once geometric construction; organic production; the beginning of new life; concrete and abstract senses of pregnancy; the emergence of a third being from an interpenetrating pair of opposites (sperm and egg); and the imagined inception of the universe.

Thought to be following a universal plastic principle, and representing an unfolding evolution from a hidden inner code that would not be rendered understandable until Gregor Mendel presented his discovery of the laws of inheritance in 1865, the incipient heart also demonstrates the simple complexities of an evolution furthered by involution. For the vesicle of the punctum saliens grows as a section of that initial line of blood and then the primitive heart tube doubles up to fold in on itself, and it continues this involution to produce the upper and lower chambers, respectively, of the right atrium, collecting the deoxygenated blood, and the right ventricle, which pumps out the blood to reoxygenate in the lungs to return to the left atrium then left ventricle, continually circulating around the body. This all develops from the punctum saliens, the leaping point where forces *intensify*, then leap out into *extension*, carrying the symbolic connotations of the initial beating of an incipient heart as the 'tiny animal', the 'animalculum' in the embryo, pulsing within the new life.

15.4 The Dynamical Theory of Matter

Coleridge can be categorized among an important group of idealist thinkers of various schools and periods for whom matter is real but not basic. These thinkers hold matter, and subsequently organic life, to derive from powers and forces, conceived as energy or spirit, with life as one of its eventual consequences. Engagement with their theories reveals the fascinating topic of how certain philosophical idealists, who held energy or spirit to have ontological precedence to matter, theorized material being. In Coleridge's thought, one finds a constellation of theories that together provide a particular form of dynamical idealism. These are: the emergence of matter from the opposition of physical, pre-material forces; the elevating circulation of energies in life and mind; and his co-option of the notion of the incipient heart as a punctum saliens for a metaphysics of emergence. In this final section, I shall contextualize these theories with reference to natural-philosophical speculations in Leibniz and Kant. The term 'dynamical' will primarily refer to power or force (Gk: *dúnamis*) both within matter and as its foundation. As various (e.g. mathematical, physical, psychological) theories of opposition hold, relative stasis can arise from the effect of opposed forces.

Coleridge formed a position that he described as 'the very rudiments of a "*Heraclitus redivivus*"', expounding his commitment to the idea of the interpenetration of opposite forces (1956–71, 4: 775 [September 1817], letter to Charles Augustus Tulk). Here, everything in nature arises in the interplay of opposed forces, which manifest an original unitary power. The drive or attraction towards resembling such unity was held to propel natural evolution and human history. In this view, matter and phenomena are at once irruptions and temporary stabilizations, surging from an originary unity as opposed forces that 'coagulate' into material forms and phenomena which mirror, at a lower, re-stabilizing level, the original, unitary power. This open-systemic flux is conveyed in an illustrative passage in *The Statesman's Manual*:

> Thus instead of the position, that all things *find*, it would be less equivocal and far more descriptive of the fact to say, that Things are always *finding* their level: which might be taken as the paraphrase or ironical definition of a storm … (Coleridge 1972 [1816], 206)

Being always in flux, the interpolar strife only reaches temporary or symbolic equilibrium when energy within the opposed poles is directed higher or lower, in a metaphysical sense. When the energy is directed hypostatically lower, from energic flux into phenomenal and material relative stability, the flux forms a synthesis of the opposed poles into a relatively concrete product, in a process of materialization: 'in all pure phænomena we behold only the copula, the balance or indifference of opposite energies' (1956–71, 4: 760). In materialization, 'the productive power [is] suspended, and as it were, quenched in the product' (1995c [1825], 1272). Coleridge saw unity in the opposite, sublimating direction as spiritualization, or a '*repotentiating*', from energic flux towards the transcendent, originary power that the opposed poles manifest.

This dynamic of a fluid balancing of opposed polarities and energies is central to Coleridge's interpretation of the dynamical theory of matter as an emergence. It also underlies his theory of the arising and evolution of organic matter, of life. In a theory uniting these dynamics in matter and life, Coleridge held that the material substance of life in flesh and blood maintains a balance between the inertia of matter and restoration as living energy through repotentiation. That is, in the living body, matter tends towards energy or spiritualization, towards return to its originary power, such that the phenomena of organic reconstruction, including growth, nourishment, and repair, demonstrate a superordinate, dynamic telos that marshals the natural processes of physical and chemical decomposition. As Coleridge saw it:

> Flesh and Blood is … in the mid or balanc~~ed~~*ing* state between fixation and revivescence…. For matter itself is but Spiritus in coagulo; and organized matter the coagulum in the act of *being* restored, i.e. it is *repotentiating*. Stop its self-destruction as matter, and you stop its self-reproduction as a vital organ. (1980–2001, 5: 545–6, annotating Jeremy Taylor)

This process of repotentiation, pulsing between living and inanimate matter, enacts the metaphysical sense of the punctum saliens. Bloodless but intellectually exhilarating, Coleridge's metaphysical new punctum saliens leaps from the nullity of the indifference-point to the absolute of the prothetic idea or power, the origin and identity of the thesis and antithesis of opposed forces.

The phenomenal world, in such a dynamical view, represents the fluxional equations of the opposed forces on local and cosmic scales. By 1817, Coleridge could explain his position on the relation of matter to spirit or energy as follows:

> It is peculiar to the [dynamic] Philosophy … to consider matter as a Product—coagulum spiritûs [a coagulation of spirit], the pause, by interpenetration, of opposite energies—and … I hold no matter as real otherwise than as the copula of these energies, consequently no matter without Spirit … (1956–71, 4: 775, to Tulk [September 1817])

Around the same time, in *Theory of Life*, and generally agreeing with Schelling, he specifies that he sees 'the construction of matter' in its geometrical, transcendental, and dynamical account 'as the product, or *tertium aliud* [third, other thing], of antagonist powers of repulsion and attraction' (1995a [1816?/1819?], 522).

Elsewhere, he had asked, 'what *is* Matter but, the synthesis of its essential component Powers, Attraction and Repulsion?' (1980–2001, 3: 851 [*c*.1812–16], discussing Moses Mendelsohn's *Morgenstunden*, 1785).

In the Prometheus essay, Coleridge attributes to Leibniz this theory of matter as the congealing of opposed energies or spiritual powers into a relative pause, when he describes material phenomena as 'formed, as it were, by the precipitation, or, if I may adopt the bold language of Leibnitz, by a coagulation of spirit' 1995c [1825], 2: 1268. Although there seems to be in Leibniz no term suggestive of 'coagulum spiritûs', Coleridge may have had in mind Leibniz's dynamical, ontological theory, namely the theory of the monads being formed by 'continual fulgurations of the Divinity from moment to moment' (Leibniz 1991 [1714], §47), that is, by emanation in constant 'lightning flashes' from God. Annotating Kant's *Metaphysical Foundations of Natural Science* (1786), Coleridge (1980–2001, 3: 282–3) commented on Leibnizian monads and the repulsion theory of matter of Roger Boscovich, who considered atoms as points of force rather than as material objects.

An editor's footnote to the Prometheus essay (1995c [1825], 2: 1268–9) states:

> As his daughter, S[ara]C[oleridge], was the first to point out, C seems to have misremembered a passage from Schelling *System des transcendentalen Idealismus* 190: .... tr. 'When Leibniz calls matter the sleep-state of the monads, or when Hemsterhuis calls it curdled spirit ['den gronnenen Geist'], there is a meaning in these expressions that ... may very easily be understood. In fact, matter is no other than spirit contemplated in the equilibrium of its activities.'

It has been hitherto unnoticed, however, that Frans Hemsterhuis seems to have repeated the notion expressed by Henry More, and expounded by Platonists and Christian Kabbalists including Anne Conway, that 'the so-called material world is formed ... [from] particles of the divine essence, which have contracted and coagulated [contractis & constipatis] into monades or physical points' (More 1677, 294). Coleridge might have coined the phrase 'coagulum spiritûs' after Hemsterhuis's 'den gronnenen Geist', likely read in Schelling, but it is likely that he also recollected his reading the concept as More's 'Divinæ Essentiæ ... constipatis' but misattributed it to Leibniz because More contracted this coagulated spirit into 'monades'.

Leibniz has been described as 'the first great protestant against the new metaphysical orthodoxy', in that he opposed the Cartesian mechanist cosmology of 'a vast clock-like machine, set in motion at the beginning by the Creator and thence running merely by operation of its own secondary causes' (Burtt 1932, 200–1). Leibniz, who coined the term 'dynamics' (Dictionnaire de l'Académie française 1762), was joined by Newton in criticizing Descartes for insufficiently considering the role of force in physics. Leibniz's dynamicist essays 'Specimen dynamicum' (1695) and 'De ipsa natura' (1698) are key works of his mature physics. 'Specimen dynamicum' comprehensively summarized his 'new science of dynamics' on the anti-Cartesian principle that 'corporeal things contain something other than ..., indeed ... prior to extension, namely the force of nature implanted in all things by the Creator ... this force ... constitutes the inmost nature of bodies. For to act is the mark of a substance' (Leibniz 1998 [1695], §2, 154). In this work, he distinguished between *vis viva*, active or living force, and *vis mortua*, dead force, which his twentieth-century translator interprets as equivalent to kinetic and potential energy (Loemker, in Leibniz 1989, 451 n. 7). *Vis viva* is energy in action, force producing change through motion. *Vis mortua*, on the other hand, is an as-yet inactive tendency toward *vis viva* and is accordingly 'only an urge (solicitation) to motion', as with 'a stone in a sling before the stone is released … and also the force of gravity or centripetal force … and the force by which a stretched elastic body tries to spring back into shape' (1998 [1695], §12, 158–9).

However pertinent the Leibnizian binary, fulguration, and aforementioned dynamical theories, a fuller explanation for Coleridge attributing to Leibniz the theory of matter as the coagulation or equilibrium of opposed powers lies, I suggest, in his reading in Kant the argument that Leibnizian 'Physical monads are … to be conceived as pointlike centres of attractive and repulsive forces, where the repulsive force, in particular, generates a region of solidity or impenetrability in the form of a tiny "sphere of activity" emanating from a central point' (Friedman 2004, ix, glossing Kant, 1786). Synthesizing the physics and the metaphysics of the previous half-century, Kant's *Physical Monadology* (1756) introduced the fundamental, opposed forces of attraction and repulsion into his explanation of how matter can be conceived in its simplest parts as immaterial points surrounded by a field of repulsion. The intersection of mathematical points and physical forces in Kant's theory of matter and its subsequent influence in physics as well as philosophy makes it the most important representative of the 'dynamical geometry, constructive physics' I discussed in Sect. 15.3. As the leading twenty-first-century interpreter of Kant's dynamical theory of matter observes, 'This theory … has played a not inconsiderable role in the development of physics and natural philosophy' and it was an 'anticipation … of the field-theoretic approach to physics' (Friedman 2013, 97).

In *Physical Monadology* (1756), Kant synthesized Leibniz's theory of matter, as composed of monadic points that are atom-like non-spatial, eternal forms, with its opponent view, held by Newton, that the infinite divisibility of space requires the divisibility of the constituent parts of matter. Kant's dynamical solution of monadic points as extensionless centres of spatially divisible regions of repulsive force reconciled the a priori divisibility of space and the at least theoretical impenetrability of minute packets of space. Kant arrived at his dynamic theory of matter independently of and earlier than Roger Boscovich, who published his theory of impenetrability as fields of force around monadic points in his *Theory of Natural Philosophy* in 1758 (Boscovich 1966). Both thinkers forged similar syntheses of Leibniz and Newton on matter. Kant explained solidity in general and the impenetrability of the monadic elements, further explaining gravitation as attraction around a point. Seven years later, Kant's dynamical approach also counterbalances logical form in his 'Attempt to Introduce the Concept of Negative Magnitudes into Philosophy' (1763), where he argued for the existence of real dynamical or qualitative and not only logical opposition, whereby 'real opposition, is that where two predicates of a thing are opposed to each other, but not through the law of contradiction' (1992 [1763], 211; Ak. 2: 171).

Kant's dynamical theory of matter was not only a pre-Critical concern, as he further developed it in his *Metaphysical Foundations of Natural Science* (1786).

There, the fundamental, opposed forces of attraction and repulsion explain how matter can be conceived in its simplest parts as immaterial points surrounded by a field of repulsion, explaining solidity in general and the impenetrability of the monadic elements, with attraction explaining gravitation around a point. Kant's dynamical solution works by affirming the non-spatiality of monadic points at the centre of spatially divisible regions of repulsive force.

Coleridge recognized that Kant had arrived at an important energic theory of material objects and, primarily, of matter itself, as arising from opposed forces. In this, Coleridge differs from Hegel (1975 [1830], §48, 78), who curiously did not recognize the extent and innovation of Kant's dynamical theory, in opining that 'Kant … never penetrated to the discovery of what the antinomies really and positively mean …: that every actual thing involves a coexistence of opposed elements. Consequently, … to comprehend an object is equivalent to being conscious of it as a concrete unity of opposed determinations'. On the contrary, Kant conceived very well how everything, in its minutest parts, subsists in the 'coexistence of opposed elements'. From Kant's dynamical theory, and stepping, with Schelling, beyond Kant's stance at the threshold before the thing-in-itself and noumenal reality, Coleridge developed his version of a dynamical idealism where matter arises and evolves in the interplay of oppositional powers. For Coleridge, the interplay of opposed forces itself arises as an expression of unitary powers, the correlates of noumenal ideas, that become imperfectly actualized through the evolution of natural and human history. Because universal and eternal noumenal powers and ideas cannot be expressed in their essential unity in a finite, temporal form—as Kant argued in the logical sphere with his antinomies of pure reason—, their expression manifests instead as opposite and interdependent forces.

Coleridge saw matter as produced according to the Kantian dynamic model, with a Leibnizian heritage, of synthesis and relative pause in the tension between opposed forces. He saw these forces from a post-Kantian perspective, as immanent expressions of an absolute unity, this being the formula that led Hegel to describe the true meaning of the Kantian antinomies as interdependent and oppositional determinations of a comprehensive and unified reality.

According to dynamical idealism, the opposed forces—arising from antecedent, unitary powers, and which in turn give rise to matter and material forms—are actualizing energies and not merely potentialities or logical possibilities. Hence, Coleridge held a dynamical philosophy in that he viewed the physical world as sustained within a cradle of ideas, powers, and forces, concretizing into syntheses that produce matter, which in turn evolves opposing forms and forces that produce further syntheses of natural, historical, and cultural forms. The ideas have an equivalence to powers, and Coleridge defines them, drawing on Kant's formal-versus-dynamical distinction, such that 'For the *Understanding*, an Idea can only be described *negatively*—as that it is super-sensuous, ιδεαι [ideas] in opposition to ειδωλα [idols (images, likenesses)]; that it is not merely *formal* but dynamic—πᾶσα

Ιδεα Δυναμις [every Idea is a Power]' (Coleridge 1972 [1816], 61 fn.). Like Plato's,[11] and following Leibniz's and Kant's, Coleridge's idealism does not refute the existence of matter. Instead, he argues for an energism in which ideas and powers precede material existence.

Since the mid-nineteenth century and the widespread acceptance of the first law of thermodynamics, and especially after Einstein's mass–energy equivalence theory, physics has increasingly supported dynamical, energic theories of matter, and has long abandoned the mechanist, corpuscular theories that were dominant in eighteenth- and early nineteenth-century science. Compared with Enlightenment expressions of materialism and mechanism, dynamical idealist theories from Leibniz to Hegel, and even the transformative theories of energy and matter prefigured in Renaissance alchemical and hermetic thinkers including Böhme and Paracelsus, seem nearer to the view of modern physics, insofar as dynamical idealists construe matter as the interrelation or 'pause' of energies. According to Max Planck (1944), for instance:

> As a man who has devoted his whole life to the most clear-headed science, to the study of matter, I can tell you as a result of my research about atoms this much: There is no matter as such. All matter originates and exists only by virtue of a force which brings the particle of an atom to vibration and holds this most minute solar system of the atom together. We must assume behind this force the existence of a conscious and intelligent spirit. This spirit is the matrix of all matter.

Rather than wrongly dismiss the dynamical idealists and their various theories as apparently denying matter, it should be better recognized that they were almost prepossessed with the problem of its nature and with its origin and subsistence in energy, whether cosmic *dýnamis*, divine spirit, or some sense of mind.

For this reason, dynamical theories of matter and the metaphysics of powers might help to make further inroads into the 'hard problem' in the philosophy of mind, namely, the ultimately ontological question of how there can be consciousness at all. This problem lies behind a current revival in theories of panpsychism, arguing that some organisms can have self-consciousness because all matter is in some elementary sense already conscious (e.g. Goff, 2017, 106–25). Dynamical idealism could rival or expand this approach by commencing not from matter plus some quality, which the panpsychists call basic or primal consciousness, but with matter as itself arising from fundamental powers, its quantitative extension evolving from the qualitative intension of powers, as the universal reason, or Logos, in Coleridge's phrase, 'whose choral Echo is the Universe' (Coleridge 1983 [1817], 2: 240). The eventual realization of consciousness in living organisms would then be theorized as the actualization of fundamental powers as 'ideas', in the sense of intellectual potencies, as has been suggested by thinkers from Plato, Plotinus, and John Scotus Eriugena to Schelling, Hegel, and Coleridge.

---

[11] e.g. the *Timaeus*, gives a constructive cosmogony, hedged as a 'likely story' (29d), of the demiurge's creation of the world as a complex living creature from originary (Platonic) ideas and geometrical forms (Platonic solids).

For Coleridge, the evolving concretes of history and experience are pressed into physical and phenomenal existence by opposed forces that manifest, through a kind of attenuated reflection, the originary power that gives rise to those forces. That power was held by Coleridge and other dynamical idealists to be an energy of intensity and ideality that they viewed as anteceding extension and materiality. Between the synthesized and evolving phenomenal level and that of transcendent, unitary powers and ideas, is a storm of opposed, immanent forces: the level of existence and manifestation. If this were the whole story, then the flow of reality in this theory would simply be a ripple of ever-attenuating emanation, bottoming out at the most basic material forms. Yet this dynamical theory, in its Coleridgean form, is circulatory and includes a pathway for ascent, directed back towards the unitary, transcendent power: for Coleridge, the prothesis or identity point, from which these forces became manifest. Whether conceiving the processes of individuating, organic life as enacting unity throughout multeity in a spiritualizing or sublimating drive toward towards the originary unity anteceding the opposed forces, or as the intellectual or aesthetic contemplation of those powers and ideas through and beyond phenomena, the pattern is one of ascent through material and phenomenal forms to sustenance of the meaning and value of the ideas or powers. At the midpoint, Coleridge drew on a biological metaphor that struck observers from Aristotle to Harvey and Harrington as having a wonderful ontological significance, the pulse and leap of the embryonic heart, the punctum saliens, to express this most crucial moment in his metaphysics.

References

Aristotle. 1999. *Metaphysics* (*c*.330 BC). Tr. Joe Sachs. Santa Fe: Green Lion Press.

———. 2002. *On the parts of the animals I–IV* (*c*.350 BC). Tr. and ed. James G. Lennox. Oxford: Oxford University Press.

———. 1910. A history of animals (c.350 BC). Tr. D'Arcy Wentworth Thompson. Oxford: Oxford University Press.

Blankaart, Steven. 1754 [1679]. Lexicon medicum renovatum ... Leuven: Joan. Franc. Van Overbeke.

———. 1684 [1679]. The physical dictionary: Wherein the terms of anatomy, the names and causes of diseases ... are accurately described: Also the names and virtues of medicinal plants ... London: S. Crouch.

Boscovich, Roger. 1966 [1758]. *Theory of natural philosophy*. Tr. J. M. Child. Cambridge, MA: MIT Press.

Briggs, Gerry, and Kiran Toor. 2017. Coleridge, the ancient mariner, and the philosopher's stone. *Journal of Romanticism* 2:1–29.

Burtt, E. A. 1932. *The metaphysical foundations of modern science*. 2nd ed. London: Kegan Paul.

Canning, George, and John Hookham Frere. 1798. 'The loves of the triangles: A mathematical and philosophical poem' (1798). In George Canning, et al. 1799. *Poetry of the Anti-Jacobin*: 108–41. London: J. Wright.

Carlisle, Anthony. 1805. The Croonian lecture: On muscular motion. *Philosophical Transactions of the Royal Society of London* 95:1–30.

Cheyne, Peter. 2020. *Coleridge's contemplative philosophy*. Oxford: Oxford University Press.

Coffmann, Ralph J. 1987. *Coleridge's library: A bibliography of books owned or read by Samuel Taylor Coleridge*. Boston, MA: G. K. Hall.

Cohen, I. Bernard. 1994. Harrington and Harvey: A theory of the state based on the new physiology. *Journal of the History of Ideas* 55 (2): 187–210.
Coleridge, Samuel Taylor. 1956–71. In *The collected letters of Samuel Taylor Coleridge*, ed. E.L. Griggs, vol. 6. Oxford: The Clarendon Press.
———. 1957–2002. *The notebooks of Samuel Taylor Coleridge*. Vols 1–3, ed. Kathleen Coburn, 1957–73; vol. 4. Ed. Kathleen Coburn and Merton Christensen, 1990; vol. 5, ed. Kathleen Coburn and Anthony Harding. Princeton: Princeton University Press.
———. 1969. *The friend* (vol. 1, 1818; vol. 2, 1808–9). Ed. Barbara E. Rooke. Princeton: Princeton University Press.
———. 1972. *The statesman's manual* (1816). In *Lay Sermons*, ed. R.J. White. Princeton: Princeton University Press.
———. 1980–2001. *Marginalia*. Vols 1 and 2 (1980–4). Ed. George Whalley, vols 3–6 (1992–2001). Ed. H. J. Jackson and George Whalley. Princeton: Princeton University Press.
———. *Logic*. 1981 [*c*.1818–22]. Ed. J. R. de J. Jackson. Princeton: Princeton University Press.
———. 1983. *Biographia literaria, or biographical sketches of my literary life and opinions* (1817). Ed. James Engell and W. Jackson Bate, 2 vols. Princeton: Princeton University Press.
———. 1987. *Lectures 1808–1819: On literature*. Ed. R. A. Foukes, 2 vols. Princeton: Princeton University Press.
———. 1993. *Aids to reflection* (1825). Ed. John Beer. Princeton: Princeton University Press.
———. 1995a. Theory of life (1816?/1819?). In *Shorter works and fragments*, ed. H.J. Jackson and J.R. de J. Jackson, vol. 1, 485–557. Princeton: Princeton University Press.
———. 1995b. 'Notes on polar logic' (1828). In *Shorter works and fragments*, ed. H.J. Jackson and J.R. de J. Jackson, vol. 2, 1383. Princeton: Princeton University Press.
———. 1995c. 'On the Prometheus of Aeschylus' (1825). In *Shorter works and fragments*, ed. H.J. Jackson and J.R. de J. Jackson, vol. 2, 1251–1301. Princeton: Princeton University Press.
———. 2000. In *Lectures 1818–1819: On the history of philosophy*, ed. J.R. de J. Jackson, vol. 2. Princeton: Princeton University Press.
———. 2001. In *Poetical works, part one: Poems (reading text)*, ed. J.C.C. Mays, vol. 2. Princeton: Princeton University Press.
Connell, Sophia M. 2001. Toward an integrated approach to Aristotle as a biological philosopher. *The Review of Metaphysics* 55 (2): 297–322.
Cudworth, Ralph. 1678. *The true intellectual system of the universe*. London: Richard Royston.
Dictionnaire de l'Académie française. 1762. 4th ed. Paris: Bernard Brunet.
Descartes, Rene. 1985a [1662; written 1632]. 'Treatise on man'. Tr. Robert Stoothoff. In *The philosophical writings of Descartes*, 99–108. Cambridge: Cambridge University Press.
———. 1985b [1664; written 1647–8]. 'Description of the human body'. Tr. John Cottingham. In *The philosophical writings of Descartes*, 314–24. Cambridge: Cambridge University Press.
Friedman, Michael. 2004. Editor's introduction to Kant, *Metaphysical foundations of natural science* (1786), vii–xxx. Cambridge: Cambridge University Press.
———. 2013. *Kant's construction of nature: A reading of the metaphysical foundations of natural science*. Cambridge: Cambridge University Press.
Goethe, J. W. 1962. Italian journey: 1786–1788. Tr. W. H. Auden and Elizabeth Mayer London: Penguin.
Goff, Phillip. 2017. Panpsychism. In *The Blackwell companion to consciousness*, ed. Susan Schneider and Max Velmans, 2nd ed. Oxford: Blackwell.
Greene, Robert A. 1962. Henry More and Robert Boyle on the spirit of nature. *Journal of the History of Ideas* 23 (4): 451–474.
Harrington, James. 1992. The commonwealth of Oceana (1656). In *The commonwealth of Oceana and a system of politics*, ed. J.G.A. Pocock. Cambridge: Cambridge University Press.
Harvey, William. 2016 [1628]. *On the motion of the heart and blood in animals*. Tr. Robert Willis (1847) and ed. Jarrett A. Carty.

———. 1989 [1651]. *On the generation of animals*. In *The works of William Harvey*. Tr. Robert Willis (1847), 145–588. University of Pennsylvania Press.
Hegel, G. W. F. 1975. *Hegel's logic: Being part one of the encyclopaedia of the philosophical sciences* (1830). Tr. William Wallace (1873). Oxford: The Clarendon Press.
Hippocratic corpus. 2014 [fourth-century BC]. 'Cord. (pe?? ?a?d???, de corde) On the heart'. Ch. 10 in *The 'Hippocratic' corpus: Content and context*. Tr. and ed. Elizabeth Craik. London: Routledge.
Kant, Immanuel. 1992 [1763]. 'Attempt to introduce the concept of negative magnitudes into philosophy' (1763). In Kant. *Theoretical Philosophy, 1755–1770*. Tr. and ed. David Walford with Ralf Meerbote, 203–42. Cambridge: Cambridge University Press.
Kragh, Helge. 2013. Big bang: The etymology of a name. *Astronomy & Geophysics* 54 (2): 28–30.
Leibniz, G. W. 1991. *G. W. Leibniz's* Monadology*: An edition for students* (1714). Tr. and ed. Nicholas Rescher. Pittsburgh: Pittsburgh University Press.
———. 1989. *Philosophical papers and letters*. Tr. and ed. Leroy E. Loemker, 2nd edn. Dordrecht: Springer.
———. 1998. 'Specimen dynamicum: An essay in dynamics' (1695). In *Philosophical texts*. Tr. and ed. R. S. Woolhouse and Richard Francks, 153–79. Oxford: Oxford University Press.
Magnuson, Paul. 1998. *Reading public romanticism*. Princeton: Princeton University Press.
More, Henry. 1677. 'Fundamenta philosophiae sive Cabbalae aeto-paedo-melissaeae'. In ed. Knorr von Rosenroth. 1677. *Kabbala Denudata*. Vol. 2 (of 2), 293–312. Sulzbach: Abraham Lichtentahler.
Mortensen, Peter. 2004. *British romanticism and continental influences: Writing in an age of Europhobia*. New York: Palgrave Macmillan.
Needham, Joseph. 1931. *Chemical embryology*. Cambridge: Cambridge University Press.
———. 1926. S. T. Coleridge as a philosophical biologist. *Science Progress in the Twentieth Century* 20 (80): 692–702.
Planck, Max. 1944. 'Das Wesen der Materie [The nature of matter]'. Speech at Florence. *Archiv zur Geschichte der Max-Planck-Gesellschaft*, Abt. Va, Rep. 11 Planck, Nr. 1797.
Plato. 1997a. Theaetetus (c.369 BC). Tr. M. J. Levett, rev. Myles Burnyeat. In *Plato, complete works*, ed. John M. Cooper and D.S. Hutchinson, 157–234. Indianapolis: Hackett.
———. 1997b. Timaeus. (c.360 BC). Tr. Donald J. Zeyl. In *Plato, complete works*, ed. John M. Cooper and D.S. Hutchinson, 1224–1291. Indianapolis: Hackett.
Schmitt, Stéphane. 2020. The description of the development of the chicken embryo by Volcher Coiter (1572): Text translated into French, annotated, with commentary. *Revue d'histoire des sciences* 73:339–361.
Southey, Robert. 2023. *The collected letters of Robert Southey*. Vol. 3 (1804–9). Ed. Carol Bolton and Tim Fulford (1 May 1805): https://romantic-circles.org/editions/southey_letters.
Swift, Jonathan. 2003. *Gulliver's travels (1726)*. London: Penguin.
Vickers, Neil. 2004. *Coleridge and the doctors*. Oxford: Oxford University Press.
Whalley, George. 1997. 'On translating Aristotle's *poetics*', in Aristotle, *poetics*. Tr. and ed. George Whalley, 1–32. Dalhousie: McGill-Queen's University Press.
Wordsworth, William. 2013. Lines composed a few miles above Tintern Abbey, on revisiting the banks of the Wye during a tour. July 13, 1798. In *Wordsworth and Coleridge, Lyrical ballads: 1798 and 1801*, ed. Fiona Stafford, 87–91. Oxford: Oxford University Press.

Peter Cheyne is professor in the School of Letters at Hiroshima University; life member at Clare Hall, Cambridge; and visiting fellow in philosophy at Durham University. He is the author of *Coleridge's Contemplative Philosophy* (Oxford UP, 2020); editor of *Imperfectionist Aesthetics in Art and Everyday Life* (Routledge, 2023) and *Coleridge and Contemplation* (Oxford UP, 2017); and co-editor of *The Philosophy of Rhythm: Aesthetics, Music, Poetics* (Oxford UP, 2020).

Chapter 16
Matter, Will, and Human Life in Coleridge and Schopenhauer

Allison Dushane

Abstract This essay explores how Samuel Taylor Coleridge and Arthur Schopenhauer develop theories of the will that are in dialogue with the work of romantic-era scientists and grounded in anti-dualist conceptions of matter, yet divergent in their philosophical conclusions. Coleridge and Schopenhauer were both interested in developing a theory of life that acknowledges and incorporates the experimental findings of science, yet they insisted that scientific methods are not the sole arbiters of truth. Coleridge began by theorizing the dynamism at work in nature to formulate positive definitions of the human will, agency, freedom, and God. Schopenhauer, on the other hand, began with an unconventional theory of the materiality of the will as the foundation of all nonhuman and human life, proposing a negative definition of freedom. These contrasting accounts of will, freedom, and human life lie at the foundation of critical conceptions of modern subjectivity and selfhood.

The ongoing search in romantic-era science to isolate and identify a formative principle of life, such as J. F. Blumenbach's *Bildungstrieb*, Erasmus Darwin's *formative nisus*, and John Hunter's vital principle, was characterized by efforts to distinguish the processes of the living from the laws of the machine. These ideas worked their way into the cultural imaginary though public discourses such as the 'vitality debates' between the surgeon John Abernethy and his student William Lawrence that took place in widely read journals.[1] Vitalist positions held that life cannot be explained by physical forces alone and must be defined through something else:

[1] See Sharon Ruston, *Shelley and Vitality* (Ruston 2005) for a comprehensive summary and analysis of the vitality debates in relation to Romantic-era literature.

A. Dushane (✉)
Department of English and Modern Languages, Angelo State University,
San Angelo, TX, USA
e-mail: Allison.Dushane@angelo.edu

P. Cheyne (ed.), *Matter and Life in Coleridge, Schelling, and Other Dynamical Idealists*, International Archives of the History of Ideas Archives internationales d'histoire des idées 253, https://doi.org/10.1007/978-3-031-78315-9_16

perhaps the soul, a more secular spirit, or another animating force.[2] The public perception of the opposing position—materialism—was shaped by the systems of several prominent Enlightenment philosophers. For example, La Mettrie (La Mettrie 1912 [1748] 148), on surveying the emerging field of physiology, ends *Man a Machine* with the invitation to 'conclude boldly that man is a machine, and that in the whole universe there is but a single substance differently modified'. Blumenbach had argued in *Über den Bildungstrieb* (1781) that 'there is no such thing in nature, as pre-existing organized germs', but rather a moment at which 'the unorganized matter of generation' takes on a 'particular action, or nisus' (Blumenbach 1791, 20). The suggestion that individual human lives may be entirely determined and directed by natural laws haunted writers who otherwise embraced the potential they saw in the Enlightenment project for human progress through a scientific program based on reason, observation, and experiment.

In an October 1804 notebook entry, Samuel Taylor Coleridge includes a fragment titled 'What is Life?' which he later transcribed in an 1819 letter to an autograph collector, 'as evidence of [his] very early Turn to metaphysics and speculative philosophy':

> Resembles Life what once was held of Light,
> Too simple in itself for human sight?
> An absolute Self—an Element ungrounded
> All that we see, all colors of all shade
> By 'incroach of darkness' made?
> Even so, is Life by *conscious* thought unbounded?
> And all the Joys and Woes of Mortal Breath,
> A War-embrace of wrestling Life and Death!
> (Coleridge 2001, 2: 767)[3]

In an extended analogy that invokes the ongoing debates about the nature of life that posit a distinguishing vital force in contrast to the previous generation's focus on physical phenomena, Coleridge alludes to the Newtonian hypothesis that light and colors are products of the arrangements of corpuscular matter. The October 1804 notebook entry includes a telling deletion in an omitted line 6: 'Is Life (in its own nature)/itself by consciousness unbounded' (1957–2002, 2: §2224, fol. 25^{v}). 'What is Life?' invokes the challenge facing philosophy as science began to apply physical laws and principles to the development of living organisms in the late eighteenth and early nineteenth centuries: if human beings are a part of the ongoing activity of material nature, then what kind of agency belongs to the human, and what are its limits? Is there such a thing as free will determined by consciousness, or is it an illusion?

---

[2] Recent studies on the centrality of vitalism and materialism to romantic-era thought include Denise Gigante, *Life: Organic Form and Romanticism* (Gigante 2009); Catherine Packham, *Eighteenth-century Vitalism: Bodies, Culture, Politics* (Packham 2012); and Robert Mitchell, *Experimental Life: Vitalism in Romantic Science and Literature* (Mitchell 2014).

[3] For the letter in which this fragment appears, see Coleridge (1956–71), 4: 937.

This essay explores how Coleridge and Arthur Schopenhauer develop theories of the will that are in dialogue with the work of Romantic-era scientists and grounded in anti-dualist conceptions of matter, yet divergent in their philosophical conclusions. Coleridge and Schopenhauer are both interested in developing a theory of life that acknowledges and incorporates the experimental findings of science, yet both also insist that scientific methods are not the sole arbiters of truth. Coleridge was always captivated by the sciences, showing an early interest in a medical career before he 'bewildered [him]self in metaphysicks', and began an unfinished degree at Cambridge (Coleridge 1983 [1817], 1:15). To further investigate the intersections of scientific, philosophical, and theological ideas, he enrolled at the University of Göttingen in 1799, where he attended lectures by Blumenbach on anatomy and physiology and quickly came to the conclusion, as he wrote in an April 1799 letter to Thomas Poole: '*Life*, *Power*, *Being*!—organization may [be] and probably *is*, their *effect*; their *cause* it *cannot* be!' (1956–71, 1: 479). Around 1816, Coleridge composed *Hints Towards the Formation of a More Comprehensive Theory of Life*, confronting the question of life (as power, being, and organization) in dialogue with physicians and scientists. *Theory of Life* lays the foundation for an anti-dualist stance that would inform, but also complicate, his later philosophical and theological voluntarism.

Schopenhauer likewise attended scientific lectures at Göttingen starting in 1809, motivated by a desire to study medicine that was overturned by an interest in philosophy (Janaway 2002, 4). His *On the Will in Nature* (1836) corroborated the universal metaphysical voluntarism (where the universe is the expression of an irrational will) that he formulated in *The World as Will and Representation* (1818) with evidence from scientific sources in an effort to publish a second version of *The World as Will* to a wider audience. Where Coleridge began by theorizing the dynamism in nature to formulate positive definitions of human will, agency, freedom, and God, Schopenhauer, by contrast, commenced from his unconventional theory of the will as the foundation of material and energic being, including all nonhuman and human life, to propose a negative definition of freedom. These contrasting accounts of will, freedom, and human life, both informed by dynamic theories of matter, lie at the foundation of critical conceptions of modern subjectivity and selfhood.

In Coleridge's later prose, he combined his formulation of a theory of 'Life' with a theory of the human will. These writings reveal a conflict between two distinct philosophical impulses: one that asserts the human will as supernaturally given, and the other that claims that the human will emerges gradually out of the ongoing activity of material nature itself. Coleridge repeatedly made anti-dualist arguments that prevented him from simply separating the soul from the material world. Instead, he attempted to link the principles of reason and conscience in human beings to the activity of 'Life' through his conception of the will.

In *Theory of Life*, Coleridge shifted the discourse on life from a question about the line separating life and death to an inquiry about the significance and distinction of the individual human life. He argued that explanations derived from physiological research in the sciences, such as John Hunter's, Abernethy's, and Lawrence's 'presuppose the arbitrary division of all that surrounds us into things with life, and things without life—a division grounded on a mere assumption' (1995 [1816?], 1: 488). Characteristically, he continually sought the most inclusive and infinitely flexible starting point from which to theorize 'Life' in the context of the experimental findings of the biological sciences. He proposed his own definition of life 'as the principle of individuation, or the power which unites a given all into a whole that is presupposed by all its parts' (510). The distinguishing feature of life for Coleridge is this 'tendency to individuation', itself manifested in degrees:

> Thus, from its utmost *latency*, in which life is one with the elementary powers of mechanism, that is, with the powers of mechanism considered as qualitative and actually synthetic, to its highest manifestation, (in which, as the *vis vitae vivida* [living life-force], or life as life, it subordinates and modifies these powers, becoming contradistinguished from mechanism, *ab extra* [from without], under the form of organization,) there is an ascending series of intermediate classes, and of analogous gradations in each class. (511)

Coleridge posited the emergence of complex and diverse life through an opposition of intrinsic agency and external forces, based on a fundamentally anti-dualist and dynamic conception of matter. Through a gradual progress, matter, whose impulses are first subject almost entirely to environmental influence, develops to a point at which the *vis vitae vivida*, the powers 'ab intra', can subordinate the forces from without, or 'ab extra'. Human beings, as the most complex—and the most individual—forms of life, occupy the pinnacle of the scale.

In a March 1819 lecture, Coleridge elaborated on these concepts of organization and individuation, contextualizing his position against the history of philosophical materialism. He began by contrasting the assumptions of what he terms the 'ancient' and 'modern' schools of materialist thought:

> It is sufficient for our purposes to know that the fundamental positions of ancient materialism were, first, that motion and sensation are properties of a specific kind of atoms and that the mind is but a species of sensation, and all the processes of perception and of reflection purely passive, and all the … phenomena or appearances of life (just as the seeming acts of a dream), are wholly mechanical or produced by necessitating antecedents. (Coleridge 2000 [March 1819], 517)

Ancient materialism, then, falters in its attribution of latent powers to specific species of matter; such a scheme, according to Coleridge, leads to a system governed by necessity and determinism that renders life a series of passive and predetermined events. However, 'the modern scheme of materialism' differs in 'representing life, mind, and will as the result of organisation, not as pre-existing in the specific [atoms so organized]' (518). Although modern materialism presents its own set of problems for Coleridge, he points to this distinction as a starting point for his continual assertion that life, mind, and will become manifest in emergent and ongoing processes.

'Organization' is a key term for Coleridge in his theorization of life. He divides the prevailing opinions of his own time into those that claim organization as the source of vitality, or life, and those that claim a superadded vital force as the source of life. His own approach is to reevaluate the assumptions behind the term:

> For what, again I say, is organisation? Not the mere arrangement of parts as means to an end, for in that sense I should call my watch organisation or a steam-engine organisation; but we agree these are machines not organisations. It appears, then, that if I am to attach any meaning at all to the word 'organisation' it must be distinct from mechanism in this, that in all machines I suppose the power to be from without, that if I take my watch there is nothing in the component parts of this watch that constitutes it peculiarly fit for a watch, or produce[s] it. (524)

Here, Coleridge invokes the analogy of man as machine to argue against a deist branch of natural history, most recognizable in William Paley's argument for a clockwork universe operating on rational principles. Opposing this, he draws on a Kantian assertion of reciprocal causation to conclude:

> Organisation, therefore, must not only be an arrangement of parts together as a means to an end, but it must be such an interdependence of parts, each of which in its turn being means to an end, as arises from within. The moment a man dies, we can scarcely say he remains organized in the proper sense. (524)

This assertion of interdependence and reciprocal determination results in a conception of nature that is dynamic, and a conception of teleology in which ends and means are reciprocally determined through an ongoing process.[4]

For Coleridge, the idea that natural processes, whether harmonious or antagonistic, could ever be devoid of, or oblivious to, the internal strivings of the individual is unthinkable. He dismantles the notion in his poem 'Human Life', first published in *Sibylline Leaves* (1817):

> If dead, we cease to be; if total gloom
> Swallow up life's brief flash for aye, we fare
> As summer-gusts, of sudden birth and doom,
> Whose sound and motion not alone declare,
> But *are* their *whole* of being! If the Breath
> Be Life itself, and not its Task and Tent,
> If even a soul like Milton's can know death;
> O Man! thou vessel purposeless, unmeant,
> Yet drone-hive strange of phantom purposes!
> (2001, 1: 887, ll. 1–9)

The poem begins as a series of propositional arguments, a repeated series of if-then statements. If there is no life after death; if physical qualities like 'sound' and 'motion' are not merely the perceived effects of, but the '*whole* of being'; if breath, or life, is an effect but not a telos, then 'Man' becomes a 'vessel', paradoxically 'purposeless' and filled with 'phantom purposes'. The stress of the passage, however, indicated by the double stress in the final line, falls on 'unmeant'. The next five lines elaborate on the nature of this un-meaning:

---

[4] Immanuel Kant, *Critique of the Power of Judgment*, tr. and ed. Paul and Eric Matthews Guyer (Kant 2001), 247–8, articulates reciprocal causation. Kant argues in paragraphs 64–6 of the third *Critique* for the existence of 'natural ends', working with a principle that states: 'an organized product of nature is that in which everything is an end and reciprocally a means as well. Nothing in it is vain, purposeless, or to be ascribed to a blind mechanism of nature'.

> Surplus of nature's dread activity,
> Which, as she gazed on some nigh-finished vase,
> Retreating slow, with meditative pause,
> She formed with restless hands unconsciously!
> Blank accident! nothing's anomaly!
> (2001, 1: 887, 10–14)

Such a depiction of nature's activity opposes the progressive vision that he set forth in *Theory of Life*, which has its telos in human life: 'In Man the centripetal and individualizing tendency of all Nature is itself concentred and individualized—he is a Revelation of Nature!' (1995 [1816?], 1: 551). This teleological trajectory is driven by the goal-directed 'individualizing tendency' of nature, as opposed to the activity of 'restless hands'. At this point, the poem resumes its original argumentative structure:

> If rootless thus, thus substanceless thy state,
> Go, weigh thy dreams, and be thy Hopes, thy Fears,
> The counter-weights!—Thy Laughter and thy Tears
> Mean but themselves, each fittest to create,
> And to repay the other! Why rejoices
> Thy heart with hollow joy for hollow good?
> Why cowl thy face beneath the Mourner's hood,
> Why waste thy sighs, and thy lamenting voices,
> Image of Image, Ghost of Ghostly Elf,
> That such a thing as thou feel'st warm or cold?
> Yet what and whence thy gain, if thou withhold
> These costless shadows of thy shadowy self?
> Be sad! be glad! be neither! seek, or shun!
> Thou hast no reason why! Thou can'st have none!
> Thy being's being is a contradiction.
> (2001, 1: 887, 15–29)

According to Coleridge, there is no being without something that is not defined solely through the physical categories of 'sound and motion', something to oppose the 'restless hands' of 'nature's dread activity'. To assert such is a contradiction of the very definition of being, and indicates the first step on the path to a hopeless relativism. Yet, at the same time, Coleridge repeatedly made anti-dualist arguments that prevent him from simply positing a soul that exists and operates separately from the material world.

To reconcile his dynamic theory of matter with the singularity of human life, Coleridge attempts to link the principles of reason and conscience in human beings to the activity of 'Life' through his conception of the will. *Aids to Reflection* (1825) is the richest source of his thinking on the will. There, the distinguishing feature of human life versus nonhuman life is spiritual, and he argues that: 'If there be aught *Spiritual* in Man, the Will must be such. *If* there be a Will, there must be a Spirituality in Man' (1993 [1825], 247). Making a distinction between the laws at work in the will and the laws at work in the natural world is a crucial part of his efforts to distinguish human life as the pinnacle of the scale of life. Thus, he asserts that:

> there is more in man that can be rationally referred to the life of Nature and the mechanism of Organization; that he has a will not included in this mechanism; and that the Will is in an especial and pre-eminent sense the spiritual part of our Humanity. (247)

Preserving the significance of the individual human life, Coleridge distinguishes between the nature of the powers at work in the will and the laws that govern the natural world. He defines his own conception of the will against its previous materialist incarnations in writers like David Hartley and Joseph Priestley, insisting on articulating it not as 'the mere result and aggregate of fibres, motions, and sensations' but a 'real, distinct, correspondent power' (247). Coleridge consistently asserts the will against the ideas of 'result' and 'aggregate', refusing to see it as an effect or a product; instead, he insists that it must be a power distinct from, yet 'correspondent' to, those in material nature.

In his preliminary remarks to the 'Aphorisms on Spiritual Religion', which make up the bulk of *Aids to Reflection*, Coleridge distinguishes the will through its power to originate, its capability for action. He argues:

> a Will conceived separately from Intelligence is a Non-entity, and a mere Phantasm of Abstraction; and that a Will, the state of which does in *no sense* originate in its own act, is an absolute contradiction. (141)

For Coleridge, the will illuminates the difference between a formative *impulse* and a formative *agency*. Something that lacks freedom of action 'might be an Instinct, an Impulse, a plastic Power, and, if accompanied with consciousness, a Desire; but a Will it *could* not be!' (141). Coleridge's thinking allows for this process of emergence in nature, but throughout the 'Aphorisms', he insists on a distinct power at work in human intelligence:

> For this is the essential attribute of a Will, and contained in the very *idea*, that whatever determines the Will acquires this power from a previous determination of the Will itself. The Will is ultimately self-determined, or it is no longer a *Will* under the law of perfect Freedom, but a *Nature* under the mechanism of Cause and Effect. (285)

Coleridge's conception of the will, then, is linked to a conception of 'perfect Freedom', which is separate from the realm of mechanical causes and effects.

The conclusion to *Aids to Reflection* elaborates further on how the agency at work in the natural world instantiates itself in living forms. He begins with the proposition that matter is characterized by inherent active properties (as in chemistry), which lead to a more conscious, collective agency (as in J. W. Goethe's *Elective Affinities*, 1809): 'perhaps the material particles possess this combining power by inherent reciprocal attractions, repulsions, and elective affinities; and are themselves the joint Artists of their own combinations?' (Coleridge 1993 [1825], 398). He continues articulate form as a characteristic acquired through an emergent physical process rather than as a given property:

> It will be sufficient to remind the thoughtful Querist, that even herein consists the essential difference … of an Organ from a Machine; that not only the characteristic Shape is evolved from the invisible central power, but that the material Mass itself is acquired by assimilation. The germinal power of the Plant transmutes the fixed air and the elementary Base of

> Water into Grass or Leaves; and on these the Organific Principle in the Ox or the Elephant exercises an Alchemy still more stupendous. (398)

The acquisition of form through an emergent process separates the 'Organ from a Machine', the living from the non-living:

> As the unseen Agency weaves its magic eddies, the foliage becomes indifferently the Bone and its Marrow, the pulpy Brain, or the solid Ivory. That what you see *is* blood, *is* flesh, is itself the work, or shall I say, the translucence, of the invisible Energy, which soon surrenders or abandons them to inferior Powers, (for there is no pause nor chasm in the activities of Nature) which repeat a similar metamorphosis according to their kind ... (398)

Coleridge includes this argument in the Conclusion as a part of his larger refutation of dualism, asserting that continued reflection upon the fruits of empirical research in the natural sciences will lead to the rejection of philosophical systems that dominated the previous century:

> the utter emptiness and unmeaningness of the vaunted Mechanico-corpuscular Philosophy, with both its twins, Materialism on the one hand, and Idealism, rightlier named *Subjective Idolism* on the other: the one obtruding on us a World of Spectres and Apparitions; the other a mazy Dream! (398–9)

This objection to the dualist 'Mechanic or corpuscular Scheme' which he says 'in its absoluteness and strict consistency was first introduced by Des Cartes' (399), echoes his argument against the variety of 'ancient' (Democritean) materialism that he criticizes in his philosophical lectures. He opposes the removal of all positive properties from corporeal substance, referring to such attempts here, and repeatedly in the philosophical lectures as a '*Fiction of Science*'. In the philosophical era prior to the reintroduction of dynamic forces into nature by the scientists that Coleridge professedly admired in *Theory of Life* (and in *Aids to Reflection*), 'Des Cartes', having been cited as the major source of modern dualism, 'propounded it [sc. matter] as *truth of fact*' and, moreover, did so while 'abstracting from corporeal substance all its *positive* properties' (1993 [1825], 399). The '*Fiction*' of science is that it is capable of illuminating everything about the world. Mistaking science for the *whole* truth rather than a particular and specialized *part* of it is a mistake with heavy metaphysical and ontological implications, reducing the world, humans included, to 'a lifeless Machine whirled about by the dust of its own Grinding' (400–401).

Yet, in his own endeavours, he finds himself unable to resolve completely his theory of life as an emergent, uninterrupted scale of natural processes, on one side, with a 'self-determining' human will, on the other. By insisting on the will as an ultimately positive, spiritual force, he must answer the question of whether he has simply reinstated the dualism that he commits himself to refuting in *Theory of Life* and the philosophical lectures. At some moments in *Aids to Reflection*, Coleridge appears to recognize this implication, writing the appearance of the will into the emergent process that it eventually transcends. In the corollary to Aphorism 15 of the 'Moral and Religious Aphorisms', for example, he contrasts will to motives (which would ostensibly fall into the category of nature), arguing: 'Motives are symptoms of weakness, and supplements for the deficient energy of the living PRINCIPLE, the LAW within us' (97). Yet, he positions motives as a foreshadowing of the will, as that in which the 'Promise and Token of a true Individuality are disclosed' (98). Coleridge attempts to articulate a key turning point in the developmental process of nature in which the individual will acquires a transcendent power:

> both the Reservoir of Sensibility and the imitative power that actuates the Organs of Motion (the Muscles) with the net-work of conductors, are all taken inward and appropriated; the Spontaneous rises into the Voluntary, and finally after various steps and a long Ascent, the Material and Animal Means and Conditions are prepared for the manifestation of a Free Will, having its Law within itself and its motive in the Law—and thus bound to originate its own Acts, not only without but even against alien Stimulants. (98)

In this passage, Coleridge attempts to write his conception of the will into his wider theorization of 'Life'. This effort to unite a conception the human will defined by freedom and originality with an account of the development of higher organisms grounded in the system of material nature preoccupied him throughout the latter stages of his career, and likely played a significant factor in his failure to complete his more ambitious synthetic projects.

Illuminating the crux of the difficulty Coleridge faced in articulating the distinction of the human individual, Schopenhauer theorized the will in opposition to freedom and individuality. His most comprehensive work, *The World as Will and Presentation* (1818–19), also begins with a dynamic and anti-dualist matter theory. He asserts that matter is 'through and through nothing but causality', and that 'its being, namely, is its effectuality'. He conceives matter as pure potential, or power to affect its surroundings, claiming that 'its entire being and essence consists only in the law-governed alteration that one of its parts brings about in another' (Schopenhauer 2008 [1818–19], 37–8). An individual in a world generated through the action of matter as causality defines itself through the will. Here, however, Schopenhauer departed significantly from Coleridge—and attempted to distinguish himself from his peers in the European philosophical community—by refusing to conceive the will as a force given outside of the action of physical nature.[5] For Schopenhauer, the will is an actuating principle that works simultaneously through the individual perceiving subject and the matter that makes up the body, insisting that 'the act of will and action of the body are not two distinct objectively cognized states that are connected by the bond of causality, do not stand in the relation of cause and effect'. Instead, they are the same light in different states of refraction: 'they are rather one and the same, only given in two entirely distinct manners: once quite immediately and once in perception for the understanding' (2008 [1818], 137). The objectification of the will is a process of individuation, which results in a gradual hierarchy of living beings:

> While every human being is then to be regarded as a particularly determined and characterized phenomenon of will, even to a certain extent as an Idea of its own, but in significance—and its trace ever more vanishes with every greater distance from human beings … (173).

[5] Schopenhauer and Coleridge were both influenced by the dynamic idealism of Jakob Böhme. See Celia Muratori and James Vigus's Chap. 2 in this book.

Like Coleridge, Schopenhauer conceived of a range of causal strength corresponding to the scale of awareness, or intelligence, that manifests itself in the material world, from inorganic crystals to plants, animals, and humans.

Introducing *On the Will in Nature*, Schopenhauer claims that his 'metaphysics proves itself to be the only one that actually has a common contact with the physical sciences'. He looks to developments in physiology as 'empirical corroborations' that support the 'paradoxical, fundamental truth' of his conception of the will (Schopenhauer 2012 [1836], 325). Yet, in an argument reminiscent of the critique of dualist matter theories in Coleridge's work, he cites the vitalist tendencies in contemporary scientific accounts of matter to argue that scientific method is inadequate for formulating a complete theory of life:

> they want to describe that which is inaccessible and unknown to physics, at which their investigations end, and which their explanations then presuppose as a given, with expressions such as natural forces, vital forces, and formative drives, etc. that is to say nothing more than x, y, z. (325–6)

Against the Cartesian privileging of cognition as the primary model of agency, he supports the reciprocal causality that drives the Kantian system, although he argues that his formulation of the will supplants the Kantian '*thing in itself*', which is 'considered absolutely unknowable', as 'the substratum of all appearances and hence all of nature … that which with we are immediately acquainted and precisely intimate, that which we find in our innermost selves as *will*' (324). Schopenhauer's argument is that:

> this will is fundamentally different from and fully independent of cognition and so can even exist and express itself apart from cognition in all of nature, from the animal on down … this will, the only thing in itself, the only true and real thing, the only original and metaphysical thing in a world where everything else is only appearance … gives to everything, whatever it may be, the power by which it can exist and have effect. (324)

His conception of the will, then, is doubly paradoxical in the context of his philosophical contemporaries. Cognition, when conceived as a secondary phenomenon that accompanies the objectification of the will, is now 'physical, not metaphysical'. Will itself is also physical in its operation, yet it remains the metaphysical anchor of Schopenhauer's system.

In *On the Will in Nature*, Schopenhauer redefines the foundation and development of human life using a chemical metaphor, arguing for the 'total separation of will from cognition.' He claims that this separation, 'this analysis of the I or soul into two heterogeneous components,' will be 'for philosophy what analysis of water has been for chemistry'. For Schopenhauer, 'what is eternal and indestructible in humans, what thus also constitutes the principle of life in humans, is not the soul, but, to use an expression from chemistry, the radical of the soul – and this is will' (339). He argues for the '*will* as *thing in itself*, something completely original', the body as 'mere visibility, objectivation', and cognition as that which is able to observe the operations of that body (40). As part of the process of the objectification of the will, human life is characterized by the development of intellectual activity that leads to a distinct form of self-awareness:

> For an intellect is only known to us from animal nature, hence, as a completely secondary and subordinate principle in the world, a product of latest origin; therefore, it can never have been the condition of the world's existence, nor can an intelligible world precede the sensible world, since it receives its material solely from the latter. An intellect has not produced nature; rather, nature the intellect. However, will—that which fills everything and immediately reveals itself in all, so that everything indicates itself as appearance of will—appears everywhere as that which is original. Therefore all teleological facts can be explained from the will of the being itself in which they are discovered. (2012 [1836], 339)

For Schopenhauer, the will is that which is *original*, but not that which is *originating*. Comparing this passage to Coleridge's account of the "Free Will" in *Aids to Reflection*, Schopenhauer's will emerges as the pinnacle of the natural process, but it retains the quality of immanence rather than aspiring for transcendence.

Like Coleridge in *Theory of Life*, Schopenhauer discards the absolute lines dividing the natural world and the human intellect, depicting instead their development as mutually reciprocal. However, Schopenhauer distinguishes between the will and 'free choice', arguing that 'the former can exist without the latter', and that will is only customarily conflated with free choice, appearing as such when 'illuminated by cognition' (2012 [1836], 341). Cognition, in Schopenhauer's philosophy, has long been mistaken for the cause of the will, and will has long been conflated with voluntary action. In his *Prize Essay on the Freedom of the Will* (1839), Schopenhauer argued for the primacy of will over self-consciousness on philosophical, rather than scientific terms, in an answer to the question, set by the Royal Norwegian Society of Sciences: 'Can the freedom of the will be proven from self-consciousness?' *On the Freedom of the Will* does away completely with the Cartesian conception of an absolutely free, self-determining, stable human self. Arguing that a free will would be one that was entirely without ground, Schopenhauer claims that an absolutely free will, or *liberum arbitrium indifferentiae*, violates the principle of sufficient reason on which all of philosophy rests. Responding to Immanuel Kant's claim that 'freedom is the power to initiate of itself a series of changes', he points to the negativity inherent in what purports to be a positive definition of freedom:

> Thus, although that definition gives the concept of freedom the appearance of being positive, on closer examination its negative nature is again apparent … a free will would be one that was determined by nothing at all. The particular manifestations of such a will (acts of will) would therefore proceed absolutely and quite originally from itself, without being brought about necessarily by antecedent conditions, and thus without being determined by anything according to a rule. (Schopenhauer 1999 [1838], 7–8)

He goes on to demonstrate that such a conception of freedom actually dissolves the distinction in character between human actors, and thus voids them of responsibility: 'if we assume the freedom of the will, it is absolutely impossible to see how virtue and vice and really supposed to originate' (48).

Instead, he theorizes that human beings have both an *individual* character, determined by the necessary and inexplicable motives of the will, and an *acquired* character, which is derived from 'a precise knowledge of his own empirical character', that is, from a knowledge of the activity of the will. The essay concludes that philosophy should 'look for the work of our *freedom* no longer in our individual actions … but in the whole being and essence (*existentia et essential*) of the human being himself' (1999 [1838], 87). This stance is consistent with his arguments about the self-origination of the will that he makes in the 'Reference to Ethics' in *On the Will in Nature*:

> Then it is still to be taken into consideration that freedom and responsibility, these two fundamental pillars of all ethics, can indeed be asserted in words, but certainly cannot be thought without presupposing the aseity of the will … Responsibility has freedom as its condition, but freedom has originality. Since I will as I am, thus I must be as I will. This aseity of the will is the first condition of a seriously thought-out ethics … To connect dependence regarding being and essence with freedom regarding action is a contradiction. (442–3)

The aseity of the will renders the kind positive definition of freedom, which Coleridge strove to articulate throughout his career, a contradiction. Christopher Janaway observes that in Schopenhauer's 'grand plan', his insistence on the aseity of the will is central to the 'culminating idea that the will to life might of its own accord give up its attachment to life, or annul itself within a human being, while there survives a pure consciousness detached from all striving and evacuated of all affect' (Janaway 1999, 158).

What Schopenhauer (2012 [1836], 168) sees as the route to 'genuine value, salvation in the will's turning against itself' is precisely the outcome of a vision of nature that, for Coleridge in 'Human Life', fears would render 'being's being a contradiction'. In an April 1827 notebook entry, Coleridge, musing on the 'the *Living Soul* distinctive of Man', remarks on the relation of human life to the ongoing activity of nonhuman nature:

> Wonderful are the efforts of Nature to reconcile chasm with continuity—to … vault and nevertheless to glide!—tho' in truth the continuity alone properly belongs to Nature, the chasms are the effect of a higher Principle, limiting the duration and regulating the retention of the Products. (Coleridge 1957–2002, 5: §5494)

The second paragraph contains an assertion that echoes the central thesis of *Theory of Life*: 'From the Vermes to the Mammalia ⟨Organic⟩ Nature is in every class and every where tending to Individuality; but Individuality actually commences in Man' (1957–2002, 5: §5494). In looking to the various forms of agency inherent in nonhuman life as continuous with the potential that actualizes itself in the distinct 'Individuality' that distinguishes the human, Coleridge's theory of the self develops out of an opposition to dualism. Yet, he also strives to articulate the 'higher Principle' that separates and distinguishes the human from the nonhuman through not only a difference in degree, but also a difference in kind.

The meeting point—and essential contrast—between Coleridge and Schopenhauer's thinking on the will as that which negotiates between human and nonhuman life as 'continuity' or 'chasm' is evident in their borrowing of a concept from embryology. The *punctum saliens* ('leaping point') was observed by Aristotle as the point at which a visible speck in the egg crosses the line from inanimate to animate existence: 'The point beats and moves as though endowed with life' (Aristotle 1984, 2: 774–993: 883). Sixteenth- and seventeenth-century physiologists such as William Harvey solidified the term, which continued to serve as both experimental descriptor and metaphor in the Romantic era.[6] In *Theory of Life*, Coleridge introduces this metaphor as he begins to discuss his doctrine of polarity, insisting that the true 'Idea of Life' as '*the tendency to individuation* … must subsist wherever the tendency is manifested. The power which comes forth and stirs abroad in the bird, must be latent in the egg'. Life is the outcome of a process that can only be perceived by the fancy, which works in concrete images, as an 'equatorial point' of 'rest, or mere neutralization' between 'counteracting forces'. Yet, Coleridge insists, what appears to be at rest is the manifestation of a transcendent and originary force: 'To the fancy alone it is the null-point, or zero, but to the reason it is the *punctum saliens*, and the power itself in its eminence' (Coleridge 1995 [1816?], 517). Schopenhauer employs the same image to disavow any suggestion of a transcendent power within his philosophy in *On the Will in Nature*:

> Therefore we first of all say that the world is not created with the aid of cognition; hence, it is not created from without, but from within, and then we are concerned to establish the salient point of the world-egg. The physico-theological idea that there must be an intellect that orders and forms nature is just as fundamentally wrong as it is suited to the uneducated understanding. (Schopenhauer 2012 [1836], 355)

Whereas Coleridge ultimately insists on a difference in kind between the nonhuman and the human, which nevertheless emerges from the continuity of nature, for Schopenhauer, the self is an emergent phenomenon that lacks a specific and goal-directed author, and the human individual differs from nonhuman nature only in degree.

A key distinction between Schopenhauer's and Coleridge's thinking on the will lies in their respective ideas about human freedom. To secure the significance of human being, Coleridge began with a positive definition of the agency in matter and the dynamism at work in nature to deduce a positive definition of human individuality, agency, freedom and God. Schopenhauer, fully committed to the latent agency of matter as will, arrived at a negative definition of human freedom. Schopenhauer's theory of the will arrives at novel conceptions of freedom and individuality. As one of his twentieth- and twenty-first-century interpreters points out, Schopenhauer 'has a radically different understanding of what is the core of the human being designated by the word self', arguing that he establishes 'two alternative but complementary conceptions of selfhood: one in which the will forms the core of the human being and one in which the human being achieves selfhood through the cultivation of the intellect' (Zöller 1999, 18–43: 19). Coleridge, however, insists on positive definitions of human freedom and selfhood. Even though his vision of nature as an emergent material process sometimes sits in conflict with his conception of will as an originating force, he continually strives to write them into coexistence.

In Fragment 4 of the extant manuscripts comprising his *Opus Maximum*, Coleridge returns to the problem of defining and explaining life in physical terms. Here, he points to what has become his now oft-repeated argument that pure

[6] For a reading of Coleridge's use of the *punctum saliens* concept, see Peter Cheyne (2020, esp. 285–6), and his Chap. 15 in this book.

physical laws are inadequate when it comes to explaining living organisms, that 'the planetary mechanism, contemplated mechanically and as inanimate, cannot causatively explain life and organisation'. Again invoking the Kantian idea of natural purpose, or teleomechanism, he reverses the question, proposing to:

> examine whether we may not be more successful in reversing the order of priority, namely, whether in life we may not find the conditions of universal organisation, and, again, in universal organisation the conditions and the solution of mechanism. (Coleridge 2002 [*c*.1819–23], 295)

Coleridge finds this 'universal organisation' in the development of the will; however, his thinking on the will poses its own set of problems. In his prolegomena to the *Opus Maximum*, Thomas McFarland argues that Coleridge's conception of the will is consistent 'with the defining emphasis of romanticism' because it 'was a dynamism, not a static thing', working with a definition of romanticism that substitutes 'process and current' for 'fixity and classification'. He also sees the will as a solution to Coleridge's problem with dualism: 'As a dynamism, will also stood outside and above both the pantheism and the "mechanico-corpuscular" conceptions that posed such a threat to Coleridge's interests' (Coleridge 2002 [*c*.1819–23], cxix–x). He continues to argue that Coleridge was invested in an account of human subjectivity that 'must somehow contain a structure grounded not in nature and its transformations but in a realm compatible with the Christian heaven and its invulnerability to change and dissipation' (cxiv).

Such compatibility between the natural and the divine, however, does not come so easily. Thomas Pfau argues that Coleridge's conception of the will as 'a primal and ineffable force that creates a new reality rather than reacting to the one given' is aligned with Gnosticism, and thus carries with it a metaphysical dilemma. According to Pfau, the assertion of a will that has the power to determine itself above and outside of nature asserts a radical—and troubling—freedom, inaugurating the modern individual as 'the perpetrator of a volitional and literally groundless skepticism that denatures and unhinges all cosmological order' (Pfau 2007, 949–1004, 988). Coleridge's conception of the will, then, in its insistence on the power of origination and 'perfect Freedom', sits in an uneasy relationship with his arguments against dualism and for an emergent system of physical nature, establishing one of the key critical challenges to articulating the nature of human agency and the significance of human life as both product of and participant in a vast and dynamic material universe (Coleridge 1993 [1825], 285).

References

Aristotle. 1984. History of animals. In *The complete works of Aristotle: the revised oxford translation*. Tr. D'Arcy Wentworth Thompson. Ed. Jonathan Barnes. Princeton: Princeton University Press.

Blumenbach, Johann Friedrich. 1791. *An essay on generation.* Tr. Alexander Crichton (1792). London: T. Cadell et al.

Cheyne, Peter. 2020. *Coleridge's contemplative philosophy*. Oxford: Oxford University Press.
Coleridge, Samuel Taylor. 1995. *Theory of life* (1816?). In *Shorter works and fragments*. Ed. J. R. de J. Jackson. 2 vols. Princeton: Princeton University Press.
———. 1983. In *Biographia literaria*, ed. James Engell and Walter Jackson Bate, vol. 2. Princeton: Princeton University Press.
———. 2000. *Lectures 1818–1819 on the history of philosophy*. Ed. J. R. de J. Jackson. Princeton: Princeton University Press.
———. 2002. *Opus maximum* (*c*.1819–23). Ed. Thomas McFarland, with the assistance of Nicholas Halmi. Princeton: Princeton University Press.
———. 1993. *Aids to reflection* (1825). Ed. John Beer. Princeton: Princeton University Press.
———. 1956–71. In *Collected letters of Samuel Taylor Coleridge*, ed. Earl Leslie Griggs, vol. 4. Oxford: Oxford University Press.
———. 2001. In *Poetical works, part one: Poems (reading text)*, ed. J.C.C. Mays, vol. 2. Princeton: Princeton University Press.
———. 1957–2002. In *The notebooks of Samuel Taylor Coleridge*, ed. Kathleen Coburn et al., vol. 5. Princeton: Princeton University Press.
Gigante, Denise. 2009. *Life: Organic form and romanticism*. New Haven: Yale University Press.
Janaway, Christopher. 1999. Will and nature. In *The Cambridge companion to Schopenhauer*, ed. Christopher Janaway, 138–170. Cambridge: Cambridge University Press.
———. 2002. *Schopenhauer: A very short introduction*. Oxford: Oxford University Press.
Kant, Immanuel. 2001. *Critique of the power of judgment* (1790). Tr. and ed. Paul and Eric Matthews Guyer. Cambridge: Cambridge University Press.
La Mettrie, Julien Offray de. 1912. *Man a machine* (1747). Tr. Gertrude C. Bussey. Chicago: Open Court.
Mitchell, Robert. 2014. *Experimental life: Vitalism in romantic science and literature*. Baltimore: Johns Hopkins University Press.
Packham, Catherine. 2012. *Eighteenth-century vitalism: Bodies, culture, politics*. New York: Palgrave.
Pfau, Thomas. 2007. The philosophy of shipwreck: Gnosticism, skepticism, and Coleridge's catastrophic modernity. *Modern Language Notes* 122:949–1004.
Ruston, Sharon. 2005. *Shelley and vitality*. New York: Palgrave.
Schopenhauer, Arthur. 2008. *The world as will and presentation* (1818–19). Tr. Richard E. Aquila with David Carus. London: Routledge.
———. 1999. *Prize essay on the freedom of the will* (1838). Tr. Eric F. J. Payne. Ed. Günter Zöller. Cambridge: Cambridge University Press.
———. 2012. On will in nature (1836). In *On the fourfold root of the principle of sufficient reason and other writings*. Tr. and ed. David E. Cartwright, Edward E. Erdmann, and Christopher Janaway. Cambridge: Cambridge University Press.
Zöller, Günter. 1999. Schopenhauer on the self. In *The Cambridge companion to Schopenhauer*, ed. Christopher Janaway, 18–43. Cambridge: Cambridge University Press.

Allison Dushane is professor of English at Angelo State University. Her research and teaching interests include eighteenth and nineteenth century literature, aesthetic theory, science studies, and posthumanism. She is co-editor of the first complete scholarly edition of Erasmus Darwin's *The Botanic Garden* (Routledge 2017) and co-editor of *Bruno Latour and Romanticism* (2024), a special issue of *Essays in Romanticism*.

Chapter 17
Life, Nature, and Race: Blumenbach's Vital Materialism in Coleridge's Later Writings (1818–28)

Kimberley Page-Jones

Abstract Using a chronological structure, this paper aims to understand the connections between Coleridge's *Theory of Life*, Blumenbach's concept of formative force and Coleridge's later writings on humans, animals, and race. Blumenbach's *nisus* appealed to Coleridge for obvious reasons but he resisted the idea of the functional adaptation of individuals to the circumstances of nature. Coleridge's later writings, and especially his contributions to Joseph Henry Green's lectures in 1828, attempted to reconcile a dynamic anthropological conception of life with the Christian interpretation of the origin of mankind. These writings not only refract the debates of the 1820s on biological life and racial theories, they also show us how they were intricately connected. They throw a fascinating light on the paradox of the 'new sciences' (anthropology, biology, chemistry) which, against a corpuscular mechanic understanding of the world, shaped a dynamic perception of nature that could have healed the 'bifurcation of nature' but for the enlightened scientist's overconfidence in man's superiority over nature.

When Seth B. Watson published Samuel Taylor Coleridge's *Hints Towards the Formation of a More Comprehensive Theory of Life* (1848 [written 1816]), he expressed some reservations about Coleridge's use of the concept of nature as seemingly endowed with an agency of its own. Yet, in the poem 'Human Life Contemplated on the Denial of Immortality' (1811–15) written a few years before *Theory of Life* was drafted in 1816, Coleridge articulated the tautological nonsense of a world governed by physical laws only, nature operating 'with restless hands unconsciously' and man, in this godless scheme, being nothing more than the 'Surplus of Nature's dread activity' (Coleridge 2001 [1811–15], 1: 887).

K. Page-Jones (✉)
Faculty of Arts and Humanities, Research Lab HCTI, University of Western Brittany, Brest, Brittany, France
e-mail: kimberley.pagejones@univ-brest.fr

P. Cheyne (ed.), *Matter and Life in Coleridge, Schelling, and Other Dynamical Idealists*, International Archives of the History of Ideas Archives internationales d'histoire des idées 253, https://doi.org/10.1007/978-3-031-78315-9_17

Theory of Life is nonetheless a curious piece of writing as Coleridge, grappling with various laws and theories related to the ideas of life and nature, takes his reader from the mineral world to the vegetal and animal realms, and demonstrates that all matter surrounding us is vital and that each natural order assimilates and perfects the lower formative powers (reproduction, irritability and sensibility). To understand the ambiguity underlying Coleridge's conception of nature and man's relationship to it, I argue in this chapter that the law of life that he defined as the tendency to individuation owes to a certain extent to the theories of physical anthropologist Johann Friedrich Blumenbach.

Writing from Göttingen in May 1799, Coleridge confided to Thomas Poole that 'nothing [could] be conceived more delightful than Blumenbach's lectures, and, in conversation, he [wa]s, indeed, a most interesting man' (Coleridge 1956–71, 1: 494). While studying in Göttingen, Coleridge, as the only British student, had indeed become a favourite of Professor Johann Friedrich Blumenbach, befriending his son who accompanied him during an excursion to the Harz mountains in May 1799. In summer 1800, Coleridge considered translating Blumenbach's *Handbuch der Naturgeschichte* (1779; 1799) into English, and later notebook entries, after he returned to England, are clear evidence of his ongoing interest for Blumenbach's dynamic conception of the human species.[1] A well-established biologist and physical anthropologist, Blumenbach was a prominent 'race scientist', internationally reputed for his craniological research and remarkable collection of skulls in the late eighteenth century. By the time Coleridge travelled to Göttingen, Blumenbach had catalogued 82 skulls detailing their 'varieties' and published the third edition of his thesis *De generis humani varietate native* (1795) which classified mankind into five races according to physical characteristics ranging from skin colour to facial configuration.

Although Blumenbach's influence on Coleridge's *Theory of Life* may seem minor compared to that of Friedrich Schelling or Henrik Steffens, I argue in this chapter that Coleridge and Blumenbach shared a worry about the potential indistinction between man and beast promoted by some late-eighteenth-century naturalists and philosophers. This period saw the emergence and proliferation of writings on the 'in-betweens' of creation—hybrids, domestic animals, wild boys—which blurred the frontier between the animal realm and humankind and paved the way for racial differentiation (Schaub and Sebastiani 2021, 319). Carl Linnaeus's 1735 classification is usually seen as the first step in the eighteenth-century process of racial thinking as he lumped humans with apes and sloths in the 'Anthropomorphia' order of the 'Quadrupeds' class. The changing Enlightenment approach to life and to historical progress, and the increasing importance given to the concepts of refinement and improvement in the redefinition of the civilized man, entailed a shift from a fixed conception of the human species to a dynamic one. Living forms and

[1] Coleridge, 1957–2002, 1: §§1738–41, 1743–8, and 1753 (December 1803) refer to Blumenbach's *Handbuch der Naturgeschichte*. Van Woudenburg, *Coleridge and Cosmopolitan Intellectualism* details Blumenbach's lecture methods, his use of natural objects, and the courses Coleridge attended at Göttingen.

creatures could possibly degenerate, affected by various natural circumstances, thus creating new 'varieties' belonging to the same species. What differentiated man from beast was perfectibility, the human capacity to transcend a natural state to become a moral and civilized being. This regenerative progress implied the domestication of nature by human labour and the spirit of invention.

Racial theories in the wake of Buffon's evolutionary approach were used in various discursive practices which promoted and amplified the idea of racial variety. Although Blumenbach disagreed with some aspects of Buffon's system,[2] the German anthropologist also relied on an evolutionary approach to life and race, emphasizing the porous borders between varieties but nonetheless freezing five human ancestral types into races (Blumenbach 1865, 303). The concept of degeneration is central to Blumenbach's interpretation of humankind: human 'varieties' (Blumenbach's term) gradually degenerated from the primeval Caucasian variety 'in respect of colour, structure, proportion and stature' (Blumenbach 1865, 263). Yet, if Blumenbach was intent upon showing that all human varieties had degenerated from the same stock and as such downplayed differences between human varieties, his research opposed the theories of evolutionary naturalists who classified humans as one species among other animal groups. Consequently, he insisted on an unbridgeable divide between humans and animals, and the superiority of man, 'lord and master of the rest of the animals' (Blumenbach 1865, 183).

Enlightenment anthropology thus effected a paradoxical movement, naturalizing human beings by submitting the body to natural circumstances, yet postulating a human dimension beyond nature thanks to uniquely species-specific rational powers (Schaub and Sebastiani 2021, 377). In the sub-section titled 'Faculties of the mind, which are peculiar to man', Blumenbach defined these faculties as 'gifts in which man so far surpasses the rest of the animals' (Blumenbach 1865, 183). This positioning of man both within and beyond the natural world troubles Coleridge's *Theory of Life* and his overall conception of nature (Modiano 1985, 139).[3] Indeed, most of his essay is concerned with the powers assimilated by the organisms of the

[2] Buffon opposed the static categories and subcategories of Linnaeus's *Systema naturae* (1735). While Linnaeus used the word 'variety', Buffon was the first to systematize the use of the term 'race', which he conceived as 'a family lineage of animals or humans that was by no means permanent or inflexible, but formed a veritable "history" of traits passed down through generations in innumerable different forms' (Jenkins, 'From "Nation" to "Race"', 253). Buffon saw nature and human nature as constantly changing in response to environmental causes (climate, diet, customs). Nonetheless, Buffon, like Blumenbach, argued for a clear distinction between men and brutes, an argument challenged by Lord Monboddo, among others, who relied on Rousseau's *Discours sur l'inégalité parmi les hommes* (1755) to delineate a new line of progress for mankind by suggesting that the orangutan belonged to the human variety (Sebastiani, 'L'orang-outang, l'esclave et l'humain').

[3] The Chapter 'Coleridge's System of *Naturphilosophie*' analyses the influence of *Naturphilosophie* (esp. Steffens and Schelling), rather than of German natural history (Blumenbach and Kant): 'From the *Naturphilosophen* Coleridge derived a dynamic model of … nature that asserted the primacy of ideal powers interacting according to a principle of polarity. He was also influenced by the view, of … Schelling …, that nature was a productive force … activated by a ceaseless movement that stemmed from an opposition of polar powers. But unlike Schelling, Coleridge could not

vegetal and animal orders, sketching an inclusive, interconnected and dynamic network of living forms. Using an upward spiral movement from the mineral to the vegetal then to animal worlds, Coleridge concluded his essay on life with mankind. Here, man occupies a limited and abstract place at the end of his essay, suggesting Coleridge's uneasiness in highlighting natural affinities between man and the 'inferior' kingdoms.

Coleridge's scepticism becomes even clearer in various fragments and marginalia written between 1818 and 1828 that addressed issues of race, varieties, and degeneration, harnessing the theories of Blumenbach, Immanuel Kant, Lorenz Oken, and Steffens and trying to reconcile them with a scriptural perspective. It is, nonetheless, in contributions to Joseph Henry Green's 1825–8 lectures that we grasp the importance of these anthropological debates for Coleridge's conception of man and nature. This chapter will thus read these contributions as a sequel to *Theory of Life*. It will show the importance of Blumenbach's formative force and racial theories for Coleridge's late involvement with the concept of nature. The latter's turn from *Naturphilosophie* and vital materialism and the endorsement of a more conservative and religious approach to life and nature can partly be understood as the result of his speculations on the fate of the human soul if given up to the biological interpretation of life as stated by German anthropologists.

17.1 Vital Materialism in Blumenbach and Coleridge: Power with Purpose

17.1.1 Biology, Aesthetics, and Moral Improvement

The originality of Blumenbach's epigenetic view of biological development is to be understood in the context of the rivalry between pre-formationism and epigeneticism. At first, he adopted Albrecht von Haller's pre-formationist account of life that saw ontogeny as the unfolding of features preformed in the sperm or the egg and created by God. This approach to life guaranteed a 'stable, ongoing process … unfolding that [divine] plan, and God took responsibility for any wayward forms of life that might be considered monstrous' (Gigante 2013, 6). Yet, the Comte de Buffon's 1745 experiments with John Needham—on what they assumed to be the divine and spontaneous generation of microscopic life in boiled chicken broth—and the publication of his theory of generation and individuation in Buffon's *Natural History* (1749–1804) sparked new interest in vital power (Gigante 2013, 16). While defending a theological notion of the generation of life, Buffon[4] also challenged

regard nature as self-subsistent, for this conflicted with the biblical account of the creation of the world' (139).

[4] Born in 1707, Georges Louis Leclerc, Comte de Buffon, studied both botanic and mathematical sciences at Angers in the late 1720s, before settling in Paris in the early 1732. He entered the

Linnaeus's static classification of man by adopting an evolutionary approach that was also theologically motivated. For Buffon, each species connected to a prototype which, 'in becoming a reality', could improve or degenerate depending on natural circumstances:

> The original form subsists entire in each individual; ... yet no two exactly resemble each other ... [or] the model from which they are sprung. This difference, which proves how far Nature is from making any thing absolutely perfect, ... is exactly the same in the human species, in all animals, and in all vegetables; and what is singular, the model of what is handsome and excellent is dispersed through all parts of the earth, and that in each climate there is a portion, thereof, which perpetually degenerates, unless united with another portion taken from a distant country ... (Buffon, 1792, 129).

Buffon's theory of a *moule intérieur* ('inner mould' or 'cast') suggested a unity in diversity, however subject to degeneration. Here, an ideal, 'original form' persisted for each individual of the same species; yet, the theory also explains irreducible differences due to degeneration. What Coleridge would relentlessly resist was the naturalist's argument that the degenerative process was to be assigned solely to natural circumstances. This notwithstanding, Buffon's belief in man's creative ability to improve himself and other species would become a remarkable feature of Coleridge's conception of the individuation process:

> Man can, therefore, not only make every individual in the universe useful to his wants, but, with the aid of time, change, modify, and improve their species; and this is the greatest power he has over Nature.' (Buffon, 1792, 313).

This new understanding of organisms endowed with energies and powers and susceptible to outward stimuli gave rise to a counter, complementing theory that explained how degeneration could be resisted, and therefore maintain or restore expression of the prototype. The moral values of refinement, cultivation, and civilization were to ensure a regenerative process, and this implied that humans, conceived as uniquely endowed with a power over nature, could actively direct their evolution (Schaub and Sebastiani 2021, 411). The importance of this double movement is conspicuous in the troublesome translation of Enlightenment anthropological ideas by late-eighteenth and early-nineteenth-century romantics. Yielding to nature brought a taint of degeneration, a passive submission 'to the mercy of nature or the elements' (Davy [1802] 1839, 319).[5]

Académie des Sciences and in 1739 joined the botanic section of the Académie as intendent of the Jardin du roi. This position enabled him to start working on his *Histoire naturelle,* publishing the first three volumes in 1749, which met with immediate success. He joined the Académie française in 1753, along with other many European academies and, by the time of his death, was the most famous naturalist in Europe. His *Natural History* was published in 36 volumes from 1749 to his death in 1788.

[5] Davy, in 'A Discourse Introductory to a Course of Lectures on Chemistry', defined man in a state of nature as 'a creature of almost pure sensation'. He lauded the benefits of science, providing man with 'powers which may be almost called creative; which have enabled him to modify and change the beings surrounding him' and allowed him 'to interrogate nature with power, not simply as a scholar, passive and seeking only to understand her operations, but rather as a master, active with

17.1.2 The Formative Force Theory

Blumenbach's anthropology owes much to Buffon's *Natural History.* However, Blumenbach objected to Buffon's conception of vitalism as 'a kind of soul superimposed on matter'. (Lenoir 1980, 84). As a consequence, he reworked Buffon's notion of *Kraft* (power) into *Bildungstrieb* (translated as *Nisus* in the English editions), and expounded his conception of vital power in his small treatise *Über den Bildungstrieb* (1781), revised in 1789 and 1791, and translated into English in 1792 by Sir Alexander Crichton. What distinguished Blumenbach's formative drive, within the context of various theories in evolutionary naturalism, was its distinction from other vital forces (contractility, irritability and sensibility), notwithstanding an interplay with these forces:

> the unorganized matter of generation ... takes on ... nisus, which ... continues to act through the whole life of the animal, and ... by it the first form of the animal, or plant is not only determined, but afterwards preserved, and when deranged, is again restored. [It] ... is as distinct from the other qualities of living bodies ... as from the common properties of dead matter: that it is the chief principle of generation, growth, nutrition, and reproduction, and that to distinguish it from all others, it may be denominated the *Formative Nisus (Bildingstrieb, or Nisus formativus)* (Blumenbach 1792, 20).

Through successive refinements of his concept, Blumenbach achieved a reconciliation of the mechanical and of the teleological, and for this his concept was praised by Immanuel Kant (Lenoir 1982, 22). Like Haller, who had left the causes of vital powers unexplained, Blumenbach used the Newtonian *qualitas occulta* argument to concentrate his research efforts on the effects of vital power rather than on its causes. Teleology suggested an interiority of purpose unfolding within the creature itself and, as such, a more perfect level of organization than what could be achieved by mechanism or chemistry. Blumenbach's teleology thus made purpose independent from external influences at the origin of life.

He expounded his theory thanks to experiments on a polyp (a small, tubular, plant-like invertebrate with many fine tentacles at its head, possessing extraordinary powers of regeneration, like the mythical Hydra): observing the regeneration of the cut tentacle, although in a depressed form, he speculated on the existence of a vital power that could restore the injured part to its original form. Grounded in matter, the *nisus,* as a teleological agent, implied a sense of purpose lacking in evolutionary theories but suggested in pre-formationism. Unsurprisingly, the idea of *nisus* captured Coleridge's imagination. Articulated as a force, tendency, drive or effort, Blumenbach's *Trieb* suggested, to a certain extent, the existence of an overarching principle working from within the organism that could ensure a sense of purpose and stability. Coleridge's interest in experiments on polyps is evident in his use of it as an image or metaphor to describe change and alteration:

---

his own instruments'. This empowering view of being was common during the revolutionary decade.

> but my Thoughts, my Pocket-book Thoughts at least, moved like a pregnant Polypus in sprouting Time, clung all over with young Polypi each of which is to be a thing of itself—and every motion out springs a new Twig of Jelly-Life. (Coleridge 1957–2002, 2, §2431).

> Nature is a wary wily long-breathed old Witch, tough-lived as a Turtle and divisible as the Polyp, repullulative in a thousand Snips and Cuttings, integra et in toto! She is sure to get the better of Lady MIND in the long run, and to take her revenge too—transforms our To Day into a Canvass dead-colored to receive the dull featureless Portrait of Yesterday (Coleridge, *Letters,* 5: 496–7).

The two excerpts, both referring to the polyp, were written 20 years apart. They show how Coleridge moved from a perception of natural life as a purposeful process (1805) to one of monstrous, uncontrolled proliferation (1825).

17.1.3 The Bildungstrieb in Coleridge's Theory of Life

Theory of Life holds a pivotal yet temporary position in Coleridge's conception of life and nature. Written hastily in November and December 1816 as an extension or digression to his essay on scrofula drafted for James Gillman,[6] his theory of life-as-individuation was as quickly discarded as unsatisfactory in the following months. Published posthumously in 1848 by Seth Watson, Coleridge's reflections on organic life stepped into the controversy sparked by William Lawrence's attack over John Abernethy's interpretation of John Hunter's theory of life. The debates revolved around the antecedence of life over organization. Lawrence argued for a vital yet purely materialistic interpretation of life, dependent only upon the structure and organization of the body, while Abernethy defended the idea of an independent vital force, pervading matter and setting it in motion. As Lawrence's position smacked of French materialism, Coleridge leaned towards Abernethy's interpretation while nonetheless arguing for a radically different perspective on life. Life indeed was to be comprehended not through the organic body as the latter was only the 'PRODUCT and representant of the power' (Coleridge 1995, 1: 533), but through the dynamic interplay of powers, tendencies, intensities, assimilations and strife binding together the different parts of an organism and creating real or imagined connections between organisms belonging to different realms:

> the metamorphoses of insects, from the egg to the maggot and caterpillar, and from these, through the nympha and aurelia [i.e. chrysalis] into the perfect insect, are but a more individuated and intenser form of a similar transformation of the plant from the seed-leaflets, or cotyledons, through the stalk, the leaves, and the calyx, into the perfect flower, the various colours of which seem made for the reflection of light, as the antecedent grade to the burnished scales and scale-like eyes of the insect (Coleridge 1995, 1: 545).

For Coleridge, the first clear manifestation of the *nisus formativus* working *ad intra* (inwards) is in the fish. In a long description, he details the complex interplay of

[6] Editors' headnote to Coleridge, *Theory of Life* (1995 [1811–16], 1: 485–557: 481).

organic forces (reproduction, irritability, and sensibility) working through ascending and retrograde movements, and its consequences on the organization and structure of the living form:

> the emergence of sensibility in the fish calls forth the opposite pole of reproduction, as a *distinct* power, and causes therefore the irritability to flow, in part, into the power of reproduction. The second result of this ascent is the direction of the organizing power, *ad intra,* with the consequent greater simplicity of the exterior form (Coleridge 1995, 1: 548).

Using the classificatory schemes of Oken and Steffens, Coleridge would regularly sketch new schemes for the organisation of living matter in his notebooks and letters: 'Nature-history on a new scheme of Classification—first, of Powers, each power subdi~~vided~~stinct into its Force & ~~all~~ each Force again according to the *proportion* of its predominance in the Species of animals subsumed' (Coleridge 1957–2002, 4, §4724 [October 1820]).[7]

Like Blumenbach, Coleridge believed that the causes of life had to be assumed but could not be explained. Yet unlike the anthropologist who followed the Newtonian strategy of refraining from inquiring into occult *causes* to focus instead on the *operation* of the mysterious formative drive, Coleridge unambiguously stated the spiritual nature of the 'moving power' of the organism using the analogy with a machine. Its inner principle is hidden to all but its Creator. Not only has he the knowledge of the machine but, Coleridge argued, he 'constitutes it' as he is 'the soul of the work', and his mind 'had pre-existed long before the machine' as well as the organization of the different parts, 'with all their bearings and relations' (Coleridge 1995, 1: 506). The analogy enabled Coleridge to distance himself from Blumenbach's biological interpretation of life, keeping the idea of movement while suggesting the existence of a divine mind empowering the machine. At this stage, Coleridge's interpretation of the purposive principle at the origin of life may have had more in common with Ralph Cudworth's plastic principle, mediating between nature and God, than with Blumenbach's *nisus* that could accommodate a divine intent but Coleridge did not explicitly mention it. In the same way, the *nisus formativus* of the organism could only be considered through its manifestations in the elementary powers (reproduction, irritability, sensibility): 'The mysterious ground, the efficient causes of vitality, and whether different lives differ absolutely or only in degree, He alone can know' (Coleridge 1995, 1: 506).

Schelling's and Steffens's ideas are the most conspicuous in *Theory of Life,* far more than those of Blumenbach. Coleridge indeed brings into play the Schellingian spiral ascent, as analysed by Heather Jackson (Jackson 1982, 848–56).

> this is one proof of the essential vitality of nature, that she does not ascend as links in a suspended chain, but as steps in a ladder; or rather she at one and the same time *ascends* as

[7]Annotating Blumenbach, Coleridge also reflects on other possible classifications which would encompass 'the *qualities* of matter'. See *Marginalia,* 1: 536. 'To elicit the *qualities* of matter, to compel the elementary stuffs to reveal their interior Being, and their finer affinities, is an evident End in the organizing process, without reference to which it would be impossible to reduce a multitude of phaenomena in the lower orders of Organic Nature to any intelligible connection'.

> by a climax, and expands as the concentric circles on the lake from the point to which the stone in its fall had given the first impulse. (Coleridge 1995, 1: 537).

However, Coleridge's later fragments on nature, life, and degeneracy point to Blumenbach's *Bildungstrieb* as a significant subtext in Coleridge's thoughts and writings on life and matter. A note written in May 1810, for instance, alludes to the drive as an 'instinctive spirit of plastic volition' (Coleridge 1957–2002, 3, §3840) allowing animals to adapt to their environment and so accounts for varieties in species. In another note from the same year, he records his wish to 'investigate connection of the Imagination with the Bildungstrieb' (1957–2002, 3, §3744). In German, the imagination is *Einbildungskraft*, and Coleridge could easily consider it to be related to *Bildungstrieb*. Though its causes remained mysterious, Blumenbach's *nisus* as a biological force shed light not just on the evolution of animals and their functional adaptation, but also on the process of degeneration, creating racial varieties. Nonetheless, Coleridge never subscribed to Blumenbach's anthropological theory of the evolution of man. For, if humans were envisaged as a product of mere biological forces, this would imply their falling prey to 'Circumstancial nature' (Coleridge 1995, 2: 1400), just like plants and animals, nature acting as a force superior both to human will and life as 'a tendency to progressive individuation' (1995, 1: 533).

17.2 Life, Degeneration, and Race in German Anthropology

17.2.1 The Formative Force in Racial Theories

Blumenbach saw no reason why man should not be studied as a 'natural product', like other species, and thus 'according to the difference of race, bodily and national peculiarities, &c.' (Blumenbach 1865, 299). Adopting a morphological conception, he assumed that human variety was due to external circumstances such as climate, diet, or mode of life. Central to his thesis however was the idea that varieties differed in degree and not in kind, that all races belonged to one and the same species, and he would relentlessly insist on the connections between varieties and moderate their distinctiveness: 'as all the differences in mankind, however surprising they may be at first glance, seem, upon a nearer inspection, to run into another by unnoticed passages and intermediate shades' (Blumenbach 1865, 303).

The *Bildungstrieb* played a fundamental part in the functional adaptation of the organism or body to its environment. In the third edition of *De Generis Humani Varietate Nativa* (1795), Blumenbach expounded on the issue by clarifying his own conception of the formative force and describing its role in the process of degeneration. Located in the 'genital liquid', the formative force ensured the development of the organism into a form determined by 'certain circumstances of maturation, mixture, place, &c.' (Blumenbach 1865, 194), its preservation and restoration in case of mutilation. Unlike other forces (contractility, irritability, etc.) the formative force was responsible for shaping the matter into form and thus propagating species. Degeneration could thus be explained thanks to the formative force:

> Now the way in which the formative force may sometimes turn aside from its determined direction and plan is principally in three forms. First by the production of monsters; then by hybrid generation … finally, by degeneration into varieties … (Blumenbach 1865, 195).

The mixing of different generations or the actions of external stimuli could deflect the pre-determined direction of the formative force. Varieties multiplied and grew more diverse due to the continuous pressure exercised on organic bodies for protracted periods. The 'deflection' of the *Bildungstrieb* 'from its accustomed path', proposed Blumenbach, was 'the most bountiful source of degeneration, and the mother of varieties properly so called' (Blumenbach 1865, 196).

Lenoir's research work has shown how Blumenbach adapted the pre-formationist Kantian concepts of *Stamm* and *Anlagen* (stock and adaptive capacities) to his epigenetic conception of life (Lenoir 1980, 77–108). Through the impulse of Kant, Blumenbach's formative force, preserving as much as degenerating species, played a crucial part in shaping racial differentiation and establishing environmental factors as causes of degeneration. The functional adaptation of the organism to its habitat modified its structure, absorbed in the energy of the formative force.

For Blumenbach, all varieties of men were degenerations of one original stock that he called 'the Caucasian race'. Although Blumenbach did not rank varieties, he nonetheless used aesthetic criteria to characterize the Caucasian race. Thus in *Beyträge Zur Naturgeschichte* (1806), he described the usefulness of his collection of skulls in determining the 'corporeal characteristics of humanity' and contended that 'according to the European conception of beauty in the countenance and shape of the skull', the Caucasian race was 'the most handsome' (Blumenbach 1865, 303–4). Diverging from Kant, who mainly used skin-colour for differentiation, Blumenbach was the first to rely on skeletal structure and posture for racial classification.

In Blumenbach's system, the biological term 'degeneration' implied no negative judgment on varieties, as his argument was free from moral connotations, and theology was absent from his scientific demonstration, although his classificatory scheme could accommodate a biblical interpretation of the origin of man. Furthermore, he repeatedly opposed polygenetic arguments that justified the racial inferiority of black populations (Blumenbach 1865, 305).

17.2.2 Humanity against Animality in Vital Materialism

On a political level, Blumenbach's theory thus vehemently opposed two ideas which were gaining ground in the late eighteenth century: that mankind was comprised of several species that originated from different stocks, and that apes and humans belonged to the same species. Both contentions carried detrimental political overtones that Coleridge would also address. For Blumenbach, the greatest names of natural history—Linnaeus, Buffon, and Haller—were unambiguous on the issue: man was a distinctive creature with no identifiable evolutionary connections with the ape. Cutting against this tradition, Blumenbach, in the first edition of his thesis, targeted the ideas of Jean-Jacques Rousseau (*Discours sur l'origine et les fondements de l'inégalité parmi les hommes*), Lord Monboddo (*The Origin and Progress of Language*), and Edward Long (*History of Jamaica*). The latter, for instance, used Buffon's theory to justify similitudes between an orangutan's brain and that of a human being: the ape's brain being a 'senseless *icon* of the human', made of the same matter but without the 'thinking principle'. That type of reasoning established a gradation in terms of intellectual faculties and a hierarchy between races. Long thus argued that the orangutan had 'in form a much nearer resemblance to the Negro race, than the latter bear to white men', bringing black populations closer to beasts and further from white men (Long 1774, 371). Quoting both William Shakespeare and Buffon, Long used the argument of matter unendued with reason to call for racial discriminations and the subordination of the 'Negro race'.

Blumenbach countered this argument on the ground that no inference that related mental faculties to bodily structure could be drawn. Man thus differed from the beast both because of his bodily structure and because of his reasoning power. For Blumenbach, man was the only living creature deprived of the natural instincts necessary for the adaption to its habitat. Matter reacting instinctively could not become reasonable or cultivated. This led Blumenbach to a curious socio-biological conception of reason as similar to a 'developing germ, which in the process of time, and by the accession of a social life and other external circumstances' could become increasingly cultivated (Blumenbach 1865, 82). In line with Enlightenment values, man 'born naked and weaponless' could only adapt and progress thanks to society and education. If humans had no natural instincts, animals, on the other hand, were destitute of reason and could never claim the same level of intelligence and dignity: '[God's] highest master-piece far excels all other animals in its noblest part, which is reason' (Blumenbach 1865, 84).

Coleridge would argue for the same clear-cut separation between men and animals in *Theory of Life*. The conclusion of that essay leaps forward to claim that analogy with the animal world cannot illuminate the distinctive qualities of human beings: 'all the analogies to ourselves, which we can discover in the quadrupeds or quadrumani, are of our vices, our follies, and our imperfections' (Coleridge 1995, 1: 550). Despite but also energized by such radical differences, Coleridge persisted in searching for polar unity, taking influence from Schelling and Steffens, among others, and this balanced his sense of human detachment from nature with the opposite tendency to connect:

> As the ideal genius and the originality, in the same proportion must be the resignation to the real world, the sympathy and the inter-communion with Nature. In the conciliating mid-point, or equator, does the Man live, and only by its equal presence in both its poles can that life be manifested! (Coleridge 1995, 1: 551).

In this passage, on man as both high point in *natura* and as standing in 'counterpoint' to the world, nature is conceptualized and as such 'de-biologized'. Coleridge moves towards a transcendental version of nature: 'in the highest realization … that last work [sc. Man], … Nature did not assist as handmaid under the eye of her sovereign Master, who made Man in his own image, by superadding self-consciousness …, and breathed into him a living soul' (Coleridge 1995, 1: 550). The biological process at work to perfect the human body reappears, when Coleridge compares the osseous structure of man to that of the insects, birds and mammalia: 'The whole force of organic power has attained an inward and centripetal direction' (550). Man is not a product of Nature but its revelation.[8] *Theory of Life* stages an unresolved tug-of-war between man-within and man-beyond nature. Yet, Steffens's idealized view of man as a unifier ensured a balance between detachment from and sympathy with Nature; a balance that would shift from late 1818, as Coleridge grew increasingly sceptical of the soundness of *Naturphilosophie*.[9]

17.3 Coleridge on 'Circumstancial Nature' and the Formative Force Theory

The next two sections will examine Coleridge's contributions to J. H. Green's 1824–8 comparative anatomy lectures at the Royal College of Surgeons. They retain aspects from Coleridge's earlier conception of nature and testify to the centrality of Blumenbach's theories of *nisus* and race, and more generally of German anthropology, in Coleridge's revision of life and nature in the 1820s (Sloane 2007, 159, 164).[10]

17.3.1 Coleridge's Contributions to Joseph Henry Green's Lectures

Joseph Henry Green,[11] then Hunterian Professor at the College, delivered 48 lectures on organic and animal life and, although it seems that he had not seen *Theory of Life* at the time of these lectures, the ideas he develops have interesting common-

[8] J. H. Green would develop a similar logic in *Vital Dynamics* to justify man's position at the apex, a 'compendium of Nature' but freed from its external powers (127).

[9] Writing to Green, Coleridge outlined how he was growing distant from Schelling and *Naturphilosophie* as he realized they saw the physical and the ideal as equal opposites, dual aspects of the same reality, whereas Coleridge insisted on the primacy of reason and spirit (1956–71, 4: 874 [30 September 1818]).

[10] Sloan has shown how Green relied on Blumenbach's formative force to explain the nature of life in his opening course. He concludes on the importance of the Royal College of Surgeons 'in shaping the "dynamics"' of the appropriation of German thought by British science in the 1820s and suggests that Coleridge's and Green's ideas provided a framework for Richard Owen 'to construct a means of relating all vertebrate forms in terms of a branching relationship' (165).

[11] On Coleridge's relationship with Green, see Jackson (1982).

alities with Coleridge's conception of life there. Coleridge drafted a series of eight essays which bear watermark dates from 1825–8 and which would have been intended for Green's final lectures delivered in March and April 1828. Green published a revised and condensed version of them in the section 'Recapitulatory Lecture' of his Coleridgean work *Vital Dynamics* (1840); the concluding section made very little use of Coleridge's contributions. Writing to Green on 26 July 1832, sometime after the lectures, Coleridge asked him for his 'two concluding Lectures of the zoological Course, on the Characteristics of Man', expressing his 'wish to look over again the passage on the *Federative* Character of the N. W. Branch of the Japetic Race'. It seems, then, that Green used Coleridge's notes on human and racial characteristics in his lectures at the Royal College of Surgeons yet left them out of *Vital Dynamics*.[12]

Coleridge reread Blumenbach for the preparation of these essays. He had been acquainted with his racial ideas as early as 1798, when he studied at Göttingen, but issues of race, evolution, and degeneration only emerged explicitly in his writings in the mid–1810s. For instance, he longed for a work that would have 'exploded the absurd notion of Pope's *Essay on Man*, Darwin, and all the countless Believers ... of Man's having progressed from an Oran Outang state'. In that letter, to William Wordsworth dated 30 May 1815, he articulated the philosophical need for:

> a general revolution in the modes of developing & disciplining the human mind by the substitution of Life, and Intelligence (considered in it's different powers from the Plant up to that state in which the difference of Degree becomes a new kind (man, self-consciousness) but yet not by essential opposition) for the philosophy of mechanism ... Facts elevated into Theory—Theory into Laws—& Laws into living & intelligent Powers—true Idealism necessarily perfecting itself in Realism & Realism refining itself into Idealism ... (Coleridge 1956–71, 4: 574–5).

The following six of Coleridge's eight essays contributing to Green's 1828 lectures demonstrate the intricate connections in the former's mind between life, nature and race: (a) 'On the *distinguishing* Characters of Man and Mankind'; (b) 'Distinction between 'species' and 'race' and 'variety''; (c) 'Distinguishing Men from Animals'; (d) '"Solitary" and "Gregarious"'; (e) 'The Races of Men'; (f) 'Origins of the Human Race' (Coleridge 1995, 2: 1388–1409).[13] Essays (a), (c), and (d) have an air of *Theory of Life*, reinstating the Steffensian idea of man as a reconciler of all the forms revealed by nature in the lower orders, yet insisting on his power to modify the circumstances of nature, 'creating for himself external influences for his Being

[12] It is yet unclear how much of Coleridge's material Green used in his own lectures. Jackson, 'Coleridge's Collaborator', quotes from two sets of papers to show how Green appropriates Coleridge's ideas and makes them more intelligible for his audience. Green's manuscript notes for his lecture 'The Distinguishing Characteristics of Mankind' (1826) draw heavily from Coleridge's 'The Races of Men' (1995 [1828], 2: 1399–409), which explains Coleridge's views on races and the existence of a 'Historic Race' as an ideal type towards which humanity should aspire. I am indebted to Roma Kail at Victoria University Libraries, University of Toronto, for helping me track Green's lecture notes for further traces of Coleridgean ideas.

[13] Haeger(1974), 'Coleridge's Speculations on Race' is the only comprehensive study of these lecture notes.

and well-being' (Coleridge 1995 ['Distinguishing Men from Animals', 1828], 2: 1391). The turning point in Coleridge's logic and where he explicitly focusses on the question of human racial distinction is also the most interesting and comprehensive of these essays: '"Solitary" and "Gregarious"'.

17.3.2 *Animals Vs Men: The Bifurcation of Nature and Culture*

'"Solitary" and "Gregarious"' best displays Coleridge's enduring interest in man's biological connections with the natural world. Using the same upward spiral movement as he does in *Theory of Life*, he briefly reflects on the way the organisms and animals of the lower orders develop their individuality while remaining connected to the group by an 'invisible copula'. Coleridge then develops the analogy between man as a 'social animal' and 'cluster-animals' (such as mussels and other of the 'lowest sorts of the Mollusca') to suggest the existence of a similar invisible copula —'the needs, affection and associations of domestic Man' binding man to a group (Coleridge 1995, 2: 1392–3). What differentiates the members of human clusters from those of animal clusters is their ability to be 'master of [their] own movements' and, as such, independent:

> the integrity of … each animal, and the co-existence of multitudes … Now under favourable circumstances, the natural increase of an animal, like man, too powerful to have that increase materially impeded by other animals or even by the agencies of nature, & at the same time by the domestic & social instincts, affections & necessities kept together, will present at once the latter [sc. co-existence]—We have only to seek for the former [sc. individual integrity]—the means of retaining his independence notwithstanding the preceding counter-powers—and this can only be found in ⟨his⟩ *reason*/and the necessary result of Reason—a moral law & a scheme of action correspondent thereto which both may be & ought to be one and the same and entire in every individual (Coleridge 1995, 2: 1393–4).

Coleridge here suggests that the *nisus* (ensuring 'natural increase'), acting on men and animals, and ensuring its gradual development into a pre-determined form (an individual), is more powerful, notwithstanding the circumstances, than external powers. Yet, in man, it is reason that will act as the greatest opposing force to preserve his individuality. Man is the only true *'federative'* being as he alone reconciles in himself, and in a more perfect form, two characters—'the solitary & the gregarious'—found in the animal realm but opposing each other. Human sociality and the formation of civil bodies are not the result of blind mechanical forces but the consequences of the 'free movements of [man']s own inward Being' (Coleridge 1995, 2: 1396), warranted by reason. Coleridge was probably here also opposing the Lamarckian idea of spontaneous self-development. Quite surprisingly, Coleridge lingers on the analogy of man with animals in this essay, playfully though safely by avoiding any reference to the mammalia category: 'that the most striking analogies or types of man & human societies are found not in the animals nearest himself, the mammalia for instance; but in far lower classes, in the Birds more than in the mammalia, and in the Insects most strikingly of all' (1995, 2: 1394).

In his thesis, Blumenbach had largely insisted on reason as the main characteristic distinguishing humans from animals. Coleridge took the anthropologist's argument one step further, to insist on what reason can produce, namely, the 'moral law'. At this point of the demonstration, Coleridge polarizes even further man and nature claiming that:

> The natural state of man is the end of his Being & every approximation to it the product of his own energies; the natural state of every Animal is the condition of ~~his~~ its existence ... (1995, 2: 1396).

Unlike animals submitted to the agencies of nature, man is not a mere synonym of his organism. As owner of his own body and in free control of his movements, he is a free and moral agent. In an 1827 notebook entry, Coleridge describes all living creatures below man as 'the joint Product of the inherent and the circumstancial Nature' (Coleridge 1957–2002, 5: §5554 [July 1827]) and simply 'recipients', 'instruments', 'Living Machines' fed by nature. Similarly, in 'The Distinguishing Characteristics of Mankind', he argues that 'for man, there exists no state, which consistently, with either fact of history or propriety of science can be entitled his *natural state*'. While the animal depends on this state for the development of its form and, as such, is a product of nature, man's humanity can only be perfected through reason and in circumstances produced by man: 'these auspicious circumstances would in the most essential points have been the product & the result of the evolving humanity, of human skill[,] virtue & enterprize' (Coleridge 1995, 2: 1414). At the end of the essay, his rejection of the orangutan hypothesis suggests the crucial importance of anthropological debates in his revision of the concept of nature. Constructing a disciplinary barrier between nature and culture, he intellectually separated the study of humanity from biology and thus widened the gap between men and animals:

> we meet with characters which can never be brought into a mere difference in degree ... to acknowledge difference of *kind* ... opposing the terms man & animal, our animal nature & our proper humanity: this too is a part of our science though the concluding part ... we deliver the torch ⟨of science⟩ for a new studium to the moralist, the political historian, or the divine: at once reverencing the boundaries of our own science, and yet sensible that the several sciences are circumscribed and kept distinct from each other not by chasm or interspace but as the ~~colours in the rainbow~~ different organs in the body by distinction of forms or as the light from a many-colored flower by the difference of the reflecting surfaces ... (Coleridge 1995, 2: 1413).

In 'The Races of Men', Coleridge also expanded the idea of human improvement as 'an affair of the Moral Will' that therefore 'belongs to *History* as distinct from Science or Insight a priori' (Coleridge 1995, 2: 1400). Paradoxically though, Blumenbach's theory of formative force, applied to organisms, animals, and racial varieties, would be the anvil on which Coleridge's theory of race would be forged. As we shall see in the final section, Coleridge distanced himself from the German conception of race as degeneration for functional adaptation elaborated by Kant and Blumenbach, and tried to articulate a system that could befit the Christian idea of mankind. However, he relied on Blumenbach's notion of *Bildungstrieb* to account for the moral degeneracy of some races, thus transferring the idea of formative force from the biological to the moral realm.

17.4 Coleridge, Nature, and Race

17.4.1 Reconciling the 'Scientific' Concept of Race with Scriptural Interpretation

As early as June 1819, Coleridge had expressed a deep interest in theories on race and a wish to coalesce anthropological theories and systems with a scriptural interpretation:

> I cherish a hope bordering on a belief, that I can explain the origination and geographical position of Blumenbach's five Races, the Caucasian with the … American and Tartar on one side, and the Malay and Negro on the other, on the Mosaic Triad, Shem Ham and Japeth—nay, what is yet more difficult, reconcile the Kantean diagnostic of *Race* with the Mosaic Documents respecting the *Deluge* (Coleridge 1957–2002, 4: §4548).

This 'hope' may have been sparked by various factors in the late 1810s that created a renewed public interest in the orangutan hypothesis: a Sumatran ape brought to London by the British naturalist Clarke Abel in 1818 and Lawrence's 1818 Royal College of Surgeons lectures on physiology, zoology, and the natural history of man, which Coleridge attended. Published in 1819, Lawrence's *Lectures* dealt with evolution and argued for a zoological study of man, laying it open to charges of atheism and blasphemy, and it was rapidly suppressed after its publication. That book, usually abbreviated to *Natural History of Man*, advanced two main arguments: the 'specific [sc. distinguishing] character of man' (Lawrence 1819, 133), and the 'different races of mankind' and whether they belonged to distinct species or to one and the same. Lawrence dedicated the book to Blumenbach and used his classification of five varieties (Caucasian, Mongolian, Ethiopan, American, and Malay). Interestingly though, he added a moral and intellectual layer on the ground that 'one or two characters [were] not sufficient for determining the race' (Lawrence 1819, 550). The 'characters' of the Caucasian variety, for instance, included skin and hair colour, cranial shape and face features but also encompassed 'Moral feelings and intellectual powers most energetic, and susceptible of the highest development and culture' (Lawrence 1819, 551). Lawrence highlighted the aesthetics of the Georgian skull (by a young Georgian woman captured in the Russo-Turkish War and who died in prison; sent to Blumenbach in 1793), adding immediately after, as if it were a causal argument, that:

> The greatest mental powers have been bestowed on this variety; so that they have discovered nearly all the arts and sciences … These nations have the most intelligent and expressive countenance, and the most beautiful body proportions … The most ancient and most early civilized nations have belonged to this division … (Lawrence 1819, 554).

Lawrence naturally concluded that the original man was Caucasian.

Coleridge's approach to racial issues would be fairly consistent from 1818–28. His essays, especially 'The Races of Men', expanding '"Solitary" and "Gregarious"', explicated his perception of degeneration as a 'separative process', while he contended nonetheless for the possibility of a 're-unitive process'. His thinking bears the marks of the Enlightenment over-confidence in man's superiority over nature. While Coleridge refused a mere biological explanation of degeneration and included, as Lawrence did, moral, intellectual and cultural specificities, his argument had nothing of Lawrence's shortcut logic. Instead, he relied on Blumenbach's five-race system because it had become the dominant model in racial thinking and because the *nisus* theory associated with degeneration enabled him to consider the issue through the perspective of forces, powers, and energies. Coleridge's dynamic system may have inspired Richard Owen, assistant conservator at the Royal College of Surgeons, then founder of the Natural History Museum in London. He attended Green's lectures as associate curator at the Hunterian Museum in 1828 (Rupke 2009, 117).

17.4.2 The Curse of Ham and Moral Depravity

Although Coleridge's deep abhorrence for slavery never wavered, his ideas on race were aligned with the dominant British prejudices inherited from French and German anthropology. His scriptural approach further tainted his perception of degeneration. As argued by Colin Kidd, if the Bible offered no racial interpretation of history, it nevertheless 'possessed a more sinister capacity to encourage the importation of divinely authorised categories of blessed and cursed—and by extension objective moral categories of good and evil—into the reading of ethically neutral differences between races' (Kidd 2006, 21). This was the path chosen by Coleridge in these essays, his hypothesis tested in earlier notebook fragments.[14] He believed that German anthropology could be re-Christianized as Blumenbach's degenerative-yet-teleological theory could accommodate the Noachic account of the origin of man.[15]

Genesis 10 traces the origin of peoples to the three sons of Noah (Shem, Ham and Japheth who themselves had sons after the flood), an account of the origin of human races which had, according to Colin Kidd, a long-lasting influence on seventeenth- and eighteenth-century naturalistic histories of humankind (Kidd 2009, 9). Ham's son, Canaan, was cursed by Noah following the shameful act of Ham who saw the nakedness of his drunk father (Genesis 9:18–25). The line spoken by Noah to Ham, 'Cursed be Canaan; a slave of slaves shall he be to his brothers', served as

[14]Coleridge, 1957–2002, 4: §§4548 (June 1819), 4722 (October 1820), 4856 (1820); 1957–2002, 5: §§5506–7 (May 1827), 5734 (January 1828), 5736 (January 1828), 5742 (January 1828).

[15]And indeed, Kidd underscores the fact that 'the spectre of Genesis haunted the birth of ethnology' (2006, 27).

'justification for Black slavery' (Goldenberg 2005, 1). It was indeed commonly held that the three sons of Noah were the ancestors of the brown race (Shem), the black race (Ham) and the white race (Japheth).

In his notebooks and essays, Coleridge adapted this common belief by adjoining it with Blumenbach's and Kant's readings of racial differentiation, thus reducing their five-race system into a two-race gradation system. In a notebook entry of 1828 (January), Coleridge suggests that the 'germs' of races were not formed by the descendants of Noah's sons but by Shem, Ham, and Japheth themselves, 'during the interval from Adam to Noah' (Coleridge 1957–2002, 5: §5734). His racial system is further reduced to two varieties as he brings together the posterities of Shem and Japheth to form one race with an intensive (−) and an expansive (+) pole. Not limited to biological differences, the 'germs' of races extend to moral and intellectual characteristics. Coleridge forged his 'ideal' being, the '*idealiter*', to which man should strive, out of this Biblical interpretation of human races. Arguing for a synthesis of Shem and Japheth that 'ought to have been the Christian', he interpreted the curse of Ham as having kindled the process of degeneracy. For each race and depending on the proportion of degeneracy due to the quantity of the 'Chamite blood', Coleridge contended that 'two states are predictable, the Minimum & the Maximum of Degeneracy, or vice versa the Maximum and Minimum of Cultivation' (Coleridge 1957–2002, 5: §5506).

His racial system was first elaborated in an 1819 notebook fragment which used the Latin adjectives 'Generosus' and 'Degener' that Coleridge defined as such:

> Generosus = in full possession of the qualities & of the degree, ~~or~~ intensive or extensive, of the qualities by which the Genus was originally formed and characterized.
>
> Degener = the contrary of Generosus, and yet contradistinguished from Bastard, Hybrid, crossed ... Apples *degenerate* where *Grafting* is not in use (Coleridge 1957–2002, 4: §4548)

In this long and intricate note, Coleridge attempts to map the different peoples of the earth according to their migration, settlement in specific places, and intermarriages, and claims to have integrated Blumenbach's biological explanation in his own system that includes 'the effects of intermarriages, and the mighty contribution of Causes from the Moral World'. Coleridge's approach used Blumenbach's biological and teleogical dynamics but gave it a substantial moral and religious twist. Coleridge thus conceived a system that opened a vista of redemption for mankind.

Unlike Lawrence, Coleridge did not position the Caucasian variety at the apex of creation as he believed it was also degenerate; he nonetheless saw it as the least degenerate and as the only variety able to lead the way to further enlightenment and cultivation:

> when we review what man has already accomplished in civilizing the surface of the Earth and indirectly the region of Air, & the climate &c or only compare what this our own country is and is rapidly becoming with swamps & inundated Savannahs of South America in the parts as yet free of man, it was at most a venial hyperbole, to entitle the working of ... historic & Scientific man on circumstancial Nature ... (Coleridge 1995, 2: 1403).

Here, Coleridge again draws on the distinction, a hallmark of his later writings, between the human striving to attain an ideal and surrendering to the contingencies of nature. Reversing the logic of biological varieties, Coleridge argued that 'historic & Scientific' man was equipped with 'Unities' (language, aesthetics, sciences, religion), which he listed in his essay 'The Races of Men' (1995, 2: 1402), and as such could effect a regenerative process to 'raise and ennoble' human nature. The degenerative process was not ineluctable since man, as a free agent, possessed an energy 'in relation to his own Will' that could resist the pressures of nature:

> The separative *Process*—which we are required to bring under the principle of modifiability and the energy that resists the modifying causes ab extra—
>
> In man, we found the latter in the highest energy far beyond all other animals—& yet at the same time the former equally so but in relation to his own Will— (1995, 2: 1399).

Again, Blumenbach's formative force theory is tangible here. Blumenbach had indeed argued that the *nisus* could curb functional adaptation. Yet Coleridge transfers this energy to the will. This moralization of biology, however, led Coleridge to a very different interpretation of racial differentiation:

> it is possible that certain climates are inhabitable without physical degeneracy only by the animalizing changes brought about by moral degeneracy ... and yet a far higher state of moral and intellectual energy in the central Race, with the scientific powers & resources of that far higher state, might enable the Masters of the world to reside unharmed on any part of their Estate (1995, 2: 1404–5).

According to David Haney, the 'central' or '"Historic race" functions for Coleridge as a kind of ideal to which no actual race fully corresponds': this ideal holds for the moral Will, a supra-individual force which makes man a 'free agent'. As he distanced himself from Blumenbach's and Kant's naturalistic interpretation of man's evolution,[16] he nonetheless used Blumenbach's formative force theory to interpret moral depravity as a force diffusing itself in the bodies, minds, habits and customs, gradually encroaching upon the formative principle. In other words, depravity was given by Coleridge a semi-biological explanation:

> Individual depravity *may* become the Fashion of a whole ... People—A Nation of depraved Individuals is a conceivable Phænomenon ... Individual Depravity in a nation of depraved Individuals may so diffuse itself thro'all the acts, faculties, affections, passions and appetites, and take such total possession of the Man, as to include in it's sphere of influence ... the reproductive ... Life itself ... concentred as the Generative and mastering the formative principle exist and act as a *Nature* in the Individual—Frightful Anomaly! When the Individual has become a Nature to the ... infection and decline of the Individuality, till the

[16] See D. Haney (2009) ' Coleridge's 'historic race': Ethical and political otherness' In *'The Races of Men'* (1995 [1828], 2: 1408), Coleridge rejects Kant's theory, and a marginalium, (1980, 1: 539 [*c*.May 1827]) attests to the limits of Blumenbach's theory: 'Without rejecting Bl.'s Pentad of Races but likewise without attributing more to it than the merit of being the most convenient Division hitherto proposed, I am ... unwilling to detach the subject from its historical Staple-ring, the Noachidæ: and would therefore class the Haupt-variatäten, into the Generous, the Degenerous, and the Mixed. If the mythic Curse had not disfraternized the second or middle Brother, ... Mankind would have been, as one Species, so only one *Race*.'

approach to the state of the lowest Orders of animal life ... Frightful perversion! (Coleridge 1957–2002, 5: §5736 [January 1828]).[17]

The note contains many crossed-out words and lines that may suggest a painstaking elaboration of a theory of national character, borrowing from the Enlightenment ideas of Montesquieu, Hume and from the Scriptures, yet also making use of Blumenbach's physical and physiological approach to mankind as species biologically adapting to the environment, to explain national depravity. Coleridge's approach to nations and races, fusing various traditional theories of national character with racial 'sciences', shows that the line between both concepts was thin in the early nineteenth century and that they could easily be confused (Hudson 1996, 256).[18]

17.5 Conclusion

This chapter has traced the presence of Blumenbach's vital materialism in Coleridge's writings spanning the decades of the 1810s and 1820s. Although Coleridge had been acquainted with the work of Blumenbach in the 1790s, he made little mention of it before raising concerns about issues of race and degeneracy while working with Green in the 1820s. *Theory of Life* though testifies to Coleridge's anxiety regarding modern anthropology and its incorporation of mankind into the natural kingdom. This chapter has brought together Coleridge's letters, marginalia and notebook writings from 1815–28 to understand his conservative shift in the late 1810s: his rejection of *Naturphilosophie* and his later approach to mankind steeped in Enlightenment anthropology and a Scriptural interpretation of races. They also show a clear shift in Coleridge's conception of the human-nature relationship as they dwell almost obsessively on the essential difference between animal and man, the latter being constituted not by instinctive volition but by an intelligent and moral will.[19] The interest of Coleridge's 1828 essays lies less in his theorisation of racial differentiation than in his use of anthropological writings to think afresh man's con-

[17] The note was used in a letter to Green, describing an 'aweful ... inference from the Phænomenon of the Races—assuming the rejection of the OO Hypothesis & that of Five Pairs of Adams & Eves', namely, 'that Individual Depravity ... may sink so deep & diffuse itself thro' all the acts, affections, faculties, passions and habits ... to master the formative principle itself ... What a dreadful inversion! A guilty, and a necessitated Individual—an Individual sunk below guilt!' (Coleridge 1956–71, 6: 723 [25 January 1828]).

[18] Hudson traces the seventeenth- and eighteenth-century meanings of 'race' and 'nation' and argues that Hume's essay 'Of National Characters' (1748) shows the importance of 'racial worldview' in the late eighteenth century.

[19] This is also the conclusion of Green in *Vital Dynamics* when he discusses animal instincts. However, he makes no use of Coleridge's demonstration using racial differentiation. Man is given as a homogeneous whole: 'But man's instincts are elevated and ennobled by the moral ends and purposes of his being. He is not destined to be the slave of blind impulses, a vessel purpose-less, unmeant. He is constituted, by his moral and intelligent will, to be the first freed being, the masterwork and the end, of nature' (95).

nection with and distinction from the animal world. As they bring together, in a kind of *bricolage,* various racial and national theories, and biblical interpretations of national varieties, they are key documents for the understanding of the intellectual debates on life, nature, race and nation in the 1820s.

Using both Blumenbach's and Coleridge's writings on life, nature and race, this chapter has also highlighted the paradox of the sciences in the making in the late eighteenth and early nineteenth centuries. Working against a corpuscular mechanic perception, German anthropology participated in the shaping of a more dynamic conception of life, laying the emphasis on mutations, alterations and variations rather than on static forms and dead matter, while nonetheless securing and thus fixing man's place above nature in a 'Supernatural' and historicized realm to ensure his moral improvement (Coleridge 1995, 2: 1405).[20] Racial writings indeed encapsulate this dialectic at play during the Romantic period between stability and change, improvement and degeneration, uniformity and diversity. Coleridge's 'metaphysicalization' of both dynamic matter and racial theories captures this intellectual drama and points at the intricate connections between the conception of man-as-a-biological-force and racial differentiation in the early nineteenth century.

References

Blumenbach, J. F. 1792. *An essay on generation*. Tr. A. Crichton. London: T. Cadell.

———. 1865. *The anthropological treatises of Johann Friedrich Blumenbach and the inaugural dissertation of John Hunter on the varieties of man*. Tr. Thomas Bendyshe. London: Longman, Roberts, & Green.

Buffon, Georges Louis Leclerc. 1792. *Buffon's natural history, containing a theory of the Earth, a general history of man, of the Brute creation, and of vegetables, minerals*, & c. 5 vols. Tr. J. S. Barr. London: H. D. Symonds.

Coleridge, Samuel Taylor. 1956–71. *The collected letters of Samuel Taylor Coleridge*. Ed. E. L. Griggs, 6 vols. Oxford: The Clarendon Press.

———. 1957–2002. *The notebooks of Samuel Taylor Coleridge*. 5 vols. Ed. Kathleen Coburn et al. Princeton: Princeton University Press.

———. 1995. In *Shorter works and fragments*. 2 vols, ed. H. J. Jackson and J. R. J. de Jackson. Jackson.

———. 1980. In *Marginalia*, ed. George Whalley, vol. 1. Princeton: Princeton University Press.

———. 2001. In *Poetical works, part one: Poems (reading text)*. 2 vols, ed. J. C. C. Mays. Princeton: Princeton University Press.

Davy, Humphry. 1839. In *The collected works of Sir Humphry Davy*, ed. John Davy. London: Smith, Elder, and Co.

Gigante, Denise. 2013. *Life: Organic form and romanticism*. New Haven: Yale University Press.

Goldenberg, David M. 2005. *The curse of ham*. Princeton: Princeton University Press.

[20] 'Man can only solve ... Nature by the Supernatural—and let it not be forgotten, that the phænomenon of Races is an historic phænomenon ⟨that the central Race is the Historical⟩ and that even the Races more or less separated (*sloughed*, as it were) by the central Race is part physical, but still in part Historical.'

Green, Joseph Henry. 1840. *Vital dynamics: The Hunterian oration before the Royal College of Surgeons in London, 14th February 1840*. London: William Pickering.

Haeger, J. H. 1974. Coleridge's speculations on race. *Studies in Romanticism* 13 (4): 333–357.

Haney, David P. 2009. Coleridge's 'historic race': Ethical and political otherness. In *Levinas and nineteenth-century literature*, ed. Donald R. Wehrs and David P. Haney, 61–81. Newark: University of Delaware Press.

Hudson, Nicholas. 1996. From 'nation' to 'race': The origin of racial classification in eighteenth-century thought. *Eighteenth-century Studies* 29 (3): 247–264.

Jackson, H. J. 1982. Coleridge's collaborator, Joseph Henry Green. *Studies in Romanticism* 21 (2): 160–179.

Kidd, Colin. 2006. *The forging of races. Race and scripture in the Protestant Atlantic world, 1600–2000*. Cambridge: Cambridge University Press.

———. 2009. *British identities before nationalism. Ethnicity and nationhood in the Atlantic world, 1600–1800*. Cambridge: Cambridge University Press.

Lawrence, William. 1819. *Lectures on physiology, zoology and the natural history of man*. London: J. Callow.

Lenoir, Timothy. 1980. Kant, Blumenbach, and vital materialism in German biology. *Isis* 71 (1): 77–108.

———. 1982. *The strategy of life: Teleology and mechanics in nineteenth-century German biology*. Dordrecht: Springer.

Linnaeus, Carl. 1735. *Systema naturae*. Leiden: Theodorum Haak.

Long, Edward. 1774. *The history of Jamaica. Or, general survey of the antient and modern state of that Island*. London: T. Lowndes.

Modiano, Raimonda. 1985. *Coleridge's system of Naturphilosophie*. Part 4 of *Coleridge and the concept of nature*, 138–206. London: Palgrave Macmillan.

Rupke, Nicolaas Adrianus. 2009. *Richard Owen: Biology without Darwin*. Chicago: University of Chicago Press.

Schaub, Jean-Frédéric, and Silvia Sebastiani. 2021. *Race et histoire dans les sociétés occidentales (XVe–XVIIIe siècle)*. Paris: Albin Michel.

Sloane, Phillip. 2007. Kant and British bioscience. In *Understanding purpose. Kant and the philosophy of biology*, ed. Philippe Huneman, vol. 8. Rochester, New York: University of Rochester Press.

Kimberley Page-Jones is senior lecturer in British studies at the University of Western Brittany. She has coordinated the H2020 DIGITENS project on British, European, and Colonial sociable practices, and has published *Energie et mélancolie: Les entrelacs de l'écriture dans les Notebooks de S.T. Coleridge* (2018); *La fabrique des sociabilités en Europe et dans les colonies. Espaces et identités (XVIIIe–XIXe siècles)* (2023) with Valérie Capdeville.

Chapter 18
Coleridge and Human Individuation: The Passions

Andrea Timár

Abstract For Coleridge, 'the riddle of humanity' is the co-existence in each individual of 'life and mind' and of what he considers our shared 'human nature' and the uniqueness of the person. He kept asking whether life can shape itself into mind, and, if it can, how human nature develops into human uniqueness. The chapter discusses Coleridge's Theory of Life, 'Human Life. On the Denial of Immortality', the fragments 'True Love Illustrated ...Geometrically' (1812) and 'On the Passions' (1828; Shorter Works, 1: 285–7; 2: 1419–53) to argue that passions play a prime role in the passage from life to mind and from human nature to human uniqueness. Re-contextualising Opus Maximum, it also shows that maternal love guides the infant from being a general member of the human species to becoming a unique human individual, a person.

18.1 From Animal to Human

In the first section of *Opus Maximum,* Coleridge examines the relationship between natural and moral necessity, between the performative use of the term 'must' and the descriptive sense, expressing physical necessity. The first belongs to the world of science, the 'is', ('If you threw me out of the window, I must go'), the second to the realm of the 'ought' ('You ought to relieve an afflicted and deserving parent', Coleridge 2002 [1823], 9). He argues that even though it is possible to transgress or disavow moral truths (unlike scientific ones), the principles of reason, including the truths of the moral world, should maintain an equality with, even a superiority to, those of science.

A. Timár (✉)
Department of English Studies, Eötvös Loránd University, Budapest, Hungary
e-mail: timar.andrea@btk.elte.hu

P. Cheyne (ed.), *Matter and Life in Coleridge, Schelling, and Other Dynamical Idealists*, International Archives of the History of Ideas Archives internationales d'histoire des idées 253, https://doi.org/10.1007/978-3-031-78315-9_18

The complex difference between science and morality, and the 'superior dignity' of the moral world then leads to one of Coleridge's many attempts to define the 'human'. On the physiological side, humans are part of the species, humankind, which he defines, after Johann Blumenbach,[1] in stating that 'all the prominent chinned, erect-walking, full-bottomed, tool-making, word-minting bimanous Bipeds of this Planet are of the one and the same Species'[2] (Coleridge 1995 [1828], 2: 1390). On the moral side, the human awareness of the moral law within, appearing in this passage as 'the riddle' of our humanity, reveals the human being as, according to Coleridge able both to acknowledge and transgress the principles of reason. For example, when someone excuses him or herself for not caring for an infirm parent by saying that a young lion does not feed its parent either, the logic is flawed, since the lion, having no moral obligations, cannot transgress. Man, on the other hand, knows of the moral imperative, and still transgresses it. In the first section of *Aids to Reflection,* man is thus distinguished from the animal (ignorant of the principles of reason), from 'the fiend' (who both transgresses and disowns them), and from 'angels' (who know and never transgress them).

The distinction and ranking between scientific and moral truths are part of two larger questions for Coleridge. First, the difference between animals and humans, and second, the difference between humankind (as a species) and the human individual (as a person). Silvia Sebastiani has shown how the Comte de Buffon's *Histoire naturelle* (10 vols, 1749–88) and Linnaeus's *Systema naturae* (1758) represented a major break in placing man within the animal kingdom and outside the biblical framework. Both 'located humankind under the same system of knowledge as the rest of nature, and classified accordingly' (Sebastiani 2021, 67). In *Coleridge's Defence of the Human* (1986), Anya Taylor describes Coleridge's various attempts to preserve the crucial difference between man and beast, enumerating the various factors that he saw threatening human dignity, such as the degrading forces of empiricism, capitalism, slavery, and even drinking. Thomas Pfau (2013, 843), in *Minding the Modern*, discusses Coleridge's endeavour to 'retrieve the human' from nature, alongside 'the extraordinary depth' of Coleridge's idea of the person as opposed to the species. In *A Modern Coleridge: Cultivation, Addiction, Habits* (2015), I focused on Coleridge's idea of opposing the dehumanizing shock of modernity with cultivation, 'the harmonious developement of those qualities and faculties that characterise our *humanity*' (Coleridge 1976 [1829], 42). From his middle years, Coleridge considers 'the science of EDUCATION … the appointed

[1] Blumenbach revised 'Linnaeus's classification of man as a mammal by introducing the further distinction of biped and bimanous, quadrupeds and quadrumanous, separating humanity from the apes (an unclear boundary for Linnaeus)' (Kitson 2007, 103). Coleridge borrows Blumenbach's definition of the human species and its widened distance from the animals.

[2] Although Coleridge's ideas on race were controversial (Taylor 1986; Kitson 2007; Page-Jones 2025 [this volume]), he continued this passage to say, 'the points of difference, by which the several Varieties are interdistinguished … shrink into insignificance when compared with the points of identity or close resemblance' (1995 [1828], 2: 1390). However, Coleridge's ideas on race are beyond the scope of this paper.

PROTOPLAST of true *humanity*' (1969 [1808–9], 1: 494, italics added). In *On the Constitution of the Church and State* (1829), he famously defines 'NATIONAL EDUCATION' as

> the *nisus formativus* of the body politic,[3] the shaping and informing spirit, which *educing*, i.e. eliciting, the latent *man* in all natives of the soil, *trains them up* to citizens of the country, free subjects of the realm. (1976 [1829], 48)

In Coleridge's ideal scheme as presented in *The Friend* and *Church and State*, national education draws humanity from its dormant state, to guide the individual from mere membership of humankind to becoming a person. In this scheme, the educator would be a male parson (Coleridge's Christianized version of the Kantian person), embodying and exemplifying God's love.

I suggest that Coleridge's onus on education can be explained by the fine boundary separating humanity from animality. I shall further elaborate on Coleridge's idea of the human in the context of what he considers 'the riddle of humanity'; not only the co-existence in each individual of life and mind but also that of human nature and human uniqueness. By human nature, I refer, following Coleridge's fragment 'True Love Illustrated Geometrically' (Coleridge 1995 [1806], 1: 184–7), to humans' shared inherent nature, regardless of education. By human uniqueness, on the other hand, I mean everything that makes one a human individual, a person, rather than a mere member of our species.[4] Secondly, I shall address the questions, which preoccupied Coleridge, of whether life can shape itself into mind, and, if it can, how human nature develops into human uniqueness. I shall discuss the fragments 'True Love Illustrated ... Geometrically' and 'On the Passions' (1995 [1812; 1828] 1: 285–7; 2: 1419–53), to argue that passions play a prime role in the passage from life to mind, as well as from human nature to human uniqueness. Finally, I will show that rather than the institution of national education, traditionally then led by a (male) parson, it is, actually, maternal love—a passion understood as both *care* and *caress*—that elicits humanity, guiding the infant from being a general member of the human species to becoming a unique human individual, a person.

One of the most significant differences Coleridge finds between the animal and the human is that humans also have *active* agency whereas animals are naturally passive. Humans share with animals the property of 'Sense', that is, 'whatever is passive in our being', the 'sensations, and impressions', but are distinguished from them by their active faculties: 'the faculty of thinking and forming *judgements* on the notices furnished by the Sense', as well as 'Practical Reason, ... the power by which we become possessed of Principles (the eternal verities of Plato and Descartes)

[3] The term 'nisus formativus' is Coleridge's own Latin translation of Blumenbach's *Bildungstrieb* (formative drive), which Coleridge renders as the 'vis plastic, or vis vitae formativus' in the 1818 (rifacimento) *Friend* (1969 [1808–9], 1: 493 n.). Blumenbach's *Bildungstrieb* informs Coleridge's *Theory of Life*, too, which, as we shall see, describes life as an active, dynamic process leading from inorganic to organic nature, from the simplest to the most complex animals, and commencing of a 'new series' to man.

[4] The distinction is also often made in social psychology, albeit with different meanings (e.g. Haslam 2006).

and of Ideas, (N.B. not images) as the ideas of a point, a line, a circle in Mathematics; and of Justice, Holiness, Free-Will, &c. in Morals' (1969 [1809], 2: 104). As Coleridge also argues in *Opus Maximum*:

> We have said that Man hath from Birth that which is common with the Animal, and that which is especially human. With the Beasts of the Field it possesses the senses, and the sensations, and the desires of self-preservation, and the impulse to pleasure from the pain of its absence. Beyond the beasts, yea, and above the nature of which they are the inmates, man possesses love, and Faith, and the sense of the permanent. (2002 [1823], 122)

Many commentators (e.g. Pfau 2013, 504–90) have shown that humans, as opposed to animals, partake of Coleridge's Christian-Kantian idea of 'Personëity'. This is what Coleridge would call the 'moral truth' about man, as distinct from the 'scientific truth'. Meanwhile, it is still difficult to see exactly where humans come to differ from animals. Whereas previous scholars have tended to emphasize those features that most clearly distinguish man from beast, in what follows, I shall concentrate on the smallest possible differences between animals and humans. To show how Coleridge attempts to resolve the riddle of the 'transfer' from life to mind, I shall emphasize that although it may be a small step on the 'ladder', 'degree' nonetheless leaps into 'kind'.

Rejecting the orang-utan hypothesis of human descent from apes (1995 [1828], 1: 1406), Coleridge spoke instead in vitalistic terms, refusing to derive mind from life. Nonetheless, he sometimes speaks about a steady 'rise', a 'long Ascent', and of some material and animal means and conditions 'preparing for' the manifestation of free will (1993 [1825], 98). Still, such *avant la lettre* quasi-Darwinian rhetoric notwithstanding, in *Theory of Life* Coleridge sustainedly emphasizes that human individuality initiates a new series in the gradual process of individuation, which is life itself.[5]

> In the lowest forms of the vegetable and animal world we perceive a totality dawning into *individuation,* while in man, as the highest of the class, the individuality is not only perfect in its corporeal sense, but *begins a new series,* beyond the appropriate limits of physiology'. (1995 [1816?/1819?], 1: 516, second italics added)

Indeed, humans possess individuality not only in the physical or biological sense, but also beyond it: they possess *both* 'Human Nature, that which in all men is the same .... wholly unindividualized, and characterless' *and* the '*characteristic* Identity of ... individual Beings', a 'particular Identity' (Coleridge 1995 [1811], 1: 286).

In the poem 'Human Life ⟨contemplated⟩ on the Denial of Immortality' (1811–15), which J. C. C. Mays elucidates at the end of this book, Coleridge explores the consequences of the levelling of human nature with human uniqueness, and the negation of this latter. The poem imagines the meaning of spirit reduced to

[5] 'By Life I everywhere mean the true Idea of Life, or that most general form under which Life manifests itself to us, which includes all its other forms. This I have stated to be the *tendency to individuation*, and the degrees or intensities of Life to consist in the progressive realization of this tendency' (Coleridge 1995 [1816], 1: 519).

breath, and the lack of higher purpose, which renders man similar to animals, a 'drone-hive strange of phantom purposes!', to some 'Blank accident',[6] the mere 'Surplus of Nature's dread activity' (2001 [1811–15], 1: 887). In the fragment of an 1828 lecture, 'Distinguishing Man from Animals', Coleridge raises the example of the beehive as something that has often been evoked by poets and philosophers to illustrate the similarity between the social organization of humankind and organization in the animal kingdom. According to Coleridge, the behaviour of insects, despite their being a far lower class than mammals, presents a livelier resemblance to the human race; in particular, he mentions their living in a 'federative' monarchy. However, whereas 'the corpus politicum' of the bees is produced by blind instinct, that of humans, argues Coleridge, results from the movements of their 'own inward Being' (1995 [1828], 2: 1396). This also helps him, implicitly, to desynonymize 'surplus' and 'addition': rather than being a mere 'Surplus of Nature's dread activity', humanity brings, as he puts it in *Opus Maximum,* (2002 [1823], 11), the 'addition' of 'goodness' (i.e. morality) to 'truth' (science, physiology).

Hence, we encounter three interrelated problems: (1) the passage from passivity (in animals) to activity (in humans) through some shared, common features; (2) the transfer from life to mind, which are logically distinct categories but in nature form a continuum (Coleridge 1995 [1828], 2: 1440), and (3) the development of the members of humankind, starting a new series, into unique human individuals or persons. For even though there is a *logical* break, a chasm separating life and mind, nature itself produces 'distinctions without Breach of continuity' (1995 [1828], 2: 1440). Indeed, what Coleridge finds most difficult, and what he tries to describe in the tiniest detail, is 'continuity', that is, the transfer from life to mind in the passage from passivity to activity. Differently put, in his *Theory of Life*, while he is 'Convinced ... by the wide chasm between man and the noblest animals of the brute creation', he still keeps pondering 'whether (or no) Life is capable of perfecting itself to Mind' (1995 [1816?/1819?], 2: 1425).

18.2 'On the Passions'

The disavowal of humanity is the theme of the poem 'Human Life ⟨contemplated⟩ on the Denial of Immortality', describing the human self-reduction to a copy of a copy, 'Image of Image', preferring a doom of meaningless existence without hope or the consciousness of some ultimate transcendence. In this state of psychological limbo, the human 'being's being a contradiction', life could be unbearable. As Coleridge movingly puts it:

> Why rejoices
> Thy heart with hollow joy for hollow good?
> Why cowl thy face beneath the mourner's hood?

[6] The term 'accident' uncannily anticipates 'mutation' in later theories of evolution.

> Why waste thy sighs, and thy lamenting voices,
> Image of Image, Ghost of Ghostly Elf
> (2001 [1811–15], 1: 887)

The quotation indicates that the question, the riddle of the co-existence of life and mind, animal and human, species and person *within* humanity has high emotional stakes. These stakes concern not only Coleridge's subjective states of mind, but also, and more importantly for the present purpose, his theory of sensations and emotions, of passions in general. For where the human approximates most the animal world is in the realm of sensibility, mediating between body and mind. In *Theory of Life* (1816?/1819?), he argues that:

> the constituent forces of life in the human living body are—first, the power of length, or REPRODUCTION [as with plants and animals]; second, the power of surface (that is, length and breadth), or IRRITABILITY [with animals, starting with the insects]; third, the power of depth, or SENSIBILITY [with the higher animals]. (1995 [1816?/1819?], 1: 557)[7]

Significantly, in 'On the Passions' (1828), Coleridge sets a corrective to the Cartesian tradition in which passions are impressions made on the mind and the brain by external agents (1995 [1828], 2: 1422). The Cartesian approach, according to Coleridge, failed to distinguish the passive sensibility of animals from the human passions, which are passive *and* active.[8] His main question thus concerns whether, and how, sensibility (shared with animals) transforms into something uniquely human.

'Life', Coleridge writes, 'has an ascension towards Mind and culminates in a *form* of Mind'. He also calls this ascension a 'transit from Sensibility' to 'the *form* of Mind.' A '*form* of Mind', however, is not yet human mind. Instead, it is 'sensual understanding', which is the highest level that life can achieve without mind, by the full development of the function of sensibility (1995 [1828], 2: 1427). This sensual understanding is also called 'Animal Mind'[9] and already partakes of the character of the passions without actually being a human passion. In fact, passions already represent something *eminently human*, offering a link between, on the one hand, what Coleridge calls appetites and impetites[10] (those forms of sensibility that can shape

[7] Meanwhile, typically opposing all static view of nature and man, he equally emphasizes that 'Life itself is not a *thing* … but an *act* and *process*' (Coleridge 1995 [1816?/1819?], 1: 557).

[8] As Ewan Jones assesses the originality of the fragment compared to Descartes's 'On the Passions of the Soul': 'The primary innovation of "On the Passions" is its specification of affect as a means of grasping this individuating impulse. Feeling … offers a counterpart to material changes [in] the organism. Yet it is also [… how] organic matter … refines itself into increasingly complex wholes. Reactive to, yet also constitutive of, the material realm, the theory of affect developed in "On the Passions" therefore represents a thoroughgoing reconceptualization of the Cartesian model, where passionate states are alternately (never concurrently) passive or active' (Jones 2014, 90).

[9] The idea of the animal mind as a part of the human being's 'Animal I' is outlined by Suzanne Webster (2017, 300–1), summarizing what humans share with animals in Coleridge's thinking (i.e. the reproductive, the instinctive and the sensible organs, and how sensibility is connected to sensible understanding, which is the first manifestation of understanding, though still in its animal form).

[10] 'This Third Power (Impetite) flowers into Sensual understanding is also called Animal Mind' The 'highest in the Class of Impetites is the articulatory Link that connects the Impetites with the

into sensual understanding or the 'Form of Mind'), and, on the other, uniquely human, always subjective affections (1995 [1828], 2: 1427–9; 1439–40). Indeed, passions mark the beginning of the 'new series' Coleridge mentions in *Theory of Life*.

> By the passions … we mean—A state of emotion, which tho' … it may have its predisposing Cause in the Body, and its occasion in external Incidents or Appearances, is yet not immediately produced by … the incidents themselves, but by the Person's Thoughts and Reflections … concerning them. Or more briefly, A Passion is a state of Emotion, … having its immediate Cause not in Things, but in our Thoughts of the Things. (1995 [1828], 2: 1423)

Passions presuppose thoughts and are predicated on the capacity of reflection. Passions cannot be 'fully explained otherwise than in connection with Thought—but only that in the first instance the modifying influence of the mind will be taken for granted *generally* (1443). Hence, in humans, these are the passions that connect body and mind, because they comprise both emotions induced by external impressions and internal thoughts or reflections.[11] This, however, equally implies that even though Coleridge's fragment 'On the Passions' sets out to determine, among other things, 'whether (or no) Life is capable of perfecting itself to Mind' (1425), the riddle, the 'transfer' from life to mind is not resolved. Coleridge's thinking, it appears, is trapped in circular logic, the logic of always already human.

Still, it is in his dealings with the passions that we find the most clues concerning human individuation, that is, the passage from the human as species to the human as person. This is especially telling as, despite the recurrence of the Kantian vocabulary of personhood, moral law, and reason, Coleridge is far from staking our humanity solely on the dignity of the Kantian person. Instead, he is deeply engaged with the idea of love, and with the specificities of human love as partaking of God's love. For example, he often opposes active love, which he importantly calls a passion, to passive sensibility, shared with animals. In his section 'On Sensibility', in *Aids to*

Passions', and can 'shape … itself into a Passion' (1995 [1828], 2: 1440) On appetites and impetites see also Jones (Jones 2014, 91): 'appetite … implies a passive relation to fundamental needs. The simple appetites … blend imperceptibly into the second experiential mode, whose instances correspond to the same respective bodily organs (stomach and chest). Coleridge terms this second category … the "impetites" …. from the Latin '*impeto*, "to rush upon, assail, assault". They are … still to some degree spontaneous, yet also imply a certain intentionality.'

I add that Coleridge, also specifies lower impetites (e.g. incidental impulses of rage and fear) and, importantly, a highest form of impetite, called the 'Desiderium', a self-impulsive sense of insufficiency also found at the origin of love (1995 [1828], 2: 1441 n. 2). This highest form of impetite can shape itself into sensual understanding, or 'Animal Mind', forming the 'articulatory link' between the impetites and the passions to provide the continuity so much sought for by Coleridge (1995 [1828], 2: 1427; 1439–40).

[11] 'In late works … including … "On the Passions", Coleridge works toward a physiological psychology that gives primacy to mind and makes the body its expression. "What is the Body", as he had written in a letter many years earlier, "but the fixture of the mind' [1956–71 (*Letters*), 3: 305] …. This primacy of mind, exerted by … the "Vital Power" that shapes and informs it, marks the absolute difference between human beings and the rest of the "Animal world" [1995 (1828), 2: 1442, 1390–1]'.

Reflection, he calls Buffon's description of the awakening of love in adolescence a 'profanation of the word Love', because the kind of sensual experience Buffon describes in *De l'homme* (an experience Coleridge equally sees represented in the sentimental novels of Laurence Sterne)[12] actually belongs to the sensibility proper to human *nature* (Coleridge, [1824] 1993, 61–2). Buffon's description of the stirrings of adolescent passion lacks, therefore, the activity characterising the kind of love that is formative of our human uniqueness. In this sense, Coleridge follows Kant's arguments against (passive) inclinations, as well as Kant's claim that duty must prevail even over sympathetic benevolence.[13] However, as opposed to Kant, he equally emphasises the eminent role love plays in our transformation from being the member of the human species into becoming uniquely human individuals.

In 'True Love Illustrated Geometrically', he explicitly differentiates between human nature as characterless, unindividualized, and common to all mankind, and the particular identity of the person as the whole of his human being. It is this unique, personal latter that, rising above the brute, yearns to be united with another 'Whole Being' in love (Coleridge 1995 [1811], 1: 285–7). That is, rather than aim solely for reproduction, following their sexual nature, ensuring the survival of the species, human beings yearn to unite in love with their entire being, through all their features, the texture of their body, their moral affections, and their understanding too.[14] The active yearning to unite with another person in love is thus different from Kantian inclination, not only in degree, but also in kind. Meanwhile, it also partakes of something entirely foreign to Kant's idea of duty and autonomy: love as yearning contains the element of bending, indeed, a movement of active (rather than passive) inclination towards the other, an inclination that is both sensual and spiritual. And if only two people, whose qualities are opposite and yet correspondent, can constitute perfection in unity, then the person in love is neither independent nor autonomous: 'Love is the yearning of the whole Being to be united with some one other Being … so that each may perfect the other' (Coleridge 1995 [1811], 1: 285). As Coleridge's geometrical illustration of love, the triangle, also testifies, the human person is not erect, but *leaning towards* another in love:

[12] Of course, sensibility has important political stakes too, as in the context of the French Revolution (Timár 2015, 7).

[13] On the difference between sympathy and love in Coleridge, see also: Timár (2015).

[14] 'Now the human Being rises above the brute in exact proportion as those sexual *opposites in correspondency* instead of being confined to the immediate purposes of the Species extend thro' the whole Being, thro' all the features, and expressions, & even in the very texture of the Body, thro' understanding, & thro' the moral Affections, all those having a common basis of Identity, but in each so modified, that in the man they shall be masculine, in the woman feminine—i.e. Opposites & yet correspondencies.—This being premised, we say that true Love is the yearning of the whole Being to be united with some one other Being … so that each may perfect the other.' (Coleridge 1995 [1811], 1: 285)

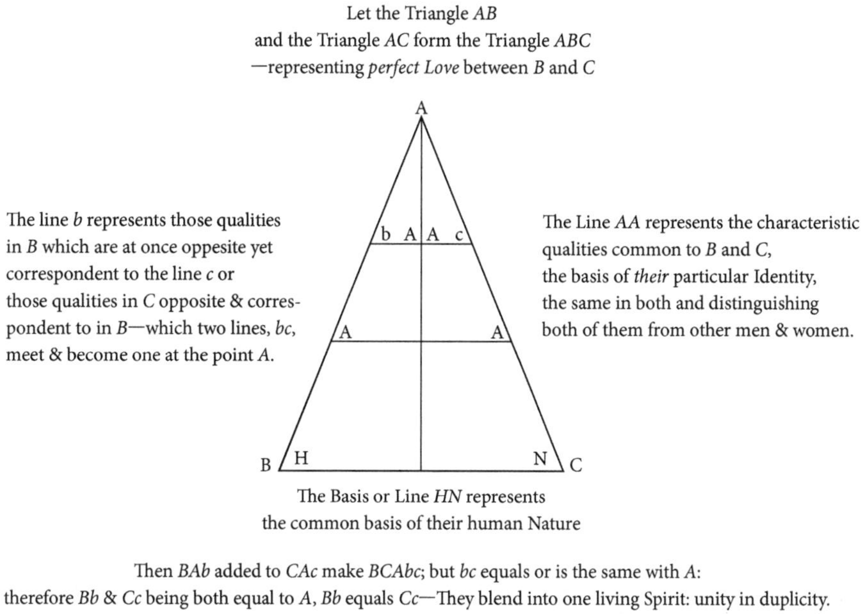

DUPLICITER UNUM, or Love's compound Triangles

Let the Triangle AB and the Triangle AC form the Triangle ABC—representing *perfect Love* between B and C.
The Basis or Line H.N. represents the common Basis of' their human Nature.
The Line AA represents the characteristic qualities common to B and C. The basis of *their* particular Identity, the same in both and distinguishing both of them from other men & women.' (1995 [1812], 1: 287)

The base of the triangle represents the common human nature of the lovers, while the two sides of the triangle, leaning towards each other, represent their individuality, their uniqueness. The stakes of this argument are especially high, because it is, as we will see, our capacity to both give and receive love that grounds our humanity, indeed, love is the very possibility condition of humanity. How can and how does love transform the characterless, unindividualized member of the human species into a unique individual, a human person, primarily defined as, precisely, being capable of love? And what becomes of 'the erect walking', 'bimanous Biped', which is not only Coleridge's biological definition of the human (as distinguished from the animal), but also the foundation of Western philosophy, then?

18.3 The Mother's Care(ss)

In *Inclinations: A Critique of Rectitude*, Adriana Cavarero shows that Western thinking is replete with the figure of the *homo erectus*. Humans are humans and not animals because they stand erect, their hands are free, and their eyes look straight ahead. Taking the example of Kant, she argues that humans in the West have always been represented as self-sufficient, and autonomous, and the (Kantian) person is not only singular but also single, giving laws for self-conduct independently of others. In this view, whenever the individual becomes prone to inclinations (physically or psychologically), duties are abandoned with the respect for the law within, which is supposed to make one human.

In the chapter 'Kant and the Newborn', Cavarero describes Kant's annoyance with mothers and children. She writes:

> They bother him because they are both borderline figures between the animal and human world: mothers beçause they pamper and care for human pups, and as such demonstrate an inclination that resembles the females of other species; children because, in essence, they still resemble animals.' (Cavarero 2016, 27–8)

According to Kant, the first cry of the newborn already signals striving for independence and autonomy and pain at being dependent on the mother, rather than the need for response, care, and nursing (Cavarero 2016, 27–8). Besides Kant, philosophers as different as Rousseau and Buffon see the mother–child relationship from a similar perspective: the mother reduced to a function, a means to an end; the child desiring independence from the mother. In Buffon, the mother is basically a womb that protects and restrains the child, who cannot wait to be born and set him or herself free.[15] As for Rousseau, although he is a revolutionary in recommending mothers breast-feed their own children, rather than employ wet nurses, what lurks behind his advice is not the acknowledgment of the prime emotional significance of mothers *per se*, but rather the utility of strengthening the intra-familial and, therefore, the social bond.[16]

In England, maternal breastfeeding had already become fashionable by the 1750s among the middle and upper classes; according to Susan Greenfield and Carol Barash, the image of the tender, full-time mother was invented in the eighteenth century.[17] Coleridge's philosophy of motherhood should thus be weighed against this background. But while his position can be considered as the continuation of an eighteenth-century ideology of domesticity, elevating the intimate bond between

---

[15] 'Hardly has the infant left the womb of its mother, and enjoyed the liberty of extending its limbs, when it is again put into a more cruel confinement. The head of the helpless infant is fixed to one position; its arms and legs put in strict bondage, and it is laced with bandages so strait as not to be able to move a single joint' (Buffon [1707–88] 1791, 3: 346).

[16] Wet nursing was a widespread custom in eighteenth century France (Brace 2007, 361–83).

[17] Yet, until the second half of the nineteenth century, English common law granted the father sole custody of children after divorce and sole right to determine their guardianship in the event of his death (Barash and Greenfield 2015, 1).

infant and mother, his foregrounding of the role of the mother in the development of humanity was also a philosophical position, to be compared with thinkers such as Kant and Rousseau. Conspicuously, for Coleridge, the mother is the prime embodiment of the love upon which our humanity is predicated, or, more precisely, maternal love is the primary condition in the process of *becoming* human, of the process of individuation of humans from species member to person.

Coleridge praises the child's earliest bond with the mother until he or she 'leaves the gentle teachings of his first home' to engage in the 'austere discipline of the understanding' with the help of male educators ([1823] 2002, 136). As I have argued elsewhere (Timár 2015, 125–40), for Coleridge, the mother serves as the initial medium between God and the child, her pious face being the first indication of God's presence: 'That which the mother is to her child, a someone unseen and yet ever present, is to all' ([1823] 2002, 126). An answer to the enigma of the child's existence, it is the mother's love that guides the child across the bridge between species and person.

In what follows, I shall suggest that Coleridge's understanding of human individuality can also be placed in the context Adriana Cavarero's philosophy of inclination, which is part of a recent tradition of feminist philosophy that stakes human uniqueness on the concept of inter-relationality. The importance of the I–Thou relationship for Coleridge, anticipating a core aspect in Martin Buber's theology and Emmanuel Levinas's phenomenology, have been discussed by David Haney, Thomas Pfau, and Tom Marshall, however, the singularly high stakes Coleridge places on motherly love as the passionate response of care and caress to the infant's vulnerability has not been emphasized enough. For Coleridge, human uniqueness emerges from the newborn's asymmetrical relationship with his mother, entailing a helpless dependency on her care and caress. Although the significance of the mother in Coleridge's scheme of education has been also touched upon by critics, including myself, the focus has generally been on how Coleridge anticipates later psychological findings (e.g. the editorial notes to *Opus Maximum* 120–32; Ruderman 2016, 86–90). What has remained unelaborated is how Coleridge grounds humanity in maternal love as a symbolic embodiment of God's love that subverts Western philosophical tradition. Through relegating Platonic 'sight',[18] Coleridge recuperates the

[18] 'Aristotle was challenging the dominant prejudice of his time. The Platonic doctrine of the Academy, as noted, held that sight was the highest sense because it was deemed the most distant and mediated; and hence the most theoretical, holding things at bay, mastering meaning from above. Touch, by contrast, was judged the lowest sense because ostensibly immediate and thus subject to pressures from the material world. Against this, Aristotle made … radical counter-claim' (Kearney 2021, 33). In *De anima*, cited by Coleridge as the forerunner of associationism in *Biographia Literaria*, Aristotle discusses touch, of which the flesh is not the organ, but the medium (Aristotle 1984 [350 BC] 2.11). Also, at 2.9, Aristotle explains human intelligence by the perfection of the sense of touch: 'It seems that there is an analogy between smell and taste, and that the species of tastes run parallel to those of smells—the only difference being that our sense of taste is more discriminating than our sense of smell, because the former is a modification of touch, which reaches in man the maximum of discriminative accuracy. *While in respect of all the other senses we fall below many species of animals, in respect of touch we far excel all other species in exact-*

Aristotelian emphasis on touch and anticipates twentieth-century feminist philosophies of care and caress.[19] For when Coleridge predicates the infant's humanity on the love of the mother and writes that 'Ere yet a conscious self exists, the love begins; and the first love is love of … another' ([1823] 2002, 121), he not only evokes the love of God radiating from the mother's face, but articulates the yearning for the 'touch' and 'caress' of the mother, which, reciprocated and responded, becomes constitutive of the child's self: 'I am not here, touch me, Mother, that I may be here!' (132). From the beginning of the love relationship between mother and child, touch grounds our humanity, and paradoxically precedes all knowledge and self-consciousness, as the child 'knows not what the Mother is, but still less does it know what itself is' (121).

The child feels God's love reflected in the mother's eyes, and embodied in their close, immediate relationship of proximity. The infant clings to the mother's body and is answered by her tender kisses and embrace. The mother's passionate response to the child's uniquely human condition of nakedness, vulnerability, exposure, and utter helplessness thus gives birth to the humanity of the child, as if it was a second birth. 'The first dawnings of [the infant's] humanity will break forth in the Eye that connects the Mother's face with the warmth of the mother's bosom' (Coleridge [1823] 2002, 121). The humanity of the child awakens in mimesis, in imitation of the mother radiating humanity responding with love and protection to the demands of the child. 'In vain would the conditions, the possibility, of the human have been inlaid, were not the human, in its full development, already there to meet and protect it' (120).

When the mother experiences the passion of love, as the unconditional acceptance and recognition of the child ('a love that asks no leave of the judgement', 121), her caress and embrace of the child are also *as if*

> unconscious. The child is looking up to the mother, stretching forward its limbs, and the mother's 'hands close as it were of th[em]selves' (121). Again we witness a circular process: the child gives birth to the mother, to the mother's love, as her care(ss) can only emerge as a response to the child. In other words, although it is already there to respond to the child's demands, motherly love is predicated on the birth of the child. In 'Ode to Georgiana, Duchess of Devonshire', Coleridge presents the virtues of the Duchess, a mother, as being born from the experience of childbirth:O Lady, nurs'd in pomp and pleasure!
> Where learnt you that heroic measure?'
>
> ….
>
> You were a Mother! at your bosom fed
> The Babes that lov'd you. You, with laughing eye,

ness of discrimination. That is why man is the most intelligent of all animals. This is confirmed by the fact that it is to differences in the organ of touch and to nothing else that the differences between man and man in respect of natural endowment are due; men whose flesh is hard are ill-endowed by nature, men whose flesh is soft, well endowed'. In 'Aristotle on the Sense of Touch' (Freedland 1979, 228) Cynthia Freeland writes that 'In Aristotle's classic catalogue of the five senses, the sense of touch is special for several reasons. It is the most basic of all the senses, the only one common to all animals (*De anima*, 413b4–10).'

[19] e.g. Irigaray (2011).

Each twilight-thought, each nascent feeling read,
Which you yourself created. Oh! Delight!
...
The Mother of your Infant's Soul!;
... from his aspect sweet
New influences in your being rose,
Blest Intuitions and Communions fleet
With living Nature, in her joys and woes!
(Coleridge 2001, 2: 612)[20]

Culturally, the idea that the child's birth awakens motherly love is associated with the Nativity, signifying the birth of love on earth. Indeed, as opposed to the maternal instincts of the animal, motherly love is also illuminated by what Coleridge calls the "'divine Fiat'.

In *Aids to Reflection,* Coleridge refers to the 'Position of the Aristotelians'—expressed in Aquinas's dictum, 'There is nothing in the mind that was not previously in the senses'—as a *half-truth* taken up by Locke, which should be complemented by the spiritual ground of our humanity, that is, by the will that is not nature (1993 [1824], 79–80), to round it out into truth. Yet, since humans do belong to nature too—'Man hath from Birth that which is common with the Animal, and that which is especially human' (Coleridge [1823] 2002, 122)—it seems that only maternal love, as spiritual *and* sensual care(ss), can respond to the riddle of our humanity, to our uniquely human predicament of being both life and mind.

In *Opus Maximum,* Coleridge describes what happens if the mother alienates the love of the child, and directs it towards what is sensuous around him. Then, the infant 'may be made to exhibit the pride of the Horse, the rage of the Turkey, the Vanity of the Peacock', that is, he may become disturbingly similar to beasts—not to mention that anthropomorphism renders beasts disturbingly similar to humans too. Earlier in the same passage, Coleridge also outlines the predicament of the new-born when deprived of the care(ss) of a human mother:

> In vain the conditions, the possibility, of the human have been inlaid, were not the human, in its full development, already there to meet and protect it. Only by disproportion of the means to the ends will the babe [survive], abandoned from its birth and suckled in the Forest by the blind and kindly instincts of nature in the Goat or the Wolf. Its eye will remain glazed; its lips, the seat of expression for appetite and rage alone. The tongue will utter the sound of Ape; the very ear will be deaf to all but the inarticulate sounds of Nature.' (Coleridge [1823] 2002, 120)

This passage suggests that maternal care(ss) is of prime significance in sustaining the difference between animals and humans: although the infant raised by animals will only be *like* an animal ('Its eye will remain glazed; its lips, the seat of expression for appetite and rage alone. The tongue will utter the sound of Ape; the very ear will be deaf to all but the inarticulate sounds of Nature'), and will 'never be indeed an animal' (Coleridge [1823] 2002, 121), in order to become an individual, a person, he must necessarily be loved, cared for, and caressed by his mother from birth.

[20] Fort the political and erotic context of the poem see: Thomson (2016, 84–6).

Deprived of maternal caress, his humanity remains *forever* in a dormant state, and his fate will be similar to that of the wild boy found in the forests of France who, despite all the efforts put in his later education, was never able to learn a language ([1823] 2002, 120 n. 155). Maternal love as both spiritual and sensual thus turns out to be the possibility condition of all (later) educative process[21]; without it, our humanity will never become manifest. In other words, so that human uniqueness can ever be realized, it is indispensable that the infant be met with maternal care(ss) at its birth. Hence, although in *Church and State* Coleridge calls for a 'NATIONAL EDUCATION' to 'elicit, the latent *man*', led by a Parson, *Opus Maximum* suggests that if the mother's love has not elicited the latent man in the first place, then national education remains groundless, even useless.

References

Aristotle. 1984. *De anima* (350 BC). Tr. J. A. Smith. In *Complete works of Aristotle, Volume 1: The revised Oxford translation*, ed. Jonathan Barnes, 641–692. Princeton University Press. http://classics.mit.edu/Aristotle/soul.2.ii.html.

Barash, Carol, and Susan C. Greenfield. 2015. *Inventing maternity: Politics, science, and literature, 1650–1865*. Lexington: Lexington.

Brace, Laura. 2007. Rousseau, maternity and the politics of emptiness. *Polity* 39 (3): 361–383.

Buffon, Count de (Georges–Louis Leclerc). [1707–88] 1791. Of infancy. In *Natural history containing a theory of the earth, a general history of man, of the brute creation, and of vegetables, minerals, &c. &c.*, Vol. 3 (of 10). Tr. William Smellie. https://www.gutenberg.org/files/45639/45639–8.txt

Cavarero, Adriana. 2016. *Inclinations: A critique of rectitude*. Stanford: Stanford University Press.

Coleridge, Samuel Taylor. 1956–71. *The collected letters of Samuel Taylor Coleridge*. Ed. E. L. Griggs, 6 vols. Oxford: The Clarendon Press.

———. 1969. In *The friend*. (vol. 1: 1817; vol. 2: 1808–9), ed. Barbara E. Rooke. Princeton: Princeton University Press.

———. [1829]1976. In *On the constitution of the church and state according to the idea of each*, ed. John Colmer. Princeton: Princeton University Press.

———. 1993. In *Aids to reflection (1825)*, ed. John Beer. Princeton: Princeton University Press.

———. 1995. [citing:1806; 1811; 1812; 1816?/1819?; 1828]. In *Shorter works and fragments. 2 vols*, ed. H. J. Jackson and J. R. Jackson. Princeton: Princeton University Press.

———. 2001. In *Poetical works, part one: Poems (reading text)*. 2 vols, ed. J. C. C. Mays. Princeton: Princeton University Press.

———. 2002. In *Opus maximum (c.1823)*, ed. Thomas McFarland . Princeton: Princeton University Press.with the assistance of Nicholas Halmi

Freeland, Cynthia. 1979. Aristotle on the sense of touch. In *Essays on Aristotle's De Anima*, ed. Martha Nussbaum and Amélie Rorty, 227–248. Oxford: Oxford University Press.

Haslam, Nick. 2006. Dehumanization: An integrative review. *Personality and Social Psychology Review* 10 (3): 252–256.

Irigaray, Luce. 2011. The fecundity of the caress: A reading of Levinas, *Totality and infinity*, "phenomenology of Eros". In *Feminist interpretations of Emmanuel Levinas*, ed. Tina Chanter, 119–144. Pennsylvania State University Press: University Park.

[21] See the discussion of *Logic* in Timár (2015, 107–24).

Jones, Ewan. 2014. *Coleridge and the philosophy of poetic form*. Cambridge: Cambridge University Press.
Kearney, Richard. 2021. Philosophies of touch: From Aristotle to phenomenology. In *Touch: Recovering our most vital sense*, 33–60. New York: Columbia University Press.
Kitson, Peter. 2007. *Romantic literature, race, and colonial encounter*. New York: Palgrave.
Page-Jones, Kimberley. 2025. Life, nature, and race: Blumenbach's vital materialism in Coleridge's later writings (1818–28). In *Matter and life in Coleridge, Schelling, and other dynamical idealists*, ed. Peter Cheyne, ch.19. Dordrecht: Springer.
Pfau, Thomas. 2013. *Minding the modern: Human agency, intellectual traditions, and responsible knowledge*. Notre Dame: University of Notre Dame Press.
Richardson, Alan. 1999. Coleridge and the dream of an embodied mind. *Romanticism* 5 (1): 1–25.
Ruderman, David. 2016. *The idea of infancy in nineteenth-century British poetry: Romanticism, subjectivity, form*. London: Routledge.
Sebastiani, Silvia. 2021. Enlightenment humanization and dehumanization. In *The Routledge handbook of dehumanization*, ed. Maria Kronfeldner, 64–83. London: Routledge.
Taylor, Anya. 1986. *Coleridge's defence of the human*. Columbus: Ohio University Press.
Thomson, Heidi. 2016. *Coleridge and the romantic newspaper*. Basingstoke: Palgrave.
Timár, Andrea. 2015. *A modern Coleridge: Cultivation, addiction, habits*. Basingstoke: Palgrave.
Webster, Suzanne E. 2017. Coleridge, contemplation, and the 'Triple Ichheit'. In *Coleridge and contemplation*, ed. Peter Cheyne, 293–309. Oxford: Oxford University Press.

Andrea Timár is associate professor in British literature and critical theory at Eötvös Loránd University, Budapest. She is the author of *A Modern Coleridge: Cultivation, Addiction, Habits* (Palgrave, 2015). Recent publications include 'Against Compassion: Post-traumatic Stories in Arendt, Benjamin, Melville, and Coleridge' (*Arendt Studies*, 2022) and 'Dehumanization in Literature and the Figure of the Perpetrator' (*The Routledge Handbook of Dehumanization*, 2021). Her forthcoming book is *Reading with Hannah Arendt: Literature, Criticism, Theory* (Bloomsbury).

Chapter 19
Coleridge's Poem 'Human Life Contemplated on the Denial of Immortality' (1811–15)

J. C. C. Mays

Abstract In this chapter, J. C. C. Mays contends that a failure to take proper notice of the hidden features of Coleridge's poem 'Human Life Contemplated on the Denial of Immortality' (1811–15) has led to an amount of confusion as to what it actually does. The author gives a more careful reading of Coleridge's late and often-neglected poem 'Human Life', which takes off from what one might call its complicated double-rhyme scheme. It turns out to be more carefully constructed than it is usually taken to be, even though it remains a poem that attempts—as Coleridge said of some of John Donne's—to say too much.

19.1 Introduction

I give two versions of the above-named poem at the beginning of what I want to say. The first version is in the form Coleridge always published it: the form in which it is usually read. The second has the same words but I have differently arranged them to clarify the additional means that carry the poem forward. Readers who are unfamiliar with the poem will find it helpful to read the poem as Coleridge published it before they read my revised arrangement, which is a construct that I hope presents new insight into what he is doing. It seems to me that a failure to take proper notice of the poem's hidden features has led to an amount of confusion as to what it actually does.

First, then, the poem as Coleridge published it and as it is printed in the Bollingen edition. (The line numbers have been retained to make the linkage to my following version easier.)

J. C. C. Mays (✉)
School of English, Drama, and Film, University College Dublin (emeritus),
Dublin, Ireland
e-mail: james.mays@ucd.ie

P. Cheyne (ed.), *Matter and Life in Coleridge, Schelling, and Other Dynamical Idealists*, International Archives of the History of Ideas Archives internationales d'histoire des idées 253, https://doi.org/10.1007/978-3-031-78315-9_19

If dead, we cease to be: if total gloom
   Swallow up life's brief flash for aye, we fare
As summer-gusts, of sudden birth and doom,
   Whose sound and motion not alone declare,
But *are* their *whole* of being! If the Breath
   Be Life itself, and not its Task and Tent,
If even a soul like Milton's can know death;
   O Man! thou vessel purposeless, unmeant,
Yet drone-hive strange of phantom purposes!
   Surplus of nature's dread activity,
Which, as she gazed on some nigh-finished vase,
Retreating slow, with meditative pause,
   She formed with restless hands unconsciously!
Blank accident! nothing's anomaly!
   If rootless thus, thus substanceless thy state,
Go, weigh thy dreams, and be thy Hopes, thy Fears,
The counter-weights!—Thy Laughter and thy Tears
   Mean but themselves, each fittest to create
And to repay the other! Why rejoices
   Thy heart with hollow joy for hollow good?
   Why cowl thy face beneath the Mourner's hood,
Why waste thy sighs, and thy lamenting voices,
   Image of Image, Ghost of Ghostly Elf,
That such a thing as thou feel'st warm or cold?
Yet what and whence thy gain, if thou withhold
   These costless shadows of thy shadowy self?
Be sad! be glad! be neither! seek, or shun!
Thou hast no reason why! Thou can'st have none!
Thy being's being is contradiction.
(2001 [1811–15], 1: 887)

A reader will quickly notice that the repeating pattern of indented lines extends through the poem but turns around at roughly mid-point, centred on the unrhymed 'Yet drone-hive strange of phantom purposes!' Following this, the reader as it were kicks off on the other foot. The original a/b indentation is replaced at the point marked by the right-edge line-up of lines 11–12. We then begin each line again with a different pattern of indentation heading-up a similar message, this time the first rhyming line being indented. Lines 16–17 vary this with a couplet to confirm that it has taken place, lines 24–25 repeat that point over again, in the same way and the last lines conclude the argument with a unique triple rhyme.

Two brief notes before I give the other version: first, more on Coleridge's use of indented lines. My impression is that he did not normally use them in any significant way. While poems like 'Fancy in Nubibus' (2001, **540**)[1] are printed with alternately indented lines, the holograph versions are more often not indented. The present poem is unusual in having all the four printed versions maintaining the same consistent pattern, although the evidence concerning its significance is slight. The second point has to do with the unmentioned presence of Donne: call it that, rather than

[1] Numbers in bold refer to the numbers given to poems in my 2001 Princeton Bollingen edition of the Collected Coleridge.

Donne's influence because it hovers over the poem in a rather peculiar way. It allowed or encouraged Coleridge's habit of association to extend beyond its ordinary boundary to become somewhat extreme. It prompted him—when he mentioned Donne as his example, as he did in the present case (see the note in Francis Wrangham's copy of *Sibylline Leaves*: 2001, 2: 1101)—to pursue connections to a limit beyond his ordinary poetry. I take it that one must be on one's guard for intellectual obscurity.

Next for the version of the poem I prepared for the present chapter. In short, I divided the poem into rhyming stanzas according to a different logic from the version Coleridge printed, which may reveal a sub-structure which creates a different sense

1. If dead, we cease to be: if total gloom	a	
Swallow up life's brief flash for aye, we fare	b	
As summer-gusts, of sudden birth and doom,	a	
Whose sound and motion not alone declare,	b	
2. But *are* their *whole* of being!		
If the Breath	a	
Be Life itself, and not its Task and Tent,	b	
If even a soul like Milton's can know death;	a	
O Man! thou vessel purposeless, unmeant,	b	
Yet drone-hive strange of phantom purposes!		**unrhymed**
3. Surplus of nature's dread activity,	a	
Which, as she gazed on some nigh-finished vase,	b	
Retreating slow, with meditative pause,	(b)	**inserted**
She formed with restless hands unconsciously!	a	
Blank accident! nothing's anomaly!	a	
4. If rootless thus, thus substanceless thy state,	a	
Go, weigh thy dreams, and be thy Hopes, thy Fears,	b	
The counter-weights!—Thy Laughter and thy Tears	b	
Mean but themselves, each fittest to create	a	
5. And to repay the other!		
Why rejoices	a	
Thy heart with hollow joy for hollow good?	b	
Why cowl thy face beneath the Mourner's hood,	b	
Why waste thy sighs, and thy lamenting voices,	a	
6. Image of Image, Ghost of Ghostly Elf,	c	
That such a thing as thou feel'st warm or cold?	d	
Yet what and whence thy gain, if thou withhold	d	
These costless shadows of thy shadowy self?	c	
7. Be sad! be glad! be neither! seek, or shun!	a	
Thou hast no reason why! Thou can'st have none!	a	
Thy being's being is contradiction.	a	

Twenty years ago, I edited this poem for the Bollingen Coleridge, recording the facts but not worrying too much about the poem as a poem. The textual record seemed fairly straightforward and not to raise any serious problems. I revisit it now because I realize it is considerably less simple than I thought. For a start, the title it bore when Coleridge first read it in public—in his 1819 philosophical lectures—is much improved by the insertion of the one word, 'contemplated'. I insert it now and it immediately limits and helps to clarify the parameters of the poem's argument. It also makes less appropriate some of the wide-ranging speculations that the poem has attracted: for instance by Anya Taylor in the *Coleridge Bulletin* (Summer 2016). The poem is about much less and paradoxically about something different from what she describes.

I should say at once that I have no wish to put down Taylor's argument, which pursues aspects of Coleridge's theme into all kinds of corners with keen sympathy. It assumes that Coleridge is meditating on the topic of human life, if immortality is denied, asking the question: Would life on such terms be human? But the problem still remains: why did Coleridge repeat his argument across such a broad front? In particular, Taylor reads the poem as if it was a single block of meaning, and does not adequately pick up on the way it changes its tune in the second half. Note how the first fourteen lines literally slow the pace as one reads; this in contrast to the bundled confusion of lines 15–29, which tumble over themselves despite summing-up twice over (in lines 22 and 26). Part 1 moves seriously, utters a grave warning, makes a complete sonnet in itself. Part 2—another sonnet extended by an extra line to finish the poem in its triplet—is a parody of meaning that accumulates only to confuse: a grotesque, fantastical replay set to an altered tune, or to no proper tune at all. The repeated warning is enlarged and destroyed. To speculate and take notice becomes to speculate even further and realize how potentially ridiculous the process has become.

I repeat that I have no wish to put down the arguments of Taylor and others.[2] I only suggest that their reading of the text does not do it justice. It is made up of two significantly different sonnets and does not hammer home the same point twice. It makes a single developing argument with elegant precision and then turns this around to demonstrate the consequences of ignoring it. It is much less and much more than what Taylor and others suggest. It is too simple to read the poem as if it was two sonnets joined together; sonnet one providing the premise and sonnet two the conclusion. We do not have to suppose the second sonnet was written first, and the first sonnet added to provide a better explanation. Both less and more is going on here than first meets our eyes. It is encouraging that Allison Dushane, Kimberley Page-Jones, and Andrea Timár, in the three preceding chapters of this book, read the poem as composed from a perspective of faith and not despair.

[2] Such as Edward Kessler (1979, 71–6) and Onita Vaz-Hooper (2009, 529–44).

19.2 A Fresh Beginning

I suggest that it is first useful to link the poem to four interrelated notebook poems by Coleridge—numbers **476–79**—which arose out of his reading of Donne's 'The Flea' in 1811 (1957–2002, 3: §§4073, 4074; 1980–2001, 2: 214–15). The first supplies the local circumstances and very nearly gets lost in them. It was never republished although it was sometimes referred to. The second—'Moles'—was first published in the 1818 *Friend* (1969 [1818], 1: 494) and, together with the third and fourth notebook entries, was republished in *Poetical Works* 1834. The present poem (**482**) was written independently, certainly after the four notebook poems in 1811 and perhaps about the same time the *Sibylline Leaves* proofs passed through Coleridge's hands. It was first published in that much-delayed collection in 1817, where it was subtitled 'A Fragment' and accompanied by remarks on its Donnean beginnings in annotated copies. He quoted it again in full in his sixth lecture on philosophy (2000, 1: 269–70), and republished it in his three last collections. It may be understood as another poem consciously in the manner of Donne or as yet another attempt by his imitator to describe, 'in a very different style', the contents of his earlier poem 'Phantom' (see **347**, and compare 2001, 1: 1091–92 and 975).

This latter connection, which Coleridge made only once, is important. When 'Phantom' was copied out for young Sara Coleridge in 1820, it was under the title 'Fragment from a Dream'. I assume that one reads this as positive and celebratory and here the poem is, like so many of Coleridge's dreams of Sara Hutchinson (Asra), a glorious fantasy.

> All Look or Likeness caught from Earth,
> All accident of Kin or Birth,
> Had pass'd away: there was no trace
> Of aught upon her brighten'd face,
> Uprais'd beneath that rifted Stone,
> But of one Image—all *her own!*
> She, She alone, and only She
> Shone thro' her body visibly—(2001, 1: 763)

This is at the same time a strange claim to make, that this ideal fabrication, this dream-fragment, should be classified by Coleridge, at the same time, as 'a dream of Purgatory, alias Limbo' (2001, 2: 1091–92). One does not expect such a description of an ideal love to be applied to a place of shadows, 'unmeaning they | As Moonlight on the Dial of Day': a place of shades more cut off than the moles and the blind men on earth, of 'dull Privation', suggesting a 'fear far worse' that 'Hell knows', of '*Positive Negation!*' ('Limbo', 2001, 1: 883, 884).

The point made by entitling the present poem 'Human Life ⟨contemplated⟩ on the Denial of Immortality' is that Coleridge's severe description is, *on his terms*, true. Not to realize that life is a stage on a longer journey, of which death is only a part, is to live a delusion. It is to surround oneself with comforting images that have no reason to exist. Asra was indeed but a shadow of his shadowy self:

> Thy Laughter and thy Tears
> Mean but themselves, each fittest to create
> And to repay the other! (**482** lines 17–19)

In short, one has to read the present poem as an alternative to poems in which hope prompts love, as they advance blindly into a glorious future, because that future is entirely the product of dreams which are 'rootless' and 'substanceless' (line 15). It is a poem of the same order as, say, 'The Picture' (**300**), in which the speaker is both the author and the knowing victim of his own fantasies.

Whether this was Coleridge's consistent argument is a moot point. 'Be sad! be glad! be neither! seek, or shun!' ('Human Life', line 27) is a comprehensive statement. But to hope till hope creates the thing it contemplates is not his assumption. He came to see that his manufactured ideal does not live in the larger, real world. A metaphysic based upon the 'dread activity' of nature (line 10) would seem to be a delusory solution: arbitrary, pointless, 'nothing's anomaly!' (line 14). And one can only admit that the situation does exist in this form if one confronts it in this way. The choice is between a shorter view deriving from immediate impulses and a longer view guided by forethought and reflection ('Method'). So the present poem, as I have interpreted it, fits the argument of the 1818–19 philosophical lectures perfectly: it embodies and exemplifies eudaemonism as conceived by Hegesias. It is again surely not fortuitous that the poem was preceded by 'A Christmas Carol' (**260**) and followed soon after by 'The Visit of the Gods' (**190**), imitated from Schiller in all Coleridge's collections from *Sibylline Leaves* onwards. Indeed, in his final collection, this 'Human Life' was immediately followed by 'Moles' (**477**), who 'See but to dread, and dread they know not why'.

The transition from a poem in the manner of Donne to one that celebrates withdrawal imaged by 'live mandrakes [sc. who] … Creep back from Light—then listen for its sound' (**477**) may be difficult to accept, but it is the accommodation Coleridge apparently reached. After many years—perhaps up to fifteen years—of meditation, he came to accept that the ideal version of Sara Hutchinson he had constructed had no real reason. In the larger scheme of things, it was a barren construction, a phantom imposition, even a selfish indulgence. He did not ponder his mistake for very long in his poetry, and he grew away from it in his later decades. 'Alice du Clós' (**655**) succinctly illustrates the cruel ending that afflicts this most innocent of his heroines. 'Human Life', on the other hand, written about the time he came to a similar conclusion, thrills forcibly to its unforgiving theme; and perhaps it is this that leads some commentators (Onita Vaz-Hooper, for example) to give the impression that the poem is as much about the necessity of an afterlife as about those who do not believe in it. In reality, the subject of Coleridge's poem is restricted to those whose field of perception is foreshortened. He presents a paradox of nullity, things turned upside down and inside out, to show how positively contradictory the speaker's blindness turns out to be. It is a rare fun-poem about a deadly subject. It juggles concepts with a mischievous ease, so that one can lose one's grip on the primary purpose of the performance.

19.3 Fail Again, Fail Better

This short poem somehow gets under under one's skin and is difficult to leave behind. It is finished yet difficult to dispatch, to leave alone, but I will try one more time.

I copied it out in seven separate stanzas at the beginning of this piece because I think the different format clarifies important features which are often overlooked. The different deployment of two consecutive sonnets, written according to different logics is primary. They each have their different rhyme-scheme, which is tailored to the purpose each serves; indeed, their own strikingly different styles attempt different things. If one does not take note of this, one is likely to be led astray into the kind of general comments that are possibly true but are not the primary basis of this short poem. Coleridge is intent on one thing here and one thing only, as if this thing is the most significant that anyone needs to know.

The poem tells us that while dreams preoccupy its speaker they are at the same time pointless: 'rootless' and 'substanceless' (line 15). They make up a dream world that has no reason, no basis in fact. And if you attempt to live the impossible contradiction, in what one supposes is the real world where people literally live and die, you fail. Your effort lacks real intent because it reaches towards a fantasy:

> unmeaning they ('Limbo' lines 7–8)
> As Moonlight on the Dial of Day

Thus, while the first sonnet-part of 'Human Life' describes an ideal through its destruction—'If even a soul like Milton's can know death'—you have to bear in mind it is a soul that knows that death leads on to another life: it is on a meaningful journey, not wasting its time on earth in a mindless, driven existence, a 'Blank accident! nothing's anomaly!' (line 14). And it is important to realize from the start of the poem that 'dead' means exactly that in this barren world. 'We cease to be' (line 1): that and no more. It deals with specifically 'human life' cut off from any everlasting—'immortal'—life promised by the great religions. This in poems like 'Moles' (**477**) and 'Ne Plus Ultra' (**479**) is about those who do not believe in saving grace, who live a life that is literally, completely, utterly terminated in death. Or one might even say it ends where they in fact begin, without faith. What they do in their blindness is to create anomalous worlds, which serve no purpose: 'Surplus of nature's dread activity' (line 10).

If the above is clear, one might be tempted to consider the present poem as like a sandwich made by the joining of two different parts, the second sonnet completing a statement separately prepared from the first. The first part reports the consequences of denying a Christian belief in life after death; the second part enlarges on the consequences of doing this in a different way. The first part moves slowly and impersonally in the voice of the thinking poet: it makes a statement that is complete in itself. The second part changes gear—*we* becomes *thou*—and it assails an unnamed possibly male auditor (line 8's 'O Man!') with a series of direct, unanswerable questions. These are posed by the poet to be sure, but they enter the mind of the person he is addressing and they accumulate until they explode in whirling contradiction. The second sonnet does not follow naturally from the first. It fits on uncomfortably, as if a frenzied person had taken control.

It is important to read the differences between the second and the first parts very carefully. They are written in different ways; they possess a different grammar and metric which only come together (almost) in a single style at the end. Part 1 is declarative, part 2 repeatedly asks questions to which there are no answers (three times 'Why' in stanza 5). That is, the second part quickens the pace, looking for a response, making pronouncements, leaving no room to think. Looking back, one appreciates more fully the space given to the reflection in the opening sonnet, which extends across four lines into the second stanza. Line 7 almost holds one up as if it were a puzzle, although it is both crystal clear and thoroughly provoking, however many times one reads it. The same again might be said of 'nature's dread activity' which fills the whole of the third stanza. Whatever the cause of such mysterious events, they are real, unexplained, and disturbing.

Indeed, it makes sense to ask how such an unsettling effect is created by such a solid structure. The first sonnet is quick to incorporate its readers into its grammar with 'we' as the third word, which it repeats as the sentence progresses. 'If' is repeated four times over in the first eight lines so that we cannot complain we are not being interrogated, over and over, to remove ambiguity which nonetheless accumulates. The second sonnet of the poem adopts a different grammar in stanza 4, but to equal effect. Beginning with a fifth conditional—in 'If rootless thus'—the speaker shifts from the shared premise of 'We' to the distanced 'Thee' and 'Thy', which recur altogether seventeen times in the final fifteen lines. However, although the identity of this third person remains distant and obscure—'rootless' and 'substanceless' (line 15)—its *a priori* limited identity is clear.

Having said so much about the components of this poem, a few words on its metrics may be apropos. It is basically constructed as two matching sets of stanzas, the first set made of three stanzas in four lines with stanzas 2 and 3 having an extra line each, and the second set of four stanzas, the first three with no extra lines and the extra fourth concluding in three rhyming lines. Seen another way, the three stanzas of the first sonnet are increasingly irregular: viz. rhyming abab, ababx, ab(b)aa (the second b line inserted); the next three stanzas (the second sonnet) rhyme abba, and the conclusion rhymes aaa. This makes compelling sense: line 9 adds an extra unrhymed, summarising line to a difficult stanza; the extra b-rhyme in line 12 adds something similar to line 11; and the couplet at lines 13–14 sums up a turning point in the poem.

At the same time, the regularity of the second sonnet differently mirrors the lack of change in the addressed person's situation. Line 18 lacks a syllable and we move more swiftly across the gap to the next stanza. The extra syllables in lines 22 and 26 are entirely appropriate in laying emphasis on what they say. If the shift of perspective between the two sonnets is ignored one has to suppose that Coleridge is repeating himself, which he is not. I repeat that the poem is emphatically conceived as whole: part 2 expands the dead conclusion it announces from the beginning. Ask where would part 2 be without the admittedly compressed part 1. Part 1 is serious, articulates a grave warning; part 2 is a parody, a grotesque fantastical replay to an altered tune.

The first part is about human life conceived as a Christian might do: the second part is about human life at a distance from this position, that is, in a way that denies (ignores, frustrates) the earlier perspective. The argument differs from that pursued in Wordsworth's great 'Ode: Intimations of Immortality', but its consequence is similar. Coleridge does not exalt childhood and expect us to live off its diminishing crumbs. Only if we and Milton's reputation die, are completely forgotten, will we cease to be; and, as no one is born and dies completely alone, so no one lives completely alone. There are chinks—indeed great gaps—in this argument, but in charity one can appreciate its direction.

19.4 In Short, An Irritating Asking Poem?

The poem is certainly a unified whole. The two sonnets are joined together and fashion a sharper larger meaning, despite its liability to be lost. For instance the grammatical run-on of stanzas 1–2 is matched in stanzas 4–5; the uneasy progression of stanzas 2 and 3 communicates the unnatural processes that are being contrived; the almost facile ease with which the subsequent rhyming verses find their rhymes is equally telling. But the question remains: is it constructed to compel a reader to read it always right? Not always, one has to admit. One needs to know that Coleridge sometimes labelled his dreams of Sara Hutchinson as empty gusts:

> Whose sound and motion not alone declare,
> But *are* their *whole* of being! (lines 4–5)

He may have distanced his subsequent address as if to a separate person, but that person did no more than he has done here, cowling his face beneath the Ancient Mariner's shroud (the original reading of line 21). Coleridge is writing to versions of himself in both parts of his poem, and no less in the second part than in the first.

There is much to give one pleasure on the way. 'Life's brief flash for aye' quickly becomes a jolly evasion as we move on to be swallowed in doom. 'Not alone declare' is explained by the following 'But', and meanwhile notice how 'declare' is emphasized to make sense of the line before it continues. 'Tent' in line 6 is an odd word until one recognizes its Biblical overtones and connections with, for instance, the tabernacle. Or might it also be related to 'try' or 'aim', stretching out as in 'intent' or 'tentative'? And the entire mysterious stanza 3 becomes no less mysterious on re-reading and re-reading it over again: the processes of nature indeed obey a law of their own, and forever escape explanation. So the poem goes, with a mounting accumulation of questions, until they only contradict one another and lie about helpless, useless, pointless. This surplus of nature's activity points up again the lost sense of purpose, the need for direction. The questions become mesmeric, the grammar becomes entangling.

I return to the fact that the earlier part of the poem is more compacted and positive than the later part. Although the details are obscure, the writer would seem firmly committed to a Christian understanding of death and the afterlife, and intellectually opposed to the workings of what he calls 'nature's dread activity'. One might indeed say that the later part of the poem has a much thinner texture. It is almost repetitive in its enumeration of what might be called the mole's view, which is the emptiness and futility of a life lived without consideration of its meaning and end.

It is important that such is the case: the poem goes no further, and certainly does not elaborate an alternative to the mole's view. It likewise says nothing at all of what lies behind the critique of the same, which lies vaguely suggested but is not stated. And, by seeking to comment on the poem by enlarging on what the poem only vaguely suggests, the remarks of some recent critics distort what Coleridge was writing about here. A mole can signify the smallest blemish as well as a small animal, but its size can multiply and become deadly.

Of course, his emblematic moles are ultimately relevant, but the point is that Coleridge did not write about them here. And he probably did not want to because the subject was one which he was uncertain about: he was in process of evolving views that were changing, emerging, and far more important than the obvious truth he could write about in the period centred on 1815–16. Whatever the dim signs of what he thought before this—multiple, often contradictory signs I hasten to add—it is probably important that he was to write his long-projected statements on 'Method' in 1817 and 1818 (1969 [1818], 1: 448–524; 1995, 1: 629–87). It is a great pity that these mind-changing essays were not printed together, along with related material, separately in the Bollingen edition, because they tend to get lost as parts of different wholes. They serve as turning-points in Coleridge's thinking about logic and the need for a prior anticipated turning-point on which such an argument must turn.

'If rootless thus, thus substanceless thy state', the second half of the poem opens, and this fundamental argument goes back to the *Biographia Literaria* and beyond. The root, the initiative, the mental antecedent, is the crux: the synderesis or whatever it is called determines the eventual discovery. Coleridge went on elaborating this insight to the end—apropos the necessary logic of the Trinity through the 1820s, for example—but the poem tries to say little more than it knew at the time. It demonstrates a startling lack on the part of humanity, to be sure, but, if we remember the whole poem, 'Human Life', as only a denial of immortality, we have totally misinterpreted it. One might argue that this is evidently such a common mistake that there must be good reason for making it. To which there is no answer other than that this is surely the point the poem wants to make. It is both more simple and more complicated than it is usually held to be. And that is characteristic of all its author's poems at their best.

I would finally emphasize that I am not aiming to replace the text of this poem. It is as Coleridge always printed it and my alternative version, broken into quatrains and the like, is not designed to replace the version printed in the Bollingen Coleridge. My purpose is simply to clarify the more fluid arrangement Coleridge settled on, and to understand how it burrows under one's skin. It ultimately reads better that way: a fluid conjugation of parts that make a more memorable whole. The parts keep one's eyes engaged on the poem, and do not allow it to wander into seas unknown.

References

Coleridge, Samuel Taylor. 1959–2002. *The collected works of Samuel Taylor Coleridge*. Ed. Kathleen Coburn, 16 titles in 33 volumes, Bollingen Series LXXV. Princeton: Princeton University Press, in which I make reference to the following:

———. 1969. *The friend.* (vol. 1, 1818; vol. 2, 1808–9). Ed. Barbara E. Rooke. Princeton: Princeton University Press.

———. 1980–2001. *Marginalia*, vols 1 and 2 (1980–4). Ed. George Whalley, vols 3–6 (1992–2001). ed. H. J. Jackson and George Whalley. Princeton: Princeton University Press.

———. 1995. *Shorter works and fragments*. 2 vols. Ed. H. J. Jackson and J. R. J. de Jackson. Princeton: Princeton University Press.

———. 2000. *Lectures 1818–1819: On the history of philosophy*. Ed. J. R. J. de Jackson. Princeton: Princeton University Press.

———. 2001. *Poetical works*. 3 double vols. Ed. J. C. C. Mays. Princeton: Princeton University Press.

———. 1957–2002. *The notebooks of Samuel Taylor Coleridge*. Ed. Kathleen Coburn et al. London/Princeton: Routledge/Princeton University Press. 6 double volumes

Kessler, Edward. 1979. *Coleridge's metaphors of being*. Princeton: Princeton University Press..

Taylor, Anya. 2016. Coleridge's worries about the afterlife: Doubts, circularities, and imaginings. *The Coleridge Bulletin* 47:1–19.

Vaz-Hooper, Onita. 2009. If dead we cease to be': The logic of immortality in Coleridge's 'human life. *European Romantic Review* 20 (4): 529–544.

J. C. C. Mays is a member of the Royal Irish Academy and professor emeritus of University College Dublin. He is the editor of Coleridge's poetry and plays in the Bollingen *Collected Works* (vol. 16, Princeton UP and Routledge). He has published numerous articles on Coleridge's verse and three books on Coleridge's poems with Palgrave Macmillan, *Coleridge's Experimental Poetics* (2013), *Coleridge's Ancient Mariner* (2016), and *Coleridge's Dejection Ode* (2019).

Correction to: Bloody Speck: How S. T. Coleridge Turned the Embryological Punctum Saliens into a Metaphysical Principle

Peter Cheyne

Correction to:

Chapter 15 in: P. Cheyne (ed.), *Matter and Life in Coleridge, Schelling, and Other Dynamical Idealists*, International Archives of the History of Ideas Archives internationales d'histoire des idées 253, https://doi.org/10.1007/978-3-031-78315-9_15

The original version of the chapter "Bloody Speck: How S. T. Coleridge Turned the Embryological Punctum Saliens into a Metaphysical Principle" was inadvertently published with the junk characters on page no. 318 (bottom) and page no. 319 (top). This has now been corrected and updated in the chapter.

The updated version of this chapter can be found at
https://doi.org/10.1007/978-3-031-78315-9_15

P. Cheyne (ed.), *Matter and Life in Coleridge, Schelling, and Other Dynamical Idealists*, International Archives of the History of Ideas Archives internationales d'histoire des idées 253, https://doi.org/10.1007/978-3-031-78315-9_20

MIX
Papier aus verantwortungsvollen Quellen
Paper from responsible sources
FSC® C105338

If you have any concerns about our products,
you can contact us on
ProductSafety@springernature.com

In case Publisher is established outside the EU,
the EU authorized representative is:
Springer Nature Customer Service Center GmbH
Europaplatz 3, 69115 Heidelberg, Germany

Printed by Libri Plureos GmbH
in Hamburg, Germany